Human Reproduction and Sexual Behavior

RICHARD E. JONES
University of Colorado

PRENTICE-HALL, INC., *Englewood Cliffs, N.J.* 07632

Library of Congress Cataloging in Publication Data

Jones, Richard E. (Richard Evan), (date)
    Human reproduction and sexual behavior.

    Includes bibliographies and index.
    1. Human reproduction. 2. Sex. I. Title.
[DNLM: 1. Reproduction. 2. Sex behavior. WQ 205 J78h]
QP251.J635 1984   612'.6   83-567
ISBN 0-13-447524-0

To kids, especially my own

Editorial/production supervision: Karen J. Clemments
Interior design: Karen J. Clemments
Cover design: Wanda Lubelska
Manufacturing buyer: John Hall

The cover art is reproduced from MAMMALIAN FERTILIZATION wallchart from the INTERNA-
TIONAL PLANNED PARENTHOOD FEDERATION'S Research in Reproduction series, edited by
Dr. R. G. Edwards.

Printed in the United States of America

10 9 8 7 6 5 4 3 2 1

ISBN   0-13-447524-0

Prentice-Hall International, Inc., *London*
Prentice-Hall of Australia Pty. Limited, *Sydney*
Editora Prentice-Hall do Brasil, Ltda., *Rio de Janeiro*
Prentice-Hall Canada Inc., *Toronto*
Prentice-Hall of India Private Limited, *New Delhi*
Prentice-Hall of Japan, Inc., *Tokyo*
Prentice-Hall of Southeast Asia Pte. Ltd., *Singapore*
Whitehall Books Limited, *Wellington, New Zealand*

Contents

PART II   PROCREATION   135

7   The Menstrual Cycle   135

8   Gamete Transport and Fertilization   159

Contents **vii**

Preface

In the spring of 1979, I decided to create a new undergraduate biology course in human reproduction at the University of Colorado. At that time, there was a course on human sexuality taught in the psychology department which was (and still is) very popular. Nevertheless, undergraduate students in my course in human anatomy, many of whom were taking the human sexuality course, came to me with questions about the biological and biomedical aspects of human reproduction. This experience, along with my research interests in reproductive endocrinology, led me to believe that there would be student interest in a course about the biology of human reproduction and that I could teach it. This course was first taught to 110 undergraduate students in the spring of 1980.

During the rigorous planning stages of this course, I soon recognized a dilemma. Although there were several books published that emphasized human sexuality, none of them gave more than lip service to the biological aspects of human reproduction and sexual behavior. What was needed was a book that (1) left the reader with a thorough understanding of the current state of knowledge about human reproductive anatomy and physiology, and (2) related this basic scientific knowledge to biomedical aspects of reproduction and to human sexual behavior. Thus, I began writing this book.

Why had such a book not been available previously? One reason is that it has not been until recently that research advances, as well as technological developments and social change, have brought the relevance of reproductive biology into focus in our minds and lives. One is hardly able to pick up a newspaper today without seeing headlines about induced abortion, contraception, sexually transmitted disease, effects of drugs and pollutants on the fetus, treatment of infertility, sex differences in brain function, test-tube babies, artificial insemination, birth defects, the population problem, and so on. Today's advances in reproductive science have brought many choices and concerns to our lives.

A second reason that such a book was not available is that teachers often feel that students are unable to grasp enough of the details of reproductive anatomy and physiology necessary to understand the more applied aspects of human reproduction. This, I feel, is an insult to the intelligence of today's students. Indeed, a great majority of the topics in this book have been integrated and understood by hundreds of undergraduate students in my course without a previous college background in biology. To ensure understanding, this book is written at the undergraduate learning and language level.

The chapters of this book are arranged into four parts, each building on knowledge learned previously. Part I, The Male and Female Reproductive Systems, gives the student a detailed knowledge of the adult male and female systems and the developmental aspects of these systems. Part II, Procreation, is concerned with aspects of conception and childbirth, beginning with the menstrual cycle and ending with the biology of the newborn child. A chapter on human genetics is included in this part for those instructors who wish to use it. All these chapters give the student a solid background for understanding Part III, Medical Aspects of Human Reproduction, in which contraception, induced abortion, infertility, and sexually transmitted diseases are discussed. Finally, Part IV, Human Sexual Behavior, introduces the student to aspects of human sexual response, and patterns, psychology, and sociology of human sexual behavior. These topics are not covered in the same depth as in human sexuality books, but do offer an overview. The final chapter, dealing with human population and family planning, is one of the most important chapters of the book.

The chapter organization begins with a list of learning goals to tell students what they will be able to do after studying each chapter. The chapter summaries also serve as reminders of the chapter coverage. I have chosen illustrations and tables that offer clear examples of phenomena discussed in the text. Furthermore, I explain each figure fully in the legend, sometimes even repeating a text discussion, thus saving searching the text for clarification. Within the text, key terms are italicized the first time they are used and are also defined in the glossary. Finally, at the end of each chapter, there is a further readings section. The readings are chosen for their relevance and because they are at approximately the same reading level as the text.

To increase flexibility of use and amplify text discussions, I have noted sources for statements in the text originating either from (1) recent publications that represent significant research advances, (2) older, landmark research publications, or (3) recent, scientific reviews that are pertinent to the discussion. These sources can be used in advanced courses as primary literature.

Writing such a book is a demanding, time-consuming, and often nerve-wracking task. At the same time, however, I have experienced the joy of learning and creating. Many people supported me during this venture, and, at the risk of forgetting someone, I must list a few: Elizabeth Owen, Dorothea Slater, Barbara Miller, Jeanie Cavanagh, Genevra Metcalf, Gretta Howell, and Pat Holman for several hours of clerical assistance; my graduate students, especially, Louis Guillette, David Duvall, Kevin Fitzgerald, Kristin Lopez, Harriet Austin, and Cliff Summers, for their interesting discussions and understanding that I was unable to give them enough attention while I was writing this book; my colleague David O. Norris (himself the author of a very successful textbook on endocrinology) and Elizabeth Guillette (an RN and Instructor of Obstetrics and Gynecology in the School of Nursing at Wichita State University) for their careful critique of each chapter of the book: Jonathan Van Vlerkom for reviewing Chapter 8; several teachers of human reproduction and sexuality around the country that reviewed this book and offered many helpful suggestions; the staff at Prentice-Hall (especially Karen J. Clemments) for their continuous encouragement and professional service; finally I owe my deepest gratitude to my wife (and favorite person), Betty, for her understanding while she had to share me with my attachment to this project.

R. E. J.

1 *Introduction to Human Reproduction and Sexual Behavior*

LEARNING GOALS

Having read this chapter, you should be able to:

1. Compare and contrast sexual and asexual reproduction, and discuss the adaptive value of sexual reproduction.
2. List the general characteristics of the male and female reproductive systems of primates.
3. Write five questions you have about human reproductive biology, and share them with your instructor or classmates.
4. Discuss the evolution of polygyny and monogamy in humans.
5. List five questions you have about human sexual behavior, and share them with your instructor or classmates.

INTRODUCTION

Like all animals, our species (*Homo sapiens*) has an evolutionary history. That is, we carry with us, and pass to our offspring, inherited structural features, physiological mechanisms, and behavioral tendencies. In addition, however, humans have a cultural history. In fact, we are unique in the degree to which learning and cultural training can interact with, modify, and even dominate our inherited characteristics. More specifically, our reproductive biology appears to be a product of our evolutionary past, whereas our sexual behavior is the result of a complex interaction between our evolutionary history and cultural training. In addition, our unique intellectual ability, through scientific advance, has resulted in a capacity to modify our reproductive biology and sexual behavior and to treat reproductive disease and illness. In this chapter, we look at the evolutionary history of our reproductive patterns, discuss how learning can modify our sexual behavior, and point out how our uniqueness and capacity to influence our reproduction have resulted in choices about the direction of our sexual lives.

THE EVOLUTIONARY HISTORY
OF HUMAN REPRODUCTIVE BIOLOGY

Sexual and Asexual Reproduction

Humans reproduce by a process of *sexual reproduction*, which is common in most animals and plants. In sexual reproduction, a female cell (egg) fuses with a male cell (sperm) to produce a new, unique individual. This is in contrast to *asexual reproduction*, in which new individuals simply bud off from a parent individual, without fusion of cells from a male and female parent. Thus, all offspring produced by asexual reproduction are identical to the parent and to each other. Asexual reproduction is common in bacteria and protozoa as well as in many plants and a few animals.

Genetic Variability and Natural Selection

Why did sexual reproduction evolve? Since a cell from each of two parents fuses to form a new organism, the offspring possesses inherited traits (genes, Chapter 12) from both parents. That is, sexual reproduction allows for a recombination of inherited traits so that the new individual is slightly different from each parent. This is why no two human beings (except identical twins) are exactly alike. This individual variability, a product of sexual reproduction, is the raw material for *natural selection*, a process by which the environment of a species exerts pressure that results in the survival and reproduction of individuals that are adapted to that environment, while other nonadapted individuals fail to survive or reproduce. A characteristic favored by natural selection is said to be an *adaptive trait*. If there are gradual or catastrophic changes in the environment, there may be a few individuals in a species that, because of their inherited make-up, are better adapted to the new conditions than other individuals. Thus, these selected individuals will survive and reproduce. In this way, genetic variability due to sexual reproduction ensures that a species will persist and will not become extinct in a changing environment. In contrast, all individuals of an asexually reproducing species may perish if the environment changes drastically, because they all are genetically similar.

Reproductive Biology of Early Humans

Our reproductive anatomy and physiology evolved to enable the sperm and egg to meet, fuse, and form and nurture a new individual. But what were the reproductive characteristics of the first humans?

Present anthropological evidence suggests that our species, *Homo sapiens*, evolved from another *Homo* species (*Homo erectus*) about 500,000 years ago.[1] These first humans were primitive hunters and food gatherers. They probably existed as small, nomadic bands of closely related individuals. If one could look at the reproductive system of these early humans, what would be found? Unfortunately, we can only make an educated guess because fossilized bones do not reveal much about physiology and anatomy of reproductive tissues. It is certain, however, that these early humans possessed reproductive systems characteristic of other mammals. The females had paired ovaries and a single uterus. Considering that these early humans evolved from another apelike primate, the females must have exhibited a monthly menstrual cycle. The males had paired external testes lodged in a scrotum. Successful mating produced fertilization inside the female reproductive tract, and the resulting fetus spent several months developing in the uterus. When born, the young nursed milk from the female's breasts.

The reproductive biology of present-day humans has not changed much from that of the first humans. What has changed, however, is the ability to manipulate our reproduction through scientific and medical advancements. This, in turn, offers us many new choices about our reproduction.

CURRENT CHOICES AND QUESTIONS CONCERNING HUMAN REPRODUCTION

Recent advances in reproductive science have been so numerous and remarkable that it is difficult for a professional, much less a layman, to keep abreast. A person now, for example, can choose between having a baby or using contraceptive devices. In addition, many contraceptive options are available. Also, infertile couples can become fertile through medical treatment. And tests can be done to determine if a fetus is normal; but with these tests goes the burden of choice to have a handicapped child or to terminate a pregnancy by induced abortion. A mother can also choose to nurse or to bottle-feed her baby.

No wonder today's youth often are confused. Every year I teach a course at the University of Colorado, "The Biology of Human Reproduction," to about 250 undergraduate students. Many of these students have had one year of college biology. Most have had sexual relationships, and many are sexually active to varying degrees. Yet, when I ask them to write down questions that they have about human reproduction and sexual behavior at the beginning of the course, it becomes clear that they are curious and confused about a lot of things. Let me give you some examples of their questions that relate to reproductive biology (not edited except for spelling).

In Chapter 2, we discuss how our brain and pituitary gland govern the function of our reproductive system. Student questions about this subject included: What hormones from the pituitary gland are necessary for ovulation? What is the role of the pineal gland in reproduction? How can the brain change our fertility? The adult female reproductive system is reviewed in Chapter 3, and some questions were: What are the effects of menopause on a woman? What makes cervical cancer so common? What does a Pap smear show? How does an egg ovulate? What organisms normally live in the vagina? What is a tilted uterus? How frequent is breast cancer, and how is it cured? How common are ovarian cysts? Questions relating to the adult male reproductive system (Chapter 4) included: At what rate are sperm produced? What is male menopause? Do males have a monthly cycle of hormone levels? Do males secrete female sex hormones?

In Chapters 5 and 6, we discuss the development and maturation of the reproductive system, and there were several student questions about these subjects: How do sex chromosomes control whether a person is male or female? Is it true that the penis and clitoris are the same? Do hormones cause sex differences in the brain? What causes a person to be hermaphroditic, and how common is it? What causes acne? What causes precocious puberty? Delayed puberty? Why is puberty happening earlier today? Why does strenuous exercise delay puberty? Why do men develop larger muscles?

The menstrual cycle (Chapter 7) was of interest to many students: Why is only one egg released if there are two ovaries? How does the egg get into the fallopian tube? How is it physically possible for a woman to get pregnant during her period? Exactly when can a woman get pregnant? Is it true that the gravitational pull of the moon influences the menstrual cycle? Why do women living together have their period at the same time? What causes premenstrual tension, and is it physical or psychological? How do hormones cause the uterus to bleed once a month? What is in menstrual flow? What are the causes of menstrual pain, and what can be done about this pain? Is having sex during menstruation dangerous? Why would a woman miss a period? What causes toxic shock, and how common is it? What are the effects of menopause on a woman? In Chapter 8, we discuss how sperm reach the egg in the female tract, and review the process of fertilization. Students wondered about the following: Is it true that a couple can determine the sex of their baby by having intercourse at a certain time of the month? How can abnormal chromosomes produce birth defects? How much research is really being done on cloning? What happens to cause twins?

Pregnancy (Chapter 9) was a major topic of interest to students: How does the embryo receive its nourishment and oxygen? What does the embryo look like early in development? How does the female body prepare for actual birth without extensive tearing of her membranes? Do the ovaries play any role in pregnancy? Can smoking harm the fetus? Does pot smoking hurt the fetus? What drugs harm the fetus? What exactly is involved with the Rh- and Rh + blood types, and why is the first baby OK but the second in danger? How can amniocentesis be used to tell if the baby is a boy or girl? What specific tests determine genetic disorders of the fetus, and how safe are they? What happens if an egg is fertilized and implants in some other place beside the uterus? Why isn't the fetus killed by the mother's immune system? How does the embryo attach itself to the uterine wall? How do hormones change at the beginning of pregnancy? How do pregnancy tests work, and are they reliable? What are the causes of miscarriage? Do twins have two placentas?

Labor and birth (Chapter 10) also were the subjects of many student questions: Why can't some women have normal births and use other means such as a C-section? Can smoking cause an abnormal birth? Is it true that the fetus determines when it is born? Does female pelvis size bear a relationship to the ease of pregnancy? What are the different methods of childbirth? Is it true that more births occur at night? Are there any bad effects of anesthetics used during birth on the infant? The students also had questions relating to the newborn and new parents (Chapter 11), such as: What are the pros and cons for circumcising little boys? Can a woman still conceive when she is breast-feeding? How is milk produced in the breasts? What are the advantages of breast-feeding? What does a PKU test detect? What causes crib death, and how frequent is it? Human genetics is reviewed in Chapter 12, and students asked these questions: What are some congenital disorders, and how often do they occur? Should we be concerned about possible selection for certain inherited traits? Is it wise to see a genetic counselor before having children? Why do some inherited traits, like color blindness, only occur in one sex? What does your blood type mean, and why is it important to know it?

Contraception (Chapter 13) is of obvious concern to university students, many of whom are sexually active but do not wish to have a child yet. Some questions were: What is a cervical cap? How do IUDs work, and do they cause infections of the uterus? What are the problems and long-term effects of the various contraceptive methods? Does the pill have long-term, adverse effects? What are possible future birth control methods? Will there be a male birth control pill? How is a vasectomy performed? How and why are a woman's tubes tied? How can cervical mucus be used as a form of birth control? Is sterilization reversible? Induced abortion (Chapter 14) also was of interest to students: What are the effects of an abortion on future pregnancies? What are the different kinds of abortion, and which are dangerous? What are the legal aspects of abortion in the United States? Infertility is the subject of Chapter 15. Items of concern for students included: What causes infertility in both men and women? What causes blockages of the female and male sex ducts? How is artificial insemination done? How is a test-tube baby made? Does freezing sperm harm them? Sexually transmitted diseases are the subject of Chapter 16: What are the dangers to the newborn baby when a pregnant women has a venereal disease? How do people know when they have gonorrhea? Is there a cure for herpes? What are crabs? How can you avoid getting venereal disease? What causes vaginal infections? How does a woman know when she has a yeast infection? Can bladder infections occur in men? What are the modern tests for syphilis?

There are answers to the above questions, and many more, in this book. To understand these answers, and to make intelligent reproductive choices, you must grasp the basic biological principles discussed in this book.

THE EVOLUTIONARY HISTORY
OF HUMAN SEXUAL BEHAVIOR

Courtship and Mating Behaviors

Sexual reproduction necessitated the evolution of sexual behaviors that serve to bring a male and female together so that sperm can be introduced by the male into the female reproductive tract (*internal fertilization*). That is, *courtship behaviors* evolved to attract the sexes to one another, and *mating behaviors* served to facilitate introduction of sperm into the female. These sexual behaviors differ among different animal species. In many mammals, the behaviors are rigid and stereotyped and are instinctive in nature. That is, there is little individual variation in these behaviors, and learning has little influence. Human sexual behavior, in contrast, is highly variable among different cultures and among individuals within a given culture. Therefore, learning and cultural training will exert great influence on any inherited behavioral tendencies that do exist. But what are these inherited tendencies, if present?

Sexual Behavior of Early Humans

How did early humans behave sexually? Again, we can only speculate on the basis of the sexual behavior and mating systems of present-day primates (monkeys, apes, and chimps). *Polygyny*, when a single male has more than one mate (either concurrently or one after the other) is the most common mating system in mammals. Primate species that are polygynous, such as the Old World monkeys, also are *sexually dimorphic*; that is, the males are larger than the females, and there are other sex differences in physical characteristics as well. [2] Those present-day primates that are not sexually dimorphic, such as the New World marmosets, are *monogamous*; that is, a male has a single female as a mate for an extended period. Today's humans are moderately sexually dimorphic (Figs. 1–1 and 1–2). Human males, for example, are about 20 percent heavier, on the average, than females. Males also have broader shoulders, wider hips, stronger muscles, a lower voice, more facial and body hair, and greater height (these are male secondary sexual characteristics). If the early humans possessed a similar sexual dimorphism, they may have had a polygynous mating system.

Is there an inherited tendency toward polygyny in today's humans? Studies of several human societies show that polygyny is common, but many cultures exhibit monogamy through the institution of marriage. [3] The tendency toward monogamy in many human cultures may be the result of cultural training and traditions such as legal marriage, rather than an inherited behavioral tendency. Still, some argue that even early humans evolved a mating system that was more monogamous in nature. They argue that monogamy was adaptive because of the fact that the human infant was, and still is, born in a relatively undeveloped state. That is, there is a long childhood period during which acquired knowledge is transferred by parents,

Figure 1-1. Chimpanzees are moderately sexually dimorphic, males being about 10 percent larger than females. These primates have a polygynous mating system. Humans are more sexually dimorphic than chimpanzees in that human males are not only about 20 percent larger than females but they also differ in other physical characteristics such as the pattern of facial hair. Reproduced from S. Ohno, "The Development of Sexual Reproduction," in *Reproduction in Mammals, Book 6, The Evolution of Reproduction,* eds. C. R. Austin and R. V. Short, pp. 1–31 (Cambridge, England: Cambridge University Press, 1976). © 1976; used with permission.

through learning, to children. They argue that this prolonged period of childhood learning in the early humans required the parents to stay together for a long period of time. That is, monogamy rather than polygyny became adaptive.

Some of our anatomical, physiological, and behavioral characteristics may indeed have evolved as adaptations that supported monogamy. In most mammals, including nonhuman primates, the females are only sexually receptive to the sexual advances of males for a brief period during their ovarian cycle, near the time of ovulation. This female receptive period is called "estrus" (Chapter 7). Such mammals, in turn, would evolve polygyny because the male, which is sexually arousable most of the time, would seek out other sexually receptive females when the female he is with is not receptive.[4] Human females, in contrast, can be sexually receptive at any time during the menstrual cycle (Chapter 17). Perhaps this evolved to help keep the mate around. In other primates, mating is from the rear, and the estrous female often develops sexual skin around her vaginal region that attracts the male. With the evolution of an upright posture and the development of language, however, human mating changed to a face-to-face position, although other coital positions are commonly practiced (Chapter 17). The human female breasts, in turn, have evolved to be fat-filled and larger than those of other primates so as to be sexually attractive to the male at all phases of the menstrual cycle (Fig. 1-2). A recent theory suggests that the increased monogamy and extended parental care in early humans led to the evolution of an upright posture so that the parents could more easily carry food to the young.[5]

Figure 1-2. The human female is the only primate in which the breasts serve as a sexual attractant and are large at times other than during pregnancy and nursing. Reproduced from S. Ohno, "The Development of Sexual Reproduction," in *Reproduction in Mammals, Book 6, The Evolution of Reproduction,* eds. C. R. Austin and R. V. Short, pp. 1–21 (Cambridge, England: Cambridge University Press, 1976). © 1976; used with permission.

CURRENT CHOICES AND QUESTIONS
CONCERNING HUMAN SEXUAL BEHAVIOR

Whatever the inherited pattern of sexual behavior, it is clear that learning and cultural training produce great human variability in how sexual behavior is expressed. The individual differences in training and beliefs, as well as possible conflicts with inherited tendencies, tend to produce many choices and questions about how we express ourselves sexually. Younger people, for example, can choose to be sexually active or to abstain from sex. They may require that they are in love before having sex, or they may not. They may believe that sex before marriage is healthy, or they may feel it is wrong. They may do what their parents or church taught them, or they may do what their peers are saying is the "in" thing to do. Therefore, it is not surprising that the students had many questions about human sexual behavior.

In Chapter 17, we review the human sexual response. Naturally, the students had questions on this subject, such as: How many ejaculations can be achieved? Is it true that males can only be aroused once, but females many times? Is a large penis more satisfying to a woman? What are some reasons why a female does not have an orgasm? What actually happens during orgasm? Do other mammals have orgasms? Are women more "horny" during menstruation, or when they ovulate? What are some physical causes of frigidity and impotence? Are there human pheromones? Do hormones play a role in sexual arousal? Do males biologically need sex more than females, or is that just an erroneous belief our society saddles us with? What is the most sexually sensitive part of a man and of a woman? What are aphrodisiacs? Does marijuana increase sex drive?

Patterns of human sexual behavior are discussed in Chapter 18. Students had several concerns about this topic, including: Is masturbation normal, and can it harm a man or woman? Do rapists have unusually high sex urges? What causes homosexuality, and is it a behavioral adaptation or a physical illness? If men feel effeminate enough to become gay, why do some assume a traditional active male role during sex? What is incest, and why is it outlawed? What causes a person to molest a child? What happens during a sex-change operation, and why does a person want such an operation? In Chapter 19, we review aspects of the psychology and sociology of human sexual behavior. Related student questions were: How do we develop to be masculine or feminine? How is women's liberation affecting men? Are teenagers having sex more often today? Are there really racial differences in sexual behavior? Should a couple be in love before they have sex? What goes on sexually in a commune? Why is divorce so common today? In Chapter 20, we discuss human population growth and family planning. Related student questions were: How fast is our population really growing, and what can be done about it? Is human population size a problem, or will science solve it? Isn't most of the population problem in poor countries?

The purpose of this book is to educate people about the current understanding of human sexual behavior and to point out unanswered questions. After studying this text, you should have the answers to the above questions as well as to many others; and it is hoped that you can then make more intelligent personal, moral, and political decisions about your sexual behavior.

CHAPTER SUMMARY

Inheritance and environment interact to influence our reproductive biology and sexual behavior. Humans exhibit sexual reproduction, and our reproductive system is adapted for internal fertilization and development of the young within the uterus. Human courtship and mating behaviors, as well as sexual dimorphism, evolved to ensure mating, fertilization, and parental care. Cultural training, however, has a great influence on human mating patterns, which range from polygyny to monogamy. Individual variability in the expression of sexual behavior is the result of a complex interaction of inheritance and learning. Our evolutionary history, as well as our cultural training and scientific advances, poses several reproductive questions, concerns, and choices to students. A grasp of the knowledge in this book will answer many of these questions and will allow the student to make intelligent choices about his/her reproduction and sexual behavior.

NOTES

1. D. C. Johanson and M. A. Edey, "Lucy: The Inside Story," *Science*, 81, no. 2 (1981), 44–55.
2. R. V. Short, "The Origin of Species," in *Reproduction in Mammals, Book 6, The Evolution of Reproduction,* eds. C. R. Austin and R. V. Short, pp. 110–48 (Cambridge, England: Cambridge Univeristy Press, 1976).
3. Ibid.
4. D. Symons, "Eros and Alley Oop," *Psychology Today*, 15, no. 2 (1981), 52–61.
5. D. C. Johanson and M. A. Edey, "How Ape Became Man," *Science,* 81, no. 3 (1981), 45–49.

FURTHER READING

OHNO, S., "The Development of Sexual Reproduction," in *Reproduction in Mammals, Book 6, The Evolution of Reproduction*, eds. C. R. Austin and R. V. Short. Cambridge, England: Cambridge University Press, 1976. Pp. 1–31.

SHORT, R. V., "The Origin of Species," in *Reproduction in Mammals, Book 6, The Evolution of Reproduction*, eds. C. R. Austin and R. V. Short. Cambridge, England: Cambridge University Press, 1976. Pp. 110–48.

SHORT, R. V., "The Origins of Human Sexuality," in *Reproduction in Mammals, Book 8, Human Sexuality*, eds. C. R. Austin and R. V. Short. Cambridge, England: Cambridge University Press, 1980.

SYMONS, D. S., *The Evolution of Human Sexuality*. London: Oxford University Press, 1979.

2 *The Brain and Pituitary Gland*

LEARNING GOALS

Having read this chapter, you should be able to:

1. List differences between endocrine and exocrine glands.
2. List 11 major components of the endocrine system.
3. Describe steps taken to detect levels of a hormone in tissue or blood, using bioassay.
4. List differences between a regular neuron and a neurosecretory neuron.
5. Describe the major structural components of the hypophysis.
6. Discuss how and where the hormones oxytocin and vasopressin are synthesized and released.
7. List the eight hormones secreted by the adenohypophysis and give their abbreviations, target tissues, and major effects on the body.
8. Describe where releasing hormones and release-inhibiting hormones are made, where they are secreted, and how they reach the adenohypophysis to exert their effects.
9. List evidence that gonadotropin-releasing hormone plays a role in human reproductive physiology.
10. Define prostaglandins, and describe evidence that they play a role in the secretion of FSH and LH from the adenohypophysis.
11. Describe the interaction of the ''surge center'' and the ''tonic center'' of the hypothalamus in controlling FSH and LH secretion.
12. List evidence that the pineal gland plays a role in human reproduction.

13. Diagram a simple feedback system, listing its component parts.

14. Describe long-loop negative and positive feedback of gonadal steroid hormones on FSH and LH secretion in both sexes, including where these feedback effects act.

15. Differentiate among long-loop, short-loop, and ultrashort-loop feedback effects on FSH and LH secretion.

16. Describe the role of the hypothalamus in controlling secretion of prolactin from the adenohypophysis.

INTRODUCTION

The Hypophysis and Reproduction

The sphenoid bone lies at the base of your skull, and in this bone is a small, cup-shaped depression, the *sella turcica* ("Turkish saddle"). Lying in this depression is a round ball of tissue, about 1.3 cm (0.5 in.) in diameter, called the *hypophysis,* or *pituitary gland* (Fig. 2–1). The word pituitary is derived from "pituita," which means "slime" or "phlegm." Many years ago (A.D. 100), Galen, a physician in Asia Minor, proposed that phlegm secreted by the pituitary gland was one of the four "humors" of the body, the others being blood, black bile, and yellow bile. When the four humors were in balance, the result was pleasure, which explains the origin of the word humorous. Phlegm was supposed to be a cold and moist fluid that caused slug-gishness; a modern meaning is that it is mucus discharged from the throat. Well, the hypophysis is not the source of phlegm, and it is now known that this gland syn-thesizes and secretes chemical messengers (hormones) that travel in the bloodstream and influence many aspects of our body. Realization of the importance of the hypophysis led early scientists to call it the "master gland." More recently, however, we have become aware that activity of this gland, which is connected to the base of the brain by a stalk, is greatly influenced by brain messages. Indeed, one might think of the brain as the conductor of a marvelous chemical symphony played by the pituitary orchestra.

If the hypophysis is removed or destroyed (an operation called *hypophysec-tomy*) our reproductive system becomes nonfunctional, and even our sexual behavior is affected. Therefore, this gland plays a very important role in our reproductive biology. The purpose of this chapter is to describe the hypophysis and its relationship to the brain. We first review the endocrine system and talk about some techniques used to study this system. Then we look at hypophysial structure and function and how our brain regulates pituitary function. Finally, we see how the brain and hypophysis are influenced by homeostatic feedback mechanisms that con-trol our reproductive function. Our discussion of the brain and pituitary gland is complex, but an understanding of this topic is essential for you to grasp the informa-tion in subsequent chapters.

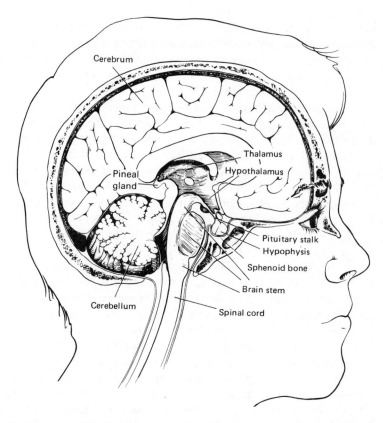

Figure 2–1. Section through the middle of the brain, showing the hypophysis, hypothalamus, and pineal gland. Note that the hypophysis (pituitary gland) rests in a depression in the sphenoid bone and is connected to the hypothalamus by the pituitary stalk. Adapted from *Principles of Human Anatomy,* 3rd ed., by Gerard J. Tortora (New York: Harper & Row, Publishers, 1983), Fig. 16–1, p. 415. © 1983 by Gerard J. Tortora. Reprinted by permission of Harper and Row Publishers, Inc., and Dr. Gerard J. Tortora.

THE ENDOCRINE SYSTEM

There are two kinds of glands in your body. *Exocrine glands* secrete substances into ducts that empty into body cavities and onto surfaces. Examples are the sweat and sebaceous glands of the skin, the salivary glands, and the mucous and digestive glands of your stomach and intestine. *Endocrine glands,* in contrast, do not secrete substances into ducts, which is why they sometimes are called "ductless glands." Instead, their secretory products, *hormones,* are secreted into the tissue spaces adjacent to the endocrine cells; from here they enter the bloodstream and are carried to other regions of the body to exert their effects. Hormones are chemical messengers in that certain tissues in the body are signaled by specific hormones to grow or change their physiological activity. Even though all body tissues are exposed to a hormone, only specific *target tissues* are responsive to a given hormone. Cells of these target tissues have specific *hormone receptors* on their surface membranes, or in their cytoplasm, that bind to a given hormone.

Hormones can have several different molecular structures. Some hormones are *proteins,* small *polypeptides,* or simple *peptides.* These three kinds of molecules are made up of chains of amino acids containing oxygen, carbon, hydrogen, and nitrogen. Other hormones are *amines,* or derivatives of amino acids; these are like polypeptides except they do not contain oxygen. *Steroid* hormones are molecules derived from cholesterol. Male sex hormones (androgens) and female sex hormones (estrogens and progestogens) are steroid hormones. An *androgen* is any substance that promotes development or function of the male reproductive structures. *Estrogens,* on the other hand, are substances that stimulate maturation and function of the female reproductive structure. *Progestogens* are substances that cause the uterus to be secretory. Even though the natural androgens, estrogens, and progestogens are steroid hormones, we will see in Chapter 13 that synthetic, nonsteroid substances can have the action of androgens, estrogens, or progestogens. Finally, some hormones are also derived from fats. We discuss several examples of these kinds of hormones in this and later chapters.

The *endocrine system* is made up of all the endocrine glands in the body. Included in this system are the hypophysis, pineal gland, gonads (testes and ovaries), and placenta—all organs of primary importance in human reproduction. In addition, the endocrine system includes the thyroid, parathyroid, and adrenal glands as well as the hormone-secreting cells of the digestive tract, kidneys, pancreas, and thymus. Figure 2–2 depicts these different components of the endocrine system.

THE SCIENCE OF ENDOCRINOLOGY

Endocrinology is the study of the endocrine glands and their secretions. Suppose you are an endocrinologist who has an idea that a particular gland has an endocrine function. What would you do to test this hypothesis? One "classical" approach, still used today, is as follows: (1) *extirpation* (removal) of the gland, (2) observation of the effects of gland removal on the body, (3) *replacement therapy,* which involves administration of a preparation of the removed gland, and (4) observation to see if the replacement therapy reverses the effects of gland removal. If the replacement therapy does reverse the effects, what could you conclude?

The technique of *bioassay* is used to detect amounts of a given hormone in glandular tissue or blood. In this method, several different amounts of a purified hormone are administered to animals, and the physiological or anatomical changes in target tissues are measured. In this way, a given degree of biological response can be associated with a given amount of hormone administered. Ideally, the response should increase in proportion to the increasing amounts of hormone administered;

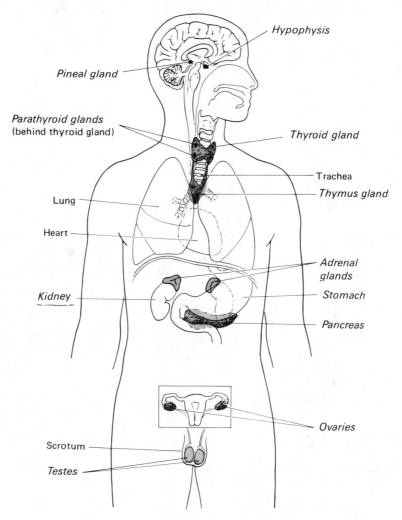

Figure 2–2. Major components of the endocrine system (italicized letters). The placenta is not shown. Adapted from *Principles of Human Anatomy,* 3rd ed., by Gerard J. Tortora (New York: Harper & Row, Publishers, 1983), Fig. 19–1, p. 504. © 1983 by Gerard J. Tortora. Reprinted by permission of Harper and Row Publishers, Inc., and Dr. Gerard J. Tortora.

that is, a *dose-response relationship,* or "standard curve," is obtained (Fig. 2–3). Then, a gland preparation or blood containing an unknown amount of this hormone is administered to other animals, and the biological response obtained is compared to the dose-response relationship. In this way, the amount of hormone in the gland or blood is determined (Fig. 2–3).

Another, more efficient and accurate way to measure the amount of a hormone in tissues or, more often, in blood is *radioimmunoassay.* Radioimmunoassay has been used to measure blood levels of hormones in humans and other animals, and it has been a valuable tool in reproductive biology and in tests for disorders of the endo-

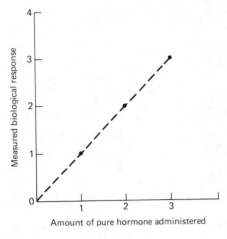

Amount of pure hormone administered

(a)

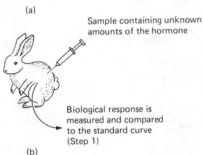

Sample containing unknown amounts of the hormone

Biological response is measured and compared to the standard curve (Step 1)

(b)

Figure 2–3. Procedure for detecting the amount of a given hormone in a sample of tissue or blood, using bioassay. (a) Step 1: Increasing amounts of pure hormone are administered to different test animals such as rabbits. Then, a specific biological response to the hormone is measured. The results are plotted as a dose-response relationship (standard curve). (b) Step 2: A sample containing an unknown amount of the hormone then is administered and the same biological response measured. Then, the degree of this response is compared to the standard curve to determine the amount of hormone in the sample.

crine system. In 1977, Rosalyn S. Yalow received the Nobel Prize in Physiology and Medicine for her development of this technique.

THE HYPOTHALAMO-NEUROHYPOPHYSIAL CONNECTION

The hypophysis has two major regions (Fig. 2–4). One is called the *adenohypophysis* (or *anterior pituitary gland*), which we discuss later in this chapter. The other is the *neurohypophysis* (or *pars nervosa, posterior pituitary gland*). The neurohypophysis is part of the brain, and it develops as an outgrowth of the portion of the embryonic brain that later will become the hypothalamus. To understand the function of the neurohypophysis, you first must realize that there are two general types of nerve cells in the body.

Most of the nerve cells, or *neurons,* in our nervous system consist of a cell body along with long extensions of the cell, dendrites, and axons (Fig. 2–5). *Dendrites* conduct a nerve impulse toward the cell body, which usually is in or near the *central nervous system* (brain and spinal cord). A sensory nerve is really a collection of long dendrites carrying messages to the central nervous system from the periphery. *Axons,* on

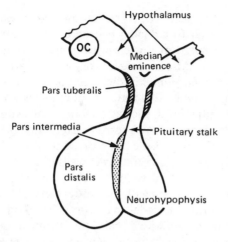

Figure 2-4. Major subdivisions of the human hypophysis (the neurohypophysis and the adenohypophysis) and their relationship to the brain. The pars tuberalis, pars distalis, and pars intermedia all are part of the adenohypophysis. OC = optic chiasma (nerves from the eyes). Adapted from E. G. Rennels and D. C. Herbert, "The Anterior Pituitary Gland—Its Cells and Hormones," *Bioscience,* 29 (7) (1979): 409. Copyright © 1979, American Institute of Biological Sciences. Adapted with permission.

the other hand, carry information away from the cell body, and motor nerves contain axons that stimulate a response in the body such as muscle contraction or glandular secretion. When one neuron connects with another, information is passed from the first to the second cell across a tiny space, the *synapse.* The first neuronal axon secretes a chemical called a *neurotransmitter,* which travels across the synapse and initiates electrochemical changes and then nerve impulses in the next neuron. Some common neurotransmitters are acetylcholine, norepinephrine, dopamine, and serotonin.

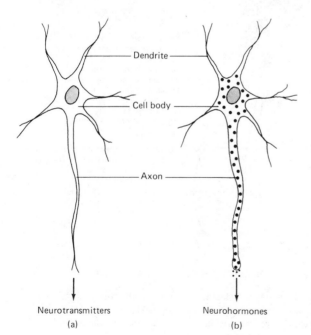

Figure 2-5. Schematic diagram of a regular neuron (a) and a neurosecretory neuron (b). Dendrites carry nerve impulses toward the cell body, whereas axons carry nerve impulses away from the cell body. Neurotransmitters are secreted by the axon endings of regular neurons, whereas neurohormones (dark dots, b) are released from the axon endings of neurosecretory neurons.

The other general kind of nerve cell is the *neurosecretory neuron* (Fig. 2–5). A neurosecretory neuron is similar to a regular neuron in that it can conduct a nerve impulse along its axon. The speed of this electrical conduction is, however, much slower than in a regular neuron. Also, neurosecretory neurons are specialized to synthesize large amounts of *neurohormones* in their cell bodies. These neurohormones then are packaged into large granules that travel in the cytoplasm down the axon and are released into the spaces adjacent to the axon ending. The neurohypophysis contains long axons of neurosecretory neurons surrounded by supporting cells. The cell bodies of these axons lie in the part of the brain called the *hypothalamus*.

The hypothalamus forms the floor and lower walls of the brain (Fig. 2–1), and contains a fluid-filled cavity, the third ventricle. This ventricle is continuous with the other ventricles in the brain and also with the central canal of the spinal cord. The fluid in the ventricles and central canal is called *cerebrospinal fluid*. The weight of the hypothalamus is only $\frac{3}{100}$ that of the whole brain, but it functions in a wide variety of physiological and behavioral mechanisms. For example, there are areas in the

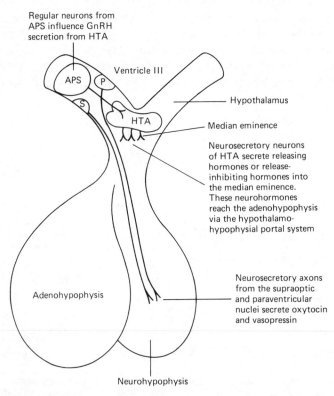

Figure 2–6. Regions of the hypothalamus involved in the function of the hypophysis: APS = anterior hypothalamic-preoptic-suprachiasmatic region; HTA = hypophysiotropic area; P = paraventricular nucleus; S = supraoptic nucleus.

hypothalamus that regulate body temperature, thirst, hunger, sleep, response to stress, and aggressive and sexual behavior.

Of importance to our discussion of the hypophysis is that the cell bodies of the neurosecretory axons in the neurohypophysis lie in paired clusters (*neurosecretory nuclei*) in the hypothalamus. More specifically, these are the supraoptic, suprachiasmatic, and paraventricular nuclei.[1] (*Note:* A neurosecretory nucleus is a group of cell bodies of neurosecretory neurons, and should not be confused with "nucleus" as meaning the body within a cell that contains hereditary material.) The axons of the neurosecretory neurons in these nuclei then pass down the *pituitary stalk* (which connects the hypophysis with the brain) and into the neurohypophysis (Fig. 2–6). The granules released by these axons contain two hormones—*oxytocin* and *vasopressin* (or *antidiuretic hormone*).[2]

Both oxytocin and vasopressin consist of nine amino acids. The two hormones differ only slightly in the kind of amino acids in their molecules, but these slight differences result in their having very different effects on our bodies. Oxytocin stimulates contractile cells of the mammary glands, so that milk is ejected from the nipples (Chapter 11). Also, oxytocin causes the smooth muscle of the uterus to contract, and thus plays a role in labor and birth (Chapter 10). Vasopressin causes the kidneys to retain water; that is, the amount of urine formed in the kidneys is reduced, and more water remains in the body. Also, vasopressin causes blood vessels to constrict and blood pressure to rise. When oxytocin and vasopressin are secreted from the axons in the neurohypophysis, these hormones enter small blood vessels (capillaries) in the neurohypophysis that drain into larger veins.

THE ADENOHYPOPHYSIS

The adenohypophysis (*adeno* = "glandular") consists of three regions: the pars distalis, the pars intermedia, and the pars tuberalis (Fig. 2–4). The *pars distalis* occupies the major portion (70 percent) of the adenohypophysis. The *pars intermedia* is a thin band of cells between the pars distalis and the neurohypophysis. In the adult human, the pars intermedia is sparse or absent. The *pars tuberalis* is a group of cells surrounding the pituitary stalk. During embryonic development, the adenohypophysis forms from an evagination (outpocketing) of the epithelial cell layer of the embryo that later will become the roof of the mouth in the adult. This evagination of epithelial cells then migrates toward the embryonic brain.

The adenohypophysis contains several types of endocrine cells. When you stain the adenohypophysis with laboratory dyes, some cells acquire a pink color. These cells are called *acidophils* (*phil* = "love") because they have an affinity for acid dyes. Some acidophils synthesize and secrete *growth hormone* (GH). This hormone is a large protein that stimulates tissue growth by causing incorporation of amino acids into proteins. The other type of acidophil in the adenohypophysis synthesizes and secretes *prolactin* (PRL), which also is a large protein. As we see in Chapter 11, prolactin acts with other hormones to cause the mammary glands in the female to become functional and secrete milk. Prolactin also may play a role in the function of the ovary (Chapter 3).

Other cells in the adenohypophysis, the *basophils,* stain darkly with basic dyes. These cells synthesize and secrete hormones that are polypeptides or glycoproteins (large proteins with attached sugar molecules). One of these hormones is the glycoprotein *thyrotropin* (TSH). This hormone causes the thyroid glands to synthesize and secrete thyroid hormones (e.g., *thyroxin),* which in turn control the rate at which our tissues use oxygen. In addition, some of the basophils in the adenohypophysis synthesize and secrete *corticotropin* (ACTH), a polypeptide hormone that travels in the blood to the adrenal glands and causes secretion of adrenal steroid hormones (*corticosteroids*) such as cortisol and cortisone. The cortisol and cortisone in turn raise sugar levels in our blood, reduce inflammation, and combat the effects of stress. Other basophils in the adenohypophysis secrete the polypeptide hormones *lipotropin* (LPH) and the *melanophore-stimulating hormone* (MSH). Lipotropin breaks fat down to fatty acid and glycerol. MSH causes synthesis of a brown pigment, *melanin,* which is present in cells called *melanophores.* Finally the basophils of the adenohypophysis secrete two kinds of natural "painkillers"—the *endorphins* and *enkephalins.*[3]

Of particular interest to our discussion of human reproductive biology are the final two hormones secreted by basophils of the pars distalis. One of these is *follicle-stimulating hormone* (FSH). We learn in Chapter 4 that FSH plays a role in sperm production in the testes. In the female, FSH stimulates the ovaries to produce mature germ cells in their enclosed tissue sacs (Chapter 3). The other hormone is *luteinizing hormone* (LH). This hormone causes interstitial cells in the testes to synthesize and secrete androgens (Chapter 4). Thus, LH sometimes is called *interstitial cell-stimulating hormone* (ICSH). In the female, LH causes the ovaries to secrete female sex hormones (estrogens and progestogens) and induces ovulation of an egg from the ovary (Chapter 3). Although most FSH and LH come from the pars distalis, there also is some evidence that the human pars tuberalis contains FSH and LH.[4] Because FSH and LH play vital roles in the function of the gonads, they are grouped under the term *gonadotropic hormones* or "gonadotropins." Figure 2-7 summarizes information about the hormones secreted by the adenohypophysis.

THE HYPOTHALAMO-ADENOHYPOPHYSIAL CONNECTION

It has been known for some time that reproductive cycles of lower animals are influenced by environmental factors such as light, nutrition, and stress, and the same appears true for humans. For example, it was observed that menstrual cycles of women often were altered or even stopped by environmental and psychological stimuli and by emotional arousal (Chapter 7). It was not until 1947, however, that J. D. Green and G. W. Harris provided anatomical evidence that neurosecretory neurons in the hypothalamus could influence the function of the adenohypophysis.[5] As mentioned above, some of the neurosecretory neurons in the hypothalamus send

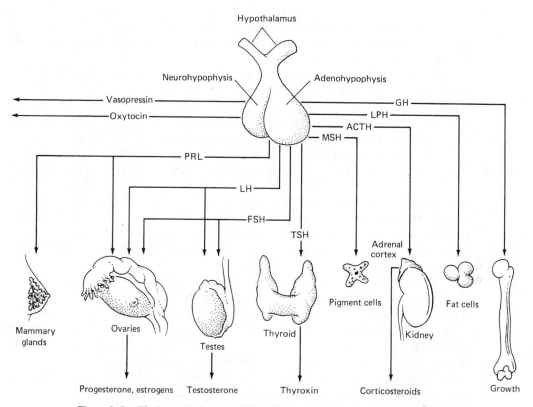

Figure 2-7. The hypophysis, connected to the hypothalamus at the base of the brain, has two lobes. The neurohypophysis stores and releases two hormones made in the hypothalamus: oxytocin and vasopressin. Oxytocin causes contraction of smooth muscle in the uterus, breast, and male reproductive tract. Vasopressin acts on the kidneys to cause water retention. The adenohypophysis secretes eight other hormones: growth hormone (GH) promotes growth; corticotropin (ACTH) causes the adrenal cortex to secrete corticosteroid hormones; follicle-stimulating hormone (FSH) and luteinizing hormone (LH) interact to regulate function of the gonads; prolactin (PRL) causes milk synthesis in the mammary glands and may influence ovarian function; thyrotropic hormone (TSH) stimulates the thyroid gland to secrete thyroxin; lipotropin (LPH) affects fat metabolism; and melanophore-stimulating hormone (MSH) stimulates melanin synthesis in pigment cells.

their axons down the pituitary stalk and into the neurohypophysis, where they release oxytocin and vasopressin. Other neurosecretory neurons existing in paired nuclei in the hypothalamus, however, do not send their axons down the stalk to the hypophysis. Instead, their axons end in an area in the floor of the hypothalamus, near the pituitary stalk, called the *median eminence. (Note:* Sometimes the median eminence is considered part of the neurohypophysis.) The cell bodies of these neurons are clustered in several pairs of nuclei in the hypothalamus, and together these nuclei are named the *hypophysiotropic area* (Fig. 2-6). These nuclei are given this name because the neurosecretory neurons in this region synthesize a family of small polypeptides (neurohormones) that either increase or decrease the secretion of hormones secreted by the adenohypophysis.

If the neurohormones controlling adenohypophysial function are released from neurosecretory neurons at the median eminence, how do they reach the adenohypophysis to influence pituitary hormone secretion? In 1930, G. T. Popa and U. Fielding described a specialized system of blood vessels extending from the median eminence to the adenohypophysis (Fig. 2–8).[6] The superior hypophysial arteries carry blood to the median eminence region. These arteries drain into a cluster of capillaries in the median eminence known as the *primary capillary plexus*. The neurohormones diffuse into these capillaries and into the blood. Then they are carried down to the adenohypophysis in small veins. These veins divide into a second capillary bed surrounding the cells of the adenohypophysis, the *secondary capillary plexus*. The neurohormones then leave the blood through the walls of these capillaries and enter the spaces between the adenohypophysial cells, where they cause these cells to either increase or decrease hormone synthesis and secretion.

A portal system is when blood flows from one capillary bed to another without going through the heart during its journey. Thus, the above vascular system connecting the median eminence with the adenohypophysis is called the *hypothalamo-hypophysial portal system,* and the small veins connecting the primary and secondary capillary plexi are the *hypophysial portal veins*. Recently, a suggestion has been made that other veins, not part of the portal system, can carry blood from the pituitary up to the hypothalamus.[7] Once the hormones of the adenohypophysis are secreted, they leave the hypophysis via the inferior hypophysial vein (Fig. 2–8).

Releasing and Release-inhibiting Hormones

The neurohormones released by the axons of the hypophysiotropic area can either increase or decrease synthesis and secretion of hormones of the adenohypophysis. When a neurohormone increases output of a particular adenohypophysial hormone, it is called a *releasing hormone* (RH). For example, the neurohormone that increases output of thyrotropin is called *thyrotropin-releasing hormone,* or TRH. And when a neurohormone lowers secretion of a particular adenohypophysial hormone, it is termed a *release-inhibiting hormone* (RIH). Thus, the neurohormone that decreases secretion of prolactin is *prolactin release-inhibiting hormone,* or PRIH.

As you can see in Table 2–1, each hormone of the adenohypophysis is controlled by a releasing hormone, and some are known to be controlled by both a releasing and a release-inhibiting hormone. Each releasing or release-inhibiting hormone probably is synthesized by a different group of neurosecretory cell bodies in the hypophysiotropic area. The chemical structures of seven of these neurohormones are known. The others are known as a result of experiments demonstrating their presence, but their chemical nature is yet to be described. In 1977, Andrew Schally and Roger Guillemin shared the Nobel Prize in Physiology and Medicine for their research on hypothalamic neurohormones.

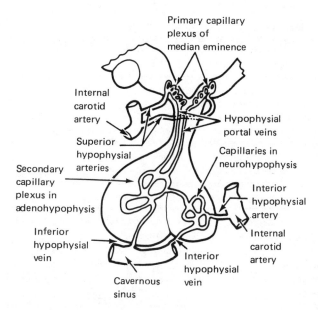

Figure 2-8. The hypothalamo-hypophysial vascular system. Arterial blood enters the median eminence and the neurohypophysis via the superior hypophysial and inferior hypophysial arteries, respectively. Both of these arteries are branches of the internal carotid arteries, major vessels supplying the brain. Neurohormones secreted into the median eminence region enter the blood in the primary capillary plexus. They pass down the hypophysial portal veins to the secondary capillary plexus in the adenohypophysis. Then, they leave the blood and cause the adenohypophysial cells to secrete or stop secreting hormones. When hormones are secreted by the adenohypophysis, they leave the hypophysis in the inferior hypophysial veins, which drain into a large vessel, the cavernous sinus. Neurohypophysial hormones enter capillaries in the neurohypophysis, which also drain into the cavernous sinus. Small blood vessels connect the capillaries of the adenohypophysis and neurohypophysis. Blood also flows from the adenohypophysis back up to the median eminence in small vessels (not shown). Adapted from E. G. Rennels and D. C. Herbert, "The Anterior Pituitary Gland—Its Cells and Hormones," *Bioscience,* 29 (7) (1979): 409. Copyright © 1979, American Institute of Biological Sciences. Adapted with permission.

Of particular interest to us are the neurohormones that control synthesis and release of FSH, LH, and PRL from the adenohypophysis. We now discuss these in some detail because research about these neurohormones has and will have profound influence in controlling human fertility and treating reproductive disorders.

Gonadotropin-releasing Hormone

Primarily through the efforts of Andrew Schally, the chemical nature of *luteinizing hormone-releasing hormone* (LHRH) is known, and it has been synthesized in the laboratory. Many thousands of hypothalami from domestic mammals were obtained from Oscar Mayer & Co. to extract and purify this and other neurohormones. LHRH

TABLE 2–1. HYPOTHALAMIC NEUROHORMONES CONTROLLING THE SYNTHESIS AND RELEASE OF HORMONES FROM THE ADENOHYPOPHYSIS

Neurohormone[a]	Abbreviation
Corticotropin-releasing hormone	CRH
Thyrotropin-releasing hormone	TRH
Gonadotropin-releasing hormone[b]	GnRH[b]
Growth hormone release-inhibiting hormone (or somatostatin)	GHRIH
Growth hormone-releasing hormone	GHRH
Prolactin release-inhibiting hormone	PRIH
Prolactin-releasing hormone	PRH
Melanophore-stimulating hormone release-inhibiting hormone	MSHRIH
Melanophore-stimulating hormone releasing hormone	MSHRH

SOURCE: Adapted from D. O. Norris, *Vertebrate Endocrinology* (Philadelphia: Lea & Febiger, 1980), p. 67. © 1980; used with permission.

[a] We refer to all of these neurohormones as "hormones." Technically, however, only seven have been isolated and identified chemically, and the others are called "factors" by some.

[b] GnRH causes release of both FSH and LH, and is identical to the LHRH discussed in the scientific literature.

is a decapeptide, consisting of ten amino acids, and has been utilized in a wide variety of research and clinical studies. Surprisingly, when LHRH is administered to humans or to laboratory mammals, both LH (Fig. 2–9) and, to a lesser degree, FSH are secreted in increased amounts into the blood,[8] and Schally concludes that there may be only one releasing hormone that controls synthesis and secretion of both LH and

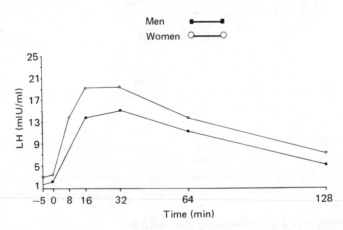

Figure 2–9. When a single injection of GnRH (LHRH) is administered to men and women at time zero, levels of LH rise in the blood, peak after 32 minutes, and then decline. Levels of FSH also rise after GnRH administration (not shown), but not as high as LH. Adapted from A. V. Schally, "Aspects of Hypothalamic Regulation of the Pituitary Gland," *Science,* 202 (1978), 18–28. © 1978 by the Nobel Foundation, Stockholm, Sweden. Used with permission.

FSH.[9] This is in spite of evidence that the hypothalamus of some mammals contains a *follicle-stimulating hormone-releasing hormone* (FSHRH) that causes release of mostly FSH and a little LH.[10] Although future research may show that the human hypothalamus secretes both LHRH and FSHRH, let us for now accept that a single releasing hormone increases both LH and FSH secretion from the pituitary; we call this *gonadotropin-releasing hormone* (GnRH), realizing that this is identical to the LHRH discussed in the scientific literature.

How certain are we that GnRH (LHRH) controls secretion of LH and FSH from the human adenohypophysis? Not only does administration of GnRH increase secretion of LH and FSH in humans, as mentioned, but GnRH levels in the blood of women increase in the middle of the menstrual cycle when a surge of LH, and to a lesser extent FSH, occurs in the blood just before ovulation.[11] Also, GnRH has been isolated from humans.[12] Therefore, this neurohormone must play an important role in human reproduction.

Because the structure of naturally occurring GnRH is known, scientists have been able to synthesize over 300 similar but not identical molecules in the laboratory; these are called *GnRH analogs*. Some of these analogs are up to 150× more potent than natural GnRH, and they cause an increase in FSH and LH secretion when taken orally, whereas natural GnRH is not active when swallowed. GnRH analogs are now used in the United States to induce ovulation in infertile women, as discussed in Chapter 15. Recent data, however, suggest that the human pituitary may become insensitive to these analogs after a short period of time.[13] Other "inhibitory" analogs of GnRH block the action of natural GnRH[14] and therefore may be used as a contraceptive (antifertility) compound in the future (Chapter 13).

Hormones called *prostaglandins* (PG) may influence GnRH secretion, but what are prostaglandins? In 1930, two New York gynecologists discovered that human semen (ejaculate) contains a substance that causes contraction of the human uterus. Shortly thereafter, it was discovered that the active substance in semen was soluble in fat, and it was named *prostaglandin* (PG) by U. S. von Euler in Sweden because it was thought to be secreted by the prostate gland into the semen. Since then, we know that the seminal vesicles (Chapter 4), not the prostate, are the major source of prostaglandins in semen, and prostaglandins also are produced in almost every tissue of the body.

Prostaglandins are a family of molecules derived from fatty acids. They all contain 20 carbon atoms. The most important kinds of prostaglandins are grouped into three categories—prostaglandins A, F, and E. As we discuss later in this book, prostaglandins play several important roles in human reproduction.[15] In fact, the 1982 Nobel Prize in Physiology and Medicine was awarded to S. K. Bergström and B. I. Samuelesson (Sweden) and J. R. Vane (England) for their research on prostaglandins. Pertinent to this discussion is that certain prostaglandins may be involved in the release or inhibition of FSH and LH secretion from the adenohypophysis, at least in rats. Some inhibitors of prostaglandin release, such as aspirin, can lower LH levels in the blood.[16] Prostaglandins probably affect LH secretion by influencing release of GnRH, although there is some evidence that they also can affect the sensitivity of the adenohypophysis to GnRH.[17] We have much more to learn about the role of prostaglandins in the control of gonadotropin secretion in humans.

OTHER BRAIN AREAS AND GONADOTROPIN SECRETION

The Gonadotropin Surge Center

In rats, the *anterior hypothalamic-preoptic-suprachiasmatic* (APS) region of the brain influences secretion of GnRH from the hypophysiotropic area (HTA) (Fig. 2–6).[18] The APS is a group of regular neurons that send their axons to the HTA. The APS is called the "surge center" because it appears to cause the HTA to release a surge of LH secretion immediately before ovulation. When the APS is not activated, the HTA releases a steady, minimal amount of GnRH, and thus it is also called the "tonic center." Rats have a four- or five-day ovarian (estrous) cycle, with ovulation occurring on the morning after the surge of LH secretion that occurred on the afternoon of the previous day. There is a critical period of about two hours on the afternoon of the LH surge when the APS sends neural signals to the HTA that cause a sudden increase in GnRH release and thus a surge of LH and FSH secretion from the adenohypophysis. The daily light cycle can influence the activity of the APS. If rats are kept on continuous light, no LH surge or ovulation occurs. The importance of the APS in the human menstrual cycle, however, is not clear. Since blind women have normal menstrual cycles (Chapter 7), daily lighting may not be as important in women as in female rats.

The Pineal Gland

The *pineal gland,* a single outpocketing from the roof of the brain (Fig. 2–1), may also influence the release of FSH and LH. In the seventeenth century, it was believed that this gland was the "seat of the soul." Now, we know that the pineal synthesizes and secretes a hormone, *melatonin,* which can inhibit the reproductive system of males and females.[19] Exposure of humans to light suppresses melatonin secretion, whereas exposure to dark increases it. Light does not directly affect the pineal. Instead, light entering the eyes causes activity of nerves in tracts (accessory optic tracts) leading to the brain. These impulses cause the part of the sympathetic nervous system that innervates the pineal gland to release the neurotransmitter *norepinephrine.* This neurotransmitter then causes a decrease in melatonin synthesis and secretion. Because of this influence of the daily light cycle on melatonin secretion, levels of melatonin in the blood exhibit a daily cycle. In humans, blood melatonin levels are highest during sleep, between 11 P.M. and 7 A.M.[20] When the daily light schedule is shifted by 12 hours, it takes four or five days for the daily rhythm in blood melatonin in humans to shift to the new light cycle.[21] Not only daily light cycles but also sleep and activity patterns can influence blood melatonin cycles in humans.

We have much more to learn about the role of the pineal in reproduction.

Analogues of melatonin have been made in the laboratory and found to inhibit the human reproductive system when given orally.[22] These analogues could act on the hypothalamus, pituitary gland, or gonads to inhibit reproduction. Melatonin also may play a role in normal puberty (Chapter 6), since levels of this hormone in the blood of young prepubertal males drop markedly just before the onset of puberty.[23] The pineal gland of laboratory mammals secretes other compounds besides melatonin, such as *methoxyindoles* and an octapeptide closely related to oxytocin called *arginine vasotocin*. However, a physiological role for these compounds in human reproduction is not clear.

FEEDBACK CONTROL OF GONADOTROPIN SECRETION

Feedback Systems

In your house or apartment, you probably have a thermostat on your wall that controls the activity of your heating system, and you can set the temperature of your room by manipulating a dial on the thermostat. Now, say that you set the thermostat at 65°F. This temperature you have chosen is called the "set point." The thermostat contains a small strip made of two metals that expands or contracts depending on the temperature. If the room temperature gets below 65°F, the thermostat sends electrical current through the wires leading to the heater, and the heater is activated. When room temperature reaches 65°F, the heater shuts off. Thus, the product of the heater (heat) influences the activity of the heater by feeding back on the device in the thermostat that controls the heater activity.

A simple feedback system is depicted in Fig. 2–10. A receptor detects changes in the system and translates this information into a message ("input"). In your thermostat, the input travels along wires from the temperature receptor (bimetal strip), and then to a "controller center." The controller center contains the set point, and it also generates an outgoing message ("output"). There are wires leading from the controller center in the thermostat to the heater that carry this output message. These output wires are activated when the room temperature is below the set point. The "effector" (heater) then responds by producing an "effect" (heat). In turn, the effect produces a change in the system (an increase in room temperature) that has a feedback effect on the controller center. This is called a *feedback loop*. In some feedback systems, other circuits ("higher centers" in Fig. 2–10) can modify the activities of the controller center by temporarily altering the set point or by inhibiting the operation of the controller center.

Many aspects of our physiology operate as feedback systems that regulate our internal environment at a steady state; this regulation is called *homeostasis*. Many homeostatic control systems in our body operate through *negative feedback*. In the case of pituitary gland function, a negative feedback system is one in which secretion of a pituitary hormone to a level above the set point causes a decrease in secretion of that same pituitary hormone into the blood. In reproductive physiology, however,

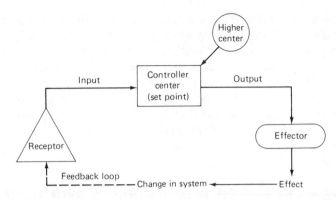

Figure 2–10. A simple feedback system. The receptor detects the level of a particular component of the system and translates it into a message (input) to the controller center. The input is compared by the controller center to its programmed set point, and this center computes whether a regulatory response is required. If necessary, the controller center generates a signal (output) that is transmitted to one or more effectors, which respond by producing some effect. This effect produces a change in the system, which then feeds back as a feedback loop on the controller center after being received by the receptor. Other circuits (higher centers) can modify the activities of the controller center by temporarily altering the set point or by inhibiting the operation of the controller center.

there are important occurrences of *positive feedback,* in which the secretion of a pituitary hormone influences the controller center so that secretion of that hormone increases even more. We discuss now in some detail the kinds of positive and negative feedback important in controlling secretion of the gonadotropins from the adenohypophysis.

Regulation of Gonadotropin Secretion by Feedback

As we discuss in detail in Chapters 3 and 4, FSH and LH cause secretion of sex hormones by the *gonads* (testes or ovaries). These steroid hormones (androgens, estrogens, progestogens) are products (effects) of the action of gonadotropins on the gonads, and it turns out that steroid hormones influence secretion of LH and FSH by having feedback effects on the systems controlling gonadotropin secretion. These feedback effects of gonadal steroid hormones are called *long-loop feedback* effects, since the effects are produced by peripheral target tissues (the gonads) and travel in the blood to the brain and pituitary gland (Fig. 2–11).

In women, administration of small amounts of estrogens will lower secretion of FSH and LH into the blood. This negative feedback effect of estrogens is even more effective when given in combination with high levels of a progestogen. In fact, this is the reason that combination contraceptive pills contain low levels of an estrogen and high levels of a progestogen (Chapter 13). In the normal menstrual cycle, high levels of a progestogen and low levels of an estrogen in the blood during the luteal phase of

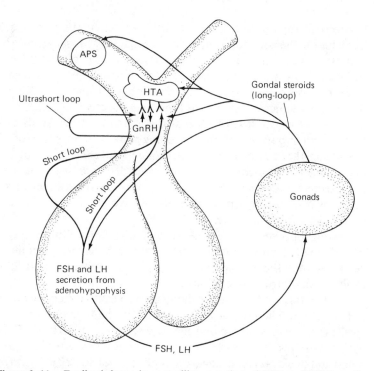

Figure 2–11. Feedback loops in controlling secretion of FSH and LH from the adenohypophysis. The gonads secrete gonadal steroid hormones (estrogens, progestogens, androgens) in response to FSH and LH, and these steroids affect activity of the anterior hypothalamic-preoptic-suprachiasmatic region (APS), hypophysiotropic area (HTA), and median eminence, thus influencing GnRH secretion by long-loop feedback. In addition, gonadal steroid hormones exert long-loop feedback on FSH and LH secretion by influencing sensitivity of FSH and LH secreting cells in the adenohypophysis to GnRH. FSH and LH also can travel in the general circulation or back up the pituitary stalk and influence secretion of GnRH at the median eminence by direct, short-loop feedback. Release of GnRH can also inhibit further release of GnRH in the median eminence by ultrashort feedback. Long-loop feedback effects can be negative or positive, depending on their sites of action and concentration in the blood. Thus far, however, short and ultrashort feedback loops have only been found to be negative.

the cycle (period between ovulation and menstruation) lower gonadotropin secretion by long-loop negative feedback and thus prevent ovulation at this time (Chapter 7).

It may help you to think that the hypothalamus contains a controller center with a *gonadostat* that works like the thermostat in your heating system. This gonadostat contains a set point for levels of estrogen and progestogen reaching it from the blood. When levels of these steroids are higher than the set point, the gonadostat signals the hypophysiotropic area to stop secreting GnRH, and thus secretion of FSH and LH from the adenohypophysis declines. If circulating levels of estrogen and progestogen are below the set point, the gonadostat signals the hypophysiotropic area to release GnRH, and circulating levels of FSH and LH rise. But where is the gonadostat in the hypothalamus? Most evidence in laboratory mammals suggests that it is in the hypophysiotropic area itself,[24] or in the median eminence.[25] In male mammals, negative feedback of androgens on FSH and LH secretion also can operate on the hypophysiotropic area[26] or median eminence.[27] Figure 2–11 depicts these sites of long-loop negative feedback.

Recent research suggests that there may be other kinds of negative feedback that play an important role in controlling gonadotropin secretion. For example, there is evidence that LH or FSH can circulate in the general vascular system back to the brain, reach the median eminence, and decrease secretion of GnRH from the neurosecretory axons.[28] This direct kind of negative feedback of a pituitary hormone on its own secretion, which does not involve gonadal steroid hormones, has been called *short-loop negative feedback* (Fig. 2–11). Finally, there may even be an *ultrashort-loop negative feedback* (Fig. 2–11) in that GnRH may directly stop secretion of GnRH from the neurosecretory axons in the median eminence.[29] We do not as yet understand the interaction of long-, short-, and ultrashort-loop feedback in controlling gonadotropin secretion.

Thus far, we have been talking about negative feedback on pituitary FSH and LH secretion. There is, however, a stage in the human menstrual cycle when high levels of estrogen in the blood increase secretion of LH and FSH from the adenohypophysis, resulting in a "surge" of these gonadotropins (primarily LH) in the blood and ovulation of an egg from the ovary (Chapter 7). Therefore, very high levels of estrogen in the blood have a positive feedback effect on gonadotropin secretion, whereas low levels of estrogen (but still above the set point) have a negative feedback effect on gonadotropin secretion. In rats, this long-loop positive feedback effect of high levels of estrogen operates by stimulating electrical activity in the APS, which then excites the hypophysiotropic area and increases GnRH secretion.[30]

This complicated story of feedback effects on pituitary LH and FSH secretion has one more chapter. Recent research has shown that the long-loop positive and negative feedback effects of gonadal steroids on LH and FSH secretion can operate directly on the adenohypophysis itself as well as on the brain (Fig. 2–11). That is, the response of FSH and LH secreting cells in the adenohypophysis to GnRH may vary depending on the kinds and amounts of steroid hormones bathing these cells. For example, it has been shown that estrogen increases the sensitivity of the pituitary to GnRH.[31] Progestogens also can increase pituitary sensitivity to GnRH in women.[32] In males, androgens have a negative feedback on gonadotropin secretion not only by influencing the brain but also by decreasing response of the adenohypophysis to GnRH.[33]

The control of prolactin (PRL) secretion by the brain differs in some respects from brain control of LH and FSH secretion. If the hypophysiotropic area of the hypothalamus is destroyed, secretion of PRL from the adenohypophysis increases,[34] whereas secretion of LH and FSH declines. Therefore, there is a prolactin release-inhibiting hormone (PRIH) secreted by neurosecretory neurons in the hypophysiotropic area that inhibits prolactin secretion. However, electrical stimulation of the medial hypothalamus can increase prolactin secretion, so it appears that there is also a PRH.[35] Finally, estrogens increase the response of prolactin-secreting cells in the adenohypophysis to PRH.[36] Knowledge of the hypothalamic control of prolactin secretion is important because of the role of this hormone in milk synthesis by the mammary glands (Chapter 11) and the association of abnormally high levels of PRL with certain kinds of infertility (Chapter 15).

CHAPTER SUMMARY

Exocrine glands secrete their products into ducts, whereas endocrine glands secrete their products (hormones) directly into the bloodstream. Specific hormones influence growth and function of certain target tissues. Hormones can be proteins, polypeptides, amines, steroids, or fats. Methods used in the science of endocrinology include gland extirpation and replacement therapy, bioassay, and radioimmunoassay.

The hypophysis has two major parts: the neurohypophysis and the adenohypophysis. Neurosecretory neurons in the hypothalamus synthesize oxytocin and vasopressin, which travel to the neurohypophysis in neurosecretory cell axons. The adenohypophysis contains three regions: the pars distalis, pars tuberalis, and pars intermedia (reduced or absent in humans). Different cells in the pars distalis secrete the hormones follicle-stimulating hormone (FSH), luteinizing hormone (LH), prolactin (PRL), corticotropin (ACTH), growth hormone (GH), thyrotropin (TSH), and lipotropin (LPH). The pars intermedia secretes melanophore-stimulating hormone (MSH), and the pars tuberalis may also secrete FSH and LH.

Neurosecretory neurons in the hypophysiotropic area (HTA) of the hypothalamus secrete releasing hormones or release-inhibiting hormones into the median eminence region at the base of the hypothalamus. Here, capillaries receive these hormones, which then travel in the blood of the hypothalamo-hypophysial portal system to the endocrine cells of the adenohypophysis. The releasing hormones then increase synthesis and secretion of specific adenohypophysial hormones, whereas the release-inhibiting hormones have the opposite effect.

Since luteinizing hormone-releasing hormone (LHRH) increases synthesis and secretion of both FSH and LH, it is also called gonadotropin-releasing hormone (GnRH). There is good evidence that GnRH plays an important role in human reproduction, and this molecule and GnRH analogues are being used to treat infertility and as possible contraceptive agents. Prostaglandins may influence GnRH secretion. In female rats, the anterior hypothalamic-preoptic-suprachiasmatic area (APS) of the hypothalamus causes a surge of LH secretion just before ovulation by increasing GnRH secretion from HTA. The pineal gland secretes the hormone melatonin, which exerts inhibitory effects on gonadotropin secretion.

Feedback systems control FSH and LH secretion from the adenohypophysis. In long-loop feedback, FSH and LH cause the gonads to secrete gonadal steroid hormones (estrogens, progestogens, androgens), which can decrease (by negative feedback) further secretion of FSH and LH. Estrogen also can have a positive feedback effect on LH secretion. The long-loop feedback effects can operate on the APS, HTA, median eminence, or adenohypophysis. Short-loop negative feedback occurs when FSH and LH directly inhibit further secretion of GnRH. Ultrashort-loop negative feedback occurs when GnRH inhibits further secretion of GnRH. Prolactin secretion from the adenohypophysis is controlled by both a prolactin-releasing hormone (PRH) and a prolactin release-inhibiting hormone (PRIH) from the hypothalamus. Estrogens increase the sensitivity of prolactin-secreting cells in the adenohypophysis to PRH.

NOTES

1. M. M. Brownstein et al., "Synthesis, Transport, and Release of Posterior Pituitary Hormones," *Science,* 207 (1980), 373–78.

2. Ibid.

3. E. G. Rennels and D. C. Herbert, "The Anterior Pituitary Gland—Its Cells and Hormones," *Bioscience,* 29 (1979), 408–13; and W. D. Odell, "Melanocyte-Stimulating Hormones, Lipotropins, and Enkephalins," in *Endocrinology,* eds. L. D. DeGroot et al., vol. 1, pp. 169–74 (New York: Grune & Stratton, Inc., 1979).

4. B. L. Baker, "Cellular Composition of the Human Pars Tuberalis as Revealed by Immunocytochemistry," *Cell and Tissue Research,* 182 (1977), 151–63.

5. J. D. Green and G. W. Harris, "The Neurovascular Link between the Neurohypophysis and Adenohypophysis," *Journal of Endocrinology,* 5 (1947), 136–46.

6. G. T. Popa and U. Fielding, "A Portal Circulation from the Pituitary to the Hypothalamic Region," *Journal of Anatomy,* 65 (1930), 88–91.

7. R. M. Bergland and R. B. Page, "Can the Pituitary Secrete Directly to the Brain? (Affirmative Anatomical Evidence)," *Endocrinology,* 102 (1978), 1325–38.

8. A. Kastin et al., "Administration of LH-Releasing Hormone of Human Origin to Man," *Journal of Clinical Endocrinology and Metabolism,* 32 (1972), 287–90; and A. Kastin et al., "Release of LH and FSH after Administration of Synthetic LH-releasing Hormone," *Journal of Clinical Endocrinology and Metabolism,* 34 (1972), 753–56.

9. A. V. Schally et al., "Hypothalamic Regulatory Hormones," *Science,* 179 (1973), 341–50; and A. V. Schally, "Aspects of Hypothalamic Regulation of the Pituitary Gland," *Science,* 202 (1978), 18–28.

10. C. Y. Bowers et al., "Biological Evidence That Separate Hypothalamic Hormones Release the Follicle-Stimulating and Luteinizing Hormone," *Biochemical and Biophysical Research Communication,* 50 (1973), 20–26.

11. J. Malacara et al., "Luteinizing Hormone Releasing Factor Activity in Peripheral Blood from Women during the Midcycle Luteinizing Hormone Ovulatory Surge," *Journal of Clinical Endocrinology and Metabolism,* 34 (1972), 271–78.

12. Schally (1978), op. cit.

13. D. Gonzalez-Barcena et al., "Stimulation of Luteinizing Release Hormone (L.H.) Release after Oral Administration of an Analogue of L.H. Releasing Hormone," *Lancet,* ii (1975),

1126–28; and Anonymous, "Clinical Use of LH–RH Analogues," *Research in Reproduction,* 11, no. 6 (1979), 2.

14. Schally (1978), op. cit.

15. V. J. Goldberg and P. W. Ramwell, "The Role of Prostaglandins in Reproduction," in *Frontiers in Reproduction and Fertility Control,* eds. R. O. Greep and M. A. Koblinsky, pp. 219–35 (Cambridge, Mass.: M.I.T. Press, 1977).

16. S. M. McCann and S. R. Ojeda, "The Role of Brain Monamines, Acetycholine, and Prostaglandins in the Control of Anterior Pituitary Function," in *Endocrinology,* eds. L. J. DeGroot et al., vol. 1, pp. 55–63 (New York: Grune & Stratton, Inc., 1979).

17. R. M. G. Nair et al., "Action of Gonadotropin-Releasing Hormone and Its Superactive Analogues on the Anterior Pituitary: The Mechanisms of Release and Synthesis of Gonadotropins," *Neuroendocrinology,* 28 (1979), 11–24.

18. B. Halasz, "Hypothalamic Mechanisms Controlling Pituitary Function," *Progress in Brain Research,* 38 (1972), 97–122.

19. R. J. Wurtman, "The Pineal Organ," in *Endocrinology,* eds. L. H. DeGroot et al., vol. 1, pp. 95–102 (New York: Grune & Stratton, Inc., 1979).

20. R. J. Wurtman and M. A. Moskowitz, "The Pineal Organ," *New England Journal of Medicine,* 296 (1977), 1329–33.

21. D. C. Jinerson et al., "Urinary Melatonin Rhythms during Sleep Deprivation in Depressed Patients and Normal," *Life Sciences,* 20 (1977), 501.

22. Wurtman and Moskowitz (1977), op. cit.

23. R. E. Silman et al., "Melatonin, the Pineal Gland, and Human Puberty," *Nature,* 282 (1972), 301–5.

24. V. D. Ramirez et al., "Effect of Estradiol Implants in the Hypothalamo-Hypophysial Region of the Rat on the Secretion of Luteinizing Hormone," *Endocrinology,* 75 (1964), 243–48.

25. F. Piva et al., "Regulation of Hypothalamic and Pituitary Function: Long, Short, and Ultrashort Feedback Loops," in *Endocrinology,* eds. L. J. De Groot et al., vol. 1, pp. 21–34 (New York: Grune & Stratton, Inc., 1979).

26. H. Sar and W. E. Stumpf, "Autoradiographic Localization of Radioactivity in the Rat Brain after the Injection of 1, 2-^{3}H-Testosterone," *Endocrinology,* 92 (1973), 251–56.

27. R. D. Lisk, "Testosterone-Sensitive Centres in the Hypothalamus of the Rat," *Acta Endocrinologica, Copenhagen,* 41 (1962), 195–204.

28. F. Fraschini et al., "A 'Short' Feedback Mechanism Controlling FSH Secretion," *Experientia,* 24 (1968), 270–71.

29. Piva et al. (1979), op. cit.

30. F. Docke and G. Dorner, "The Mechanism of the Induction of Ovulation by Estrogens, *Journal of Endocrinology,* 33 (1965), 491–99.

31. Anonymous, "Varying Sensitivity of the Pituitary to LH–RH during the Oestrous Cycle," *Research in Reproduction,* 6, no. 5 (1974), 2; R. B. Jaffe and W. R. Keys, Jr., "Estradiol Augmentation of Pituitary Responsiveness to Gonadotropin-Releasing Hormone in Women," *Journal of Clinical Endocrinology and Metabolism,* 39 (1974), 850; and Y. Nakai et al., "On the Sites of the Negative and Positive Feedback Actions of Estradiol in the Control of Gonadotropin Secretion in the Rhesus Monkey," *Endocrinology,* 102 (1978), 1008–14.

32. S. J. Nillius and L. Wide, "Progesterone-Induced Augmentation of Pituitary Gonadotrophic Responses to Luteinizing Hormone-Releasing Hormone in Oestrogen-Pretreated Amenorrhoeic Women," *Acta Endocrinologica, Copenhagen,* 83 (1976), 684–91.

33. J. W. Everett, "Central Neural Control of Reproductive Functions of the Adenohypophysis," *Physiological Reviews,* 44 (1964), 373–431.

34. J. W. Everett, "Luteotrophic Function of Autografts of the Rat Hypophysis," *Endocrinology,* 54 (1954), 786–96.

35. J. W. Everett and D. L. Quinn, "Differential Hypothalamic Mechanisms Inciting Ovulation and Pseudopregnancy in the Rat," *Endocrinology,* 78 (1966), 141–50.

36. Anonymous, "The Nature of Prolactin Secretion," *Research in Reproduction,* 6, no. 5 (1974), 2.

FURTHER READING

ARIMURA, A., and A. FINDLAY, "Hypothalamic Map for the Regulation of Gonadotropin Release, Based Mainly on Data Obtained in the Rat," *Research in Reproduction,* vol. 3, no. 1 (1974), map.

BAIRD, D. T., "Reproductive Hormones," in *Reproduction in Mammals, Book 3, Hormones in Reproduction,* eds. C. R. Austin and R. V. Short. Cambridge, England: Cambridge University Press, 1972. Pp. 1–28.

CROSS, B. A., "The Hypothalamus," in *Reproduction in Mammals, Book 3, Hormones in Reproduction,* eds. C. R. Austin and R. V. Short. Cambridge, England: Cambridge University Press, 1972. Pp. 29–41.

GUILLEMIN, R., and R. BURGUS, "The Hormones of the Hypothalamus," *Scientific American,* 227 (1973), 24–33.

RENNELS, E. G., and D. C. HERBERT, "The Anterior Pituitary Gland—Its Cells and Hormones," *Bioscience,* 29 (1979), 408–13.

3   *The Adult Female Reproductive System*

LEARNING GOALS

Having read this chapter, you should be able to:

1. List the major structural components of the ovary.
2. Describe the structure of primordial, primary, secondary, and tertiary ovarian follicles.
3. List the major pathways of steroidogenesis within the ovary, and describe the sites of steroid hormone secretion within the ovary.
4. Trace the interaction of hormones in controlling ovarian follicular growth.
5. List the steps of oocyte meiosis; describe when they occur and the stimuli needed for each step.
6. Propose a general model of the process of ovulation and its control.
7. Describe how a corpus luteum is formed.
8. Distinguish among the terms cancer, neoplasm, benign, malignant, and tumor.
9. List the regions of each oviduct, and describe adaptations of oviduct structure that facilitate egg, sperm, and embryo transport.
10. Describe the major regions of the uterus and the structure of the uterine wall.
11. Define cervical and endometrial cancer, and discuss some factors that may relate to their incidence.
12. Describe how the uterus can be out of position, and discuss the causes and treatments of these conditions.
13. Discuss how the acidic vaginal environment usually is maintained.

14. List the major components of the female external genitalia, and describe their location.

15. Describe in detail the glandular and ductal components of the mammary gland.

16. Describe early signs of breast cancer, and some methods used to detect and treat this disease.

INTRODUCTION

The adult female reproductive system consists of the paired ovaries and oviducts, the uterus, the vagina, the external genitalia, and the mammary glands. All of these structures have evolved for the primary functions of ovulation, fertilization of an ovum by a sperm, and birth and care of a child. The components of this system are integrated structurally and physiologically to these ends.

In standard terminology, the female *primary sexual characteristics* are the internal structures that are involved in reproduction in the female and include the ovaries as well as the oviducts, uterus, and vagina (female *sex accessory ducts*), and the external genitalia (Fig. 3–1). Female *secondary sexual characteristics,* on the other hand, include all those external features (except external genitalia) that distinguish an adult female from an adult male. These include enlarged breasts and the characteristic distribution of fat in the torso.

In this chapter, we look at the anatomy, endocrinology, and disorders of the adult female reproductive system, whereas the menstrual cycle will be covered in Chapter 7.

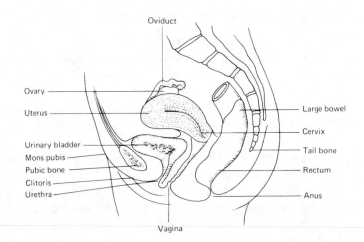

Figure 3–1. Side view of the female pelvic region, showing some major components of the reproductive system. Adapted from *Woman to Woman* by Lucienne Lanson and illustrated by Anita Karl (New York: Alfred A. Knopf, Inc., 1975). © 1975 by Lucienne Lanson and Alfred A. Knopf, Inc. Reprinted by permission of Alfred A. Knopf, Inc., and Penguin Books Ltd., Australia and England.

THE OVARIES

Ovarian Gross Anatomy

The *ovaries,* or gonads of the female, are paired structures lying on each side of the upper pelvic cavity, up against the pelvic wall and near the uterus (Figs. 3-1 and 3-2). Each ovary is about 2 to 4 cm (¾ to 1½ in.) in length and is connected to the uterus and pelvic wall by supportive ligaments. The ovary is one of the most vascular organs of the body.[1] An ovarian artery leaves the abdominal aorta on each side. Before this artery enters the ovary, it is joined by a branch of a uterine artery. The common artery then enters the ovary at an indentation on its surface. Blood drains from each ovary in the ovarian vein. This vein joins with a branch of a uterine vein to form the utero-ovarian vein, which then enters the inferior vena cava, a major vessel that takes blood to the heart.

Ovarian Microanatomy

If a slice is cut out of the ovary and this section is examined with a microscope, it will be clear that the ovary is a very complex structure (Fig. 3-3). The external surface of each ovary is covered with a thin sheet of tissue, the *surface epithelium*. This epithelium once was called the "germinal epithelium" because it was thought that the female germ cells were derived from this tissue. Now it is known, however, that germ cells originate outside the ovary during embryonic development, as we discuss in Chapter 5. Underneath the surface epithelium is a connective tissue framework (the *ovarian stroma*), which is divided into a more dense, outer *ovarian cortex* and a less dense, central *ovarian medulla*. The ovarian medulla contains large, spirally arranged blood vessels, lymphatic vessels, and nerves. The ovarian cortex contains the female germ cells.

Each female germ cell, or *oocyte,* is enclosed in a tissue sac, the *ovarian follicle* (*follicle* is Latin for "little bag"). Between the oocyte and the follicular wall is a thin transparent membrane, the *zona pellucida*. The origin of the zona pellucida is not clear; it could be secreted by the follicular wall, oocyte, or both.[2]

There are many follicles in the ovary.[3] At birth, each ovary of a newborn girl contains about 500,000 follicles. This number is all a female will have in her life; no new follicles are made after birth. Before puberty, many follicles degenerate. Death of a follicle, a process called *atresia,* can occur at any stage of follicular growth. Degenerating follicles are called *corpora atretica*. At puberty, each ovary contains only about 83,000 follicles. The process of atresia continues throughout a woman's reproductive life. An average of 50 to 75 percent of follicles are atretic at any age.[4] At 35 years of age, each ovary contains only about 30,000 follicles. Because of follicular atresia, as well as the loss of about 400 to 500 follicles through ovulation, ovaries in women 50 years or older contain less than 1000 follicles.[5] Tissue derived from atretic follicles produces clumps of endocrine cells dispersed throughout the stroma. These

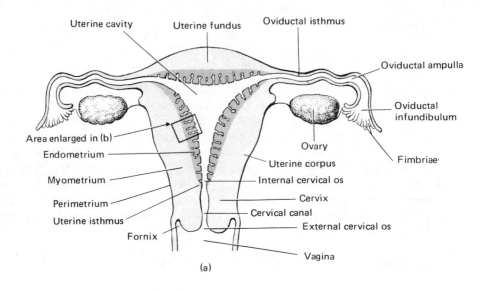

(a)

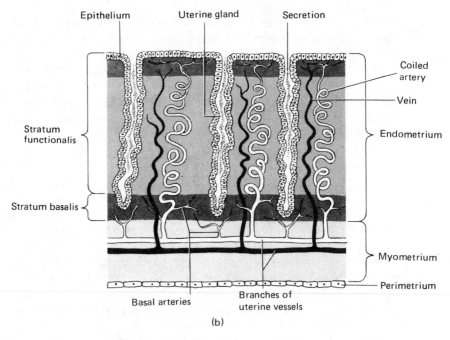

(b)

Figure 3–2. A diagrammatic section through the uterus and oviducts. Figure B is an enlargement of the box in A, showing the structure of the uterine wall. Adapted from K. L. Moore, *The Developing Human. Clinically Oriented Embryology,* 3rd ed. (Philadelphia: W. B. Saunders Company, 1982). © 1982; used with permission.

are called *ovarian interstitial cells.* We now look at the stages of growth of an individual ovarian follicle.

Stages of Follicular Growth

Most follicles within the adult ovary are very small (50 μm in diameter). These *primordial follicles* consist of an oocyte surrounded by a single layer of flattened, *granulosa cells,* the *membrana granulosa.* The primordial follicles lie in the periphery of the ovarian cortex (Fig. 3-3). Relatively few primordial follicles initiate growth at any given time. Those that do grow become *primary follicles,* which are about 100 μm in diameter. The granulosa is still a single layer of cells in primary follicles, but these cells are now cube-shaped instead of flattened (Fig. 3-3). In addition, primary follicles acquire a connective tissue covering around the granulosa. This covering, which contains small blood vessels, is the *theca.* As a primary follicle continues to grow, the granulosa cells divide mitotically, so that *secondary follicles* have a membrana granulosa consisting of two to six cell layers (Fig. 3-3). The theca is still a single layer in secondary follicles.

A few secondary follicles that avoid becoming atretic grow to a more advanced stage to become *tertiary follicles* (Fig. 3-3). First, the granulosa cells secrete fluid that accumulates between the cells (in follicles about 200 μm in diameter). These fluid spaces then join, and a lot of additional fluid leaves pores in the thecal blood vessels and is added to the fluid space. This fluid-filled space is the *antral cavity,* and the fluid is called *antral fluid (follicular fluid).* About 80 percent of the proteins present in the plasma of the blood in thecal vessels can leave the vessels and diffuse through the follicular wall and into the antral cavity. Antral fluid contains steroid and protein hormones, anticoagulants (substances that prevent blood clotting), enzymes, and electrolytes (ions with positive or negative charges).[6]

Tertiary follicles have a membrana granulosa of three or four cell layers, and the theca now is differentiated into an inner *theca interna,* containing glandular cells and many small blood vessels, and a *theca externa,* consisting of dense connective tissue and larger blood vessels (Fig. 3-3). The oocyte in a tertiary follicle, which now has enlarged to about 100 to 130 μm in diameter, is suspended in antral fluid by a stalk of granulosa cells, the *cumulus oophorus.* Immediately surrounding the oocyte is a thin ring of granulosa cells, the *corona radiata.* Figure 3-4 is a picture of such an oocyte.

Tertiary follicles can be divided further into categories based on size. Resting tertiary follicles are 1 to 9 mm in diameter, ripe tertiary follicles about 10 to 14 mm in diameter, and graafian follicles about 15 to 25 mm in diameter. The theca of the most mature graafian follicle is extremely vascular. One graafian follicle during each menstrual cycle (just before ovulation) becomes so large that it forms a blisterlike bulge on the surface of the ovary.

In both ovaries of a woman from 16 to 25 years of age, there are, on the average, 94 tertiary follicles, 6600 primary and secondary follicles, and 159,000 primordial follicles.[7]

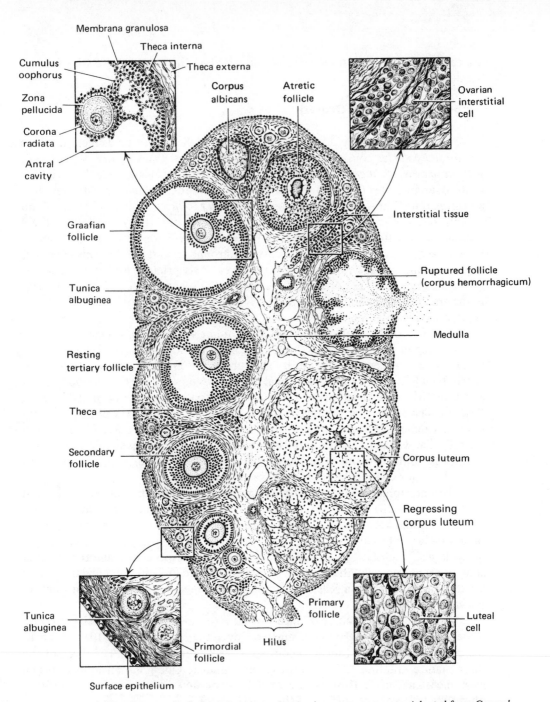

Figure 3–3. Section through the ovary, showing its component parts. Adapted from *General Endocrinology,* 6th ed., by C. Donald Turner and Joseph T. Bagnara (Philadelphia: W. B. Saunders Company, 1976). © 1976 by the W. B. Saunders Company. Adapted by permission of Holt, Rinehart and Winston.

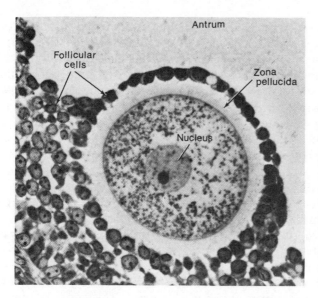

Figure 3–4. Photomicrograph of a region from a human tertiary follicle, showing the oocyte surrounded by follicular (granulosa) cells. The ring of cells immediately next to the oocyte is the corona radiata. The stalk of cells connecting the oocyte to the follicular wall is the cumulus oophorus. At the lower left corner are some cells of the theca. Reproduced from W. Bloom and D. W. Fawcett, *A Textbook of Histology,* 10th ed. (Philadelphia: W. B. Saunders Company, 1975). © 1975; used with permission.

Hormonal Control of Follicular Growth

Ovarian follicles synthesize and secrete steroid hormones that play essential roles in the female. Before discussing follicular secretion of these hormones, however, we first must see how steroid hormones are synthesized.[8] *Cholesterol* is the precursor steroid for all steroid hormones secreted by the ovary. You have heard about the harmful effects of cholesterol in the body, but this is an example of a very important role of this substance. Cholesterol is converted by enzymes to *pregnenolone,* another precursor steroid. *Steroidogenesis* (biosynthesis of steroids) within the ovary then can follow one of two pathways (Fig. 3–5). In the $\triangle^5$ ("delta 5") pathway, which occurs

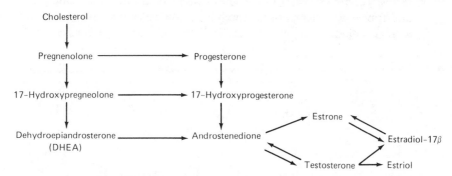

Figure 3–5. Steroidogenesis within the ovary. Conversion of pregnenolone to DHEA, then further synthesis of other androgens and estrogens, is the $\triangle^5$ pathway, predominant in ovarian follicles. Conversion of pregnenolone to progesterone and then to androgens and estrogens is the $\triangle^4$ pathway, predominant in the corpus luteum.

predominantly in large tertiary follicles, pregnenolone is converted to 17-hydroxy pregnenolone and then to a weak androgen, *dehydroepiandrosterone* (DHEA). A weak androgen is one that is not very potent in stimulating male tissues. In the second, or $\triangle^4$ pathway, pregnenolone is converted to *progesterone,* a progestogen that not only serves as a precursor for other steroids but enters the female's blood and acts as a hormone on target tissues such as the uterus and mammary glands. Within the ovary, progesterone is converted to 17-hydroxyprogesterone, which then is changed to *androstenedione,* another weak androgen. This $\triangle^4$ pathway is predominant in the corpus luteum, which as we see later is an endocrine gland formed from the follicle after it ovulates.

From here on, both the $\triangle^5$ and $\triangle^4$ pathways can produce estrogens (Fig. 3–5). DHEA from the $\triangle^5$ pathway is converted to androstenedione or *testosterone* (a potent androgen). Androstenedione and testosterone from both pathways then can be converted to *estradiol-17β,* the major estrogenic hormone secreted by ovarian follicles and the corpus luteum. Two other estrogens, *estriol* and *estrone,* also can be synthesized in both pathways.

In summary, the ovarian follicles, through the $\triangle^5$ pathway, produce and secrete estradiol-17β and a trace of the other two estrogens. The corpus luteum, in contrast, synthesizes and secretes both progesterone and estradiol through the $\triangle^4$ pathway. It is interesting to note that, in either pathway, androgens are precursors for estrogens in the female. In the testis, as we see in Chapter 4, steroidogenesis favors synthesis and secretion of androgens; very little estrogen is produced. Figure 3–6 illustrates the basic structure of some of these steroid hormones.

In humans, it appears that the theca interna is capable of synthesizing some steroid hormones, but not until the follicle has grown to the tertiary stage.[9] Beginning at this time, the theca interna glandular cells synthesize a weak androgen (DHEA), which then diffuses into the membrana granulosa. The granulosa cells then convert the weak androgen to testosterone, and the granulosa and thecal cells convert the testosterone to estradiol-17β.[10] The estradiol then enters the blood vessels in the theca and travels to other parts of the body. Low levels of testosterone are present in a woman's blood, secreted from the follicles and clusters of *hilus cells* in the ovary near

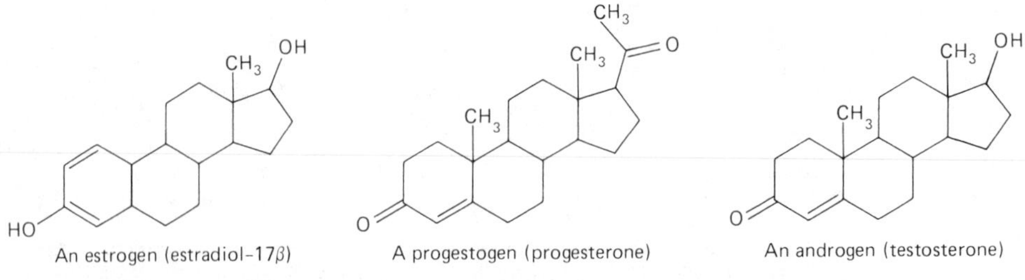

An estrogen (estradiol–17β)   A progestogen (progesterone)   An androgen (testosterone)

Figure 3–6. Chemical structures of three steroid hormones. Note that although these molecules are all derived from cholesterol and differ only slightly in structure, they have quite different actions on tissues.

where the blood vessels enter and leave. Weak androgens, however, are the predominant androgens in female blood; most of these come from the adrenal glands (Chapter 6).

Follicular growth and steroid hormone secretion are controlled by the two pituitary gonadotropins—follicle-stimulating hormone (FSH) and luteinizing hormone (LH).[11] Remember from Chapter 2 that cells in target tissues for hormones have specific hormone receptor molecules on their cell membranes. There are FSH receptors on the granulosa cells of primordial, primary, and secondary follicles, and FSH causes growth of these follicles by stimulating division of granulosa cells. The thecal cells of late secondary follicles acquire receptors for LH, and LH causes these cells to secrete DHEA, which diffuses to the granulosa cells. The latter cells are stimulated to convert DHEA to testosterone by FSH, and then the granulosa and thecal cells convert testosterone to estradiol-17β. Granulosa cells of larger tertiary follicles acquire LH receptors, and LH causes these granulosa cells to transform just before ovulation so that they begin to secrete progesterone through the delta 4 pathway, a change called *luteinization*. The interaction of hormones and their receptors in controlling follicular growth and steroid secretion is summarized in Fig. 3–7.

In the human menstrual cycle, about 20 large tertiary follicles form in both ovaries just before ovulation. Only one of these in one ovary, however, ovulates.[12] Therefore, about 19 tertiary follicles in both ovaries grow large and then undergo atresia. What causes this atresia is not clear, but one theory argues that the favored follicle is the only one that accumulates high amounts of FSH in its antral fluid. Meanwhile, estrogen secretion increases and inhibits FSH secretion from the pituitary through negative feedback (Chapter 7). Thus, the other tertiary follicles become deprived of FSH, leading to their degeneration.[13] A surprising new discovery is that ovarian follicles contain GnRH, and that this substance inhibits follicular function.[14] Whether ovarian GnRH plays a role in follicular atresia is not clear.

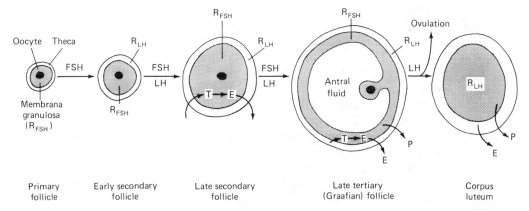

Figure 3-7. General model for the control of ovarian follicular growth, ovulation, and corpus luteum function in the human ovary: R_{FSH} = FSH receptors; R_{LH} = luteinizing hormone receptors; E = estradiol -17β; T = testosterone; P = progesterone.

Oocyte Maturation and Ovulation

Before discussing the process of *ovulation* (extrusion of an oocyte from its follicle), we first must digress for a moment and review a special kind of cell division found in germ cells called *meiosis*. The nucleus of most human cells contains 46 chromosomes, or 23 pairs of homologous chromosomes. This condition is called the *diploid* condition, and is symbolized by "2N." When oocytes first appear in the ovaries of female fetuses (Chapter 5), they begin the process of meiosis that ultimately produces cells that have only 23 chromosomes, the *haploid* or "N" condition. When a haploid ovum fuses with a haploid sperm, the diploid ("2N") number of chromosomes is restored (Chapter 8).

The process of *oogenesis* results in production of mature female *gametes* ("sex cells"). At birth of a female infant, the ovaries contain *primary oocytes* in primordial follicles. These primary oocytes are arrested in the *first meiotic division,* and this *first meiotic arrest* is maintained in the oocytes of follicles of all stages except the most mature. In the menstrual cycle, only one follicle reaches the large Graafian stage every month. Then, as we see in Chapter 7, a surge of LH secretion from the pituitary gland occurs. This LH diffuses from the theca into the antral fluid, reaches the oocyte, and reinitiates the first meiotic division.

As a first step in the reinitiation of meiosis, the membrane of the oocyte nucleus (germinal vesicle) disintegrates, a process called *germinal vesicle breakdown*. Then, the first meiotic division is completed. This is the *reduction division* of meiosis and produces two haploid daughter cells, still within the follicle (Fig. 3–8). Actually, the reduction division is unequal and produces a large haploid *secondary oocyte* (about 190 μm in diameter) and a tiny haploid *first polar body*. This polar body can divide again or remain single; in either case, it degenerates. Then, the secondary oocyte begins the *second meiotic division,* but this division is again arrested while the oocyte is still within the follicle; this is the *second meiotic arrest*. As we see in Chapter 8, the second meiotic division is not completed (resulting in a haploid *ootid* and *second polar body*) until after sperm penetration of the ovum, which occurs in the oviduct. Thus, the female germ cell is ovulated as a secondary ooctye (*ovum*) in the second meiotic arrest. The process of oocyte meiosis is summarized in Fig. 3–8.

In the human, germinal vesicle breakdown occurs about 25 hours after the beginning of the LH surge during the menstrual cycle. Much research has been conducted on how LH causes resumption of the second meiotic division. One theory relates to the fact that LH also causes the granulosa cells of the preovulatory Graafian follicle to luteinize and secrete progesterone. In lower animals such as frogs, it is progesterone that mediates the effect of LH on oocyte maturation, but it does not appear that progesterone does so in humans. Instead, this steroid hormone simply may prime the oocyte so that it can respond to LH.[15] Another factor that may play a role in oocyte maturation is an *oocyte maturation inhibitor,* which is present in the antral fluid.[16] This inhibitor may prevent premature oocyte maturation until the LH surge occurs.

Now that we have seen how an oocyte begins to mature just before ovulation,

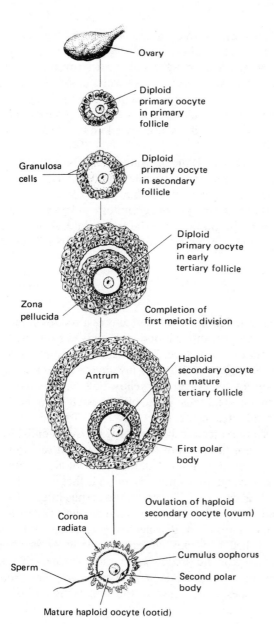

Figure 3-8. The process of oogenesis, which is not completed until a sperm penetrates an ovulated egg. The theca of the follicles is not included in this illustration. Adapted from K. L. Moore, *The Developing Human. Clinically Oriented Embryology,* 3rd ed. (Philadelphia: W. B. Saunders Company, 1982). © 1982; used with permission.

how does it escape from the follicle? The general theory is as follows: [17] Just before ovulation, a small, pale (avascular) region, the *stigma,* appears on the follicular surface. Underneath this region, the surface epithelium and thecal layers of the follicle become thinner and dissociated, and the wall in this region exhibits a reduction in tensile strength. Also, the membrana granulosa degenerates in this region. This thinning and weakening of the follicular wall at the stigma appear to be caused by estrogen-induced stimulation of the production of an enzyme (*collagenase*) from the connective tissue cells in this area. Breakdown products of the destroyed connective tissue induce an inflammatory response, with migration of white blood cells and secretion of prostaglandins in this region. The prostaglandins probably facilitate ovulation by constricting blood vessels and reducing blood supply to the degenerating tissue. Administration of inhibitors of prostaglandin synthesis is known to block ovulation in rats.[18]

After the follicular wall is thinned, the pressure within the antral cavity causes the wall to tear at the stigma. There are contractile, smooth musclelike cells in the follicular wall, but there is no clarity on what role they play in ovulation. The oocyte, which has become detached from the membrana granulosa and now is floating freely with its corona radiata and cumulus oophorus in the antral fluid, then oozes out with the escaping fluid through the tear in the follicular wall (Fig. 3-9). Thus, ovulation has occurred.

The Corpus Luteum

Once ovulation has occurred, the follicular wall remains as a collapsed sac. It then is called the *corpus hemorrhagicum* because a blood clot, derived from the torn blood vessels of the stigma, appears on its surface. The luteinized granulosa cells (*luteal cells*) in this collapsed follicle begin to divide and invade the old antral cavity (Fig. 3-3), thus forming the *corpus luteum* (Latin = "yellow body"). It is yellow because of the presence of pigment in the luteal cells. Then, blood vessels from the thecal layer grow and penetrate the central luteal cell mass.

The cells of the corpus luteum secrete high levels of both estradiol-17β and progesterone. Secretion of both hormones from the corpus luteum requires LH (Fig. 3-7). The pituitary hormone prolactin, along with LH, is essential for corpus luteum function in laboratory mammals, but in women it is only LH that maintains the function of the corpus luteum.[19] As we see in Chapter 7, the corpus luteum forms in the second half of the menstrual cycle, reaches a maximum diameter of 10 to 20 mm, and then degenerates before menstruation. The degenerated corpus luteum fills with connective tissue and then is called a *corpus albicans* (Latin = "white body"). In Chapter 9, we discuss how estradiol and progesterone prime the uterus for establishment of pregnancy if the ovum is fertilized. If pregnancy occurs, the corpus luteum does not die but instead survives to function in the first trimester of pregnancy (Chapter 9).

Ovarian Disorders

There are two major kinds of *ovarian cysts*. *Cystic follicles* are large fluid-filled sacs that are formed from unovulated follicles. *Luteinized cysts* are solid masses filled with luteal cells. Both kinds of cysts are common, and often they disappear spontaneously. In some cases, however, they can persist, and they will secrete abnormal

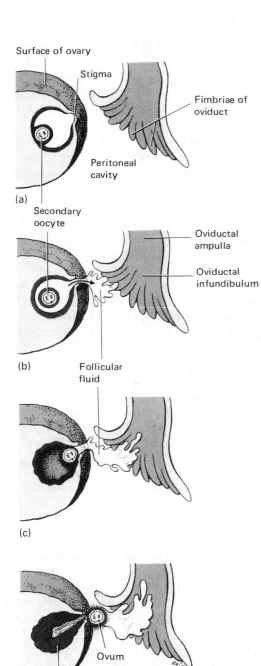

Surface of ovary

Stigma

Fimbriae of
oviduct

Peritoneal
cavity

(a)

Secondary
oocyte

Oviductal
ampulla

Oviductal
infundibulum

(b)

Follicular
fluid

(c)

(d)   Developing corpus luteum

Ovum

Figure 3–9. Diagrams illustrating ovulation. The stigma (a) ruptures (b), and the ovum is expelled along with follicular (antral) fluid (b, c). The ovulated ovum is arrested in the second meiotic division (d) until it is penetrated by a sperm (Chapter 8). Adapted from K. L. Moore, *The Developing Human. Clinically Oriented Embryology,* 3rd ed. (Philadelphia: W. B. Saunders Company, 1982). © 1982; used with permission.

amounts of steroid hormones and interfere with fertility. These cysts often must be removed surgically.

When a woman's ovaries contain many small cysts, it is called the *polycystic ovarian syndrome.* Luteinized cysts are common in women with *Stein-Levanthal syndrome,* a form of polycystic ovarian syndrome. In this syndrome, the follicles are unable to convert androgens to estrogens, so high levels of FSH and LH occur in the blood (because of lack of negative feedback of estrogen on their secretion). Androgen levels are also high in these women, whereas estrogen levels are low. Associated physical symptoms include lack of ovulation and menstruation, obesity, and growth of malelike body hair (*hirsutism*).

A rare ovarian condition involves formation of *dermoid cysts,* which contain tissues such as hair, nails, and skin glands. One theory is that these cysts result from fertilization of an oocyte while it is still in the follicle, the embryo then partially developing within the follicle.

Sometimes, body cells can multiply and form a lump or mass called a *tumor* or *neoplasm.* If the cells in such a mass are normal, it is a *benign tumor.* Examples of benign tumors are the aforementioned ovarian cysts. If, however, the cells lose their ability to control their growth and multiplication, the lump is a *cancer* or *malignancy.* Cancerous cells can remain in place and may offer little danger. If, however, cancerous cells break loose from the tumor and travel in the blood or lymphatic system to other parts of the body, a process called *metastasis,* new cancers can appear elsewhere in the body and soon will interfere with an important bodily function and can be fatal. Ovarian cancer is relatively rare among women in the United States. The annual incidence is about 15.5 per 100,000 women.[20] However, even though its incidence is low, ovarian cancer is a dangerous disease. Of the approximately 17,000 new cases of ovarian cancer a year in the United States, about 11,200 women will die. There is evidence that use of the combination contraceptive pill lowers the incidence of ovarian cancer (Chapter 13).

THE OVIDUCTS

The *oviducts,* or *uterine tubes,* are paired tubes extending from near each ovary to the top of the uterus (Fig. 3-2). They also are called *fallopian tubes,* after the sixteenth-century anatomist Fallopius. Each oviduct is about 10 cm (4 in.) long and has a diameter about the size of a drinking straw. Each oviduct can be divided into several regions along its length (Fig. 3-2). Nearest the ovary is a funnel-shaped portion, the *oviductal infundibulum.* The opening into the infundibulum, into which the ovulated ovum enters, is the *ostium.* The edges of the infundibulum have fingerlike projections, the *fimbriae.* When ovulation occurs, the ovum is captured by the infundibulum and enters the oviduct because of cilia on the oviductal lining that beat in a uterine direction. Proceeding along the tube toward the uterus, there is a wide *oviductal ampulla,* then a narrower *oviductal isthmus,* and finally the part of the oviduct that is embedded within the uterine wall, the *intramural oviduct.* The point where the oviduct empties into the uterine cavity is termed the *uterotubal junction.*

Each oviduct in cross section has three layers: (1) On the outside is a thin membrane, the *oviductal serosa.* (2) The middle layer consists of smooth muscle (the *oviductal muscularis*), which is divided into an inner circular and outer longitudinal layer. Contraction of this muscle helps transport the ovum and, later, the early embryo, toward the uterus. Near the time of ovulation, these contractions are 4 to 8 sec apart, whereas they are less frequent during other stages of the menstrual cycle. (3) The third layer, the internal lining of the oviduct, has many folds, and there are ciliated and nonciliated cells as well as mucous glands on the internal surface.[21] Ciliary beating and mucous secretion play a role in ovum and embryo transportation and in sperm movement up the oviduct (Chapter 8). Hormones play a role in oviduct function. Estrogens cause secretion of mucus in the oviducts; they also cause the oviductal cilia to beat faster and the smooth muscle to contract more frequently. Progesterone, on the other hand, reduces mucous secretion and reduces muscular contraction.

Inflammation of the oviducts, or *salpingitis,* can lead to blockage by scar tissue and subsequent infertililty (Chapter 15). This term comes from the Greek word *salpinx,* which means "tube." Tubal blockage can be repaired in some cases by microsurgery. *Salpingectomy* (surgical removal of the oviducts) or *tubal sterilization* (surgical cutting, tying off, or blocking the oviducts) can be used as a contraceptive measure (Chapter 13).

THE UTERUS

Uterine Functional Anatomy

The *uterus* (womb), is a single, inverted, pear-shaped organ situated in the pelvic cavity above the urinary bladder and in front of the rectum (Figs. 3–1 and 3–2). In *nulliparous* women (those who never have born a child), the uterus is about 7.5 cm (3 in.) long, 5 cm (2 in.) wide, and 1.75 cm (1 in.) thick. In *multiparous* women (those who have previously born children), the uterus is larger and its shape is more variable.

The uterus is supported by bands or cords of tissue, the *uterine ligaments.* First, there are a pair of *broad ligaments* that attach the uterus to the pelvic wall on each side. Then, paired *uterosacral ligaments* attach the lower end of the uterus to the tail bone. Next, the *lateral cervical ligaments* connect the cervix and vagina to the pelvic wall. Finally, the paired, cordlike *round ligaments* attach on one end to the uterus near the entrance of the oviducts and on the other end to the lower pelvic wall. These ligaments, besides carrying blood vessels and nerves to the uterus, also serve to support the uterus and other pelvic organs in their normal position. We see later in this chapter what happens to the uterus when these ligaments are not doing their job.

If a longitudinal section is made through the uterus, several distinct regions are revealed (Fig. 3–2). The dome-shaped region above the points of entrance of the oviducts is the *uterine fundus*. The *uterine corpus* or "body" is the tapering central portion, which ends at an external constriction, the *uterine isthmus*. A narrow region, the *uterine cervix,* is adjacent to the vagina.

The wall of the uterine fundus and corpus has three layers of tissue (Fig. 3–2). The external surface of the uterus is covered by a thin serosal membrane, the *perimetrium*. Under the perimetrium is a thick layer of smooth muscle, the *myometrium,* which is thickest in the fundus region. The myometrium is capable of very strong contractions during labor, as we discuss in Chapter 10. Its muscle fibers increase in length and number during pregnancy. Internal to the myometrium is the layer of the uterus that lines the uterine cavity. This layer, the *endometrium,* can be divided further into an internal layer, the *stratum functionalis,* and a deeper layer, the *stratum basilis* (Fig. 3–2). The stratum functionalis, consisting of a lining epithelium and *uterine glands,* is shed during menstruation. The underlying stratum basilis is not shed during menstruation but contains the bleeding vessels that produce part of the menstrual flow. After menstruation, the stratum basilis gives rise to a new stratum functionalis. Thus the endometrium undergoes marked changes in structure and function during the menstrual cycle, and these changes are under hormonal control (Chapter 7).

The layers of the uterine cervix are similar to those of the uterine fundus and corpus, with a few important exceptions. The cervical myometrium is thinner, and the cervical endometrium is not shed during menstruation. Glands in the lining of the cervix secrete mucus to varying degrees during the menstrual cycle (Chapter 7). In some stages of the cycle, this mucus forms a plug that retards sperm movement through the cervical canal and prevents infectious organisms from entering the uterine cavity. The *cervical canal* is the small channel within the cervix that connects the vagina with the uterine cavity. The opening of the cervical canal to the uterine cavity is called the *internal cervical os,* whereas the opening of the cervical canal into the vagina is the *external cervical os*. This external os is about the diameter of the head of a kitchen match. The cervix viewed through the vagina appears as a dome, 1 to 2 in. in diameter.

Uterine Disorders

Pelvic infection. The uterus is subject to infectious organisms entering through the vagina. Some bacteria, such as those associated with sexually-transmitted diseases (Chapter 16), can cause uterine infection and inflammation that can spread to the oviducts. Such an infection can be dangerous to the fetus during pregnancy and can lead to oviductal scarring and infertility (Chapters 9 and 15).

Incidence and cause of cervical cancer. Cervical cancer is the eighth most frequent cancer-caused fatality among women.[22] The annual incidence of cervical cancer in the United States is about 11.4 per 100,000 women.[23] Before cancer develops, cervical cells exhibit a precancerous condition called *cervical dysplasia.* These cells have an abnormal appearance but are not yet malignant. True cancer of the cervix can take many years to develop. When these cancerous cells begin to multiply, they cause lesions in the cervical wall accompanied by intermenstrual bleeding and vaginal discharge. If not treated, the cancer can then spread to the uterine myometrium, an invasion that can take up to 35 years.[24] Cervical cancer is less common in celibate women, suggesting that sperm or *smegma* (sloughed cells and secretions from the skin of the penis present under the foreskin) may be involved. The incidence of cervical cancer is higher in women who first had coitus as an adolescent and in those with a greater number of sexual partners.[25] Also, some venereal diseases, such as Herpes virus Type 2 (Chapter 16), have been implicated in the development of cervical cancer.[26] For some reason, cervical cancer is much less common in the Jewish race. Recently, it has been found that the risk of cervical cancer is four times higher in women whose partners have had a vasectomy.[27] Use of the combination pill, however, has not been shown to increase cervical cancer (Chapter 13).

Detection and treatment of cervical cancer. Because of the danger of cervical cancer, the cervical cells should be examined using the *Papanicolaou Test,* or *Pap smear,* named after Dr. George Papanicolaou, who developed it in 1942. It usually is recommended that a Pap smear be done once a year. Since the development of cervical cancer can take many years, however, it recently has been argued that a Pap smear needs to be done only every two or three years.[28] In this painless test, a few cells from the cervix are removed with a cotton swab or fine wooden spatula, and these cells are examined under a microscope. Detection of cervical precancerous or malignant cells with a Pap smear is greater than 90 percent reliable. A small piece of cervical tissue also can be removed and examined, a procedure called *cervical biopsy.* If cancer is present, a woman can be treated with radiation therapy or can undergo surgical removal of the cervix and uterus, an operation called a *hysterectomy.*

Cervical cysts and polyps. The cervical tissue can also form noncancerous growths as the result of past infection or irritation. *Cervical cysts,* for example, are very common. These cysts look like pea-sized, whitish pimples. They usually are not a problem, but if they become so, they can be removed by burning them out (heat cauterization) or by freezing the tissue (cryosurgery). Recently, laser beams also are being used to remove abnormal tissue in the female reproductive tract.[29] *Cervical polyps* sometimes appear as tear-shaped growths extending into the cervical canal. These polyps are benign. If extensive, however, they should be removed because they can interfere with fertility or coitus.

Endometrial cancer. Cancer also can develop in the body of the uterus. This *endometrial cancer* is more common in older women, in nulliparous women, and in those that are obese or diabetic. Pap smears detect the presence of endometrial cancer only about half the time. Treatment of menopausal symptoms with estrogens has been shown to increase the incidence of endometrial cancer (Chapter 7), but use of the combination pill has shown no influence (Chapter 13). The annual incidence of endometrial cancer in the United States is about 22.4 per 100,000 women.[30] Usually the first symptom is irregular menstrual bleeding. An *endometrial biopsy* can be performed to see if cancer is present. If the cancer is in early stages, the diseased endometrium can be removed using cryosurgery. Uterine cancer also can be treated with radiation therapy, chemotherapy, or hysterectomy.

Endometriosis. Sometimes, endometrial tissue can detach from the uterus and lodge in other regions of the body such as the ovaries, oviducts, or uterine ligaments. This condition, *endometriosis,* can be painful because the misplaced tissue expands and contracts (with bleeding) during the menstrual cycle. About 10 to 15 per 100 women will develop endometriosis during their reproductive years, so it is fairly common. In about 40 percent of cases, endometriosis can lead to infertility because scar tissue formed can interfere with oviductal or ovarian function. Endometriosis is more common in women who do not become pregnant early in their reproductive life and who have fewer offspring, and in those between 20 and 40 years of age. Use of tampons or diaphragms does not increase the risk of endometriosis. Women with endometriosis can be treated with drugs such as Danazol to halt their menstrual cycles for a few months to reduce the pain and inhibit growth of endometrial cells. This drug inhibits FSH and LH secretion. The misplaced endometrial tissue also can be removed surgically.

Endometrial polyps, hyperplasia, and fibroids. The endometrium can form abnormal growths that are not cancerous. *Endometrial polyps* are mushroomlike growths that extend into the uterine cavity. *Endometrial hyperplasia* is an excessive multiplication of endometrial cells. Both of these conditions can be caused by chronic exposure of the endometrium to estrogen without the occurrence of menstruation. *Fibroids* are noncancerous tumors of the smooth muscle layer of the uterus. These benign tumors are common in older women before they reach menopause. All of the above types of abnormal growth may cause irregular menstrual bleeding and can interfere with fertility. They can be removed by heat cauterization, cryosurgery, or laser beams.

Tilted uterus. The position of the uterus can vary in different women. In most cases, the uterus is in an *anteflexed* position, being tilted forward at right angles to the vagina (Fig. 3–10). In about one of five women, however, the uterus is in a *retroflexed* position, being tilted backward instead of forward (Fig. 3–10). Fortunately, a retroflexed uterus only rarely causes pain or difficulties with pregnancy.

Prolapsed uterus. A *prolapsed uterus* is when this organ slips down toward the vagina. This often occurs because the uterine ligaments are too relaxed or stretched. Delivery of a large infant, or breech or forceps deliveries (Chapter 10), sometimes can cause uterine prolapse by stretching or damaging the uterine ligaments. The degree of prolapse can be anywhere from slight to a severe protrusion of the uterus from the vagina. In the latter case, surgery is required to correct the prolapse or remove the uterus.

Flabbiness or overstretching of the *pubococcygeus muscle* can also lead to pro-

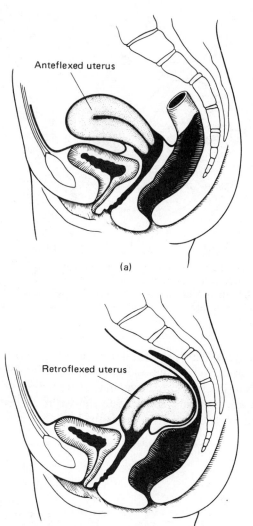

Anteflexed uterus

(a)

Retroflexed uterus

(b)

Figure 3–10. Usual and unusual positions of the uterus. In (a), the uterus is in the usual anteflexed position, whereas in (b) the uterus is tilted backward in a retroflexed position. Adapted from *From Woman to Woman* by Lucienne Lanson and illustrated by Anita Karl (New York: Alfred A. Knopf, Inc., 1975). © 1975 by Lucienne Lanson and Alfred A. Knopf, Inc. Reprinted by permission of Alfred A. Knopf, Inc., and Penguin Books Ltd., Australia and England.

lapse of the uterus and other pelvic organs. This muscle is a slinglike band of tissue that connects the pubic bone with the tail bone and forms the floor of the pelvic cavity (Fig. 3–11). Thus, it supports the uterus, part of the vagina, urinary bladder, urethra, and rectum. If this muscle is not taut, the uterus and vagina can sag, and there can be uncontrollable leakage of urine from the urethra (*urinary incontinence*). What can a woman with a sagging pelvic organ do about her problem? One answer is the *Kegel exercise*. A physician can show her how to exercise and strengthen the pubococcygeus muscle. To do this exercise, a woman first finds this muscle by sitting on a toilet, urinating, and stopping urine flow voluntarily. The muscle that stops

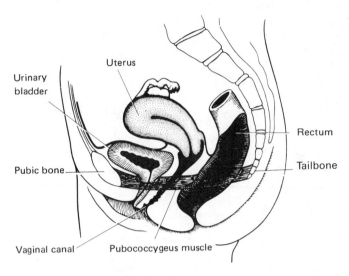

Figure 3–11. A diagram indicating position of the pubococcygeus muscle. Adapted from *From Woman to Woman* by Lucienne Lanson and illustrated by Anita Karl (New York: Alfred A. Knopf, Inc., 1975). © 1975 by Lucienne Lanson and Alfred A. Knopf, Inc. Reprinted by permission of Alfred A. Knopf, Inc., and Penguin Books Ltd., Australia and England.

urine flow is the pubococcygeus. Then, she should voluntarily contract this muscle ten times, and repeat this six or more times daily. Also, a rubber or plastic device called a *vaginal pessary* can be inserted into the vagina to support this organ as well as the uterus. Vaginal pessaries, however, can irritate the vagina and interfere with coitus, and surgery to repair pelvic support often is the best answer.

THE VAGINA

Functional Anatomy

The *vagina* is a 10–cm (4–in.)-long tube that lies between the urinary bladder and rectum (Figs. 3–1 and 3–2). It is normally collapsed. As discussed in Chapter 17, the length and diameter of the vagina increase during sexual arousal, and the diameter of the vaginal canal and opening can be controlled voluntarily to some extent by use of the muscles of the pelvic floor. As we see in Chapter 17, the vagina is lubricated during sexual arousal by fluid that leaks from blood vessels in the vaginal wall. The vagina functions as a passageway for menstrual flow, as a receptacle for the penis during coitus, and as a part of the birth canal. The opening of the external cervical os into the vagina is circumscribed by a recess, the *fornix*. This recess allows support for a diaphragm contraceptive (Chapter 13).

Structure

The muscular layer of the vagina has transverse folds (*vaginal rugae*) that allow the vagina to stretch during coitus or childbirth. The epithelial lining of the vagina consists of many layers of flattened cells. Changes in the condition of these cells during the menstrual cycle can be detected by swabbing the lining and looking at the cells under a microscope. This is called a *vaginal smear*. During periods when estrogen is high, the cells become more hardened with the protein keratin, and the cell nuclei degenerate. The vaginal wall contains only a few touch and pressure receptors, but it does possess some free sensory nerve endings located in its outer one-third.

The Vaginal Environment

The vagina is the natural home for several microorganisms (Table 3–1). Some of these bacteria, fungi, and protozoa play important roles in maintaining the vaginal environment. Others are potential disease microorganisms, which normally do not cause problems unless they multiply rapidly. Some of these organisms can be sexually transmitted (Chapter 16).

Cells of the vaginal epithelium accumulate large amounts of *glycogen* (a sugar) under the influence of estrogen. As these cells die and are sloughed into the vaginal cavity, they release this glycogen. Certain bacteria present within the vagina then metabolize the glycogen to lactic acid, rendering the vaginal environment acidic, and this acidic condition retards yeast infection. If, however, a woman takes certain antibiotics, the bacteria are destroyed. The vaginal environment then becomes more basic, which may lead to yeast infection (Chapter 16). The vaginal acidity also kills sperm. As we discuss in Chapter 8, semen deposition into the vagina during coitus

TABLE 3-1. NORMAL FLORA AND FAUNA OF THE VAGINA

Lactobacillus acidophilus	Coliform bacilli
Staphylococcus aureus	*Proteus* species
Staphylococcus epidermidis	*Mima polymorpha*
Fecal Streptococci	*Clostridium* species
Streptococcus viridans	*Bacteroides* species
Anaerobic streptococci	*Fusobacterium* species
Neisseria species (other than	*Mycoplasma*
N. gonorrhoeae and *N. meningitidis*)	*Candida* species
Diphtheroids	*Trichomonas vaginalis*
Corynebacterium species	
Hemophilus vaginalis	

Source: Adapted from R. C. Kolodny et al., *Textbook of Human Sexuality for Nurses* (Boston: Little, Brown & Company, 1979), p. 154. With permission of the Masters and Johnson Institute, ©1979.

Note: Not all these microorganisms are present in a single female at one time.

changes the vaginal environment to a more basic condition and allows sperm to survive and move up the female tract.

THE FEMALE EXTERNAL GENITALIA

The female *external genitalia* include the mons pubis, labia majora, labia minora, vaginal introitus, hymen, and clitoris (Fig. 3–12). These organs are collectively called the *vulva.*

Mons pubis. The *mons pubis,* or *mons veneris* (*mons* = "mountain"; *veneris* = "of love"), is a cushion of fatty tissue, covered by skin and pubic hair, that lies over the pubic symphysis. The skin of this area has many touch receptors and only a few pressure receptors. The distribution and amount of pubic hair will vary in different individuals. Usually the pubic hair forms the shape of an inverted pyramid. In about 25 percent of women, this hair extends in a line up to the navel.

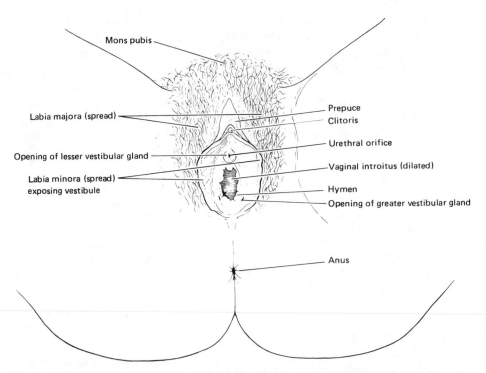

Figure 3–12. The female external genitalia, or vulva. Adapted from *Principles of Human Anatomy,* 3rd ed., by Gerard J. Tortora (New York: Harper & Row, Publishers, 1983), Fig. 23–17, p. 649. © 1983 by Gerard J. Tortora. Reprinted by permission of Harper and Row Publishers, Inc., and Dr. Gerard J. Tortora.

Labia majora. The *labia majora* ("major lips") are fleshy folds of tissue that extend down from the mons pubis and surround the vaginal and urethral orifices. These lips contain fat, and the pigmented skin has some pubic hair, sweat and oil glands, and fewer touch and pressure receptors than does the mons pubis. The labia majora are homologous to the male scrotum; that is, they are derived embryologically from the same tissue (Chapter 5).

Labia minora. The *labia minora* ("minor lips") are paired folds of smooth tissue under the labia majora. They range from light pink to brownish black in color in different individuals. In a sexually unstimulated condition, these lips cover the vaginal and urethral opening, but upon sexual arousal they become more open (Chapter 7). The hairless skin of the labia minora has oil glands (but no sweat glands) and a few touch and pressure receptors. In older women, or in women who have low estrogen levels, the skin of the labia minora becomes thinner, sensitive, and itchy.

Vestibule. The cavity between the labia minora is the *vestibule.* Most of this cavity is occupied by the opening of the vagina, the *vaginal introitus.* In women who have not previously had coitus, the introitus often is covered partially by a membrane of connective tissue, the *hymen.* This tissue often is torn during first coitus, accompanied by minor pain and bleeding. However, it also can be broken by a sudden fall or jolt, by insertion of a vaginal tampon, or by active participation in sports such as horseback riding and bicycling. Thus, the presence of a hymen is not a reliable indicator of virginity. In some women, the hymen can persist even after coitus, especially if the tissue is flexible. In rare cases, a wall of tissue completely blocks the introitus, a condition called *imperforate hymen.* The condition is present in about 1 out of 2000 young women. Since an imperforate hymen can block menstrual flow, surgery is required to alleviate the problem.

Urethral orifice. Above the introitus is the *urethral orifice.* This is where urine passes from the body. Below and to either side of the urethral orifice are openings of two small ducts leading to the paired *lesser vestibular glands* (*Skene's glands*). These glands are homologous to the male prostate gland (that is, the two gland types are derived from the same structure in the embryo; see Chapter 5) and secrete a small amount of fluid. At each side of the introitus are openings of another pair of glands, the *greater vestibular glands* (*Bartholin's glands*). These glands secrete mucus and are homologous to the bulbourethral glands of the male. Sometimes, the Bartholin's glands can form a cyst or abscess as the result of an infection (Chapter 16).

Clitoris. The *clitoris* lies at the upper junction of the two labia minora, above the urethral orifice and at the lower border of the pubic bone. Its average length is about 2.5 cm (1 in.), and it is about 5 mm (½ in.) in diameter. This cylindrical structure has a shaft and glans (enlarged end). It is homologous to the penis. The *clitoral shaft,* as does the shaft of the penis (Chapter 4), contains a pair of *corpora cavernosa,* spongy cylinders of tissue that fill with blood and cause the clitoris to erect slightly during sexual arousal (Chapter 17). Another spongy cylinder present in the penis, the *corpus spongiosum,* is not found in the clitoris; this tissue in the female is represented by the labia minora (Chapter 5). The *clitoral glans* is partially covered by the *clitoral prepuce,* which is homologous to a similar structure covering the glans of the penis (Chapter 4). The clitoris is rich in deep pressure and temperature receptors, but it has only a few touch receptors. We discuss the role of the clitoris and other structures of the female vulva in the female sexual response in Chapter 17.

THE MAMMARY GLANDS

The *mammary glands,* also called the *breasts* or *mammae,* are paired structures on the chest. Their function is to secrete milk during breast-feeding and also to serve as a stimulus for sexual arousal in both males and females. These glands evolved from sweat glands, and so milk is really modified sweat! Embryologically, the mammary glands developed from a *milk line,* which is a chain of potential mammary glands extending from the arm buds to the leg buds of the embryo. In mammals other than humans, several pairs of glands persist in the adult female, depending on the litter size. In humans, usually only a single pair persists, but in some individuals more than one pair are present, a condition called *polythelia.* There are even records of women having up to eight pairs of functional breasts, and it is estimated that 1 out of 20 males has an extra nipple. (Speaking of males, the male mammary glands are usually quiescent, but these are capable of growing and even secreting milk if properly stimulated by certain hormones. Such development of the breasts in males is called *gynecomastia.*)

Mammary Gland Functional Anatomy

Each human female breast is covered by skin and contains a variable amount of fat and the actual mammary gland tissue. The variation in breast size and shape is due to differences in amount and distribution of fat. However, large or small breasts do not differ in their ability to secrete milk. Each breast contains glandular tissue divided into 15 to 20 lobes separated by fat and ligamentous tissue (Fig. 3–13). The latter tissue (*suspensory ligaments of Cooper*) provides support for the breast, but it tends to be less effective in older women. Each mammary lobe, in turn, contains grapelike clusters of glandular cells, the *mammary alveoli* (Figs. 3–13 and 3–14). Milk is synthesized and secreted from these alveoli into *secondary mammary tubules.* The secondary tubules from each lobe join to form a *mammary duct,* which empties into a wider *mammary ampulla,* where milk can be stored. The ampulla then empties into a *lactiferous duct,* which opens into the *nipple.* There is one lactiferous duct for each lobe, but some ducts may join before reaching the nipple. Surrounding the nipple is a ring of pigmented skin, the *areola,* which contains oil glands.

Hormonal control of mammary gland function. In prepubertal females, the mammary gland tissue is inactive, but when the ovaries begin to secrete estrogen, the alveoli and ducts develop (Chapter 6). During the menstrual cycle, the glandular

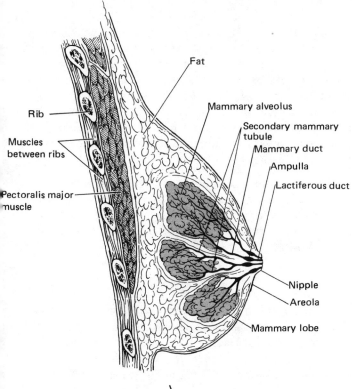

Rib

Muscles
between ribs

Pectoralis major
muscle

Fat

Mammary alveolus

Secondary mammary
tubule

Mammary duct

Ampulla

Lactiferous duct

Nipple

Areola

Mammary lobe

Figure 3–13. Section through an adult female breast. Adapted from *Principles of Human Anatomy and Physiology,* 3rd ed., by Gerard J. Tortora and Nicholas P. Anagnostakos (New York: Harper & Row, Publishers, 1981), Fig. 28–23, p. 735. © 1981 by Gerard J. Tortora and Nicholas P. Anagnostakos. Reprinted by permission of Harper & Row, Publishers, Inc., and Dr. Gerard J. Tortora.

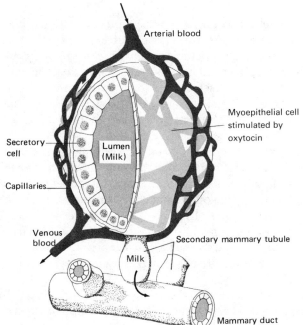

Arterial blood

Myoepithelial cell
stimulated by
oxytocin

Secretory
cell

Lumen
(Milk)

Capillaries

Venous
blood

Milk

Secondary mammary tubule

Mammary duct

Figure 3–14. Diagrammatic view of a single mammary alveolus. Milk is secreted by the secretory cells into a cavity (lumen) within the alveolus. From here, the milk is carried to the nipple through a series of ducts. The blood supply to the alveolus is important because of its support of alveolus growth and secretion. Also, the blood carries hormones to the alveolus that stimulate its secretion and cause contraction of the myoepithelial cells, leading to milk ejection. Adapted from A. L. R. Findlay, "Lactation," *Research in Reproduction,* vol. 6, no. 6 (1974), map. © 1974; used with permission of the International Planned Parenthood Federation and Dr. A. L. R. Findlay.

tissue undergoes enlargement when estrogen and progesterone levels are high, so that the breasts may feel full or tender (Chapter 7). During pregnancy, rising estrogen and progesterone in the blood, along with adrenal hormones and growth hormone, cause the glandular tissue to enlarge and the ducts to branch.[31] In the last third of pregnancy, the glandular tissue is not yet secreting milk, but it does secrete a clear fluid (*colostrum*), which, besides containing water, is high in protein and rich in maternal antibodies. Colostrum is swallowed by a nursing infant before true milk is secreted in the breasts. It provides nutrients and protects the infant against certain infections, especially those of the digestive tract (Chapter 11).

Suckling. In Chapter 11, we see how suckling of the nipple maintains prolactin secretion, which causes milk secretion in previously primed mammary tissue. We also see how nursing stimulates release of oxytocin, which causes ejection of milk from the nipple. The latter is accomplished by the contraction of *myoepithelial cells* that surround each alveolus (Fig. 3–14).

Breast Disorders

Breast cancer. The annual incidence of breast cancer in the United States is about 89.9 per 100,000 women,[32] and breast cancer has one of the highest fatality rates of all female cancers. This condition is the leading cause of death in women between 40 and 44, and is second among all age groups. About one in 18 women will die of breast cancer.[33] It is most common in women over 55, and there is an inherited tendency to develop this disease.

There are several myths about breast cancer. For example, there is no evidence that sexual activity, blows to the breast, having or nursing babies, entrance of germs into the nipple, or touching another cancer victim can cause breast cancer. In addition, there is no evidence that taking the combination pill increases the incidence of breast cancer (Chapter 13). Deposition of semen, for some unknown reason, may lower the incidence of breast cancer. Therefore, the use of contraceptive measures that prevent semen from entering the female tract, such as a condom or diaphragm, can slightly increase the risk of breast cancer.[34]

Tumors of the breast can be benign or malignant. In early stages of development, breast cancer is painless, and its discovery during these early stages is critical. If detected early, over 85 percent of patients survive the disease with proper treatment. Any lump in the breast, as well as puckering of the skin or discharge from the nipple, no matter how slight, should be reported to a physician. About 95 percent of cases of breast cancer are first discovered by the woman herself. Every woman should have a medical breast exam at least once a year and should palpate her breast for lumps once a month about a week after the end of her period (or in postmenopausal women, on the first day of each month).

There also are more sophisticated ways to examine breasts for tumors. Examination by X-rays (*mammography*) is 80 percent reliable in detection of abnormal growth and can be used to guide later breast surgery. A new method, called *xeroradiography,* uses X-ray to produce images on paper, not film. This method gives excellent detail of soft tissues, and a lower dose of radiation is required. Another method, *thermography,* detects local regions of high temperature on the breast. Benign or malignant tumors produce more heat than do surrounding tissues.

Noncancerous disorders.   Only about 30 percent of lumps in the breast are cancerous. The most common noncancerous lumps are fluid-filled cysts, a condition termed *cystic mastitis.* About one in five women between ages 25 and 55 will develop these cysts. More rarely, noncancerous solid lumps will appear.

Treatment of Breast Disorders

If a benign cyst or lump occurs, it can be removed from the breast. If the tumor is malignant but has not spread, a *simple mastectomy* (surgical removal of the breast) often is done. If, however, the tumor has spread to surrounding tissues, a *radical mastectomy* involving removal of the breast, underlying pectoral muscle, and axillary lymph nodes is performed. Recently, however, there is an awareness that radical mastectomies are not always necessary.[35] Radiation therapy or chemotherapy also can be used in conjunction with these operations.

CHAPTER SUMMARY

The adult female reproductive system consists of the ovaries, oviducts, uterus, vagina, external genitalia, and mammary glands. Each ovary consists of an outer ovarian cortex and an inner ovarian medulla. Within the cortex are ovarian follicles in various stages of growth, as well as atretic follicles. Each follicle contains a female germ cell (oocyte) surrounded by a follicular wall consisting of a membrana granulosa and the theca. A membrane, the zona pellucida, separates the granulosa from the oocyte. The number of follicles is fixed at birth, and this number steadily decreases with age because of follicular atresia and ovulation. Follicular growth first involves formation of a primary follicle from a primordial follicle. The primary follicle transforms to a secondary and then a tertiary follicle; the latter contains a fluid-filled antral cavity. The wall of tertiary follicles synthesizes estradiol-17ß and progesterone under the influence of FSH and LH. Growth and maturation of follicles involve a complex interaction of FSH, LH, and their receptors on thecal and granulosa cells. The corpus luteum, formed from the ovulated follicle, secretes estradiol-17ß and progesterone under the influence of LH.

The diploid primary oocyte contained in most ovarian follicles is arrested in the first meiotic division. The surge of LH at the middle of the menstrual cycle causes completion of the first meiotic division, producing a secondary haploid oocyte and a first polar body. The secondary oocyte begins the second meiotic division while still in the follicle, but does not complete this division until after ovulation, when the ovum is penetrated by a sperm. Ovulation is caused by LH, and involves a local degradation of the follicular wall. A tear develops at the stigma, and the ovum oozes out with antral fluid. Abnormal ovarian cysts and ovarian cancer can interfere with ovarian function, and can be dangerous.

The oviducts are paired tubes extending from near the ovaries to the uterus. An enlarged oviductal region at the ovarian end, the infundibulum, captures the ovulated egg. Sperm are transported up the oviduct. Fertilization occurs in the oviduct, and the embryo is transported down the oviduct to the uterus. Transport of these cells is aided by contraction of the smooth muscle and beating of oviductal cilia, both of which are influenced by levels of estradiol-17ß and progesterone.

The uterus is a pear-shaped, muscular organ lying in the pelvic cavity. It is divided into three regions—the fundus, corpus, and cervix, and the wall of each region has three layers. The innermost layer, the uterine endometrium, exhibits marked structural changes during the menstrual cycle. Cervical cancer can be detected by a Pap test, and endometrial cancer by endometrial biopsy. Both forms of uterine cancer can be treated by radiation, chemotherapy, or hysterectomy. Other abnormal uterine growths include cervical cysts and polyps, endometrial polyps and hyperplasia, and uterine fibroids. Disorders of uterine position include retroflexion and prolapse.

The vagina serves as a passageway for menstrual flow, as a receptacle for the penis during coitus, and as a part of the birth canal. The vaginal environment is maintained acidic by the activity of bacteria in the vaginal lumen.

The female external genitalia (vulva) include the mons pubis, labia majora, labia minora, vaginal introitus, hymen, and clitoris. These structures, and the glands associated with them, are homologous to certain structures in the adult male. The mammary glands (breasts) consist of glandular tissue and associated ducts embedded in fatty tissue. Growth and function of the glandular tissue are controlled by hormones. Breast cancer is a relatively common disease. Its early detection by breast palpation, X-ray, or thermography can eliminate much of the fatality associated with this conditon.

NOTES

1. W. E. Ellinwood et al., "Ovarian Vasculature: Structure and Function," in *The Vertebrate Ovary—Comparative Biology and Evolution,* ed. R. E. Jones, pp. 583–614 (New York: Plenum Publishing Corporation, 1978).

2. A. Tsafriri, "Ooctye Maturation in Mammals," in *The Vertebrate Ovary—Comparative Biology and Evolution,* ed. R. E. Jones, pp. 409–42 (New York: Plenum Publishing Corporation, 1978).

3. E. Block, "Quantitative Morphological Investigations of the Follicular System in Women: Variations at Different Ages," *Acta Anatomica,* 14 (1952), 108–23.

4. Ibid.

5. G. T. Ross and R. L. Vande Wiele, "The Ovaries," in *Textbook of Endocrinology,* 5th ed., ed. R. H. Williams, pp. 368–422 (Philadelphia: W. B. Saunders Company, 1974).

6. K. P. McNatty, "Follicular Fluid," in *The Vertebrate Ovary—Comparative Biology and Evolution,* ed. R. E. Jones, pp. 215–59 (New York: Plenum Publishing Corporation, 1978).

7. Block (1952), op. cit.

8. G. E. Abraham, "Ovarian Metabolism of Steroid Hormones," *Research in Reproduction,* vol. 3, no. 5 (1971), map.

9. F. Friedrich et al., "The Progesterone Content of the Fluid and the Activity of the Steroid-3ß-ol-dehydrogenase within the Wall of the Ovarian Follicles," *Acta Endocrinologica,* 76 (1974), 343–52.

10. L. Bjersing, "Maturation, Morphology, and Endocrine Function of the Follicular Wall in Mammals," in *The Vertebrate Ovary—Comparative Biology and Evolution,* ed. R. E. Jones, pp. 181–214 (New York: Plenum Publishing Corporation, 1978).

11. J. S. Richards, "Hormonal Control of Follicular Growth and Maturation in Mammals," in *The Vertebrate Ovary—Comparative Biology and Evolution,* ed. R. E. Jones, pp. 331–60 (New York: Plenum Publishing Corporation, 1978).

12. E. Block, "Quantitative Investigations of the Follicular System in Women: Variations in the Different Phases of the Sexual Cycle," *Acta Endocrinologica,* 8 (1951), 3–54.

13. Bjersing (1978), op. cit.

14. Anonymous, "LH-RH and the Ovary," *Research in Reproduction,* 13, no. 4 (1981), 3–4.

15. Tsafriri (1978), op. cit.

16. Ibid.

17. L. L. Espey, "Ovulation," in *The Vertebrate Ovary—Comparative Biology and Evolution,* ed. R. E. Jones, pp. 503–32 (New York: Plenum Publishing Corporation, 1978).

18. D. T. Armstrong and D. L. Grenwich, "Blockade of Spontaneous and LH-Induced Ovulation in Rats by Indomethacin, an Inhibitor of Prostaglandin Biosynthesis," *Prostaglandins,* 1 (1972), 21–28.

19. Ellinwood et al. (1978), op. cit.

20. W. Rinehart and J. C. Felt, "Debate on Oral Contraceptives and Neoplasia Continues; Answers Remain Elusive," *Population Reports,* ser. A., no. 4 (1977), pp. 69–100.

21. A. Ferenczy et al., "Scanning Electron Microscopy of the Human Fallopian Tube," *Science,* 175 (1972), 783–84.

22. J. L. Marx, "The Annual Pap Smear: An Idea Whose Time Has Gone?" *Science,* 205 (1980), 204–5.

23. Rinehart and Felt (1977), op. cit.

24. Marx (1980), op cit.

25. Ibid.

26. Ibid.

27. S. H. Swan and W. L. Brown, "Vasectomy and Cancer," *New England Journal of Medicine,* 301 (1979), 46.

28. Marx (1980), op. cit.

29. D. E. Thomsen, "Lasers versus Female Complaints," *Science News,* 119, no. 6 (1981), 90–91.

30. Rinehart and Felt (1977), op. cit.

31. R. Buchanan, "Breast-Feeding: Aid to Infant Health and Fertility Control," *Population Reports,* ser. J, no. 4 (1975), pp. 46–67.

32. Rinehart and Felt (1977), op. cit.

33. W. Rinehart and P. T. Piotrow, "OCs-Update on Usage, Safety and Side Effects," *Population Reports,* ser. A, no. 5 (1979), pp. 133–86.

34. A. N. Gjorgov, "Barrier Contraception and Breast Cancer," *Karger Gazette,* April 1980, pp. 1–2.

35. Anonymous, "New Treatments for Breast Cancer," *Newsweek,* July 13, 1981, p. 55.

FURTHER READING

ABRAHAM, G. E., "Ovarian Metabolism of Steroid Hormones," *Research in Reproduction,* vol. 3, no. 5 (1971), map.

COWIE, A. T., "Lactation and Its Hormonal Control," in *Reproduction in Mammals, Book 3, Hormones in Reproduction,* eds. C. R. Austin and R. V. Short. Cambridge, England: Cambridge University Press, 1972. Pp. 106–43.

FINDLAY, A. L. R., "Lactation," *Research in Reproduction,* vol. 6, no. 6 (1974), map.

JONES, R. E., ed., *The Vertebrate Ovary—Comparative Biology and Evolution.* New York: Plenum Publishing Corporation, 1978.

4   *The Adult Male Reproductive System*

LEARNING GOALS

Having read this chapter, you should be able to:

1. List the major structural components of the testis.
2. Describe the processes of spermatogenesis, spermiogenesis, and spermiation within a seminiferous tubule.
3. Name the components of the spermatic cord.
4. List functions of Sertoli cells in the testis.
5. List the hormones secreted by the testis.
6. Discuss the role of pituitary gonadotropins and androgens in controlling spermatogenesis, spermiogenesis, and spermiation.
7. Discuss the role of androgens, estrogens, and inhibin in controlling pituitary gonadotropin secretion in males.
8. Discuss the effects of stress, pollutants, and light on spermatogenesis.
9. Describe the causes and related symptoms of the male climacteric.
10. List the ducts that a sperm goes through as it travels from the seminiferous tubules to the outside.
11. Describe the location of the male sex accessory glands.
12. List the major structural components of the penis.
13. Describe the anatomy of the scrotum, and discuss the role of the scrotum in controlling testicular temperature.

INTRODUCTION

The male reproductive tract includes a pair of testes *(testicles),* which produce the male sex cells (gametes) called *spermatozoa* (sing., *spermatazoon),* or *sperm.* The word *testes* (sing., *testis)* is derived from the Latin word for ''witness'' or ''testify.'' (In ancient times, men put one hand over their genitals when taking an oath.) The testes, the male sex accessory ducts (which receive, store, and transport sperm), and the sex accessory glands (which add substances to the ducts), as well as the external genitalia, are all male primary sexual characteristics. The general location of these structures is shown in Fig. 4–1. The male secondary sexual characteristics include all those external features (except the external genitalia) that distinguish an adult man from an adult woman. These include larger muscles, greater height, facial hair, and a lower voice.

THE TESTES

Each testis is oval-shaped, with a length of about 4.0 cm (1.5 in.) and a width of 2.5 cm (1.0 in.). On the outside of each testis is a shiny covering, the *tunica vaginalis* (Fig. 4–2). Immediately under the tunica vaginalis is a thin, dense covering of the testis itself, the *tunica albuginea.* Inside each testis are about 250 compartments, *testicular lobules,* which are separated from each other by septa (tissue barriers). Each lobule contains one to three highly coiled *seminiferous tubules* (Fig. 4–2). If a single seminiferous tubule were stretched to its maximal length, it would measure about 30–91 cm (1–3 ft), and the total length of all the tubules would be longer than a football field! The male germ cells lie next to the inner wall of each tubule, and sperm are produced in the seminiferous tubules. The testes also contain some of the male sex accessory ducts that we discuss later in this chapter.

Each testis is suspended from the body wall by a *spermatic cord* (Fig. 4–2), which penetrates into the pelvic cavity through the *inguinal canal,* the route through which the testes originally descended into the scrotum from the pelvic cavity before birth (see Chapter 5). Each spermatic cord contains one of the accessory ducts (the vas deferens), a testicular nerve, and three coiled blood vessels (the *testicular artery,* or spermatic artery, carrying blood to the testis and epididymis, and two *testicular veins,* or spermatic veins, carrying blood away from the testis).

The Seminiferous Tubules

Each seminiferous tubule is lined on its inside by the *seminiferous epithelium,* which contains two kinds of cells—male germ cells and Sertoli cells. First we look at how sperm are produced from the male germ cells, and then the functions of Sertoli cells are discussed.

Spermatogenesis and Spermiogenesis. The processes by which immature male germ cells produce sperm are called *spermatogenesis* and *spermiogenesis.* First,

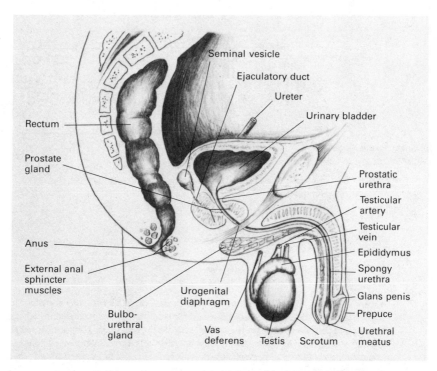

Figure 4-1. A schematic representation of the male pelvic region, showing the reproductive organs.

Labels in figure:
Seminal vesicle
Ejaculatory duct
Ureter
Urinary bladder
Rectum
Prostatic urethra
Prostate gland
Testicular artery
Testicular vein
Anus
Epididymus
External anal sphincter muscles
Spongy urethra
Urogenital diaphragm
Glans penis
Bulbo-urethral gland
Prepuce
Vas deferens
Testis
Scrotum
Urethral meatus

let us observe spermatogenesis, the process by which a diploid *spermatogonium* (pl., *spermatogonia)* transforms into four haploid spermatids. The immature germ cells lying next to the wall *(basement membrane;* see, Fig. 4–3) of each seminiferous tubule are the spermatogonia. These cells multiply by *mitosis,* a process of cell division by which two diploid daughter cells are derived from a diploid parental cell. Each spermatogonium is diploid and has 46 chromosomes (2 each of 23 different chromosomes). During mitosis, each chromosome duplicates itself, and each member of a duplicated pair passes to a separate progeny cell. Thus, the normal diploid chromosomal make-up is maintained in each new spermatogonium.

As the spermatogonia continue to divide mitotically, some change in appearance and begin a different kind of cell division, meiosis (Figs. 4–3 and 4–4). After a spermatogonium enters meiosis, it becomes a *primary spermatocyte.* A primary spermatocyte then completes the first meiotic division, which is the reduction division because the resultant daughter cells now have only 23 chromosomes (one member of each original pair) instead of 46. After reduction division, the cells thus are haploid (1N) instead of diploid (2N). The haploid cells, called *secondary spermatocytes,* then undergo the second meiotic division to produce four haploid cells, the *spermatids.* Now the process of spermatogenesis is complete. The spermatids are then trans-

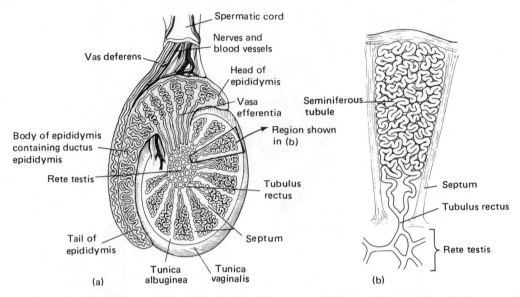

Figure 4–2. A, section of the testis, epididymis, vas deferens, and spermatic cord. B, a seminiferous tubule within a single testicular lobule. Sperm are produced in the seminiferous tubules and then move through the tubuli recti into the rete testis. They are carried through the vasa efferentia into the ductus epididymis and then into the vas deferens. The spermatic cord contains the vas deferens, nerves, and blood vessels. Reprinted by permission from Alexander P. Spence and Elliott B. Mason, *Human Anatomy and Physiology*, 1st ed., Fig. 28.3, p. 760 (Menlo Park, Calif.: Benjamin/Cummings Publishing Company, Inc., 1979).

formed into sperm, a process called *spermiogenesis.* The sperm heads then embed in the Sertoli cells; later the heads are released from these cells, an event called *spermiation.* Figure 4–4 illustrates the above-described events within the seminiferous tubules. In humans, the entire process from spermatogonia to spermiation takes about 16 days.[1] If you look at a given region of one seminiferous tubule, you will see that not all areas of the seminiferous epithelium are in the same stage of spermatogenesis (Fig. 4–3), so a male constantly is producing new sperm.

    Sertoli Cells. *Sertoli cells,* also called *sustentacular cells,* are pyramid-shaped cells lying within the seminiferous epithelium (Figs. 4–3 and 4–5). Their bases lie against the basement membrane of each tubule, and their tips point toward the cavity

Figure 4–3. (a) Cross section through a region of the human testis, showing several seminiferous tubules and interstitial tissue. Note that not all areas in one tubule are in a similar stage of spermatogenesis. The interstitial spaces contain androgen-secreting Leydig (interstitial) cells, connective tissue, and blood vessels. Adapted from W. Bloom and D. W. Fawcett, *A Textbook of Histology,* 10th ed. (Philadelphia: W. B. Saunders Company, 1975). © 1975; reprinted by permission. (b) One region of a cross section of one seminiferous tubule, showing stages of spermatogenesis. Note spermatogonium dividing by mitosis, and the different cell stages occurring during meiosis (number of chromosomes in parentheses). Also note spermatozoa with their heads embedded in Sertoli cells. Adapted from L. B. Arey, *Developmental Anatomy* (Philadelphia: W. B. Saunders Company, 1965). © 1965; used with permission.

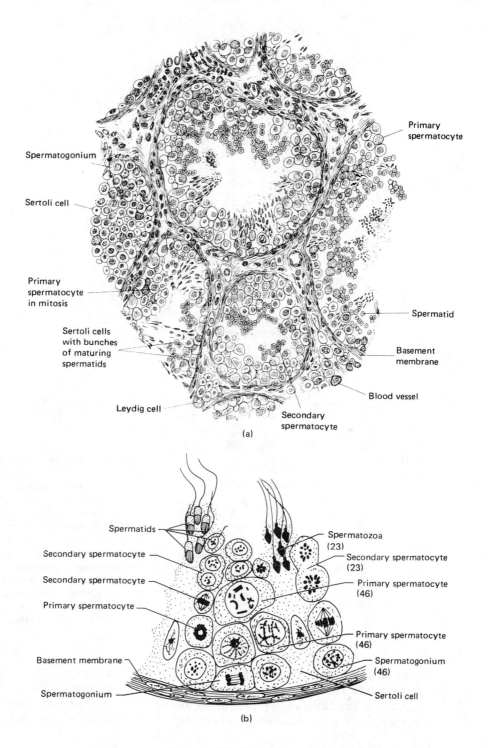

Primary
spermatocyte

Spermatogonium

Sertoli cell

Primary
spermatocyte
in mitosis

Sertoli cells
with bunches
of maturing
spermatids

Leydig cell

Spermatid

Basement
membrane

Blood vessel

Secondary
spermatocyte

(a)

Spermatids

Secondary spermatocyte

Secondary spermatocyte

Primary spermatocyte

Basement membrane

Spermatogonium

Spermatozoa
(23)

Secondary spermatocyte
(23)

Primary spermatocyte
(46)

Primary spermatocyte
(46)

Spermatogonium
(46)

Sertoli cell

(b)

Chap. 4   The Adult Male Reproductive System   **69**

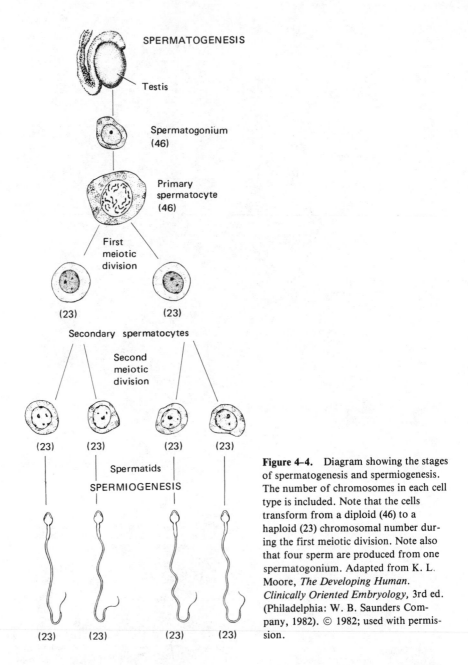

SPERMATOGENESIS

Testis

Spermatogonium
(46)

Primary
spermatocyte
(46)

First
meiotic
division

(23)   (23)
Secondary spermatocytes

Second
meiotic
division

(23)   (23)   (23)   (23)
Spermatids

SPERMIOGENESIS

(23)   (23)   (23)   (23)

Figure 4-4. Diagram showing the stages of spermatogenesis and spermiogenesis. The number of chromosomes in each cell type is included. Note that the cells transform from a diploid (46) to a haploid (23) chromosomal number during the first meiotic division. Note also that four sperm are produced from one spermatogonium. Adapted from K. L. Moore, *The Developing Human. Clinically Oriented Embryology,* 3rd ed. (Philadelphia: W. B. Saunders Company, 1982). © 1982; used with permission.

in the middle of each tubule. As mentioned, the heads of maturing sperm embed in the Sertoli cells, and these cells play a role in nurturing the sperm and in spermiation. Additional functions of Sertoli cells are the secretion of testicular fluid into the tubular cavity and *phagocytosis* (engulfing) of the remains of degenerated germ

Figure 4-5. A section through a seminiferous tubule, taken with the scanning electron microscope. The entire tubule is seen. Note the sperm tails in the center of the tubule, and Leydig cells between adjacent tubules (lower left). Reproduced from *Tissues and Organs: A Text-Atlas of Scanning Electron Microscopy* by Richard G. Kessel and Randy H. Kardon (San Francisco: W. H. Freeman & Company, Publishers, 1979). © 1979; used with permission. Photograph courtesy of Dr. Richard G. Kessel.

cells.[2] Sertoli cells also may play an important role in the hormonal control of spermatogenesis, as described below.

Testicular Interstitial Tissue

Since seminiferous tubules are circular in cross section, regions exist outside the tubules. These regions, the interstitial spaces (Fig. 4–3), contain structures vital to testicular function. Small arteries, capillaries, and small veins are present in the interstitial spaces. This is where blood-borne products (oxygen, glucose) leave the blood and diffuse into the tubules. These products are important for the germ cells since there are no blood vessels within the tubules themselves. Hormones can also leave the interstitial blood vessels, and they enter the seminiferous tubules through their basement membrane. The basement membrane, however, acts as a barrier to the movement of large molecules from the blood into the tubules. Finally, waste products produced by the cells of the tubules pass through the basement membrane and are carried away by the small veins in the interstitial spaces.

The interstitial spaces contain the *Leydig cells,* or testicular *interstitial cells* (Fig. 4–2), which synthesize and secrete androgenic steroid hormones.[3] An androgen is any molecule that will stimulate growth of male tissues. If levels of steroid hormones in the blood of an adult male are measured, the following androgens are found (listed in μg/100 ml of blood plasma): testosterone (0.7); 5α-dihydrotestosterone, or DHT (0.05); androstenedione (0.1); and dihydroepiandrosterone, or DHEA (0.5).[4] Testosterone and DHT are potent androgens in that they stimulate androgen-dependent structures in very low dosages. Levels of DHT in the blood, however, are very low, so that testosterone is the most important androgen in the blood of human males. About 95 percent of the testosterone comes from the Leydig cells of the testes, with the remainder coming from the adrenal glands. Androstenedione and dehydro-epiandrosterone are "weak" (not very potent) androgens produced by the adrenal glands, and their role in male reproductive function is not clear. Many changes caused by testosterone in the male, such as appearance of facial hair and growth of the penis and scrotum, are caused by DHT. That is, these target tissues convert testosterone to DHT, and it is the DHT that is really stimulating the male tissues (Chapter 5).

Human males also have estrogens, or "female sex hormones," in their blood.[5] You will recall from Chapter 3 that estrogens are important sex hormones in females. Blood levels of estradiol-17β (an estrogen) in human males are very low (only 0.003 μg/100 ml blood) in relation to androgen levels. The testicular source of estradiol in males is the Leydig cells, the Sertoli cells, or both. Also, some estrogen in males results from conversion of testosterone to estrogens in tissues other than the testis. Thus, just as human females produce both "male" and "female" sex hormones (Chapter 3), both androgens and estrogens are present in the blood of males. However, the estrogens in human male blood do not "feminize" most men because the androgens are present in relatively higher proportion.

Hormonal Control of Testicular Function

The hypothalamus produces a single gonadotropin-releasing hormone (GnRH), which causes the synthesis and secretion of two gonadotropins: follicle-stimulating hormone (FSH) and luteinizing hormone (LH) from the anterior pituitary gland (Chapter 2). If the anterior pituitary gland is removed from an adult male (hypophysectomy), or if the anterior pituitary gland is rendered nonfunctional because of tissue damage, the testes will degenerate; no more sperm will be produced, and levels of androgen in the blood will drop considerably because the Leydig cells cease functioning. Similarly, if the testes are removed (an operation called *orchidectomy*) or are nonfunctional for some other reason, androgen levels will be very low. Restoration of full fertility and androgen production in hypophysectomized men requires administration of both FSH and LH; either gonadotropin alone will not work.[6] If both gonadotropins are required for full testicular function, what are the specific roles of each hormone? We do not know much about the answer to this question in humans, so much of what we have to say about this is based on research using laboratory animals such as rats.

In humans and other mammals, it is LH that stimulates the Leydig cells to secrete testosterone.[7] This androgen not only enters the general blood circulation and goes to other areas of the body but also diffuses from the testicular interstitial spaces through the basement membrane and into the seminiferous tubules, where it enters the Sertoli cells. The Sertoli cells of the rat can convert testosterone to DHT.[8] Testosterone and DHT then leave the Sertoli cells and enter the fluid around the germ cells.

It is testosterone (and possibly DHT) that stimulates certain phases of spermatogenesis, not LH directly. For example, testosterone in the rat stimulates the first meiotic division (Fig. 4–6), during which diploid primary spermatocytes are converted to haploid secondary spermatocytes.[9] Thus, LH only indirectly causes this division by causing secretion of testosterone from the Leydig cells. The only direct influence of LH within the seminiferous tubule is to stimulate spermiation (Fig. 4–6).

Recently, it has been discovered that Sertoli cells secrete *androgen-binding protein* (ABP).[10] This protein is secreted into the fluid in the tubular lumen, and there it combines with testosterone and DHT. The role of ABP may be to concentrate androgens around the germ cells so that these steroids can better influence spermatogenesis. Furthermore, the androgens are kept from diffusing out of the tubules because adjacent Sertoli cells and germ cells are connected to one another, and this presents a barrier to passage of the ABP-androgen complex out of the tubule.

Why is FSH needed along with LH for complete testicular function? The answer to this question is not known for humans, but there appear to be at least three important functions of FSH in male rats. First, FSH (along with testosterone) may stimulate mitosis of spermatogonia. Second, FSH is necessary for the transformation of spermatids to spermatozoa (spermiogenesis) in the seminiferous tubules. Third, FSH causes the Sertoli cells to secrete ABP. Figures 4–6 and 4–7 summarize what is known about the hormonal control of spermatogenesis.

Control of Gonadotropin Secretion in the Male

Blood levels of FSH and LH in men remain relatively constant, although there is some evidence that levels of LH and testosterone exhibit a daily cycle, being 20 percent higher at night. There also is a monthly cycle in hormone levels in men. E. Ramey studied levels of male sex hormones in the urine of men over a period of 16 years and found 30-day cycles.[11] Men also exhibit emotional cycles of a little longer

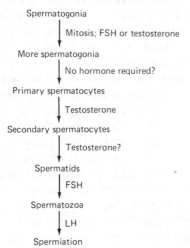

Spermatogonia

↓ Mitosis; FSH or testosterone

More spermatogonia

↓ No hormone required?

Primary spermatocytes

↓ Testosterone

Secondary spermatocytes

↓ Testosterone?

Spermatids

↓ FSH

Spermatozoa

↓ LH

Spermiation

Figure 4–6. Possible hormonal control of spermatogenesis, spermiogenesis, and spermiation.

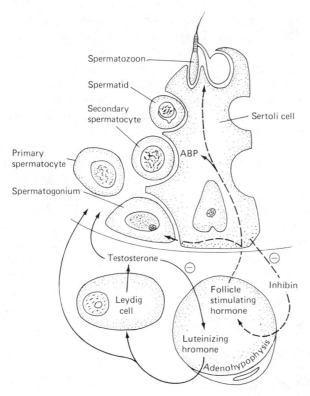

Labels on figure:
Spermatozoon
Spermatid
Secondary spermatocyte
Primary spermatocyte
Spermatogonium
Sertoli cell
ABP
Testosterone
Leydig cell
Follicle stimulating hormone
Luteinizing hromone
Inhibin
Adenohypophysis

Figure 4–7. Summary of hormonal control of testicular function and feedback control of gonadotropin secretion. For hormonal control of steps in spermatogenesis and spermiogenesis, see Fig. 4–6. Testosterone has a negative feedback effect on LH secretion, as does inhibin on FSH secretion. Adapted from W. B. Neaves, "Leydig Cells," in *Frontiers in Reproduction and Fertility Control,* eds. R. O. Greep and M. A. Koblinsky, p. 328 (Cambridge, Mass.: M.I.T. Press, 1977). By permission of the M.I.T. Press, Cambridge, Mass. Copyright © 1977 by The Ford Foundation.

than a month. In the "low" period, they tend to be apathetic, indifferent, and they tend to magnify small problems. In the high period, they have a sense of well-being and a lot of energy.[12] It is not known if the monthly sex hormone cycles in men influence or cause these cyclic mood swings. Finally, there is a seasonal cycle of blood testosterone in men, with a peak in autumn.[13] Whether this seasonal cycle influences the seasonal variation in frequency of coitus (Chapter 10) is not clear.

There are feedback effects of steroid hormones that regulate LH levels in men. Orchidectomized men exhibit abnormally high levels of FSH and LH in their blood, suggesting that testicular substances have a negative feedback effect on gonadotropin secretion from the adenohypophysis. If testosterone is given to an adult man, LH levels in the blood go down, but FSH levels do not change much.[14] Thus, testosterone exerts a negative feedback at least on LH secretion. Under normal conditions, therefore, when testosterone levels drop, LH levels rise and stimulate the Leydig cells to secrete more testosterone; the opposite occurs when testosterone levels rise. When excess androgens are administered to adult men, they can become temporarily infertile (Why?), but their sex drive is not reduced.[15] Testosterone regulates LH secretion by influencing GnRH release from the hypothalamus and by affecting the sensitivity of the LH-secreting cells in the anterior pituitary to GnRH.[16]

What feedback systems regulate FSH secretion in men? One possibility is that the low levels of estradiol-17β in male blood inhibit FSH secretion.[17] If estrogen is administered to male rats, both FSH and LH secretion are inhibited. Thus, if men for some reason take their sexual partner's oral contraceptive pills (which contain an estrogen), their levels of FSH and LH will drop, and they may suffer temporary sterility and lowering of androgen levels. Although temporarily effective, this definitely is not recommended as a method of male contraception since male secondary sexual characteristics will fade and female characteristics (such as breast enlargement) will develop!

Another possible way the FSH secretion is controlled in males is that *inhibin,* a polypeptide present in the testes, plays a role in controlling FSH secretion.[18] Inhibin has been found in the testicular fluid and sex accessory duct fluid of domestic and laboratory mammals. In rats, inhibin suppresses FSH (but not LH) secretion, possibly by decreasing the sensitivity of FSH-secreting cells in the anterior pituitary to GnRH. Inhibin has not yet been completely characterized chemically, and its cellular source (germ cells or Sertoli cells) is not known. Any role of inhibin in human male reproduction is purely conjectural at this time.

Other Factors That Influence Testicular Function

Besides androgens and estrogens, other kinds of steroids may influence testicular function. For example, progesterone inhibits spermatogenesis by inhibiting gonadotropin secretion. Emotional stress also is known to inhibit testicular function in men,[19] and one way that this could happen is that adrenal hormones (corticosteroids), which are secreted under stress, may inhibit gonadotropin secretion.[20] We know very little about the role of prolactin (PRL) in human testicular function. This pituitary hormone is secreted, however, by the male anterior pituitary gland (Chapter 15).

Chemical, radiation, and viral factors. Certain chemicals, if ingested, can damage the testes. For example, cadmium salts cause complete and irreversible sterility because they block blood flow to the testes. Environmental pollutants also can have adverse effects on testicular function. In 1929, the median sperm count of American males was about 90 million sperm in 1 ml of semen. In a sample of college men in 1979, however, the sperm count averaged only 60 million/ml, and 23 percent of these men had sperm counts of 10 million/ml, an amount that indicates functional sterility (Chapter 8). In addition, the testes of these students contained very high levels of *polychlorinated biphenyls* (PCBs) (pollutants), which along with smoking, may be responsible for this decline in sperm count.[21] Ionizing radiation, if at high levels, can also cause temporary sterility and can increase the mutation rate in germ cells. X-ray exposure can destroy germ cells temporarily, although Sertoli cells remain unaffected.

Some infections also can damage the testes. *Mumps,* for example, is a viral disease that causes painful inflammation and enlargement of the salivary glands. If this virus infects an adult male's testes, sterility may occur.

Light cycles and nutrition. As discussed in Chapter 7, daily light cycles can influence many aspects of reproduction in mammals, but human reproduction appears to be unaffected by changes in day length. For example, blind men have normal FSH, LH, and testosterone levels in their blood.[22] Nutrition can also influence testicular function in that malnutrition may lead to temporary sterility. The mechanisms by which malnutrition influences testicular function are not clear, although we do know that a severe deficiency of certain vitamins (A, D, E) can decrease male fertility. This does not mean, however, that taking excessive amounts of these vitamins can increase fertility or male sex drive.

Testicular Cancer. Testicular cancer is rare, less than 5200 cases occurring each year. Since 1972, however, there has been a 72 percent increase in the incidence of this disease. Although testicular cancer is responsible for 13 percent of all cancer deaths in American males age 20 to 34, most cases are curable (by radiation or chemotherapy or by surgically removing the diseased testis) if detected in their initial stages.[23] Thus, a man should periodically check his testes for any small lumps.

Testicular Function in Old Age

Some men undergo a phenomenon that is similar to menopause in women (Chapter 7): in fact, it is called the *male climacteric.* In women, menopause usually happens over a period of a few months (Chapter 7), but male climacteric occurs slowly, over a long period of time; it can begin any time in the forties or later. One sign of male climacteric is that erections may take longer to achieve and require more active stimulation of foreplay to enable a couple to have sexual intercourse (Chapter 18). These and other symptoms are at least partly caused by a decline in testosterone levels in some older men. Male sexual characteristics also may change: the voice can rise in pitch, facial hair growth can decrease, and the scrotum and penis may shrink. Sex accessory structures, such as the prostate and other glands, may reduce in size. Because of these changes, some older men sometimes become depressed and irritable. Men over 70 who are sexually active have higher blood testosterone levels than do those abstaining from sex, but it is not clear if the high levels in these men increased their sexual motivation, or if their sexual activity increased their testosterone levels.[24] An older man's sex life should not suffer if he accepts male climacteric in the same positive way that a woman is encouraged to accept menopause (see Chapter 7).

A primary cause of these changes in some older males appears to be a decrease in the ability of the testes to respond to pituitary gonadotropins. This is shown by the fact that older men tend to have lower blood levels of testosterone and therefore higher LH levels because of the decrease in negative feedback of testosterone on LH secretion. Blood levels of estrogens also rise in older men.[25] The testes may be less functional because their blood supply is, for some unknown reason, reduced. The seminiferous tubules show damage in some older men, but other men continue to be fertile in very old age.

Treatment of older men with an androgen, often along with short-term psychological counseling, can reverse the symptoms associated with the reduction in testosterone secretion. That is, the secondary sex characteristics and sex accessory structures return to their normal condition. In addition, sex drive and ability to have an erection improve. However, degeneration of the seminiferous tubules cannot be helped by testosterone treatment because the blood supply to the testes is minimal.

THE MALE SEX ACCESSORY DUCTS AND GLANDS

Sex Accessory Ducts

The male sex accessory ducts include the tubuli recti, rete testis, vasa efferentia, ductus epididymis, vas deferens, ejaculatory duct, and urethra (Fig. 4–8). These ducts serve to nurture and transport sperm.

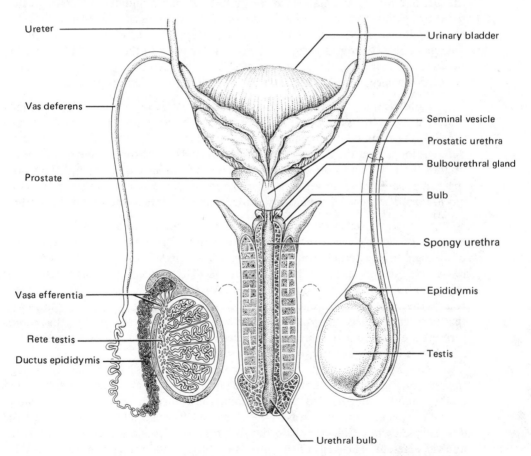

Figure 4–8. The male sex accessory ducts and glands. The structure of the penis is shown in Fig. 4–9. Adapted from L. Weiss and R. O. Greep, *Histology,* 4th ed. (New York: McGraw-Hill Book Company, 1977). © 1977; used with permission.

Mature sperm, suspended in testicular fluid, leave the seminiferous tubules and enter the *tubuli recti*. These tubules, in turn, join a network of tubules still within the testis, the *rete testis*. Then the sperm enter ducts that eventually leave the testis, the *vasa efferentia* (Fig. 4–2). The vasa efferentia then enter an organ lying outside the testis, the epididymis.

Epididymis. The *epididymis* (pl., *epididymides*) is a comma-shaped structure, about 3.8 cm (1.5 in.) long, that lies along the posterior surface of each testis (Figs. 4–2 and 4–8). As the vasa efferentia leave the testis, they enter the larger, upper portion ("head") of the epididymis. The vasa efferentia then join to form a single *ductus epididymis* in the middle region ("body") of the epididymis. This duct then enlarges to form the beginning of the vas deferens in the "tail" region of the epididymis. The tubules within the epididymis secrete important substances that help the sperm to survive and mature. While in the body of the epididymis, sperm are nurtured by epididymal secretions and undergo further stages of their maturation. For example, human sperm taken from the head portion of the epididymis can swim, but only in a circle. In contrast, those taken from the body of the epididymis can move forward by swimming in a spiral path. The ductus epididymis secretes *sperm forward-mobility protein,* which causes this change in movement pattern.[26]

Vas deferens. The vas deferens (or *ductus deferens*) is a 45-cm (18-in.) long tube that ascends on the posterior border of each testis, penetrates the body wall through the inguinal canal, and enters the pelvic cavity. Once inside, each vas deferens loops over the urinary bladder and extends down toward the region of the urethra (Figs. 4–1 and 4–8). The end of each vas deferens has an expanded portion, the "ampulla," which serves as a reservoir for sperm. Each vas deferens enters an *ejaculatory duct,* which is 2 cm (1 in.) long. These short ducts then enter the urethra.

Urethra. The *urethra* is a tube extending from the urinary bladder, through the floor of the pelvic cavity ("urogenital diaphragm"), and then through the length of the penis to its external opening (the *urethral meatus).* Thus, the urethra serves as a passageway for urine. Since, however, the ejaculatory ducts enter the urethra, this tube also transports sperm to the outside. As we see below, the prostate gland surrounds the point where the ejaculatory ducts enter the urethra (Fig. 4–8); that is why this portion is called the *prostatic urethra.* The region of the urethra that passes through the urogenital diaphragm is the *membranous urethra,* and when in the penis this duct is the *spongy* (or *cavernous) urethra.*

How do sperm move through these sex accessory ducts? Their journey through the seminiferous tubules, tubuli recti, rete testis, and vasa efferentia is passive; that is, they do not swim. Instead, the cells lining these ducts have cilia, and the beating of these hairlike processes moves the fluid and its suspended sperm toward the ductus epididymis. Cilia also help the sperm move through the ductus epididymis and vas deferens. The walls of the latter two ducts contain smooth muscle, and this tissue contracts in waves to propel the sperm into the urethra during ejaculation (Chapter 17).

Sex Accessory Glands

The male *sex accessory glands* include the seminal vesicles, prostate gland, and bulbourethral glands (Fig. 4–8). These glands secrete substances into ducts that join the sex accessory ducts. Thus, the secretion of these glands, *seminal plasma,* mixes with the sperm to form *semen* or *seminal fluid.*

Seminal vesicles. The *seminal vesicles* are paired pouchlike structures, about 5 cm (2 in.) long, that lie at the base of the urinary bladder. Each seminal vesicle joins the ampulla of the vas deferens to form the ejaculatory duct (Fig. 4–8). These glands secrete an alkaline viscous fluid rich in the sugar fructose, an important nutrient for sperm (Chapter 8). A majority of the seminal plasma is secreted by the seminal vesicles.

Prostate gland. The *prostate gland* is a single doughnut-shaped organ about the size of a chestnut. It lies below the urinary bladder and surrounds the prostatic urethra (Fig. 4–8). The alkaline secretion of this gland makes up about 13 to 33 percent of seminal plasma. The secretion enters the prostatic urethra through many (up to a dozen) tiny ducts.

Bulbourethral glands. The paired *bulbourethral glands* (or *Cowper's glands*), each about the size of a small pea, lie on either side of the membranous urethra (Fig. 4–8). Their ducts empty into the spongy urethra. These glands secrete mucus that lubricates the urethra during ejaculation.

The male sex accessory glands, because of their connections to the urethra, can become infected by microorganisms that invade the urinary tract. In fact, bacterial infection of the prostate gland *(bacterial prostatitis)* is common in sexually mature males (Chapter 16). Once infected, the prostate can enlarge and ejaculation can be painful (Chapter 17). Other possible symptoms of this condition include fever, chills, rectal discomfort, lower back pain, and an increased urgency of urination (because the enlarged prostate is pressing on the urethra and bladder). This type of prostatitis can be cured with antibiotics. The prostate gland, however, also can enlarge and become inflamed without infection by bacteria. This *nonbacterial prostatitis* is most common in men in their middle and late years; about one-third of men over 60 develop this condition. Unfortunately, the prostate can also develop benign or malignant tumors. Prostatic cancer is the second leading cause of cancer death in men in the United States; about 19,000 die each year of this affliction. Surgery or radiation therapy often is necessary for this prostatic condition.

Hormonal Control of Sex Accessory Structures

The maintenance and function of male sex accessory ducts and glands are under the control of androgens. Testosterone, for example, stimulates secretion of sperm nutrients by the epididymis. If a man is orchidectomized, the sex accessory structures will soon atrophy. There also is some evidence in laboratory mammals that prolactin acts with testosterone to control the function of the sex accessory ducts and glands. Finally, oxytocin and prostaglandins may stimulate contraction of the smooth muscle in these glands and ducts during ejaculation (Chapter 17).

THE PENIS

The *penis* has an enlarged acorn-shaped end called the *glans penis* and a *penile shaft, or body* (Fig. 4–9). The rounded ridge at the back end of the glans penis is called the *corona glandis.* The skin covering the shaft extends in a loose fold over the glans; this is the *penile prepuce, or foreskin.* In circumcised men, much of the prepuce has been removed surgically (Chapter 11). There are small glands under the foreskin that secrete a cheesy substance, smegma. Retraction of the prepuce to remove the smegma with soap and water is important, especially in uncircumcised men, because bacteria can thrive in the secretion.

The body of the penis contains three cylindrical masses of spongy tissue filled with blood sinuses (Fig. 4–9). Two of these cylinders, one on each side of the top of the body, are the corpora cavernosa. The single cylinder at the bottom of the shaft is the corpus spongiosum; the spongy urethra runs through this cylinder. The spongy cylinders fill with blood during an erection (Chapter 17), and the corpus spongiosum can be felt as a raised ridge on the bottom of the erect penis. The body of the penis attaches to the pelvic wall; extensions of the spongy tissue actually are fused to the pelvic bone.

Men (and women) often are curious or concerned about penis size. The length of the nonerect (flaccid) penis of most men ranges from 8.5 to 10.5 cm (3.3 to 4.1 in.), with an average length of 9.5 cm (3.7 in.). The average length of the erect penis is 16 cm (6.3 in.), with a range of 12.0 to 23.5 cm (4.7 to 9.2 in.). The average circumference of an erect penis at its thickest point is 13.2 cm (5.2 in.).[27] Contrary to popular belief, there is no correlation between a man's skeletal make-up and his penis size. When a relatively small flaccid penis becomes erect, it enlarges more than does a relatively large flaccid penis. Even if a man does have a large penis, it does not necessarily mean that he can better satisfy a sexual partner. In Chapter 17, we review the human sexual response, including what happens to the penis during male sexual arousal.

THE SCROTUM

The *scrotum* is a pouch, suspended from the groin, which contains the testes and some of the male sex accessory ducts. The pouch has two compartments, each housing one testis; the left testis usually hangs lower than the right. The skin of the scrotum is relatively dark and hairless and is separated in the midline by a ridge called the *raphe.* Underneath the skin of the scrotum is a layer of involuntary smooth

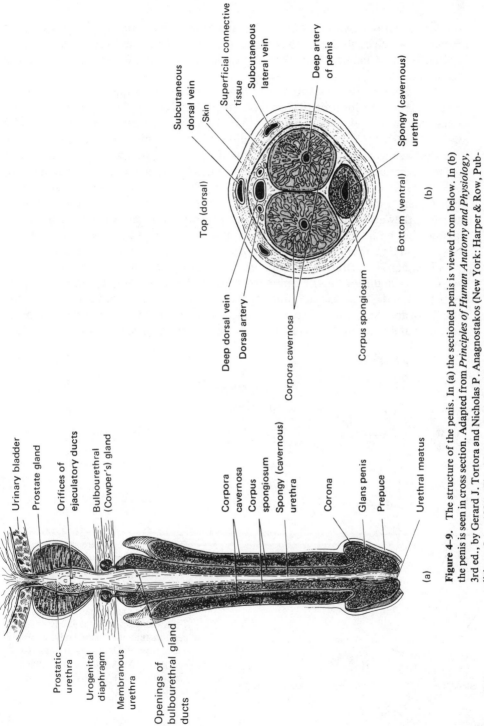

Urinary bladder

Prostate gland

Orifices of
ejaculatory ducts

Bulbourethral
(Cowper's) gland

Prostatic
urethra

Urogenital
diaphragm

Membranous
urethra

Openings of
bulbourethral gland
ducts

Corpora
cavernosa

Corpus
spongiosum

Spongy (cavernous)
urethra

Corona

Glans penis

Prepuce

Urethral meatus

(a)

Subcutaneous
dorsal vein

Superficial connective
tissue

Skin

Subcutaneous
lateral vein

Deep artery
of penis

Top (dorsal)

Deep dorsal vein

Dorsal artery

Corpora cavernosa

Corpus spongiosum

Spongy (cavernous)
urethra

Bottom (ventral)

(b)

Figure 4–9. The structure of the penis. In (a) the sectioned penis is viewed from below. In (b) the penis is seen in cross section. Adapted from *Principles of Human Anatomy and Physiology*, 3rd ed., by Gerard J. Tortora and Nicholas P. Anagnostakos (New York: Harper & Row, Publishers, Inc., 1981), Fig. 28–8, p. 719. © 1981 by Gerard J. Tortora and Nicholas P. Anagnostakos. Reprinted by permission of Harper & Row, Publishers, Inc., and Dr. Gerard J. Tortora.

muscle, the *tunica dartos.* Just under the tunica dartos is another layer of muscle, the *cremaster,* which contracts when the inner thigh is stroked; this "cremasteric reflex" is discussed in Chapter 17. Unlike the tunica dartos, the cremaster is voluntary striated muscle.

One function of the scrotum is to help maintain the temperature necessary for spermatogenesis by locating the testes out of the body cavity. The temperature in the human scrotum is 3.1°C (5.6°F) lower than the normal temperature within the body (37°C = 98.6°F).[28] Spermatogenesis requires this lower temperature, and an increase in scrotal temperature of 3°C (5.4°F) caused by wearing of tight underclothing can lead to a decrease in viable sperm and even an increase in mutation rate in male germ cells.[29] The testes normally descend from the pelvic cavity into the scrotum before birth (Chapter 5). If they fail to descend by the time spermatogenesis begins in the child, they are damaged by the high temperature in the body cavity (Chapter 6). Taking a hot bath or sauna can lead to temporary problems with sperm production. However, taking a series of hot baths is not a reliable contraceptive measure.

There are temperature receptors in the scrotum. When the temperature is too low, the tunica dartos muscle contracts, causing the scrotal skin to wrinkle and the testes to ascend. Thus, the scrotum has a smaller surface area for heat loss, and the testis now are closer to the warmer groin region. It should be noted that the other muscle in the scrotum, the cremaster, is not involved in temperature regulation.

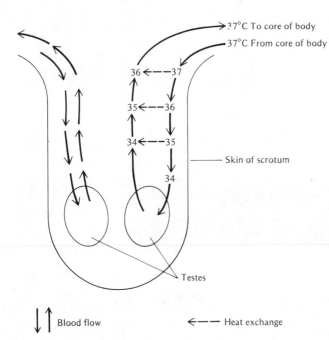

Figure 4–10. Countercurrent heat exchange in the spermatic cord. The testicular artery and veins lie close together in the spermatic cord. Blood flows in opposite directions, and heat leaves the artery and enters the venous blood. Thus, the testes are kept at a lower temperature than that in the body. Reproduced from P. Goldstein, *Human Sexuality* (New York: McGraw-Hill Book Company, 1976). © 1976; used with permission.

However, this muscle contracts in times of sexual excitement, fear, or anxiety. Numerous sweat glands in the scrotal skin secrete sweat when the scrotal temperature is high, and evaporation of this fluid also helps cool the testes.

There is another means whereby the testes are cooled. The testicular artery and veins in each spermatic cord lie very close together, and the blood flow runs in opposite directions. As warm blood from the body flows down the testicular artery, heat is lost to the cooler blood flowing up the testicular veins (Fig. 4–10). This *countercurrent heat exchange* system thus helps keep the testes cooler than the body temperature.[30]

CHAPTER SUMMARY

The male reproductive tract consists of the testes, the sex accessory ducts and glands, and the penis and scrotum. Each testis is subdivided into testicular lobules, and each lobule contains seminiferous tubules, which are the site of sperm production (spermatogenesis and spermiogenesis). Sertoli cells are also present within the seminiferous tubules. The interstitial spaces between tubules contain blood vessels and Leydig cells, the source of androgens.

The anterior pituitary gland secretes two gonadotropic hormones, which are both necessary for testicular function. Luteinizing hormone (LH) stimulates secretion of androgens from the Leydig cells, and these androgens control some stages of spermatogenesis. Follicle-stimulating hormone (FSH) stimulates Sertoli cells in the seminiferous tubules to secrete androgen-binding protein, which concentrates androgens around the germ cells. Sertoli cells also support and nourish the male germ cells. Secretion of FSH and LH is controlled by feedback effects of androgens, estrogens, and possibly inhibin produced by the testes. Stress, nutrition, chemicals, and ionizing radiation can influence testicular function. Testicular cancer, although rare, is becoming more common. The testes of older men may decline in function, and the resultant hormonal changes can influence some men physiologically and psychologically, a process called the male climacteric. Treatment with testosterone often can reverse some of these symptoms.

Sperm produced in the seminiferous tubules travel through a series of male sex accessory ducts: the tubuli recti, rete testis, vasa efferentia, ductus epididymus, vas deferens, ejaculatory duct, and urethra. Male sex accessory glands (seminal vesicles, prostate, and bulbourethral glands) produce seminal plasma that enters the urethra and mixes with sperm during ejaculation. The male sex accessory structures are under androgenic control.

The penis consists of a glans penis, shaft, and prepuce. The shaft contains spongy tissue that fills with blood during an erection. The scrotum is a pouch of skin and muscle that contains both testes. This sac plays a role in controlling testicular temperature.

NOTES

1. Y. Clermont, "Kinetics of the Human Seminiferous Epithelium," *Research in Reproduction,* 4 (1972), 3–4.

2. D. W. Fawcett, "The Ultrastructure and Functions of the Sertoli Cells," in *Frontiers in Reproduction and Fertility Control,* eds. R. O. Greep and M. A. Koblinsky, pp. 321–37 (Cambridge, Mass.: M.I.T. Press, 1977).

3. M. B. Lipsett, "Steroid Secretion by the Human Testis," in *The Human Testis,* eds. E. Rosenberg et al., pp. 407–18. (New York: Plenum Publishing Corporation, 1970).

4. B. A. Cook et al., "The Testis," *Research in Reproduction,* vol. 5, no. 6 (1973), map.

5. R. Rubens et al., "Further Studies on Leydig Cell Function in Old Age," *Journal of Clinical Endocrinology and Metabolism,* 39 (1974), 40–45.

6. R. E. Mancini et al., "Effect of Human Urinary FSH and LH on the Recovery of Spermatogenesis in Hypohysectomized Patients," *Journal of Clinical Endocrinology and Metabolism,* 33 (1971), 888–97.

7. E. S. Steinberger et al., "The Control of Testicular Function," in *Frontiers in Reproduction and Fertility Control,* eds. R. O. Greep and M. A. Kolbinsky, pp. 264–92. (Cambridge, Mass.: M.I.T. Press, 1977).

8. Fawcett (1977), op. cit.

9. Steinberger et al. (1977), op. cit.

10. S. C. Weddington et al., "Sertoli Cell Function after Hypophysectomy," *Nature, London,* 254 (1975), 145–46; and D. M. DeKrester et al., "Peritubular Tissue in the Normal and Pathological Human Testis: An Ultrastructural Study," *Biology of Reproduction,* 12 (1975), 317–24.

11. E. Ramey, "Men's Cycles (They Have Them Too, You Know)," *Ms. Magazine,* Spring 1972, pp. 8–14.

12. Ibid.

13. S. G. H. Smals et al., "Circannual Cycle in Plasma Testosterone Levels in Man," *Journal of Clinical Endocrinology and Metabolism,* 42 (1976), 979–82.

14. P. A. Lee et al., "Regulation of Human Gonadotropins. VIII. Suppression of Serum LH and FSH in Adult Males Following Exogenous Testosterone Administration," *Journal of Clinical Endocrinology and Metabolism,* 35 (1972), 636–64; and R. J. Sherins and D. L. Loriaux, "Studies in the Role of Sex Steroids in the Feedback Control of FSH Concentrations in Man," *Journal of Clinical Endocrinology and Metabolism,* 36 (1973), 886–93.

15. P. R. K. Reddy and J. M. Rao, "Reversible Antifertility Action of Testosterone Propionate in Human Males," *Contraception* 5 (1972), 255–301.

16. Steinberger et al. (1977), op. cit.

17. Ibid.

18. Y. Folman et al., "Production and Secretion of 5α-Dihydrotestosterone by the Dog Testis," *American Journal of Physiology,* 222 (1972), 653–56.

19. A. Krenz et al., "Suppression of Plasma Testosterone Levels and Psychological Stress," *Archives of General Psychiatry,* 26 (1972), 479.

20. R. E. Mancini et al., "Effects of Prednisolone upon Normal and Pathological Human Spermatogenesis," *Fertility and Sterility,* 17 (1966), 500.

21. Anonymous, "PCBs Linked to Male Sterility," *Science News,* 116, no. 11 (1979), 183.

22. S. Bodenheimer et al., "Diurnal Rhythms of Serum Gonadotropins, Testosterone, Estradiol and Cortisol in Blind Men," *Journal of Clinical Endocrinology and Metabolism,* 37 (1973), 472–75.

23. R. M. Lane and S. M. Nagelschmidt," Testicular Threat," *Omni,* 16 (July 1981), 123.

24. Anonymous, "Testosterone and Male Sex Drive," *Science News,* 116, no. 6 (1979), 104.

25. Rubens et al. (1974), op. cit.

26. T. S. Acott et al., "Sperm Forward Mobility Protein: Tissue Distribution and Species Cross-Reactivity," *Biology of Reproduction,* 20 (1979), 247–52.

27. R. Bennett, "Forum Penile Study," *Forum: The International Journal of Human Relations,* 1 (1972), 36–41.

28. A. M. Tessler and H. P. Krahn, "Varicocele and Testicular Temperature," *Fertility and Sterility,* 17 (1966), 201–3.

29. L. Ehrenberg et al., "Gonadal Temperature and Spontaneous Mutation Rate in Man," *Nature, London,* 180 (1957), 1433–34.

30. G. M. H. Waites, "Temperature Regulation and the Testis," in *The Testis,* eds. A. D. Johnson et al., vol. 1, pp. 241–79 (New York: Academic Press, Inc., 1970).

FURTHER READING

COOK, B. A., H. J. VAN DER MOLEN, and B. P. SETCHELL, "The Testis," *Research in Reproduction,* vol. 5, no. 6 (1973), map.

FAWCETT, D. W., "The Male Reproductive System," in *Reproduction and Human Welfare: A Challenge to Research,* eds. R. O. Greep et al. Cambridge, Mass.: M.I.T. Press, 1976. Pp. 165–277.

MONESI, V., "Spermatogenesis and the Spermatozoa," in *Reproduction of Mammals, Book 1, Germ Cells and Fertilization,* eds. C. R. Austin and R. V. Short. Cambridge, England: Cambridge University Press, 1972. Pp. 46–84.

SETCHELL, B. P., and B. A. COOKE, "The Action of Androgens," *Research in Reproduction,* vol. 11, no. 5 (1979), map.

SHORT, R. V., "Role of Hormones in Sex Cycles," in *Reproduction in Mammals, Book 3, Hormones in Reproduction,* eds. C. R. Austin and R. V. Short. Cambridge, England: Cambridge University Press, 1972. Pp. 42–72.

STEINBERGER, E., "Hormonal Control of Mammalian Spermatogenesis," *Physiological Reviews,* 51 (1971), 1–22.

5 *Sexual Differentiation and Development*

LEARNING GOALS

Having read this chapter, you should be able to:

1. Define and discuss the adaptive significance of X-chromosome inactivation.
2. Describe the condition of the gonads, sex accessory ducts, and external genitalia during the sexually indifferent stage.
3. Describe the differentiation of testes or ovaries from the indifferent gonads.
4. Discuss how the presence of a Y chromosome masculinizes the male reproductive tract.
5. Describe the differentiation of the male and female sex accessory ducts and glands.
6. Describe how the male and female external genitalia are derived from the indifferent external genitalia.
7. Explain experiments in laboratory mammals that show that testosterone masculinizes the brain. Then, discuss the evidence for or against a similar phenomenon in humans.
8. List homologous structures of the male and female reproductive tract.
9. Compare and contrast the condition of the reproductive system in a newborn human male and female.
10. Discuss the symptoms and causes of the testicular feminization syndrome, the "Guevedoce" syndrome, the adrenogenital syndrome, Turner's syndrome, and Klinefelter's syndrome.
11. State how a sex-change operation can change male into female genitalia.

INTRODUCTION

In Chapters 3 and 4, we reviewed the structures of the adult male and female reproductive systems. All of us were just a single fertilized cell at one time. From these previous two chapters, we know that the adult male and female reproductive systems differ to a great degree. Therefore, the reproductive systems had to undergo remarkable differentiation and development into the adult form. Even more remarkable, the reproductive system of genetically male and female embryos, are, at an early stage of development, identical! How this identical stage differentiates into the typical (and sometimes atypical) male and female reproductive systems during embryonic and fetal development is the subject of this chapter.

CHROMOSOMAL SEX

Sex is determined by chromosomes and is established at the time of fertilization of an egg by a sperm (Chapter 8). In females, each diploid body cell has 46 chromosomes (23 pairs), including 22 pairs of *autosomes* (nonsex chromosomes) and 1 pair of *sex chromosomes*. The sex chromosomes in females are XX, so the chromosomal type of female cells is designated as 46;XX. Thus, the female is called the *homogametic sex* because there are two X chromosomes. This occurs when a haploid (23;X) egg is fertilized by a haploid (23;X) sperm. Male somatic cells are 46;XY (Fig. 5-1), which occurs when a 23;X egg is fertilized by a 23;Y sperm. Thus the male mammal is the *heterogametic sex* because there is both an X and a Y chromosome.

At least 100 genes are present on the human X chromosome. These genes are sex-linked to the X chromosome, and they control such things as color vision, blood clotting, and diseases such as muscular dystrophy (Chapter 12). The Y chromosome gets short-changed, however, since the only sex-linked gene it has controls hairy ears, and even this is in doubt. Therefore, a problem confronts the cells of a female since they have a double dose of genes on the X chromosome. In 1949, M. Barr discovered that female cells, but not male cells, contain a small dot of condensed material that is a remnant of one of the X chromosomes that has been inactivated *(X-chromosome inactivation)*. This inclusion is called the *Barr body* or *sex chromatin* (Fig. 5-2). In 1961, Mary Lyon proposed that one X chromosome is inactivated randomly in female cells, so that a double gene dosage is avoided. All cells that follow a particular cell line, however, have the same X chromosome inactivated. If more than two X chromosomes are present, as in some conditions produced by errors of fertilization (Chapter 8), all of the X chromosomes are inactivated except one (Fig. 5-2).

The inactivation of an X chromosome is even more interesting when you realize that, as somatic cells divide, the inactivated chromosome must also replicate and divide. The presence of the Barr body is utilized to determine the sex of fetuses using the procedure called amniocentesis (Chapter 9). Also, cells from the lining of the

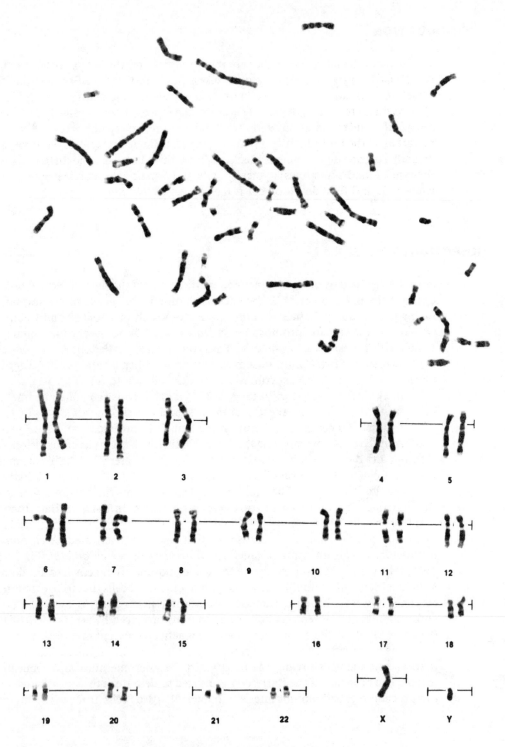

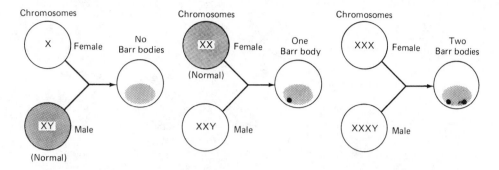

Figure 5–2. Variations in sex chromosomes and the resultant number of sex chromatins, or Barr bodies. Note that the number of Barr bodies is one less than the number of X chormosomes.

mouth membrane can be checked for the presence of a Barr body (the *buccal smear test).* The fluorescent dye *quinacrine* can be used to identify Y chromosomes.[1]

THE SEXUALLY INDIFFERENT STAGE

There is a period during embryonic development when the gonads, sex accessory ducts, and external genitalia of both sexes are identical; this is the *sexually indifferent stage.* The only way that one can determine the sex of embryos at this stage is to look for Barr bodies or Y chromosomes in the cells. The gonads develop as paired bulges of tissue near the midline at the back of the abdominal cavity. These are the *genital ridges,* which are similar in both sexes and do not as yet contain germ cells. These ridges then differentiate into *indifferent gonads,* consisting of an outer cortex and an inner medulla (Fig. 5-3). These indifferent gonads appear during the fifth week of gestation, when the embryo measures about 6 mm from the top of the head to its rump (the *crown-rump length,* or CR). The gonadal indifferent stage lasts until the end of the seventh week of gestation, when the embryo is 12 mm CR.

During the sixth week of gestation, the embryo also has indifferent sex accessory ducts, consisting of two pairs (Fig. 5-4). One pair, the *wolffian ducts (mesonephric ducts),* is derived from the ducts of the primitive embryonic kidney, the *mesonephros.* The other pair, the *müllerian ducts (paramesonephric ducts),* exists separately from the mesonephros. This indifferent duct stage lasts until the seventh week of gestation. Before the seventh week of gestation, the external genitalia are indifferent in both sexes (Fig. 5-5). These indifferent external structures consist of the *genital tubercle (phallus),* paired *urogenital folds,* and a *labioscrotal swelling.*

Figure 5–1. Human chromosomes (top) which have been arranged in homologous pairs (bottom). These are from a cell of the author's son (Evan Pryor Jones), which was obtained while he was still a fetus by a process called amniocentesis (Chapter 9). Note the X and Y chromosomes. Courtesy of Dr. George Henry and the Reproductive Genetics Center, Denver, Co.

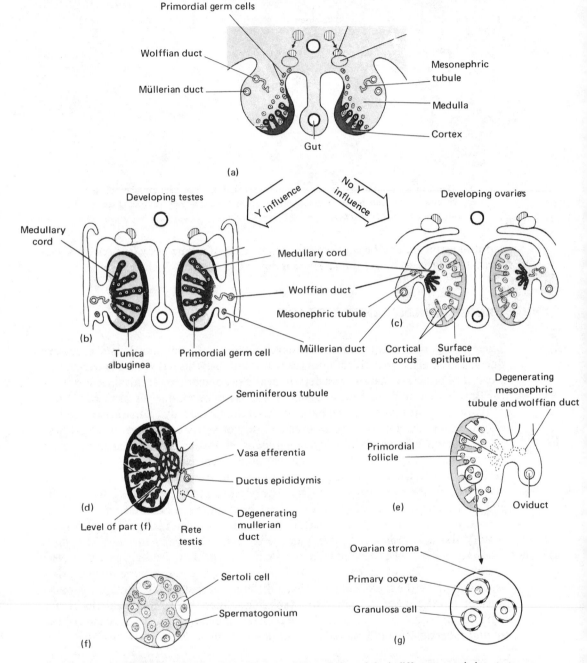

Figure 5–3. Diagram illustrating the differentiation of the indifferent gonads into testes or ovaries: (a) Indifferent gonads from 6-week-old fetus; (b) at seven weeks, showing testes developing under influence of the Y chromosome (H-Y antigen); (c) 12 weeks, showing ovaries developing in absence of the Y chromosome; (d) testis at 20 weeks, showing the rete testis and

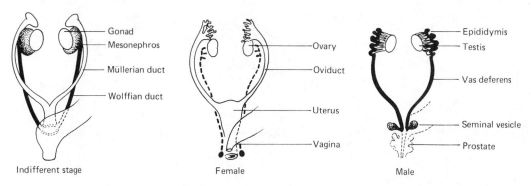

| Indifferent stage | Female | Male |

Gonad
Mesonephros
Müllerian duct
Wolffian duct

Ovary
Oviduct
Uterus
Vagina

Epididymis
Testis
Vas deferens
Seminal vesicle
Prostate

Figure 5-4. Differentiation of the sex accessory structures in male and female embryos. Adapted from J. D. Wilson et al., "The Hormonal Control of Sexual Development," *Science,* 211 (1981), 1278-84. © 1981 by the American Association for the Advancement of Science; used with permission.

GONADAL SEX DETERMINATION

During the fourth to fifth week of gestation, germ cells begin to arrive at the indifferent gonads and to penetrate into these organs. These *primordial germ cells* arise from the epithelium of an extraembryonic membrane, the yolk sac (Chapter 9), and then migrate by amoeboid movement along the embryonic gut to the indifferent gonads. In humans, there are about 100 of the primordial germ cells that start the journey, but by the time they arrive at the gonads, they number about 1700 because they proliferate en route.[2] You should remember that, in genetic males, the primordial germ cells and the tissue of the indifferent gonads are 46;XY, whereas these same structures are 46;XX in genetic females.

Ovarian Development

If a Y chromosome is not present, the cortex of the indifferent gonad forms ingrowths called *cortical cords.* The primordial germ cells concentrate in the cortical cords, and the medulla degenerates (Fig. 5-3). Differentiation of ovaries occurs a few weeks later than differentiation of testes. The remnants of the medullary cords persist as *rete ovarii* in the adult ovary. The enlarging cortex becomes an ovary, which contains *oogonia,* cells derived from the primordial germ cells. The oogonia then undergo mitotic proliferation so that there are about 600,000 by eight weeks of gestation and 7,000,000 by the fifth month of gestation.[3] Most of these oogonia

the seminiferous tubules derived from medullary cords; (e) ovary at 20 weeks, showing primordial follicles; (f) section of a seminiferous tubule from a 20-week fetus; (g) section from the ovarian cortex of a 20-week fetus, showing three primordial follicles. Adapted from K. L. Moore, *The Developing Human. Clinically Oriented Embryology,* 3rd ed. (Philadelphia: W. B. Saunders Company, 1982). © 1982; used with permission.

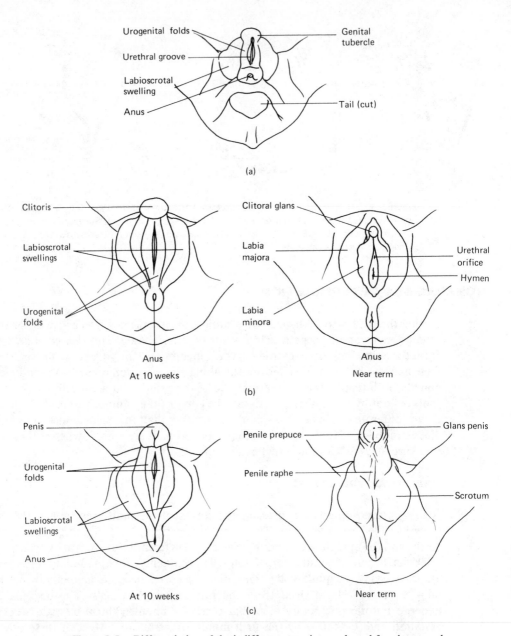

Figure 5-5. Differentiation of the indifferent stage into male and female external genitalia. (a) Indifferent stage, (b) female development, (c) male development. Adapted from J. D. Wilson et al., "The Hormonal Control of Sexual Development," *Science* 211 (1981), 1278–84. © 1981 by the American Association for the Advancement of Science; used with permission.

degenerate, but some live, enter meiosis, and become oocytes, which are surrounded by a single layer of granulosa cells. Thus, primordial follicles are formed (Fig. 5-3). At birth, all of the female germ cells are oocytes in follicles. Some of these follicles grow to reach the tertiary stage, but then they become atretic. Each ovary contains about 500,000 follicles at birth, and this is all that a female will have for the rest of her life.[4] [This is in contrast to the testes, which produce many new germ cells (spermatogonia) throughout a male's reproductive life.] From the seventh to the ninth month of fetal life, the ovaries descend from an abdominal position to reside in the pelvic cavity.

Testicular Development

If the primordial germ cells and gonadal cells are XY, testes develop from the indifferent gonads (Fig. 5-3). This event occurs at about 43 to 50 days of gestation, when the male embryo is 15 to 17 mm in CR (Fig. 5-5). If a Y chromosome is present, the primordial germ cells concentrate in the medulla of the gonad, and the cortex regresses. *Medullary cords* develop within the medulla (Fig. 5-3). These cords contain cells derived from the primordial germ cells, the spermatogonia, and also contain Sertoli cells. Thus, the medullary cords become the seminiferous tubules of the testes (Fig. 5-3). Remnants of these cords also form the rete testis. The testosterone-secreting Leydig cells begin to appear between the cords on about day 60 of gestation (30 mm CR), and these fetal Leydig cells may be stimulated to secrete testosterone by activity of the fetal anterior pituitary gland, which begins secreting at this time. Also, the placenta secretes an LH-like gonadotropin called *human chorionic gonadotropin,* or *hCG* (Chapter 9), which may pass into the male fetus and cause its Leydig cells to begin secreting testosterone. During the second to the ninth month of fetal life, the testes descend from an abdominal position into the scrotum (Fig. 5-6). Their journey is stimulated by testosterone. If the testes fail to descend, *cryptorchid testes* are present (Chapter 6).

H-Y Antigen

How does the presence of a Y chromosome control gonadal sex determination? We need to know more about this. One theory, based on good evidence in laboratory mammals, is that a gene (or genes) on the Y chromosome causes formation of a protein on the cell membrane of male cells called the *H-Y antigen,* this antigen then causing testes to form. Appearance of this protein is not dependent on testosterone but is an inherited characteristic.[5] It now appears that the gene on the Y-chromosome coding for production of H-Y antigen is not the only gene required for gonadal sex determination. The others are (1) a gene that activates the H-Y gene, (2) a gene that codes for receptors for the H-Y antigen, and (3) a gene controlling spermatogenesis.[6] Some believe that the H-Y antigen only switches on a sex-determining gene present in the X chromosome, but there is not much evidence for this.[7]

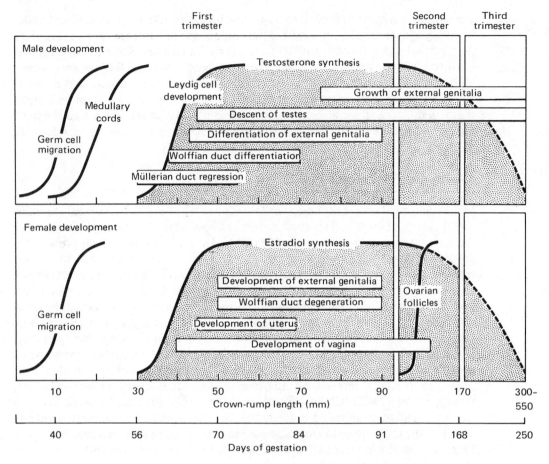

Figure 5-6. Relationship between differentiation of the gonads and sex accessory ducts in male and female human embryos. The shaded areas represent the degree of testosterone secretion from the embryonic testes or estradiol-17β secretion from the embryonic ovaries. Adapted from J. D. Wilson et al., "The Hormonal Control of Sexual Development," *Science,* 211 (1981), 1278–84. © 1981 by the American Association for the Advancement of Science; used with permission.

Differentiation of ovaries is not only a matter of the absence of a Y chromosome, and therefore of the H-Y antigen. For example, individuals with *Turner's syndrome* (45;XO cells) are sterile; the ovaries have no germ cells and are filled with connective tissue. Thus, both X chromosomes are necessary for germ cell survival and ovarian development.[8] If this is true, how is X-chromosome inactivation compatible with normal ovarian development? Well, it turns out that one X chromosome is not inactivated in the oogonia and oocytes in the fetal ovaries until after at least week 21 of gestation.[9]

DIFFERENTIATION OF THE SEX ACCESSORY DUCTS AND GLANDS

As mentioned, embryos of each sex possess both wolffian and müllerian duct systems until the seventh week of gestation (Fig. 5-4). If testes develop, they secrete testosterone, which causes development of the wolffian duct system into the epididymis, vas deferens, seminal vesicles, and ejaculatory duct.[10] The vasa efferentia form from the *mesonephric tubules,* which lead into the mesonephric duct. (The prostate gland and bulbourethral glands develop from part of the urethra, not from the wolffian duct system.) It appears that this effect of testosterone is local and not through the general circulation. That is, the right testis secretes testosterone, which diffuses to adjacent wolffian structures and stimulates their development only on that same side; the same thing happens on the other side. The embryonic testes also secrete a protein from the Sertoli cells that causes regression of the müllerian duct system.[11] This protein is called *müllerian duct-inhibiting factor.* A small, pouch-like remnant of the müllerian ducts in males is present within the prostate gland. This remnant is the *prostatic utricle.* Development of the male sex accessory ducts is depicted in Fig. 5-4.

If ovaries develop, no testosterone is secreted, and the wolffian ducts regress. Also no müllerian duct-inhibiting factor is secreted, so the müllerian ducts develop into a pair of oviducts as well as into the uterus, cervix, and the upper one-third of the vagina (Fig. 5-4). The lower two-thirds of the vagina forms from the *urogenital sinus,* which also gives rise to the urinary bladder and urethra. The greater and lesser vestibular glands of the female are derived from urethral tissue.

Given the above information, you should be able to explain the results of the following experiments on rats: (1) If a male fetus is orchidectomized (its testes removed), it develops female sex accessory organs derived from the müllerian duct system. (2) If an orchidectomized male fetus is given testosterone, both the wolffian and müllerian duct systems develop. (3) If an intact male fetus is given an antiandrogen that blocks the effects of testosterone, neither duct system develops.

DIFFERENTIATION OF THE EXTERNAL GENITALIA

If testes are present in an embryo, the male external genitalia develop during the eighth week of gestation (Fig. 5-6). That is, the genital tubercle differentiates into most of the penis, the urogenital folds form the ventral aspect of the penis shaft, and the labioscrotal swelling becomes the scrotum (Fig. 5-5). These structures develop because testosterone, secreted by the embryonic testes, is converted to 5α-dihydrotestosterone (another androgen) by an enzyme *(5α-reductase)* present in the cells of these tissues.[12] Thus, it is 5α-dihydrotestosterone that actually stimulates development of the male external genitalia as well as the bulbourethral and prostate glands. This is in contrast to the wolffian duct derivatives, which contain no 5α-reductase and respond directly to testosterone.

If no testes are present to secrete testosterone (that is, ovaries are present), the genital tubercle differentiates into the clitoris, the urogenital folds into the labia minora, and the labioscrotal swelling into the mons pubis and labia majora (Fig. 5-5). Thus, the neutral condition again is female, and it takes testosterone to masculinize the external genitalia.

SEXUAL DIFFERENCES IN BRAIN DEVELOPMENT

Differences in Hypothalamic Function

The brain of adult mammals exhibits several sex differences. For example, the hypothalamus of females is programmed to release GnRH in a cyclic manner, resulting in estrous cycles in most mammals, or menstrual cycles in humans and other primates (Chapter 7). In contrast, the hypothalamus of males exhibits very little cyclicity in GnRH release, although some is present (Chapter 4). This sex difference is not due to inherent cyclicity in the ovaries or anterior pituitary gland. That is, if one transplants ovaries into an adult male mammal, the ovaries do not exhibit cyclicity. Likewise, if one transplants the anterior pituitary gland of an adult female into the sella turcica of an adult male, no cyclic output of FSH and LH occurs. If, however, one transplants the pituitary gland of an adult male into the sella turcica of an adult female, cyclic release of FSH and LH follows.

Coincident with the inherent cyclicity of GnRH secretion by the female brain is cyclicity in female sexual behavior. That is, females of many mammalian species are only sexually receptive to male sexual advances at a time in the ovarian cycle when estrogen secretion from their ovaries is high, right around the time of ovulation (Chapter 7). Males, in contrast, show no such cyclicity in sexual motivation. Furthermore, it is difficult to elicit female sexual behavior in adult males by administering an estrogen. Similarly, it is difficult to get females to perform male sexual behavior by administering testosterone. Therefore, the brains of adult males and females of most mammals differ in their responsiveness to hormones that influence sexual behavior. The evidence for hormonal control of human sexual behavior is discussed in Chapter 17.

Hormonal Influence on Brain

It is interesting that, in many mammals, brain differences between sexes require exposure to certain hormones at critical periods in development.[13] Removal of the testes of newborn male rats, mice, and hamsters causes them to have a cyclic female-like hypothalamus as adults, and to respond to estrogen by exhibiting female sexual behavior (Fig. 5-7). Administration of testosterone to newborn females of these

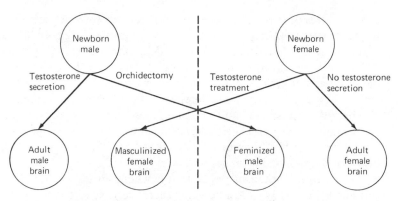

Figure 5-7. Sexual differentiation of the rat brain takes place soon after birth. In newborn males, testosterone secreted by the testes is converted by target cells in the brain into estradiol-17β, which gives rise to a permanent "male" pattern of brain activity. If the male is orchidectomized at birth, however, sexual differentiation of nerve circuits in the brain fails to take place, and the brain retains a "female" pattern. Administration of testosterone to a newborn female rat evokes a "male" pattern of brain circuitry as a result of the testosterone being converted intracellularly into estradiol. Unlike in rats, the critical period for androgenic masculinization of the human brain occurs before birth.

animals during a day or two after birth causes their hypothalamus to exhibit a male noncyclic pattern as adults and to respond to testosterone by exhibiting male sexual behaviors (Fig. 5-7). When mothers of male rat fetuses are stressed, it results in lower testosterone secretion and feminization of the males' behavior as adults. [14]

Further experiments in laboratory mammals have shown that testosterone exerts its masculinizing effect on the hypothalamus indirectly. That is, testosterone is enzymatically converted to estradiol-17β in the hypothalamus, and it is this estrogen that imprints the brain in a male direction! [15] This explains why administration of estradiol can masculinize newborn females. Baby female rats are protected against this masculinizing effect of estrogen by a protein in their blood *(alpha-fetoprotein)* that binds to the estrogen and keeps it from reaching the brain. Such a role for this protein in humans is not clear, however. [16]

In species such as guinea pigs, sheep, and rhesus monkeys, the critical developmental period for masculinizing the brain with testosterone occurs before birth, and the same may be true for humans (Chapter 18). Exposure of female monkey fetuses to androgen masculinizes their external genitalia and causes them to exhibit malelike behavior. [17] On the other hand, such masculinization of their external genitalia does not prohibit ovulation after puberty, suggesting that early exposure to testosterone may not masculinize the part of the hypothalamus controlling gonadotropin secretion.

Are there sex differences in the human brain that are programmed by hormones during fetal development? This is a controversial question. Some argue that any biologic basis of such differences implies an unacceptable "sexist" position. In reality, some interpretations of research in this area may have been influenced by the fact that the scientists were male.[18] Nevertheless, one can ask, "Are there sex differences in human brain function and behavior?" And if so, "Do these differences have a biologic basis or are they the product of experience?"

Certainly, there are sex differences in the function of the human hypothalamus. The hypothalamus in the adult human female regulates gonadotropin secretion in a cyclic manner, whereas a male's hypothalamus is more or less noncyclic in this regard (Chapters 3 and 4). The controversy arises when we discuss possible sex differences in behavior at birth and infancy.

Past research had suggested that newborn girls and boys differ in brain structure and behavior. We know now that there are no sex differences in overall brain size or proportions at birth,[19] but there are sex differences in the behavior of very young infants and children.[20] For example, male newborns sleep less and are more irritable than females and exhibit more distress if they are separated from the mother. Furthermore, the left cerebral hemisphere develops a dominance over the right earlier in girls than in boys, which leads to the fact that, overall, young girls learn to talk sooner. Boys, on the other hand, have spatial and mathematical skills on the average that exceed those of young girls.[21] In addition, young boys spend more energy in aggressive play than do young girls. Girls, on the other hand, have more fantasies and play behaviors related to mothering.[22]

There is some evidence that testosterone secretion from the testes of male human fetuses could be responsible for some of these early sex differences in behavior. For example, unfortunate "natural experiments" occur when female human fetuses are exposed to androgens, e.g., the androgenital syndrome and progestogen-induced masculinization (as discussed later), and the behavior of these exposed young females after birth is interesting. Although these girls are physically masculinized, they usually are raised as girls. Studies indicate, however, that they are "tomboyish" in their behavior, attitudes, and preferences. They usually have fewer maternal fantasies and are more aggressive than are other girls.[23] Similarly, female fetuses of Rhesus monkeys exposed to androgens exhibit malelike play and aggressive behavior after birth.[24] When human fetuses are exposed to progestogens that have androgenic activity, both males and females exhibit more aggressive behavior as children.[25]

Concluding, however, that early exposure of females to androgen when they were fetuses produces malelike behaviors is premature. Rather, it is possible that their early postnatal experience, especially the behavior of parents toward the infants, could have produced their tomboyish behavior.[26] The reality of the situation is that all human behavior is an interaction of biological maturation and experience, and it may prove very difficult to separate the biological influences from those resulting from childhood experiences. What is important is that any sex differences in human brain function, regardless of how they develop, should form the basis for understanding, and not suppression of, the human rights of either sex.

SUMMARY OF SEXUAL DETERMINATION
AND DEVELOPMENT

From our discussion, you should realize that the Y chromosome initiates events that establish a masculine condition and suppress a basically neutral feminine condition (Fig. 5-8). You also should realize that the reproductive tract of females and males develops from common embryonic tissue; these homologies are listed in Table 5-1. Scientists in this field generally assume that development of the female primary sex

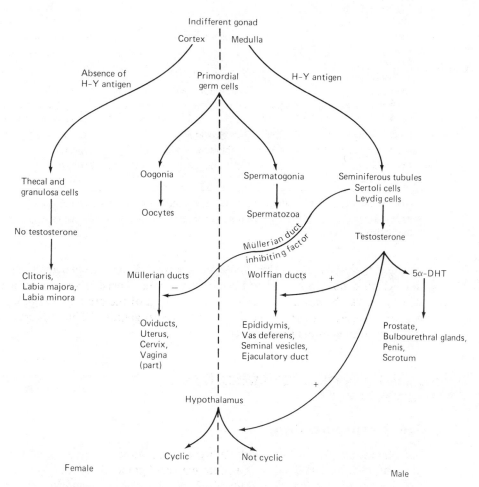

Figure 5–8. Summary of sex determination and development. Adapted from P. J. Hogarth, *Biology of Reproduction: Tertiary Level Biology* (New York: John Wiley & Sons, Inc., 1978). Reprinted by permission of Blackie and Sons, Ltd., and Dr. Peter J. Hogarth; © 1978.

TABLE 5-1. ADULT DERIVATIVES OF EMBRYONIC STRUCTURES, SHOWING HOMOLOGIES AMONG PORTIONS OF THE FEMALE AND MALE REPRODUCTIVE SYSTEMS

Female Structure		Indifferent Embryonic Structure		Male Structure
Ovary (from cortex)	←	Indifferent gonad	→	Testis (from medulla)
No functional structures	←	Mesonephric tubules	→	Vasa efferentia
No functional structures	←	Mesonephric ducts (wolffian ducts)	→	Epididymis; vas deferens; ejaculatory duct; seminal vesicles
Oviducts Uterus Cervix Vagina (upper third)[a]	←	Müllerian ducts	→	No functional structures
Clitoris (glans; corpora cavernosa)	←	Genital tubercle (phallus)	→	Penis (glans; corpora cavernosa; corpus spongiosum)
Labia minora	←	Urogenital folds	→	Ventral aspect of penis
Labia majora; mons pubis	←	Labioscrotal swellings	→	Scrotum
Lesser and greater vestibular glands	←	Urethral tissue	→	Prostate; bulbourethral glands

Note: Structures derived from the same indifferent embryonic tissue are homologous.

[a]The lower two-thirds of the vagina are derived from the urogenital sinus, which also gives rise to the urinary bladder and urethra.

organs (ovaries) as well as the female sex accessory ducts, external genitalia, and sexual behavior does not require hormonal priming. There is some evidence in the laboratory rat, however, that at least development of the brain in a female direction requires progesterone from the newborn's adrenal glands.[27] Thus, we have more to learn about this fascinating subject.

DISORDERS OF SEXUAL DETERMINATION AND DEVELOPMENT

True Hermaphroditism

Sometimes, people are born with ambiguous reproductive systems, a phenomenon generally called *intersexuality*. If a combination of gonadal tissue is present, a person is a *true hermaphrodite*. These individuals often possess an *ovotestis* on one or both sides, which is a gonad that contains a combination of seminiferous tubules and ovarian follicles. This sometimes happens because of an error in fertilization (described in Chapter 12) that results in one-half of the individual's cells being 46;XX and the other half 46;XY. Sometimes, true hermaphrodites will have an ovary on one side and a testis on the other side, and this person is a *gynandromorph*. On the side with the ovary, there are müllerian duct derivatives; and on the side with the testis, there are wolffian duct derivatives. This is further evidence that the effects of the testes on development of the wolffian ducts and regression of the müllerian ducts are local phenomena.

Pseudohermaphroditism

A *pseudohermaphrodite* is a person whose gonads agree with chromosomal sex but who has external genitalia of the opposite sex. Male pseudohermaphrodites have normal testes but incomplete masculinization of the wolffian duct system and external genitalia. One form of this condition is the inherited disorder, *testicular feminization syndrome* (Fig. 5-9). These people have normal testes and male chromosomes (46;XY), and their testes secrete normal amounts of testosterone. However, they have a genetic defect in the receptors for testosterone in target tissues.[28] Thus, they develop femalelike external genitalia, and the testes are not descended. No müllerian duct derivatives are present because their testes secreted müllerian duct-inhibiting factor during development.

Another form of male pseudohermaphroditism has been described in the

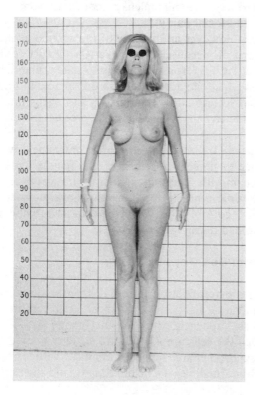

Figure 5-9. A photograph of a person with testicular feminization syndrome, a form of male pseudohermaphroditism. Photograph courtesy of Dr. Howard W. Jones.

populations of some small villages in the Dominican Republic. In this condition, genetic males (46;XY) are born with normal but undescended testes. Testosterone secretion from these testes also is normal. These individuals have a small clitorislike penis and a blind vaginal pouch but normal wolffian duct derivatives. This inherited condition develops as a deficiency of the enzyme 5α-reductase in the cells of external genitalia tissue.[29] Thus, testosterone can cause development of the wolffian duct tissues but not the tissues dependent on 5α-dihydrotestosterone. Surprisingly, when these males reach puberty, their penises grow and they develop pubic hair (but not facial hair); they are, however, infertile. The villagers call these people *Guevedoces,* which means "penis at 12." In Chapter 19, we discuss the interesting fact that these adult Guevedoces exhibit normal, heterosexual male behavior and a typical male sex role, even though they were raised as girls! This suggests that the early exposure of their brains to testosterone imprinted them with a male sexual behavior pattern.

Female pseudohermaphroditism occurs when normal ovaries are present but the body is partially masculinized. Examples are individuals with *adrenogenital syndrome,* also known as *congenital adrenal hyperplasia* (Fig. 5-10). This is an inherited disorder that can develop after, but usually occurs before, birth. It accounts for about one-half of all cases of human intersexuality. Genetic females (46;XX) with this disorder have masculinized external genitalia, and pubic and axillary hair develops only a few years after birth. (Males also can inherit this disorder. Soon after birth, these males can begin to acquire adult malelike secondary sexual characteristics such as penile enlargement and facial hair.)

Adrenogenital syndrome involves an inability of the adrenal glands to secrete steroid hormones such as cortisol. Because cortisol exerts a negative feedback effect on secretion of corticotropin (ACTH) from the anterior pituitary gland, blood levels of ACTH are very high. But since the adrenal glands are unable to synthesize and secrete cortisol, they instead secrete large amounts of androgens into the blood, which serve to masculinize the female body and cause precocious sexual development in the male.

If a woman is administered a progestogen during pregnancy, which used to be a fairly common practice to prevent miscarriage, the female fetus can exhibit female pseudohermaphroditism with symptoms similar to those of the adrenogenital syndrome. This is because some synthetic progestogens have androgenic activity. Therefore, combination birth control pills, which contain a synthetic progestogen (Chapter 13), should be avoided during pregnancy.

Chromosomal Errors and Sex Determination

Other genetic disorders of sex determination can result from errors of fertilization. That is, the cells of an individual may end up with an abnormal number of sex chromosomes. How this can happen is discussed in Chapter 12. An example is Turner's syndrome, in which the cells are 45;XO and no Barr body is present. Individuals with Turner's syndrome are sexually infantile with female external genitalia. They have facial abnormalities, a shieldlike chest, and a neck that is short,

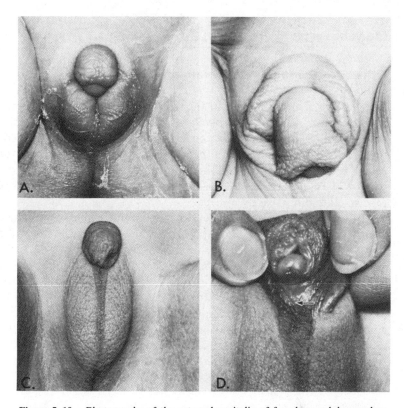

Figure 5-10. Photographs of the external genitalia of female pseudohermaphrodites with adrenogenital syndrome: (A) Newborn female, exhibiting enlargement of the clitoris and fusion of the labia majora; (B) female infant, showing considerable enlargement of the clitoris; (C) and (D) six-year-old girl, showing an enlarged clitoris and fusion of the labia majora to form a scrotum. In (D), note the clitoral glans and the opening of the urogenital sinus (arrow). Reproduced from K. L. Moore, *The Developing Human. Clinically Oriented Embryology,* 3rd ed. (Philadelphia: W. B. Saunders Company, 1982). © 1982; used with permission.

broad, and webbed. They also tend to have cardiovascular and kidney disorders. Their "ovaries" are sterile, consisting mostly of connective tissue; no germ cells are present.

Another error of sex determination resulting from an error in fertilization produces individuals with 47;XXY cells. This is *Klinefelter's syndrome,* which occurs in 1 out of 700 males, and may not be manifested until puberty. These individuals have small external genitalia, undescended testes, and are mentally retarded. Their urethras may also fail to close during development, resulting in an orifice on the lower surface of the penis, a condition called *hypospadias.* Their testes are infertile. Other, more extreme fertilization errors can produce conditions similar to Klinefelter's syndrome. These individuals have cells that are 48;XXYY, 48;XXXY, or 49:XXXYY. How many Barr bodies would there be in cells with each of these chromosomal types?

Other Problems in Sex Development

Developmental abnormalities of the reproductive tract, other than those mentioned above, are rare. One of these, *penile agenesis,* is the absence of a penis due to the fact that a genital tubercle was not formed. In another condition, called *double penis,* there are two penises because of the earlier formation of two genital tubercles. Or the genital tubercle branches, resulting in a forked, or *bifid, penis.* And if the pituitary gland of the male fetus is underdeveloped, this can result in a *micropenis,* which often is so small that it is barely noticed.

Rare abnormalities in the development of the female reproductive tract also can occur. Since the müllerian ducts initially are paired structures, their middle and lower portions must fuse to form the unpaired uterus and vagina. Partial or complete failure of müllerian duct fusion occurs in about 1 in every 1500 newborn females. At one extreme is the *arcuate uterus,* in which almost normal fusion occurred, but a slight nonfusion of the uterus produces a "dent" in the fundus of the uterus (Fig. 5-11). At the other extreme is complete nonfusion of the müllerian ducts, resulting in a *double uterus* as well as two vaginas (Fig. 5-11). Several degrees of nonfusion, between these two extremes, can occur (Fig. 5-11). For example, the uterus can be forked at the top *(bicornuate uterus).* Many of these conditions do not cause infertility, but they are related to an increase in the risk of miscarriage (Chapter 9).

SEXUAL CONDITION OF THE NEWBORN

In the way of a summary, let us look at the sexual condition of a normal newborn male and female. Despite the complexities of sex determination and development from an indifferent stage, the system works perfectly in most of us.

The newborn male child has a pair of testes that usually have descended into the scrotum. The seminiferous tubules of the testes contain spermatogonia and Sertoli cells but no sperm. The Leydig cells of the testes, after being active in secreting testosterone in the fetal period, are now inactive. The hypothalamus probably has been primed by fetal testosterone to be noncyclic after puberty. The scrotum and the penis are fully formed but small. The male sex accessory ducts and glands are fully formed but inactive. The male secondary sexual characteristics will not appear until puberty, as we see in the next chapter.

In newborn females, the ovaries contain a fixed number of oocytes in resting or growing follicles. None of the follicles will ovulate until puberty. The ovaries are now in the pelvic cavity. The female sex accessory ducts are fully formed but inactive. Similarly, the female external genitalia are fully formed but relatively small. The female secondary sexual characteristics (e.g., breasts) are present but will not mature

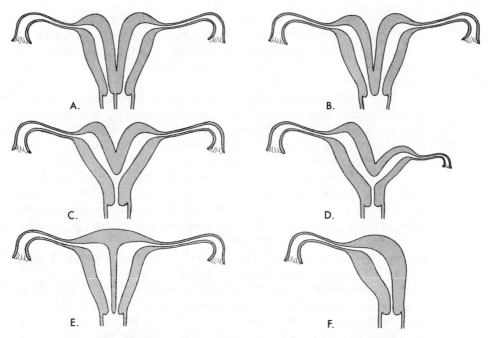

Figure 5-11. Illustrations of various types of congenital uterine abnormalities: (A) double uterus and double vagina; (B) double uterus with single vagina; (C) branched (bicornuate) uterus; (D) bicornuate uterus with a small left branch; (E) septate uterus (divided in the middle); (F) uterus formed from müllerian duct on one side only. Reproduced from K. L. Moore, *The Developing Human. Clinically Oriented Embryology;* 3rd ed. (Philadelphia: W. B. Saunders Company, 1982). © 1982; used with permission.

until puberty. The female's hypothalamus is programmed genetically to become cyclic after puberty.

CHAPTER SUMMARY

The chromosomes of diploid female cells are 46;XX, whereas those of diploid male cells are 46;XY. One X chromosome of the female cells is inactivated, producing a Barr body. Development of the reproductive tract begins with the sexually indifferent stage. The primordial germ cells migrate to the indifferent gonads. If a Y chromosome is present, H-Y antigen causes the medulla of the indifferent gonads to form testes. If a Y chromosome is absent, the cortex of the indifferent gonads forms ovaries.

In male embryos, testosterone secreted by the testes stimulates development of the wolffian ducts into the male sex accessory ducts (epididymis, vas deferens, ejaculatory duct) and the seminal vesicles. Secretion of the müllerian duct-inhibiting factor causes regression of the müllerian ducts. In female embryos, the absence of testosterone and of the müllerian duct-inhibiting factor allows the wolffian ducts to regress and the müllerian ducts to develop into the oviducts, uterus, and part of the vagina.

The indifferent external genitalia of the embryo develop into male external genitalia (penis, scrotum) if testosterone is present. This androgen is converted to another androgen, 5α-dihydrotestosterone, by tissues of the external genitalia. If testosterone is not present, the female external genitalia develop into a clitoris, labia majora, labia minora, and mons pubis.

In laboratory mammals, a burst of testosterone secretion from the testes of fetal or newborn males masculinizes the brain, especially the parts of the hypothalamus controlling adult reproductive physiology and sexual behavior. If testosterone is not present, the brain develops a cyclic female pattern of control. Some evidence suggests that a similar hormonal priming of the brain occurs in humans, but it has proven difficult to separate biological and experiential influences controlling sex differences in human brain function.

Intersexual individuals are born with ambiguous reproductive tracts. True hermaphrodites have ambiguous gonads and external genitalia, whereas pseudohermaphrodites have normal gonads and ambiguous external genitalia. Examples of male pseudohermaphrodites are people with testicular feminization syndrome and the "Guevedoces" syndrome. Examples of female pseudohermaphrodites are those with adrenogenital syndrome and progestogen-induced masculinization. Other chromosomal abnormalities also can produce developmental abnormalities of the reproductive tract. These include Turner's syndrome (46;XO) and Klinefelter's syndrome (47;XXY). Developmental errors can result in penile agenesis, double penis, bifid penis, micropenis, arcuate uterus, bicornuate uterus, or double uterus and vagina.

NOTES

1. P. L. Pearson et al., "Techniques for Identifying the Y Chromosome in Human Nuclei," *Nature,* 266 (1970), 78.
2. M. W. Hardisty, "Primordial Germ Cells and the Vertebrate Germ Line," in *The Vertebrate Ovary—Comparative Biology and Evolution,* ed. R. E. Jones, pp. 1–45 (New York: Plenum Publishing Corporation, 1978).
3. T. G. Baker and W. W. O, "Development of the Ovary and Oogenesis," *Clinical Obstetrics and Gynecology,* 3 (1976), 3–26.
4. E. Block, "Quantitative Morphological Investigations of the Follicular System in Women: Variations at Different Ages," *Acta Anatomica,* 14 (1952), 108–23.
5. G. C. Koo et al., "H-Y Antigen: Expression in Human Subjects with the Testicular Feminization Syndrome," *Science,* 196 (1977), 655–56.

6. S. S. Wachtel, "H-Y Antigen and the Genetics of Sex Determination," *Science,* 198 (1977), 797–99; W. K. Silvers and S. S. Wachtel, "H-Y Antigen: Behavior and Function," *Science,* 195 (1977), 956–60; and J. W. Gordon and F. H. Ruddle, "Mammalian Gonadal Determination and Gametogenesis," *Science,* 211 (1981), 1265–71.

7. Gordon and Ruddle (1981), op. cit.

8. F. P. Haseltine and S. Ohno, "Mechanisms of Gonadal Differentiation," *Science,* 211 (1981), 1272–78.

9. Ibid.

10. J. D. Wilson et al., "The Hormonal Control of Sexual Development," *Science,* 211 (1981), 1278–84.

11. M. Blanchard and N. Josso, "Source of Anti-Müllerian Hormone Synthesized by the Fetal Testis," *Pediatric Research,* 8 (1974), 968; and Wilson et al., (1981), op. cit.

12. Wilson et al. (1981), op. cit.

13. G. B. Kolata, "Sex Hormones and Brain Development," *Science,* 205 (1979), 985–87; A. P. Arnold, "Sexual Differences in the Brain," *American Scientist,* 68 (1980), 164–74; and N. J. MacLusky and F. Naftolin, "Sexual Differentiation of the Central Nervous System," *Science,* 211 (1981), 1294–1303.

14. I. L. Ward, "Prenatal Stress Feminizes and Demasculinizes the Behavior of Males," *Science,* 175 (1972), 82–84; and I. L. Ward and J. Weisz, "Maternal Stress Alters Plasma Testosterone in Fetal Males," *Science,* 207 (1980), 328–29.

15. F. Naftolin et al., "The Formation of Estrogens by Central Neuroendocrine Tissues," *Recent Progress in Hormone Research,* 31 (1975), 295.

16. H. L. Lau and S. E. Linkins, "Alpha-Fetoprotein," *American Journal of Obstetrics and Gynecology,* 124 (1976), 533–49; and MacLusky and Naftolin (1981), op. cit.

17. R. W. Gay, "Role of Androgens in the Establishment of Behavioral Sex Differences," *Journal of Animal Science,* 25 (1966), 21–35.

18. M. B. Parlee, "The Sexes under Scrutiny: From Old Biases to New Theories," *Psychology Today,* 12, no. 6 (1978), 62–69.

19. Ibid.

20. M. Lewis and M. Weintraub, "Origins of Sex Role Development," *Sex Roles,* 5 (1979), 135–49; and A. A. Ehrhardt and H. F. L. Meyer-Bahlburg, "Effects of Prenatal Sex Hormones on Gender-Related Behavior," *Science,* 211 (1981), 1312–18.

21. D. Goleman, "Special Abilities of the Sexes: Do They Begin in the Brain?" *Psychology Today,* 12, no. 6 (1978), 48–59, 120; and C. P. Benbow and J. C. Stanley, "Sex Differences in Mathematical Ability: Fact or Artifact?" *Science,* 210 (1980), 1262–64.

22. Parlee (1978), op. cit.

23. A. A. Ehrhardt and S. S. Baker, "Fetal Androgens, Human Central Nervous System Differentiation, and Behavior Sex Differences," in *Sex Differences in Behavior,* eds. R. Friedman and R. Vandewelle, pp. 33–51 (New York: John Wiley & Sons, Inc., 1974); and Ehrhardt and Meyer-Bahlburg, (1981), op. cit.

24. R. W. Gay and J. A. Resko, "Gonadal Hormones and Behavior of Normal and Pseudohermaphroditic Nonhuman Female Primates," *Recent Progress in Hormone Research,* 28 (1972), 707–33.

25. J. M. Reinisch, "Prenatal Exposure to Synthetic Progestins Increases Potential for Aggression in Humans," *Science,* 211 (1981), 1171–73.

26. D. M. Quadango et al., "Effect of Perinatal Gonadal Hormones on Selected Nonsexual Behavior Patterns: A Critical Assessment of the Nonhuman and Human Literature," *Psychological Bulletin,* 84 (1977), 62–80.

27. B. H. Shapiro et al., "Is Feminine Determination of the Brain Hormonally Controlled?" *Experientia,* 32 (1976), 650–51.

28. B. S. Kennan et al., "Syndrome of Androgen Insensitivity in Man," *Journal of Clinical Endocrinology and Metabolism,* 38 (1974), 1143–46.

29. J. Imperato-McGinley et al., "Steroid 5α-Reductase Deficiency in Man: An Inherited Form of Male Pseudohermaphroditism," *Science,* 186 (1974), 1213–15.

FURTHER READING

GOLEMAN, D., "Special Abilities of the Sexes: Do They Begin in the Brain?" *Psychology Today,* 12, no. 6 (1978), 48–59, 120.

HERBERT, J., "Hormonal Basis of Sex Differences in Rats, Monkeys, and Humans," *New Scientist,* 70 (1976), 284–86.

KOLATA, G. B., "Sex Hormones and Brain Development," *Science,* 205 (1979), 985–87.

MITTWOCH, U., "Sex Differences in Cells," *Scientific American,* 209, no. 1 (1963), 54–62.

SHORT, R. V., "Sex Determination and Differentiation," in *Reproduction in Mammals, Book 2, Embryonic and Fetal Development,* C. R. Austin and R. V. Short, eds. Cambridge, England: Cambridge University Press, 1972. Pp. 43–71.

6 *Puberty*

LEARNING GOALS

Having read this chapter, you should be able to:

1. Name the five general stages of the human life cycle.
2. Discuss the variation in pubertal processes among individuals.
3. List processes that tend to occur during earlier and later stages of puberty.
4. List major anatomical changes that occur during puberty in males and females.
5. Discuss characteristics of the initial menstrual cycles of adolescent females.
6. Define acne: How is it formed, what are some myths surrounding its appearance, and what are some effective treatments?
7. Describe how the ovaries and testes change from birth to puberty.
8. Describe changes in FSH, LH, and steroid hormone secretion from birth to puberty.
9. List evidence that negative and positive feedback of steroid hormones on pituitary gonadotropin secretion control the onset of puberty.
10. Discuss the theory that a female must have a critical amount of fat in her body before she exhibits menarche.
11. List possible influences of climate, day length, stressors, and social interaction on puberty.
12. Describe how studies of twins indicate that our genetic make-up influences the timing of puberty.
13. Discuss why body image during puberty is a cause of concern in some males and females.

14. Describe how undergoing early or delayed puberty can influence one's self-esteem and future psychosexual development.

15. Discuss society's treatment of adolescents, and how this treatment produces barriers to their progress to adulthood.

INTRODUCTION

The human life cycle can be divided into five stages: (1) embryonic and fetal existence, (2) infancy and childhood, (3) puberty and adolescence, (4) early and middle adulthood, and (5) late adulthood and old age. In this chapter, we discuss the process of puberty, the internal and environmental factors that influence this process, and the psychological and social adjustments teenagers must make during puberty and adolescence.

PUBERTY AND ITS TIMING

Puberty (or *sexual maturation*) is the transition period that takes a person from being a sexually immature child to a sexually mature, reproductively fertile adult. The word puberty has the root in the Latin word *pubes*, which means "hair." *Pubescence* is the state of the child between the onset of pubertal changes and the completion of sexual maturation. Much is known about timing of the biological events during puberty, yet no "typical" pattern exists. The ages at which the various changes occur vary greatly among individuals. Figure 6–1 illustrates this individual variability in the pubertal sequence. Puberty involves growth and maturation of several body systems and tissues; and the timing of a change in one part of the body (for example, maturation of the breasts in a female) in relation to a change in another part of the body (for example, appearance of pubic hair in a female) can vary among individuals.[1]

The Pubertal Process

In Table 6–1, the sequences of pubertal change in a "typical" male and female are summarized. The ages given for the various changes are averages, but these can vary greatly from what is presented; a wide range of pubertal patterns is considered normal. The earlier puberty is initiated, the earlier sexual maturity is reached. Also, the pubertal sequence begins earlier and ends earlier in females than in males. For example, the body growth spurt begins about two years earlier in females, and females become fertile about a year earlier than males (Table 6–1).

Pubertal Changes in Females

The first menstruation in a female (or *menarche*) occurs at an average age of 13 years in the Northern Hemisphere, with 95 percent of females reaching menarche between age 11 and 15. More specifically, menarche occurs in the United States at an average age of 12.3 years. A female's first ovulation can be at the time of menarche, but

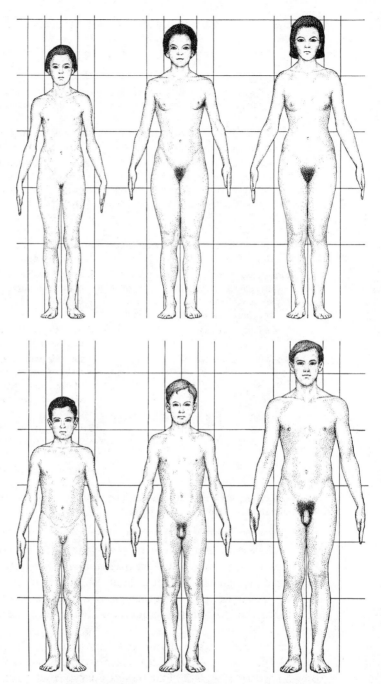

Figure 6–1. Different degrees of sexual maturation at the same age. The males are all 14.75 years old, and the females are all 12.75 years old. This illustrates that the degree of sexual maturity can vary widely in relation to age. Reproduced from J. M. Tanner, "Growing Up," *Scientific American* 229, no. 3 (1973), 34–43. Copyright © by Scientific American, Inc.; all rights reserved. Used with permission.

TABLE 6-1. "TYPICAL" SEQUENCE OF EVENTS DURING PUBERTY IN A FEMALE AND MALE

Age	Female	Male	Age
10	Growth spurt begins; initial increase in height and fat deposition	Initial stages of spermatogenesis; Leydig cells appear and begin to secrete androgens	10
11	Initial breast development; pelvis widens; pubic hair begins to appear	Subcutaneous fat deposition increases; testes begin to enlarge	11
12	Maturation and growth of internal reproductive organs (ovaries, oviducts, uterus, vagina); maturation of external genitalia; areola becomes pigmented	Increase in scrotum and penis size; increase in spontaneous erection frequency; first signs of pubic hair; growth of seminal vesicles and prostate gland; skeletal growth-spurt begins	12
13	Filling in of breasts; menarche; axillary hair appears	Pubic hair more apparent; nocturnal emissions begin	13
14	First ovulation occurs; skeletal growth-rate declines; breast maturation complete; sweat and sebaceous gland development, sometimes with acne	Larynx growth and deepening of voice occur; hair appears in axilla and on upper lip	14
15	Voice slightly deepens	First fertile ejaculation; slight breast enlargement in some individuals	15
16	Adult stature reached	Adult hair pattern, including indentation in front hairline and appearance of chest hair; sweat and sebaceous glands develop, often with acne; loss of body fat occurs	16
		Broadening of shoulders; muscle growth and increased muscle strength	17
		Adult stature reached now or later	18

Note: This table depicts a general sequence, and individuals can vary greatly in the timing of this sequence and still be within the normal range.

usually is some months (or as much as two years) after menarche. The first menstrual cycles in pubertal females, thus, are often *anovulatory* (infertile). Even though this is the case in most females, there is a slight risk of pregnancy in some who have just experienced menarche, and there are cases of females who have become pregnant before menarche. Also, for several months after menarche, a female may miss some periods.

Menarche and first ovulation are only steps in a series of changes in pubescent females.[2] The ovaries and female sex accessory structures (oviducts, uterus, vagina) grow and mature (Fig. 6-2). Female secondary sexual characteristics appear, such as growth and widening of the bony pelvis and appearance of pubic and axillary (armpit) hair. Also, the breasts grow and mature; a girl becomes more aware of her breast

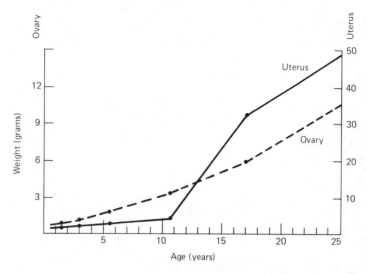

Figure 6-2. Changes in weight of an ovary and the uterus with age. Adapted from B. T. Donovan and J. J. Van der Werff Ten Bosch, *Physiology of Puberty,* in *Monographs of the Physiological Society,* no. 15, eds. H. Barcroft et al. (Baltimore: The Williams & Wilkens Company, 1965). © 1965 by B. T. Donovan and J. J. Van der Werff Ten Bosch; used with permission of the co-authors and the Editorial Board for Monographs of the Physiological Society.

size not only because of the physical change but also because of social attitudes toward breast size. *Sweat* and *sebaceous glands* in the skin become more active, and some females (and males) develop *acne* due to increased activity and inflammation of the sebaceous glands. The *linea nigra* appears as a dark band on the lower abdominal wall; it is especially pronounced in brunettes and dark-skinned races. There is also a slight lowering of voice in pubescent females, but not as much as in males. *Metabolic rate* (rate at which tissues use oxygen), as well as heart rate and blood pressure, increases in females at puberty.

Many pubescent females deposit fat under the skin, which leads to a heavier appearance. This fat becomes distributed in the hips and breasts as a secondary sexual characteristic. Also, a skeletal growth spurt occurs in pubescent females, and both height and weight increase. This bone growth usually ends at about age 16 in females, although some females will continue to grow in height until age 20 or later. High levels of estrogens secreted by the pubescent female eventually halt bone growth.

Pubertal Changes in Males

In males, pubertal changes include growth and maturation of the testes and male sex accessory structures (vas deferens, seminal vesicles, prostate gland; Fig. 6-3). In addition, male secondary sexual characteristics begin to appear. Pubic hair develops, and hair also appears in the axilla, face, chest, and extremities. Sweat glands develop

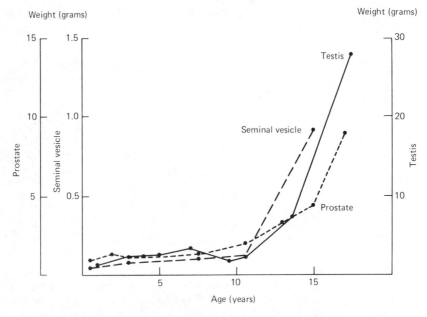

Figure 6–3. Changes in weight of a single testis, seminal vesicle, and the prostate gland with age. Adapted from B. T. Donovan and J. J. Van der Werff Ten Bosch, *Physiology of Puberty,* in *Monographs of the Physiological Society,* no. 15, eds. H. Barcroft et al. (Baltimore: The Williams & Wilkens Company, 1965). © 1965 by B. T. Donovan and J. J. Van der Werff Ten Bosch; used with permission of the co-authors and the Editorial Board for Monographs of the Physiological Society.

in the axilla, and these produce an odor because of bacteria present in the oily secretion. The sebaceous glands become active in the skin of the scrotum, face, back, and chest; and acne appears in some individuals. In pubescent males, the nipple becomes pigmented and the areola darkens and widens. The vocal cords in the *larynx* ("voice box") double in length in pubertal males, and this results in a dramatic lowering of voice. The pitch of the adult male voice is about a full octave below that of adult females.

The scrotum and penis of pubescent males grow markedly. *Spontaneous erections,* which occur in infants and even in male fetuses, increase in frequency during puberty (Chapter 18). These erections, which can be disturbing to the individual or his parents, occur in response to stressful or emotionally charged stimuli seemingly not related to sexual matters; this is quite normal and does not mean the boy is "oversexed." *Nocturnal emissions* (or "wet dreams") begin in pubescent males. These emissions tend to happen during sleep or right after waking up. Initial ejaculations of pubescent males (during nocturnal emissions, masturbation, or coitus) are relatively sperm-free, but the possibility always exists for at least some fertile sperm to be present. Both spontaneous erections and nocturnal emissions tend to decrease in frequency after puberty.

Pubescent males exhibit a spurt in body growth. Fat is deposited early in the pubertal sequence, without much skeletal or muscle growth. In fact, females aged 10 to 13 usually are taller than males of a similar age.[3] In later stages of the pubertal sequence, however, marked skeletal and muscle growth occurs in males because of the stimulatory effects of testosterone on bone growth. Thus, the average adult height of American males (177.6 cm, or 69.9 in.) is greater than that of females (169.2 cm, or 66.6 in.).

Classification of Pubertal Changes

It often is useful for physicians and others to be able to look at a teenager and determine where he or she is in the pubertal sequence. For this purpose, some anatomical characters that change in a predictable order have been classified into five stages.[4] These characters include the development of pubic hair, penis, and testes in males and pubic hair and breasts in females. Table 6-2 describes these stages, and they are pictured in Figs. 6-4, 6-5, and 6-6.

TABLE 6-2. PUBERTAL DEVELOPMENT OF CERTAIN ANATOMICAL CHARACTERS IN FEMALES AND MALES BY STAGE

Females		
Stage	Pubic hair	Breasts
1	Prepubertal	Prepubertal
2	Sparse, lightly pigmented, straight; medial border of labia	Breast and papilla elevated as small mound; areolar diameter increased
3	Darker; beginning to curl; increased amount	Breast and areola enlarged; no contour separation
4	Coarse, curly, abundant; but amount less than in adult	Areola and papilla form secondary mound
5	Adult feminine triangle spread to medial surface of thighs	Mature; nipple projects; areola part of general breast contour

Males			
Stage	Pubic hair	Penis	Testes
1	Prepubertal (none)	Prepubertal	Prepubertal
2	Scanty, long, slightly pigmented	Slight enlargement	Enlarged scrotum; pink; texture altered
3	Darker; starts to curl; small amount	Penis longer	Larger
4	Resembles adult type, but less in quantity; coarse, curly	Larger; glans breadth increased	Larger; scrotum dark
5	Adult distribution; spread to medial surface of thighs	Adult	Adult

Source: Adapted from J. M. Tanner, *Growth at Adolescence,* 2nd ed. (Oxford, England: Blackwell Scientific Publications, 1973). © 1962; reprinted with permission of the Institute of Child Health, University of London, and Dr. J. M. Tanner.

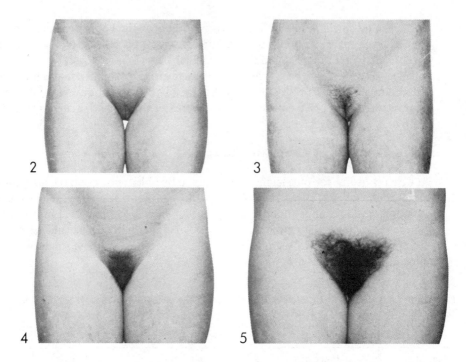

Figure 6–4. Stages of pubic hair development in females. Infantile pattern (stage 1) is not shown. Refer to the description of these stages in Table 6–2. Reproduced from J. M. Tanner, *Growth at Adolescence,* 2nd ed. (Oxford, England: Blackwell Scientific Publications, 1973). © 1962; reprinted with permission of Blackwell Scientific Publications and Dr. J. M. Tanner.

GONADAL CHANGES FROM BIRTH TO PUBERTY

Ovarian Changes from Birth to Puberty

The ovaries of a newborn female are about 2 cm in length, and each ovary contains about 500,000 ovarian follicles.[5] No new oocytes or follicles will appear after this time. Most of the oocytes are contained in primordial follicles, but several are in growing follicles; 60 percent of the ovaries in newborns contain some tertiary follicles. By one year of age, the ovaries of all females contain tertiary follicles. No follicles ovulate during childhood, and many growing follicles become cystic or atretic (Chapter 3). Because of this massive follicular death in the ovaries of female children, the number of follicles remaining in each ovary immediately before puberty

Figure 6–5. Stages of female breast development. Refer to the description of these stages in Table 6–2. Reproduced from J. M. Tanner, *Growth at Adolescence,* 2nd ed. (Oxford, England: Blackwell Scientific Publications, 1973). © 1962; reprinted with permission of Blackwell Scientific Publications and Dr. J. M. Tanner.

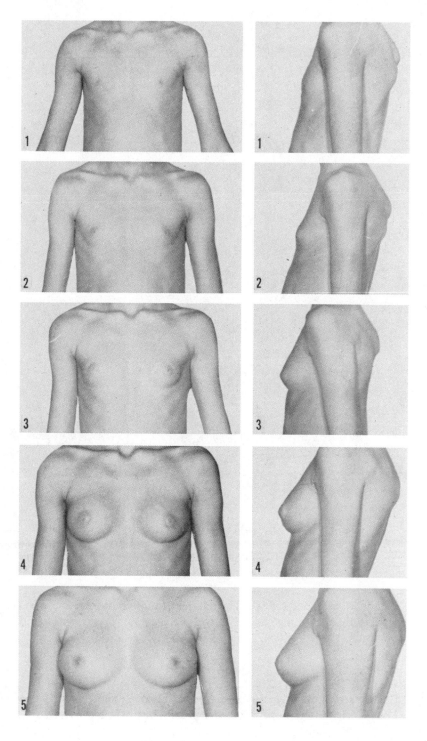

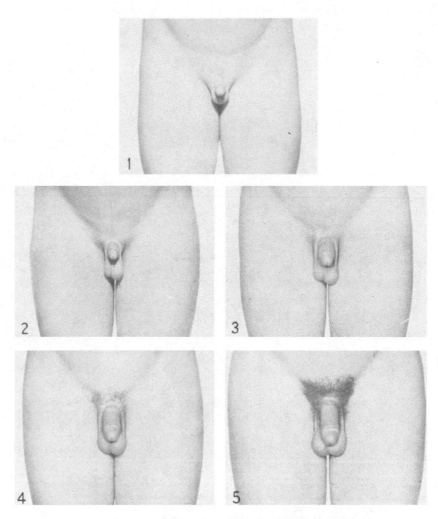

Figure 6–6. Stages of male genital development during puberty. Refer to the description of these stages in Table 6–2. Reproduced from J. M. Tanner *Growth at Adolescence,* 2nd ed. (Oxford, England: Blackwell Scientific Publications, 1973). © 1962; reprinted with permission of Blackwell Scientific Publications and Dr. J. M. Tanner.

is reduced to about 83,000.[6] The ovaries of pubescent females, however, weigh more than those of a young child (Fig. 6–2) because some remaining healthy follicles have enlarged.

Testicular Changes from Birth to Puberty

Seminiferous tubules in the testes of a newborn male contain only spermatogonia and Sertoli cells. Some Leydig cells are present at birth, but by six months of age these cells are almost invisible in the testes. At about nine years of age, spermatogenesis begins in the tubules, and the Leydig cells again are visible. Mature spermatozoa are not produced, however, until the fourteenth or fifteenth year. At the onset of spermatogenesis, the testes enlarge markedly (Fig. 6–3). The volume of a single testis of a one-year-old boy is about 0.7 ml, that of an eight-year-old boy about 0.8 ml, and that of a pubescent boy about 3.0 ml. Testicular volume is 16.5 ml in an adult male, approximately a 24-fold increase!

During the seventh or eighth month of fetal life, the testes descend from the abdominal cavity into the scrotum. They are "pulled" down by a ligament (the *gubernaculum*) that is attached on one end to the testes and at the other end to the scrotum. During their journey, the testes pass through the *inguinal canals* (two natural openings in the lower abdominal wall), and when they come to rest in the scrotum, the vasa deferentia extend into these canals. The canals then become closed and filled with connective tissue. Incomplete closure of an inguinal canal after testicular descent often can lead to an *inguinal hernia,* a condition where a weakness in the abdominal wall remains; a portion of the internal organs may extend through this weakened region into the scrotum. *Hydrocele* occurs when the canal remains slightly open, permitting drainage of abdominal fluid into the scrotal sac. Surgery often is needed to close the canal in such cases.

Cryptorchid Testes

In about 10 percent of males, the testes fail to descend by birth. By one year, about 1 percent of males still have testes within the pelvic cavity, and in 0.3 percent of cases the testes fail to descend by the time spermatogenesis begins (about age nine). Failure of one testis to descend is about five times as common as failure of both to descend. The process of spermatogenesis requires a temperature 3.1°C (5.6°F) lower than that in the abdominal cavity (Chapter 4). If spermatogenesis begins while the testes remain in the body cavity (about age ten), temperature damage occurs in these cryptorchid testes (*crypt* = "hidden"; *orchid* = "testis"); only Sertoli and Leydig cells remain. Thus, these males are sterile but have fully developed male secondary sex structures because androgen secretion from Leydig cells is normal. Cryptorchid testes often are removed surgically because they tend to develop cancerous tissue. Surgery or treatment with gonadotropins, androgens, or GnRH usually can cause testicular descent before any damage is done.

HORMONE LEVELS FROM BIRTH TO PUBERTY

In Chapters 3 and 4, we saw that the pituitary gonadotropins—follicle-stimulating hormone (FSH) and luteinizing hormone (LH)—are both necessary for complete function of the ovaries and testes. In young children of both sexes, blood levels of FSH and LH are not high enough to initiate gonadal function. At nine to twelve years

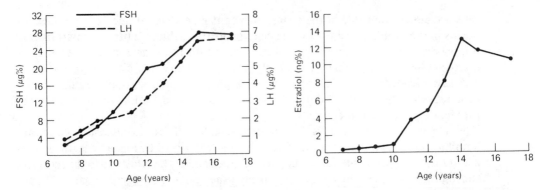

Figure 6–7. Changes in blood levels of FSH, LH, and estradiol in females of different age. Blood androgen levels are not depicted, but these also rise as puberty nears. Note that the values for hormone levels are in nannograms percent (ng %) or micrograms percent (μg %). A nannogram is V_{1000} of a microgram, and one ng percent is equivalent to 1.0 ng in 100 ml of blood plasma. Thus, not many hormone molecules are needed to change biological function. Adapted from M. M. Grumbach et al., "Hypothalamic-Pituitary Regulation of Puberty: Evidence and Concepts Derived from Clinical Research," in *Control of the Onset of Puberty,* eds. M. M. Grumbach et al. (New York: John Wiley & Sons, Inc., 1974). © 1974; used with permission.

of age, however, levels of first FSH and then LH begin to rise in the blood; these increases usually occur one or two years sooner in females (Figs. 6–7 and 6–8). Not only do LH levels begin to rise in pubescent males and females, but the degree of the hour-to-hour changes in secretion to this gonadotropin increases in the day and especially during sleep (Fig. 6–9).[7] In adults, LH levels fluctuate less markedly. The role of these

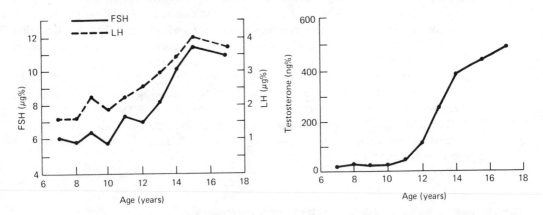

Figure 6–8. Changes in blood levels of FSH, LH, and testosterone in males of different age. Blood estrogen levels are not depicted, but these also rise slightly as puberty nears. For an explanation of ng % and μg %, see Fig. 6–7. Adapted from M. M. Grumbach et al., "Hypothalamic-Pituitary Regulation of Puberty: Evidence and Concepts Derived from Clinical Research," in *Control of the Onset of Puberty,* eds. M. M. Grumbach et al. (New York: John Wiley & Sons, Inc., 1974). © 1974; used with permission.

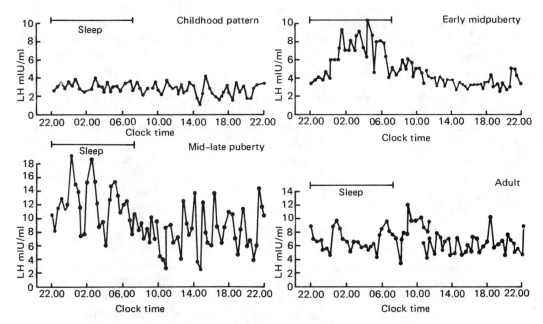

Figure 6–9. Daily patterns of LH levels in blood of females at various stages of sexual maturation. Note that during puberty, LH levels rise during sleep and exhibit marked hour-to-hour fluctuations. Average levels and fluctuations in both day and night then increase during late puberty, before decreasing somewhat in the adult. Adapted from E. D. Weitzman et al., "The Relationship of Sleep and Sleep Stages to Neuroendocrine Secretion and Biological Rhythms in Man," in *Recent Progress in Hormone Research,* 31 (1975), 399–446, ed. R. O. Greep (New York: Academic Press, Inc., 1975). © 1975; used with permission.

short-term changes in levels of LH during puberty is not understood. They may play a role in final gonadal development.

Other hormones exhibit an increase in blood levels in prepubertal males and females. Growth hormone (GH) is secreted by the anterior pituitary gland in greater amounts near puberty and is responsible, along with androgens, for the growth of long bones and other tissues (Chapter 2). Growth hormone also has a major effect on protein synthesis and elevation of blood sugar levels. Children with a GH deficiency exhibit delayed sexual maturation as well as short stature. An increase in secretion of thyroid hormones from the thyroid gland, caused by greater secretion of thyrotropic hormone (TSH) from the pituitary (Chapter 2), may account for the rise in metabolic rate in both sexes. Thyroid hormones also are essential for body growth.

Hormones in Females

Menarche and first ovulation. Near the time of first ovulation in females, there is a sudden surge of LH and, to a lesser extent, of FSH. This first preovulatory surge of gonadotropin secretion may not be sufficient to cause ovulation, but it does cause enough cyclic variation in ovarian estrogen production so that menstruation occurs (Chapter 7). Thus, as mentioned earlier, the initial menstrual cycle of pubescent females often is anovulatory. Eventually, the endocrine system and brain mature so that the LH surge is sufficient to cause the first ovulation.

Estrogens. Levels of estrogens (estradiol-17β and estrone) begin to rise in pubescent females, causing the onset of breast maturation (Table 6-1), and they continue to increase as the female approaches menarche (Fig. 6-7). The major source of these estrogens is the large, growing follicles in the ovary, maturing under the influence of FSH and LH. Estrogens initiate development of the mammary gland, cause growth of the bony pelvis, and help to deposit subcutaneous fat. These steroid hormones also cause growth of the external genitalia and of the vagina, oviduct, and uterus.

Androgens. Levels of the androgens androstenedione and dehydroepiandrosterone also increase in the blood of pubescent females, but the ovary is not the main source of these so-called "weak" androgens. The adrenal glands secrete large quantities of these hormones during puberty. Adrenal androgens are responsible for some of the changes in the female during sexual maturation, including growth of pubic and axillary hair, slight lowering of the voice, development of sebaceous glands and acne, and growth of long bones. In addition, these androgens may increase sex drive in pubescent girls (Chapter 17). When these weak androgens arrive at some target tissues of the female, they are converted by enzymes to testosterone (Chapter 3), which then makes the tissues grow and mature. In this way, masculinization of women by high blood levels of the potent androgen testosterone is avoided.

Hormone Levels in Males

Gonadotropins and androgens. The increase in secretion of FSH and LH in males at about eleven years of age is responsible for the onset of spermatogenesis and the rise in androgen secretion (mainly testosterone) from the testes (Fig. 6-8). Testicular androgens, along with adrenal androgens, cause growth of the sex accessory structures (seminal vesicles, prostate gland, penis, scrotum) and secondary sexual characteristics (pubic, axillary, and facial hair; and larynx growth). Also, testosterone causes retention of nitrogen, calcium, and phosphorus in the body, thus supporting the bone and muscle growth seen in sexually maturing males. Spontaneous penile erections become more frequent as puberty approches, due to the effects of androgens on growth and touch sensitivity of the penis and on the male sex drive (Chapter 17).

Estrogens. Blood estrogen levels also rise slightly in males before and during puberty. The source of these estrogens is probably the Sertoli cells of the testes (Chapter 4). The presence of estrogens in males near puberty explains why some pubertal boys exhibit slight growth of the mammary glands (gynecomastia). In this condition, a small (1 to 2 cm in diameter) lump appears behind the nipple, but it usually goes away in about two years.[8] Thus, both male and female sex hormones occur in both sexes, and the relative levels of the two hormones determine if secondary sexual characteristics develop in a male or female direction. The interaction of androgens and estrogens in sexual maturation remains an exciting area of research.

Androgens and Acne

Androgens are responsible for the increase in secretion of sebaceous glands in the skin of pubescent females and males.[9] In about 30 percent of pubertal males and females, these glands, which normally secrete an oily substance called *sebum,* become clogged and infected, producing pimples and blackheads on the face, chest, or back (acne). Young people experiencing this condition can suffer from embarrassment and loss of self-esteem. Contrary to common belief, acne is not caused by improper hygiene, spread by touch, associated with long hair, caused by masturbation, or cured by frequent sex.[10] Sleep loss, anxiety, and emotional stressors can aggravate acne. Some treatments used to alleviate acne include medicines containing salycylic acid, sulfur, benzoyl peroxide, vitamin A-acid, or antibiotics. Exposure to ultraviolet light also can clear up sores. Acne can take up to ten years or more to run its course, and a few people develop pronounced scars from it. In some cases, these scars can be removed surgically.

WHAT MECHANISMS CAUSE PUBERTY?

To understand the mechanisms that cause puberty, we must first look at the main causative agent: the prepubertal rise in blood levels of FSH and LH. The increased secretion of these gonadotropins ultimately is responsible for the onset of gonadal activity and the resultant rise in blood levels of gonadal steroid hormones. What causes the prepubertal rise in gonadotropin secretion?

Estrogens and androgens exert negative feedback on the secretion of gonadotropins from the anterior pituitary gland in certain situations (Chapter 2). This negative feedback occurs because of inhibitory effects of these steroid hormones on the production of GnRH (gonadotropin-releasing hormone) from the hypothalamus. Steroid hormones circulating in the blood of children exert this negative feedback on the secretion of pituitary gonadotropins. Children born without functional gonads (as in Turner's syndrome, Chapter 5) have relatively high levels of FSH and LH in their blood.[11] Therefore, even the low levels of androgens or estrogens in young children are enough to effectively regulate gonadotropin secretion.

We observed in Chapter 2 that negative feedback effects of gonadal steroids operate in relation to a set point in the "gonadostat" of the brain (hypothalamus). A major hypothesis explaining the prepubertal increase in gonadotropin secretion in both sexes is that the sensitivity of the hypothalamus to steroidal negative feedback decreases as puberty approaches.[12] That is, the set point for negative feedback increases so that higher concentrations of steroid hormones in the blood are required to decrease gonadotropin secretion from the anterior pituitary gland. (Similarly, as you increase the set point of the thermostat in your house, it takes a higher room temperature before the heater shuts off.) Thus, even though circulating levels of steroids rise because of the increase in gonadotropin secretion, the levels of these steroids are not high enough to block a further rise in gonadotropin secretion. It has been hypothesized that the hypothalamus of children is 6 to 15 times more sensitive to steroidal negative feedback than is that of an adult. How is this known? One piece of evidence is that it takes a much smaller amount of injected estrogen to lower blood gonadotropin levels in girls than in adult women.[13]

What controls the change in the set point of the gonadostat? Not much is known about this, but it does appear that other brain areas may influence the ability of the hypothalamus to secrete GnRH. For example, lesions or tumors of specific brain regions in female rats cause early puberty, which suggests that these brain areas normally inhibit secretion of GnRH until puberty by influencing the gonadostat.[14] But what causes maturation and changes in activity of these brain areas? Ultimately, perhaps, their maturation is genetically programmed in each individual, just as is maturation of many organs. Their rate of maturation may also be influenced by environmental factors.

Changes in sensitivity of the brain to negative feedback are not the only factors involved in the onset of puberty. For example, the pituitary gland itself is more sensitive to GnRH in adults than in children. That is, administration of a given amount of GnRH produces a greater secretion of FSH and LH in adults than in children. Also, the gonads of young children respond less to injected gonadotropins than do those of children approaching puberty.[15] Thus, maturation of the pituitary gland and the gonads, as well as the brain, interact to produce the onset of puberty.

As mentioned above, the set point in the gonadostat increases in late prepubertal years so that negative feedback of a given level of steroid hormone is less effective. Following this change, however, the hypothalamus of females changes in a qualitative manner so that immediately before puberty, estrogens have a positive feedback effect on gonadotropin secretion. High blood estrogen levels cause a surge of LH and FSH secretion that causes ovulation in adult women (Chapter 7). In females immediately before their first ovulation, the surge center in the hypothalamus has matured to the extent that estrogens cause this surge of gonadotropin secretion. The positive feedback effect of estrogen is not present in younger females, possibly because their hypothalamus is not yet mature.[16] If estrogens are administered to young females, no gonadotropin surge follows, but if given to girls just before puberty, an FSH and LH surge follows immediately. Therefore, the positive feedback effect of estrogen begins to appear just before menarche and is fully operable late in the pubertal sequence; it is responsible for the first ovulation. The interaction between changes in brain and endocrine function in the regulation of puberty in females is summarized in Fig. 6–10. Less is known about the regulation of puberty in males. It appears to be similar in both sexes, except that no gonadotropin surge occurs in males.

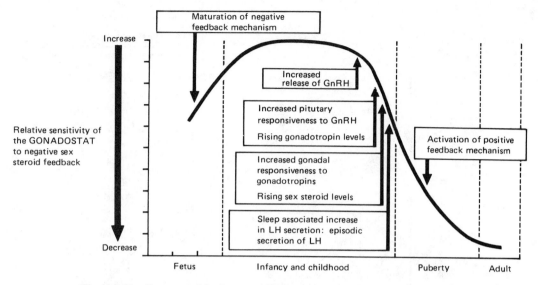

Figure 6–10. Summary of the factors regulating the change in pituitary gonadotropin secretion during female sexual maturation. The pattern is similar in males except that no activation of a positive feedback mechanism develops. Adapted from M. M. Grumbach et al., "Hypothalamic-Pituitary Regulation of Puberty: Evidence and Concepts Derived from Clinical Research," in *Control of the Onset of Puberty,* eds. M. M. Grumbach et al. (New York: John Wiley & Sons, Inc., 1974). © 1974; used with permission.

ENVIRONMENTAL FACTORS AND PUBERTY

Environmental factors that influence physiology of the child also can affect the sequence of sexual maturation and puberty.[17] Of the many possible factors, we examine nutrition, day length, stress, climate, and social interaction. You should be aware, however, that it is very difficult to separate the effects of one environmental factor from those of other factors that vary along with it.

Nutrition

The age of menarche in females generally has shown a steady decline worldwide, at a rate of about 3 months per decade over the last century (Fig. 6–11), although this rate of decline may be an overestimate.[18] In the United States, the average age of menarche was 14.2 years in 1900, but today it is 12.3 (range, 9–17) years. There also is evidence that puberty in males now occurs at an earlier age.[19] Pubertal lowering of voice in the Bach Boy's Choir in Leipzig occurred at age 18 in the eighteenth century, but this event occurs much earlier today.[20] In all probability, there is a biological limit to this decrease, and this limit may have already been reached in some countries. The decline in age at puberty has profound implications for human reproductive biology

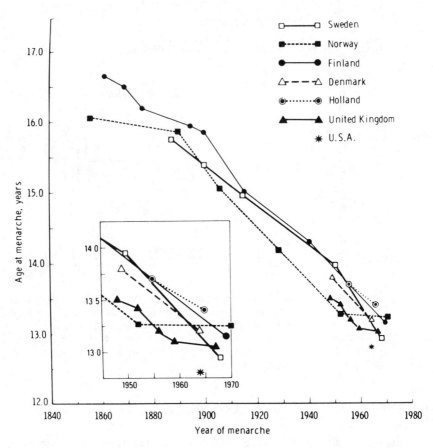

Figure 6–11. Decline in age of menarche in different countries in recent years. Reproduced from J. M. Tanner, "Foetus Into Man: Physical Growth From Conception to Maturity" (Cambridge, Mass.: Harvard University Press, 1978). © 1978 by J. M. Tanner. Reprinted by permission of Dr. J. M. Tanner, Harvard University Press, and Open Books Ltd.

and sexuality in that the period of adolescent sexuality is lengthened. We discuss the impact of this prolonged adolescence on sexual behavior patterns in Chapter 18.

What is responsible for the decline in the age of puberty? There is some evidence that improved nutrition and health in childhood are factors. Females in countries with poorer nutrition tend to exhibit a relatively delayed puberty. In fact, acute starvation prevents puberty. Also, poorer nutrition in rural versus urban regions, in lower social classes, and in larger families may account for the later age at puberty seen in these groups. It is well known that teenage girls with *anorexia nervosa,* a disorder where food intake is greatly reduced, reach puberty at a relatively late age, as do lean female athletes and ballet dancers.[21]

How does nutrition affect the mechanisms for regulating the onset of sexual maturation? One suggestion is that a certain amount of fat is needed in the body of females before menarche can occur. Obese girls tend to reach puberty at an earlier age than other girls.[22] There even is some evidence that a girl must be at a critical weight of about 45 kg (104 lb), and 11 kg (24 lb) of this must be fat (17 percent) before sexual maturation can begin.[23] It may be that, through natural selection, females have evolved a mechanism ensuring that they have enough energy stored to give birth successfully. It has been estimated that pregnancy costs from 27,000 to 80,000 calories. Thus, the 11 kg of fat would contain enough stored energy (about 99,000 calories) for pregnancy and a few months of nursing (at about 1000 calories per day).[24] Much controversy exists about this *critical body fat hypothesis,*[25] and some think that a critical metabolic rate, not a critical body weight or composition, permits puberty.

If the critical body fat hypothesis is true, how does the amount of fat in the body signal the pituitary to begin secreting gonadotropins? One idea is that the fat level in the blood is sensed by the brain. When this level reaches a threshold value, the brain somehow signals the hypothalamus to release GnRH. There is, however, scant evidence for this idea, and a recent finding suggests another possible explanation. It appears that fat cells in human females contain the enzymes necessary to convert androgens to estrogens.[26] Thus, fat may contribute to the high circulating estrogen levels needed to cause the preovulatory surge of gonadotropin secretion and the resultant first ovulation. This interesting hypothesis requires confirmation.

Day Length and Season

Recall that darkness may increase the synthesis and secretion of melatonin and other pineal gland secretions, and these chemicals may then inhibit reproductive function (Chapter 2). Light has the opposite effect. Therefore, do light-dark cycles, through their influence on pineal gland function, play a role in the onset of puberty? And has the man-made increase in day length in recent years, due to increased use of artificial lighting, been at least partially responsible for the recent decline in age of puberty? It appears that children who have lesions of the pineal gland, rendering it nonfunctional, undergo precocious puberty. The absence of circulating pineal secretions in these children may explain their predicament. In contrast, about one-third of children with secretory pineal tumors exhibit delayed puberty, and perhaps their relatively high levels of pineal secretions inhibit sexual maturation.[27] Paradoxically, however, blind children with limited light perception reach puberty at an earlier age than do people with normal vision. And females with no light perception exhibit the earliest menarche.[28] Theoretically, at least, since darkness causes melatonin secretion, these blind children should exhibit delayed maturation, but this is not the case. Deaf children also tend to reach puberty early; thus sensory deprivation in general may accelerate sexual maturity. Obviously, we have much to learn about the possible role of the day length and other environmental factors in regulating puberty in humans. There is some evidence that more females reach menarche at a certain time of year, this time varying among geographic regions. Whether or not day length influences this seasonality of puberty is not known.

Stressors

In laboratory animals, various kinds of environmental stressors (emotional or physical) can delay sexual maturity. A *stressor* is any set of circumstances that disturbs the normal homeostasis in the body. *Stress,* on the other hand, is the physiological effect of a stressor. Any kind of long-term stressor produces a set of physiological stress responses that Hans Selye has termed the *General Adaptation Syndrome.*[29] Although this syndrome encompasses a variety of responses, a major component is enlargement of the adrenal glands and increased secretion of adrenal steroid hormones. Some believe that this increased adrenal function may inhibit the reproductive system. The evidence for this is controversial. There is some information, however, that stressors (emotional or physical) can delay puberty in humans. However, there also is evidence that stress during infancy can accelerate puberty. For example, J. W. M. Whiting studied 50 societies and determined if infant stress was present. Kinds of stressors included were mother-infant separation, pain due to ornamentation or scraping with objects, and exposure to extreme temperature or loud noise. He found that the age of menarche in societies that stressed their infants was 12.75 years, whereas that of those unstressed was 14.0 years.[30] The effects among kinds of stressors on puberty, and when they occur during the life cycle, should be examined in more detail and should be evaluated for other contributing factors such as diet.

Climate

It used to be thought that children in warmer regions reached puberty at an earlier age than those in colder areas. But careful statistical comparisons have shown that these differences are due to other factors besides temperature. Altitude, however, appears to have an effect on age of puberty. Indeed, for each 100 meters increase in altitude, puberty is delayed by about three months. The mechanisms by which high altitude suppresses the sexual maturation process are not clearly known and may be related to poorer nutrition or greater energy expenditure at higher elevations.

Social Interaction

The degree of sexual and social interaction among older children may influence their age of puberty. Interestingly, girls of the *Oneida Community* (a religious colony of the last century in upstate New York that practiced group sex) participated in frequent sexual intercourse before puberty, and they reached puberty about two years earlier than did girls in surrounding regions. In general, however, there is little sound evidence suggesting a causative link between social interaction and age of sexual maturation in humans, although it has been established in many lower animals that courtship activity of males accelerates sexual maturation in females.

INHERITANCE AND AGE OF PUBERTY

Undoubtedly, inheritance interacts with environmental factors to determine the age of sexual maturation. In fact, the differences in age of menarche among different countries (Fig. 6–11) may be related to genetic differences in populations as well as to differences in childhood health and nutrition. It has been estimated that 15 percent of the variance in age of puberty is due to genetic differences. For example, age at menarche tends to be similar in mothers and daughters.[31] Moreover, in identical twin females (identical genotypes), the age of menarche usually differs by only a maximum of two months. Any difference here is probably a measure of slight differences in birth weight or in exposure to slightly different social or physical environments during childhood. In contrast, the difference between age at menarche in fraternal twin girls (with different genetic make-ups) can be up to eight months.[32] Caucasian and Negro races reach puberty at a similar age if raised in similar environmental conditions. But some groups exhibit a genotype favoring early puberty, including the Chinese. The genotype of other races, however, may favor late puberty.[33]

PUBERTY AND PSYCHOSOCIAL ADJUSTMENT

Young people must adjust to remarkable physiological, anatomical, and psychological transformations during the process of puberty. Their bodies change rapidly, and thus their *body image* (how they perceive their bodies, especially in relation to the bodies of their peers) also changes. Any deviations of their bodies from what they or their peers consider "normal" can lead to low self-esteem. For example, females will worry about the size of their breasts and other aspects of their figures, and will not feel feminine if these are not in line with the norm of their peers. Similarly, a male may worry about the size of his penis or his physique; and if these are not within the norm, he will feel that something is wrong with him. Young people should realize that there is considerable variation in the timing of the stages of puberty among different individuals, and that most people develop into "normal" adults in the course of their sexual and physical maturation.

Sexual desire increases during puberty, and sexual relationships become a primary concern. Because of the lack of heterosexual outlets, pubescent males and females may engage in sexual fantasies and sexual activities with the same sex. Masturbation also increases markedly during puberty. Teenagers often feel that they are abnormal if they display these behaviors, and they should recognize that what they do is a normal phase of their psychosexual development (Chapter 18).

It can be painful psychologically when an individual reaches puberty later or earlier than his or her peers. If not handled in a constructive fashion, early or late puberty can lead to poor self-esteem and problems in sexual and other areas of life. Medically speaking, puberty is considered *precocious* (abnormally early) if this process begins before age 8 in females and 10 in males. On the other hand, puberty is considered *delayed* in males if no testicular growth has occurred by age 14 or if no skeletal growth spurt has happened by age 18. In females, puberty is delayed if no breast growth occurs by age 14 or if no skeletal growth spurt appears by age 15.[34] A person experiencing precocious or delayed puberty should be examined by a physician to ascertain cause and possible treatment.

Even if a person has a "normal" puberty from a medical standpoint, it still can be earlier or later than most of his or her peers, and this can affect self-esteem. For example, late-developing males may suffer from a poor self-image, and this can influence them in later life. They tend to have a lower occupational attainment, get paid less, marry later, and have fewer children than other men of the same adult height.[35] Early maturing males, on the other hand, can have an easier time of it. They tend to be held in higher prestige by their peers because of their broad shoulders and masculine physique. These males also are more outgoing than those with frail physiques.[36] Early maturing females suffer more then early maturing males. Because of their mature bodies (e.g., large breasts), their peers make the assumption that they are sexually experienced and sexually "easy"; whereas in actuality, early maturing females tend to be submissive, socially indifferent, and low in popularity.[37]

Adolescence is the period between puberty and adulthood when a good deal of social learning takes place. The length of this period of youth is socially determined. In the United States, it is being extended because of the decrease in age of biological puberty and the lengthening in educational training before adulthood. In some primitive societies, however, the period of adolescence is brief or nonexistent; teenagers are expected to be adults in sexual matters and in other areas of life. In the United States, being an adult can mean several things. Biologically, teenagers are adult after they have reached puberty, when they are capable of having children. Economically, they are adult when they can support themselves and/or family. Morally, they are adult when they are responsible for their actions, can express love in a mature manner, and can have productive and meaningful relationships. During adolescence, teenagers must achieve economic and moral adulthood, deal with separation from family, develop realistic vocational goals, and come to terms with their emerging sexuality. In respect to sexuality, they must accept their sexual fantasies, ambivalence about intercourse, concerns about sexual adequacy, and confusion about love (Chapter 19). Not only that, they must achieve these adjustments in a society that expects adult behavior from them but often does not treat them as adults. Thus, it is an attestation to human understanding and flexibility that most adolescents become mature, capable members of society.

CHAPTER SUMMARY

The human life cycle can be divided into five stages: (1) embryonic and fetal existence, (2) infancy and childhood, (3) puberty and adolescence, (4) early and middle adulthood, and (5) late adulthood and old age. Puberty is the transition from a child to a reproductively mature adult. This process involves growth and development of male or female gonads, sex accessory structures, and secondary sexual characteristics. The timing of puberty varies greatly among different individuals.

In both sexes, levels of steroid hormones (androgens and estrogens) and of pituitary gonadotropins (FSH and LH) rise before puberty. The rising levels of steroid hormones cause the physiological and anatomical changes of puberty. Variations in the negative and the positive feedback of steroid hormones on pituitary secretion of gonadotropins control the onset of puberty. The brain also influences the onset of puberty through genetically programmed development interacting with environmental factors such as climate, nutrition, day length, stressors, health, and social interaction. The great decline in age of puberty is thought to be caused by improved nutrition and health care.

A teenager's self-esteem and adjustment are influenced by his or her body image. During the period of adolescence, teenagers must develop economic and moral adulthood, deal with separation from family, plan their vocation, and come to terms with their sexuality. Adolescents must also adjust to conflicting standards of society about how they should behave and what is considered to be adult behavior.

NOTES

1. W. A. Marshall and J. M. Tanner, "Variation in the Pattern of Pubertal Changes in Girls," *Archives of Disease in Childhood,* 44 (1969), 291–303; and W. A. Marshall and J. M. Tanner, "Variation in the Pattern of Pubertal Changes in Boys," *Archives of Disease in Childhood,* 45 (1970), 13–23.

2. W. A. Daniel, Jr., and R. T. Brown, "Adolescent Sexual Maturation," *Journal of Current Adolescent Research* (June 1979), pp. 27–30.

3. W. A. Marshall, "Growth and Sexual Maturation in Normal Puberty," *Clinics in Endocrinology and Metabolism,* 4 (1975), 3–25.

4. J. M. Tanner, *Growth at Adolescence,* 2nd ed. (Oxford, England: Blackwell Scientific Publications, 1973).

5. T. G. A. Baker, "A Quantitative and Cytological Study of Germ Cells in Human Ovaries," *Proceedings, Royal Society of London, ser. B,* 158 (1963), 417–33.

6. E. Block, "Quantitative Morphological Investigations of the Follicular System in Women. III. Variations in the Different Phases of the Sexual Cycle," *Acta Endocrinologica,* 8 (1952), 33–54.

7. E. D. Weitzman et al., "The Relationship of Sleep and Sleep Stages to Neuroendocrine Secretion and Biological Rhythms in Man," *Recent Progress in Hormone Research,* 31 (1975), 399–441.

8. D. Knorr and F. Bidlingmaier, "Gynecomastia in Male Adolescents," *Clinics in Endocrinology and Metabolism,* 4 (1975), 157–71; and M. Nydick et al., "Gynecomastia in Adolescent Boys," *Journal of the American Medical Association,* 178 (1961), 449–54.

9. J. B. Hamilton, "Male Hormone Substance: A Prime Factor in Acne," *Journal of Clinical Endocrinology,* 1 (1941), 570–92.

10. J. E. Jelinek, "Acne: 10 Common Myths—and the Facts," *Consultant* (May 1979), pp. 55–59.

11. J. S. D. Winter and C. Faiman, "Serum Gonadotropin Levels in Agonadal Children and Adults," *Journal of Clinical Endocrinology and Metabolism,* 35 (1972), 561–64.

12. S. R. Ojeda et al., "Recent Advances in the Endocrinology of Puberty," *Endocrine Reviews,* 1 (1980), 228–57.

13. R. P. Kelch et al., "Suppression of Urinary and Plasma Follicle-Stimulating Hormone by Exogenous Estrogens in Prepubertal and Pubertal Children," *Journal of Clinical Investigation,* 52 (1973), 1122–28.

14. M. M. Grumbach et al., "Hypothalamic-Pituitary Regulation of Puberty: Evidence and Concepts Derived from Clinical Research," in *Control of the Onset of Puberty,* eds. M. M. Grumbach, G. D. Grave, and F. E. Mayer (New York: John Wiley & Sons, Inc., 1974).

15. Ibid.

16. E. O. Reiter et al., "The Absence of Positive Feedback between Estrogen and Luteinizing Hormone in Sexually Immature Girls," *Pediatric Research,* 8 (1974), 740–48.

17. L. Zacharias and R. J. Wurtman, "Age at Menarche: Genetic and Environmental Influences," *New England Journal of Medicine,* 280 (1969), 868–75.

18. V. L. Bullough, "Age at Menarche: A Misunderstanding," *Science,* 213 (1981), 365–66.

19. A. H. Steinhaus, *Toward an Understanding of Health and Physical Education* (Dubuque, Iowa: William C. Brown Company, Publishers, 1963).

20. W. Sullivan, "Boys and Girls Are Now Maturing Earlier," *New York Times,* January 24, 1971.

21. Anonymous, "Delayed Menarche and Amenorrhea in Ballet Dancers," *Research in Reproduction,* 13, no. 1 (1981), 2–3.

22. C. A. McNeill and N. Linson, "Maturation Rate and Body Build in Women," *Child Development,* 34 (1963), 25–32.

23. R. E. Frisch and R. Revelle, "Height and Weight at Menarche and a Hypothesis of Critical Body Weights and Adolescent Events," *Science,* 169 (1970), 397–99; and R. E. Frisch and J. W. McArthur, "Menstrual Cycles: Fatness as a Determinant of Minimum Body Weight for Height Necessary for Their Maintenance and Onset," *Science,* 185 (1974), 949–51.

24. P. J. Hogarth, *Biology of Reproduction: Tertiary Level Biology* (New York: John Wiley & Sons, Inc., 1978).

25. W. Z. Billewica et al., "Comments on the Critical Metabolic Mass and the Age of Menarche," *Annals of Human Biology,* 3 (1976), 51–59.

26. A. Nimrod and K. J. Ryan, "Aromatization of Androgen by Human Abdominal and Breast Fat Tissue," *Journal of Clinical Endocrinology and Metabolism,* 40 (1975), 367–72.

27. R. J. Wurtman et al., *The Pineal* (New York: Academic Press, Inc., 1968).

28. L. Zacharias and R. J. Wurtman, "Blindness: Its Relation to Age at Menarche," *Science,* 144 (1964), 1154–55.

29. H. Selye, "The Evolution of the Stress Concept," *American Scientist,* 61 (1974), 692–99.

30. J. W. M. Whiting, "Menarcheal Age and Infant Stress in Humans," in *Sex and Behavior,* ed. F. A. Beach (New York: John Wiley & Sons, Inc., 1965).

31. Zacharias and Wurtman (1969), op. cit.

32. Tanner (1973), op. cit.

33. J. M. Tanner, "Puberty," in *Advances in Reproductive Physiology,* ed. A. McLaren, vol. 2 (New York: Academic Press, Inc., 1967).

34. J. A. Clausen, "The Social Meaning of Differential Physical and Sexual Maturation," in *Adolescence in the Life Cycle,* eds. S. E. Dragastin and G. H. Elder, Jr. (New York: John Wiley & Sons, Inc., 1975).

35. R. Ames, "Physical Maturing among Boys as Related to Adult Social Behavior," *California Journal of Educational Research,* 8 (1957), 69–75; and H. Peskin, "Pubertal Onset and Ego Functioning," *Journal of Abnormal Psychology,* 72 (1967), 1–15.

36. J. B. Cortes and F. M. Gatti, "Physique and Propensity," *Psychology Today,* 4, no. 5 (1970), 82–84.

37. D. A. Hamburg and D. T. Lunde, "Sex Hormones in the Development of Sex Differences in Human Behavior," in *The Development of Sex Differences,* ed. E. E. Maccoby (Stanford, Calif.: Stanford University Press, 1966).

FURTHER READING

BERENBERG, S. R., ed., *Puberty: Biologic and Psychological Components.* Leiden: Stenfert Kroese, 1975.

CORTES, J. B., and E. M. GATTI, "Physique and Propensity," *Psychology Today,* 4, no. 5 (1970), 82–84.

DRAGASTIN, S. E., and G. H. ELDER, JR., eds., *Adolescence in the Life Cycle.* New York: John Wiley & Sons, Inc., 1975.

GRUMBACH, M. M., et al., eds., *Control of the Onset of Puberty.* New York: John Wiley & Sons, Inc., 1974.

HAFEZ, E. S. E., "Reproductive Life Cycle," in *Human Reproduction: Conception and Contraception,* eds. E. S. E. Hafez and T. M. Evans. New York: Harper & Row, Publishers, Inc., 1973. Pp. 157–200.

MARSHALL, W. A., "The Relationship of Puberty to Other Maturity Indicators and Body Composition in Man," *Journal of Reproduction and Fertility,* 52 (1978), 437–43.

TANNER, J. M., *Growth at Adolescence,* 2nd ed. Oxford, England: Blackwell Scientific Publications, 1973.

TANNER, J. M., "Earlier Maturation in Man," *Scientific American,* 218 (1968), 21–27.

TANNER, J. M., "Growing Up," *Scientific American,* 229 (September 1973), 34–43.

WERFF TEN BOSCH, J. J. VAN DER, "Indices of Human Puberty," *Journal of Reproduction and Fertility,* suppl. 6 (1969), pp. 67–76.

WERFF TEN BOSCH, J. J. VAN DER, and A. BOT, "Puberty," *Research in Reproduction,* vol. 8, no. 6 (1976), map.

ZACHARIAS, L. et al., "Sexual Maturation in Contemporary American Girls," *American Journal of Obstetrics and Gynecology,* 108 (1970), 833–46.

7 *The Menstrual Cycle*

LEARNING GOALS

Having read this chapter, you should be able to:

1. List differences between estrous cycles and menstrual cycles.
2. Define monestrous and polyestrous species.
3. Compare induced and spontaneous ovulation.
4. Give examples of how environment and social interaction influence reproductive cyclicity in women.
5. List factors that influence length of menstrual cycles, and explain the influence of each factor.
6. Describe major hormonal, ovarian, and uterine changes during the menstrual cycle.
7. Discuss the role of negative and positive feedback in regulating the menstrual cycle.
8. List five hypotheses that attempt to explain what causes the corpus luteum to die at the end of the human menstrual cycle.
9. List methods used for detecting when ovulation has occurred or will occur.
10. List physical and psychological symptoms of the premenstrual syndrome, and discuss possible causes of this syndrome.
11. List possible causes of dysmenorrhea.
12. Describe the relationship of tampons to toxic shock syndrome.
13. List some causes of amenorrhea, and differentiate between primary and secondary amenorrhea.

14. Discuss past and present beliefs about menstruation.

15. Describe the symptoms, causes, and treatment of menopause.

INTRODUCTION

In Chapters 2 through 4, we saw how the brain influences secretion of gonadotropic hormones (FSH and LH) from the anterior pituitary gland. These gonadotropic hormones in turn regulate ovarian activity by controlling ovarian steroid hormone secretion and by causing maturation of follicles and oocytes and inducing ovulation. Ovarian steroid hormones, in turn, regulate the function of the female sex accessory structures and secondary sexual characteristics. But the whole is greater than the sum of its parts, and the menstrual cycle is the result of a finely tuned interaction among the brain, pituitary gland, ovaries, and uterus.

REPRODUCTIVE CYCLES IN OTHER MAMMALS

Many mammals exhibit seasonal reproductive cycles. That is, ovulation occurs only at a certain predictable time of year, and the ovaries are inactive for the remainder of the year. In some of these species, seasonal changes in day length influence reproductive cycles. There is no strong tendency for seasonality, however, in human reproduction, although there is slight seasonal variation in the number of births (see Chapter 10).

In most mammals, females are sexually receptive to males only around the time of ovulation, thus ensuring a greater chance for fertilization and pregnancy. Female sexual receptive behavior occurring around the time of ovulation, when estrogen levels are high, is called *estrous behavior* (Greek *oestros* = "mad desire") or "heat," and animals that exhibit cyclic estrous behavior are said to have *estrous cycles*. (*Note: Estrus* is the noun, whereas *estrous* is the adjective.) In some seasonally breeding mammals, there is only one period of estrus a year (deer, for example), and these are called *monestrous species*. The remainder of the year they are said to be *anestrous* ("without estrus"). Other seasonally breeding animals, such as wild mice, have several estrous cycles within the breeding season, and these are called *polyestrous species*. Some tropical mammals exhibit estrous cycles all year, and humans are similar to these in being reproductively active all year. *Menstrual cycles* (monthly cycles) are found only in humans and certain other primates, such as baboons, apes, and monkeys; and uterine bleeding (*menstruation*) occurs at the end of each cycle. There is controversy, however, as to whether human females exhibit an estrous as well as a menstrual cycle, that is, if women are more receptive to sexual intercourse at the time of ovulation than at other times in the menstrual cycle. In Chapter 17, we look into this controversy.

In some mammals, females are continuously receptive to males but do not ovulate unless stimuli associated with copulation or other male behaviors are present. This phenomenon is called *induced ovulation* in contrast to *spontaneous ovulation*, which is not dependent on male behavior or copulation. In induced ovulators such as rabbits, camels, raccoons, and cats, the act of copulation stimulates sensory nerves in the vagina, the neural message is relayed to the brain, the brain causes luteinizing hormone (LH) secretion from the anterior pituitary gland, and this hormone causes ovulation. In fact, mere stimulation of the vagina with a glass rod can cause ovulation in these animals. In sexually experienced rabbits, moreover, presence of a male without copulation is enough to trigger ovulation!

Human females are spontaneous ovulators in that ovulation happens periodically whether *coitus* (copulation) occurs or not. However, there is some controversial evidence that coitus can induce ovulation during this spontaneous cycle. For example, more pregnancies result from rape than occur by chance, and it is thought that aggressive coitus causes induced ovulation in these cases. Also, the high failure rate of the "rhythm method" of contraception (discussed in Chapter 13) could partly be due to coitus-induced ovulation.[1]

Daily light cycles have a profound influence on the estrous cycle of some mammals.[2] For example, laboratory rats with a 4- or 5-day estrous cycle that are kept on a 12-hours light:12-hours dark daily light cycle, the lights being turned on at 6.00 A.M., ovulate at 2:00 A.M. on the morning of estrus. Now, if the 12-hours light:12-hours dark photoperiod is reversed so that the lights turn on at 6:00 P.M., the estrous cycles will shift in a few days so that they follow the reversed light cycle. If the rats are kept in constant light, they will be in estrus continuously, and if they are kept in constant darkness, they will be in continuous anestrus. There is no evidence, however, that the human menstrual cycle is influenced by the daily light cycle. Blind women, for example, have normal menstrual cycles.

Is it a coincidence that the average menstrual cycle length is 29.5 days, the same as the lunar cycle length? Maybe not! Perhaps women evolved a cycle length that, for some reason, was adaptive to the lunar cycle. A recent study did show that more women than would be predicted by chance ovulate during the dark (new moon) phase of the lunar cycle.[3]

Social interactions also can influence estrous cycles in some mammals. If female rats are kept together without males, their estrous cycles will be irregular and not synchronized. However, if females are grouped together with males, most of the females eventually will reach estrus on the same day, the so-called *Whitten effect*. Remarkably, an analogous situation can occur in humans, and it has been called the "dormitory effect." M. K. McClintock has shown that women grouped together for a period of time in dormitories soon will have synchronized menstrual cycles; that is, most menstruate at the same time.[4] Also, an increase in dating activities decreases menstrual cycle length. A recent study shows that the menstrual cycles of college women who have coitus frequently are less variable in length than are those in women who abstain from sex.[5] Some of these social effects on menstrual cycles may be mediated by pheromones (Chapter 17).

MAJOR EVENTS IN THE MENSTRUAL CYCLE

The menstrual cycle functions so that ovulation of a single egg occurs, and the uterus is prepared to receive the embryo after fertilization. The word *menstrual* has its root in the Latin word *mensis*, which means "month"; the menstrual cycle averages the length of a lunar month (29.5 days).[6] Only about 10 to 15 percent of menstrual cycles, however, are exactly 29.5 days. Cycle length can vary greatly in a single woman and even more so among different women. Most cycles are 25 to 30 days long, but some last less than 25 or greater than 30 days. Younger women tend to have longer cycles than do older women. For example, average cycle length is 35 days in 15- to 19-year-olds, 30 days in 30-year-olds, and 28 days in 35-year-olds.[7] Cycle length is more variable in teenage and older women than in women in their peak reproductive years. With a cycle length of 28 days, a women would ovulate about 13 ova per year, or 481 ova in the 37 years from puberty to the end of her reproductive (childbearing) life. Pregnancy and lactation, however, would reduce this total number because ovulation is inhibited during these times (Chapters 9 and 11).

Each menstrual cycle can be divided into three main phases. These are: (1) the *menstrual* (or "destructive") *phase*, also called *menses*, (2) the *follicular* (or "proliferative," "estrogenic") *phase,* and (3) the *luteal* (or "secretory," "progestational") *phase* (Fig. 7–1). It is customary to designate the first day of the cycle (day 1) to be the first day of the menstrual phase. This is the day on which menstruation begins, and it is chosen as a starting point because a woman is more aware

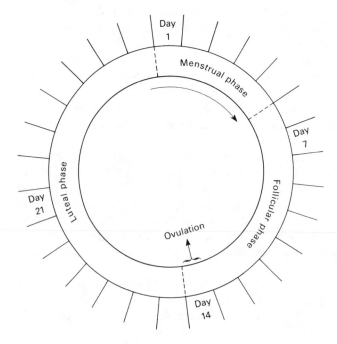

Figure 7–1. The three phases of a 28-day menstrual cycle. Day 1 is the beginning of the menstrual phase, which lasts about 5 days. The menstrual phase is followed by the follicular phase, which ends at ovulation on about day 14. Next is the luteal phase, ending 14 days later with the beginning of menstruation. Adapted from *The Cycling Female: Her Menstrual Rhythm* by Allen Lein (San Francisco: W. H. Freeman & Company, Publishers, 1979). © 1979; used with permission.

of this day than any other. The menstrual phase can last 1 to 6 days, but usually lasts 4 or 5 days. The follicular phase begins at the end of menstruation, or about day 6 if menstruation lasts 5 days. During the follicular phase, the ovaries continue growth that began during the menstrual phase, and then secrete estrogens. Estrogens in turn cause growth of the uterine endometrium. This phase continues until ovulation, which occurs on about day 14 of a 28-day cycle and day 16 of a 30-day cycle. Thus, ovulation usually occurs 14 days before the onset of the next menstruation (day 1 of the next cycle). The luteal phase lasts from ovulation to the beginning of menstruation, or about 14 days. It is during this phase that the corpus luteum, which is formed from the wall of the ovulated follicle (Chapter 3), secretes estradiol-17β and progesterone, which prepare the uterus for arrival of an embryo. The ovum disintegrates if fertilization does not occur, and its remains are washed out of the body in the menstrual flow. If fertilization does occur, the embryo implants (embeds) in the uterine lining (Chapter 9), and further menstrual cycles are inhibited during pregnancy. Changes in the oviducts during the menstrual cycle in relation to gamete transport and fertilization are discussed in the next chapter.

THE MENSTRUAL CYCLE IN DETAIL

Menstrual Phase

During menstruation ("the period"), part of the lining of the uterus (endometrium) degenerates because the blood vessels that supply it with nutrients and oxygen constrict. These vessels, although constricting, bleed as part of the lining degenerates: 33 to 267 ml (1–8 oz) of blood can be lost during menstruation, but most women lose 33 to 83 ml (1 to 2.5 oz). When a woman suspects that more than 300 ml (about a cup) of blood is lost, she should supplement her diet with iron to prevent possible iron-deficiency anemia (iron is an important component of red blood cells). Women using intrauterine contraceptive devices tend to have a heavier than average menstrual flow, whereas those using the combination pill tend to have less than average flow (Chapter 13). Excessive menstrual bleeding should be reported to a physician.

Menstrual discharge contains not only blood but also other uterine fluids along with debris from the sloughing endometrium and some cells that have been lost from the lining of the vagina. The debris from the endometrium looks like "clots" and often is mistaken for blood clots. Actually, the uterine blood clots immediately but then liquifies and leaves the body in the menstrual flow.[8] Thus, menstrual fluid really is like blood serum. The characteristic odor of menstruation is caused by the action of bacteria on the menstrual flow.

The corpus luteum of the previous cycle has regressed by day 1, and the resultant decrease in blood levels of estradiol and progesterone causes the stratum functionalis of the uterine endometrium to degenerate (Fig. 7–2). The ovaries at day 1 contain only small tertiary follicles, less than 5 mm in diameter, along with several

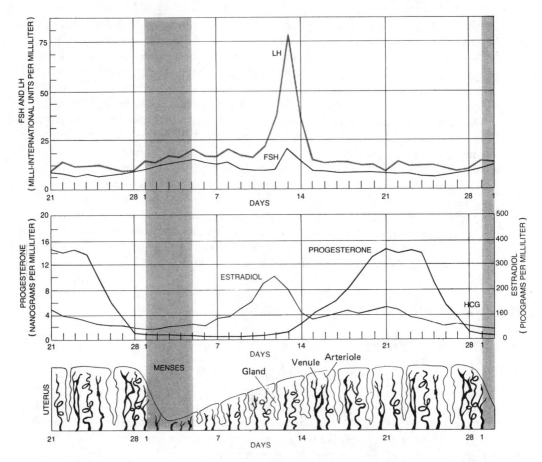

Figure 7-2. Blood levels of hormones during a 28-day menstrual cycle, along with changes in the uterine endometrium. Note that an estradiol-17β peak precedes the LH surge. Note also that the ratio of estradiol to progesterone is high during the follicular phase and lower during the luteal phase. The ovaries also secrete small amounts of another estrogen (estrone), androgens, and 17-hydroxyprogesterone (not shown). Levels of progesterone are in nannograms (one thousandth of a milligram) and of estradiol in picograms (one thousandth of a nannogram). Levels of FSH and LH are in milli-international units (one thousandth of an international unit). An international unit is an amount of a hormone that produces a given biological response in a target tissue. Adapted from C. Grobstein, "External Human Fertilization," *Scientific American,* 240, no. 6 (1979), 57–67. Copyright © 1979 by Scientific American, Inc.; all rights reserved.

atretic follicles and thousands of smaller follicles (see Chapter 3). By day 3, some of the small tertiary follicles have enlarged to about 10 mm in diameter.

Figure 7–2 summarizes changes in blood levels of pituitary hormones (FSH and LH) and steroid hormones (estradiol-17β and progesterone) during the menstrual cycle. Study this figure closely before you proceed. Note that on day 1 of the cycle, levels of all four hormones are low. By day 3, however, FSH and LH begin to rise slowly, and these hormones are responsible for the aforementioned increase in size of some follicles at this time. As these follicles grow, they begin to secrete estradiol-17β, so that estradiol levels also begin to rise by day 3. Levels of progesterone, however, remain low during the menstrual phase.

Follicular Phase

The uterine endometrium begins to thicken during the follicular phase, and the uterine glands begin to enlarge (Fig. 7–2). Also, the endometrium becomes more richly supplied with blood vessels, and water accumulates between cells in the tissues, a condition known as *edema*. In addition, the smooth muscle of the myometrium begins to contract mildly in a rhythmic fashion, although a woman usually is not aware of these slight contractions.

Some of the tertiary follicles in each ovary increase in size during the follicular phase, and a few reach 14 to 21 mm in diameter by days 10 to 12. Other follicles that have previously undergone rapid growth become atretic (degenerate). By day 13, only one large antral follicle, 20 to 25 mm in diameter, is present in only one ovary; this large follicle appears like a blister on the surface of one ovary. The remaining large follicles in each ovary become atretic by day 13.

Estradiol-17β is secreted by the larger follicles in each ovary (including those that may later become atretic), and levels of this hormone in the blood continue to rise throughout most of the follicular phase, reaching a maximum (peak) on day 12 or 13 (Fig. 7–2). About 24 hours after this peak in estradiol, a surge (marked and rapid increase) of LH occurs (Fig. 7–2), which initiates resumption of meiosis in the oocyte within the largest follicle and ovulation of this follicle (Chapter 3). Ovulation occurs about nine hours after the peak of this LH surge.[9] Progesterone and FSH levels remain low in the follicular phase until just before ovulation. At this time, a small FSH surge accompanies the greater LH surge, and progesterone levels rise slightly just before ovulation (Fig. 7–2). This progesterone comes from the wall of the largest follicle under the influence of the LH surge (Chapter 3) and may (along with estradiol) be important in regulating the LH surge.

Intricate feedback mechanisms operate during the follicular phase. Let us first look at the feedback effects of estradiol-17β on secretion of gonadotropin-releasing hormone (GnRH) from the hypothalamus. Remember from Chapter 2 that the hypothalamus of the brain may secrete only a single GnRH, which controls secretion of both FSH and LH from the pituitary gland. Now, look at Fig. 7–3, and you see that moderate levels of estradiol have a negative feedback effect on GnRH secretion. At the beginning of the follicular phase, when estradiol levels are low, FSH levels rise because of the absence of negative feedback by estradiol and GnRH secretion. After estradiol reaches moderately high levels in the midfollicular phase, it exerts negative feedback on GnRH secretion, and FSH levels begin to decline. Human ovarian follicles contain a substance (inhibin) that inhibits FSH (but not LH) secretion,[10] and this substance may also exert negative feedback effects. In the last days of the

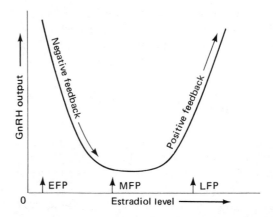

Figure 7–3. The effect of increasing blood levels of estradiol-17β during the follicular phase of the menstrual cycle on secretion of GnRH from the hypothalamus. Note that in the early follicular phase (EFP), estradiol levels are low and GnRH output is high due to the absence of negative feedback of estradiol on GnRH secretion from the hypothalamus. In the midfollicular phase (MFP), estradiol levels are moderately high and exert negative feedback on GnRH secretion. In the late follicular phase (LFP), the very high estradiol levels now exert a positive feedback on GnRH secretion, resulting in the LH surge. Adapted from *The Cycling Female: Her Menstrual Rhythm* by Allen Lein (San Francisco: W. H. Freeman & Company, Publishers, 1979). © 1979 by W. H. Freeman & Company.

follicular phase, estradiol levels are very high and now exert a positive feedback on GnRH secretion; this increase in GnRH causes the LH surge on day 14.

Many questions remain to be answered about the control of FSH and LH secretion during the follicular phase. For example, why is it that FSH and LH do not cycle in a similar manner (Fig. 7–2) if there is only one GnRH? The answer is not clear, but a possible explanation is that changing estradiol-17β levels not only influence secretion of GnRH from the hypothalamus but also change the sensitivity of the anterior pituitary gland to GnRH.[11] More specifically, it seems that a given level of estradiol may affect the sensitivity of cells in the pituitary that secrete FSH and LH in a different manner. For example, high levels of estradiol in the late follicular phase may have a positive effect on GnRH secretion and on the sensitivity of LH-secreting cells to GnRH but simultaneously decrease the sensitivity of FSH-secreting cells to GnRH. Why is there a small FSH surge along with the LH surge on day 14? It may be that the pituitary is unable to release that much LH without also releasing some FSH, or inhibin may selectively inhibit FSH and not LH secretion (Chapter 3). If an injection of estradiol is given to a woman, an LH surge follows. If, however, this estradiol injection is followed by an injection of progesterone, both FSH and LH surge in the blood.[12] Progesterone alone does not cause either gonadotropin to surge. Finally, can estrogen cause an LH surge by directly acting on the pituitary gland alone instead of on the hypothalamus, as occurs in the Rhesus monkey?[13] Future research on how steroid hormones cause the gonadotropin changes that lead to ovulation is a subject of great interest to clinical scientists because they may be able to use this knowledge to help infertile women to ovulate (Chapter 15).

Look at Fig. 7–2 and notice that estradiol-17β levels drop rapidly just before ovulation. For some reason, the largest follicle in the ovary stops secreting estradiol and begins to produce progesterone immediately before ovulation. This drop in estradiol can lead to mild uterine bleeding at ovulation, called *spotting* or *breakthrough bleeding,* a somewhat rare phenomenon.

There are also many mysteries about the ovarian changes that occur in the follicular phase. For example, we do not understand why, although many growing follicles are present in an ovary, eventually only one is selected for ovulation. The other follicles, although growing and initially secreting estrogens, become atretic. We do know that if the large follicle that is going to ovulate is removed before it does so, another follicle will take its place. Thus, a follicle that was programmed to die now remains viable and ovulates. We also do not understand the mechanisms controlling alternation of ovulation. If a woman's right ovary ovulates one egg in a particular month, chances are that her left ovary will ovulate the next time around. How this alternation occurs is not clear, but we do know that in women who have one ovary removed (*ovariectomy*), the remaining ovary will ovulate every month.

Luteal Phase

During the luteal phase, the uterine endometrium becomes thick and spongy (Fig. 7–2), and its glands secrete nutrients that will be used by the embryo—if one is conceived (Chapter 9). During this phase, the uterine smooth muscle contracts less frequently, but each contraction is more intense than during the follicular phase.

After ovulation, a corpus luteum is formed from the wall of the follicle that ovulated (Chapter 3). This structure then begins to secrete estradiol-17β and progesterone. The levels of these two hormones rise in the middle of the luteal phase (Fig. 7–2). About four days before menstruation begins, the corpus luteum begins to degenerate, and levels of these two steroid hormones decline (Fig. 7–2). It is the combination of high levels of estradiol and progesterone during the luteal phase that maintains the uterus in its secretory condition; when blood levels of these hormones decrease, the endometrium begins degeneration, resulting in the onset of menstruation.

The presence of estradiol-17β and progesterone in the luteal phase results in negative feedback on both FSH and LH secretion;[14] and because of this negative feedback, the levels of FSH and LH are low in the luteal phase (Fig. 7–2). You may wonder why a combination of estradiol and progesterone causes release of LH before ovulation but inhibits secretion of FSH and LH by negative feedback during the luteal phase. Apparently, GnRH is released when progesterone levels are low as compared to estradiol levels (day 13); but when progesterone levels are relatively high as compared to estradiol in the luteal phase, GnRH secretion is inhibited. Combination pill oral contraceptives contain a little estrogen and a large amount of progestogen because this combination mimicks the hormonal condition in the luteal phase and thus prevents ovulation (Chapter 13). When estradiol and progesterone levels begin to decline at the end of the luteal phase, secretion of GnRH is no longer inhibited, and a new cycle begins.

In humans, LH is necessary for formation of the corpus luteum and for its secretory function in the luteal phase. In other words, the low levels of LH in the luteal phase (Fig. 7–2), although not high enough to cause ovulation, are enough to maintain the corpus luteum. In some other mammals (such as the rat), LH also causes formation of the corpus luteum, but another pituitary hormone, prolactin (PRL), is necessary for maintenance of the corpus luteum. There is little evidence, however, that PRL is necessary for function of the human corpus luteum. But human PRL was only isolated in 1971, so we have much to learn about its role in the menstrual cycle.[15]

What causes the corpus luteum to die at the end of each menstrual cycle? There are several suggested reasons. In sheep, for example, a prostaglandin secreted by the uterus causes the corpus luteum to regress.[16] There is no evidence, however, that uterine prostaglandin is a *luteolytic* ("corpus luteum killing") *factor* in humans, although prostaglandins are present in the human endometrium near the end of the luteal phase, and the human corpus luteum cells have prostaglandin receptors.[17] Women who have had their uterus removed for medical reasons have normal hormone cycles, and the life span of their corpus luteum is similar to that in women with a uterus.[18] Therefore, prostaglandins secreted by the uterus probably do not kill the human corpus luteum.

Perhaps the slight drop in LH levels near the end of the luteal phase (Fig. 7–2) causes the corpus luteum to die. Injection of LH into women can prolong the life of their corpus luteum for a few days, but then it dies anyway.[19] The number of LH receptors in the human corpus luteum drops as this structure ages, and this may play a role in its death.[20] Changes in secretion of prolactin could also influence the life of the human corpus luteum, but as was said above, we know little about the role of this hormone in the menstrual cycle. Perhaps secretion of an estrogen by the corpus luteum leads to its degeneration because injection of an estrogen directly into the human corpus luteum causes it to die.[21] For some reason, administration of an inhibitory GnRH analog causes the death of the corpus luteum in women, but the natural role of GnRH in maintaining the corpus luteum is unknown.[22] Finally, it may be that the human corpus luteum has a programmed life cycle that is not influenced by other internal factors.

About a week prior to menstruation, the breasts become larger and more sensitive (maybe even painful) if touched. This change in sensitivity is brought about by the high levels of estradiol-17β and progesterone in the luteal phase, which cause edema (water retention) in the breast tissue. No milk is secreted, however, probably because not enough prolactin is present (Chapter 11).

Table 7–1 summarizes the events of the three phases of the menstrual cycle.

TABLE 7-1. SUMMARY OF EVENTS IN THE MENSTRUAL CYCLE, LISTED IN ORDER OF OCCURRENCE

1. FSH ↑ = follicular growth and estradiol secretion from follicles.
2. Estradiol ↑ = proliferative build-up of endometrium.
   = inhibition of FSH.
   = stimulation of LH surge.
3. LH surge and smaller increase in FSH = ovulation and corpus luteum formation.
4. Corpus luteum secretes estradiol and progesterone.
5. Estradiol and progesterone ↑ = inhibits FSH and LH.
   = secretory phase of uterus.
6. Corpus luteum degenerates if fertilization doesn't take place.
7. Estradiol ↓ Progesterone ↓ = menstrual flow.
8. Estradiol ↓ = FSH ↑ ; cycle begins again.

Source: Adapted from B. Goldstein, *Human Sexuality* (New York: McGraw-Hill Book Company, 1976), p. 52. © 1976 by McGraw-Hill; used with permission.

VARIATIONS IN LENGTH OF MENSTRUAL CYCLE PHASES

Figure 7–4 depicts the length of each phase for menstrual cycles that vary in total length (25, 28, 35 days). Note that it is the length of the follicular phase that accounts for most of the variation in total length of the cycle. Although the length of the menstrual phase is pictured in Fig. 7–4 as constant, it too can vary among women. In contrast, the length of the luteal phase usually is 14 days, regardless of the length of the other phases. Thus ovulation usually occurs 14 days before the onset of menstruation. Recent evidence, however, suggests that in some women the luteal phase length can also vary, and this is important for women using the "calendar method" of contraception (see Chapter 13). In one recent study of menstrual cycles in women 18 to 30 years old, the mean cycle length was about 30 days, with the follicular phase averaging 16.9 days and the secretory phase 12.9 days.[23]

METHODS FOR DETECTING OVULATION

Several methods for detecting ovulation have been used or proposed. Although reliable methods are very important for those practicing the "rhythm method" of contraception and for determining causes of infertility in women (Chapters 13 and 14), none so far devised can actually pinpoint the exact time of ovulation in advance. First, we look at those methods that can be done by a women at home and then at those more readily used in the laboratory.

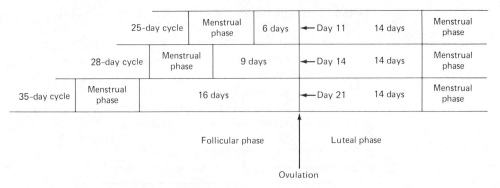

25-day cycle		Menstrual phase	6 days	← Day 11	14 days	Menstrual phase
28-day cycle	Menstrual phase		9 days	← Day 14	14 days	Menstrual phase
35-day cycle	Menstrual phase		16 days	← Day 21	14 days	Menstrual phase

Follicular phase   Luteal phase

Ovulation

Figure 7–4. Variations in length of menstrual cycle phases. Note that the length of the luteal phase is usually 14 days. The lengths of the follicular phases of these cycles are 6, 9, and 16 days. The length of the menstrual phase, although shown as constant here, can vary.

Home Methods

Ovulation usually occurs 10 to 16 days before menstruation begins. If a woman keeps an accurate record of her cycle lengths for at least a year, she may be able to predict the time in her cycle that it is relatively safe to have coitus, taking into account the two to three days that sperm would survive and the two days that the ovum is fertilizable (Chapter 8). Details of the calandar method of detecting time of ovulation and determining the "safe" period are presented in Chapter 13. Suffice it to say here that given the wide variation in total cycle length and in the length of each cycle phase, the calendar method is not very reliable in detecting the time of ovulation.

Many women feel pain in the abdomen around the time of ovulation. This *Mittelschmerz* (German for "midpain") usually is mild and probably is caused by irritation of the abdominal wall by blood and follicular fluid escaping from the ruptured follicle at ovulation. The pain can last for a few minutes or up to a couple of hours. Because of its mild nature, Mittelschmerz often goes undetected and is not a reliable method for determining time of ovulation.

Basal (resting) body temperature, after a slight decrease at ovulation, rises 0.3 to 0.5°C (0.5 to 1.0°F) during the luteal phase of the cycle (Fig. 7–5). This rise is caused by the increase of progesterone in the blood during this phase. Because of difficulty in detecting this rise in body temperature and controlling for other variables such as time of day and level of activity (see Chapter 13), this method is not totally reliable for detecting time of ovulation. This partly accounts for the high failure rate of the rhythm method of contraception when body temperature is used as an indicator.

Around ovulation, the opening from the cervix to the vagina (external cervical os) widens for a short time, and a woman can insert a speculum and mirror to examine the size of this opening. This method is awkward and may be unappealing to

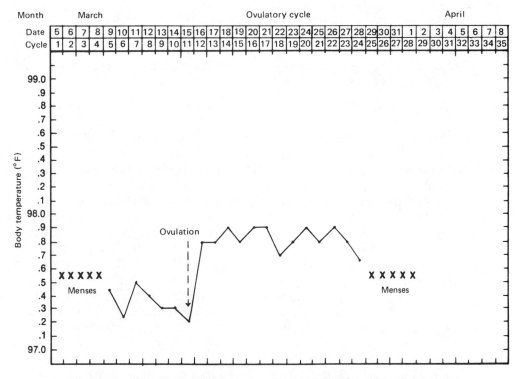

Figure 7–5. A basal body temperature chart. Note the brief drop and then sharp rise in temperature at the time of ovulation. An anovulatory cycle would show no such rise; if pregnancy occurs, the higher, luteal phase temperature will continue. Adapted from *From Woman to Woman* by Lucienne Lanson and illustrated by Anita Karl (New York: Alfred A. Knopf, Inc., 1975). © 1975 by Lucienne Lanson and Alfred A. Knopf, Inc. Reprinted by permission of Alfred A. Knopf, Inc., and Penguin Books Ltd., Australia and England.

some women. Therefore, it is not a good method for detecting ovulation in the home, although it may be used along with other cervical tests in the laboratory.

The cervical *mucus*, which is thick and sticky five to seven days before and one to three days after ovulation, becomes more abundant, watery, and stringy near the time of ovulation.[24] Women often can notice an increase in mucous flow and a moistening of the vagina near the time of ovulation. Generally, laboratory tests for these changes in mucus are more reliable than personal observation for detecting ovulation time.

Some commercial kits may soon become available for detection of ovulation. These tests, which rely on chemically treated papers that change color when a critical physiological event occurs, can be used for detecting such things as chloride in the cervical mucus. Another paper method detects an enzyme, alkaline phosphatase, in the saliva. The level of this enzyme in the saliva increases 48 to 72 hours prior to ovulation. Another potential method may involve detecting another enzyme, urokinase, in the urine. This enzyme, which is produced by cells lining the urinary bladder and urethra, increases in the urine three to four days prior to ovulation.[25] We know little about how these enzymes relate to reproduction, but their detection could be useful in predicting ovulation.

Laboratory Methods

Laboratories can take blood samples and determine levels of LH or steroid hormones. Detection of the LH surge or the increase in blood progesterone levels in the luteal phase can determine if ovulation will occur or has occurred. A breakdown product of progesterone, *pregnanediol*, can also be detected in urine.

As mentioned above, tests for the condition of cervical mucus can be performed in the laboratory. The mucus around the time of ovulation exhibits a fernlike pattern of crystals of sodium and potassium chloride when dried. This *Fern Test* for ovulation is combined with a test for *Spinnbarheit* (threadability) of the mucus and degree of opening of the cervix, to obtain an overall "cervical score." This score is highest when blood estradiol-17β is high and lowest when blood progesterone is high.

The cytology of cells lining the human vagina also changes near the time of ovulation. That is, cells sloughing from the vagina are more cornified (hardened with the protein keratin) near the time of ovulation when blood estradiol-17β levels are high. Microscopic examination of vaginal smears, therefore, can be an indication of ovulation. Finally, a piece of the endometrium can be removed by a physician. It then can be examined to see if it is in the proliferative or secretory phase. If in the latter, ovulation has occurred. This is called an endometrial biopsy (Chapter 3).

Reliability of Methods
for Ovulation Detection

Many of the above methods have drawbacks. That is, most are relatively inaccurate and detect ovulation only after it occurs. More to the point, there is no accurate method of predicting the exact time of ovulation, although there are means of detecting if ovulation has occurred. The latter are useful in studying causes of infertility in women (see Chapter 15).

THE PREMENSTRUAL SYNDROME

Symptoms

About 60 percent of all women experience some physical and emotional difficulties before menstruation begins, although these symptoms are severe and debilitating in only 5 percent. These difficulties are called the *premenstrual syndrome* (a syndrome is a group of symptoms that occur together), and they usually begin three to ten days before menstruation. Various physical symptoms are reported, including heart palpitations, breast tenderness, unpleasant tingling or swelling in the hands and feet, an increase in urination and body weight (due to water retention), acne, and migraine headaches. Emotional components of this syndrome (which are grouped within the term *premenstrual tension*) can include fatigue, irritability, depression or emotional instability, and fits of crying.

Possible Causes

Suggested causes of the premenstrual syndrome include hormonal changes, dietary deficiencies, and psychologic factors, all of which probably interact. For example, the changes in estradiol-17β and progesterone levels at the end of the luteal phase may be related to water retention, which in turn can be a factor in the occurrence of headaches, edema, and breast tenderness. Treatment with progesterone can relieve some of these complaints in certain women. Secretion of the steroid hormone *aldosterone* from the adrenal glands increases in the luteal phase, and this hormone increases water retention. Some drugs that increase urination (*diuretics*), such as *spironolactone*, antagonize aldosterone and thus can relieve some symptoms of the premenstrual syndrome. Restriction of salt and water intake also can help. Some scientists feel that dietary deficiencies, especially a lack of *vitamin B$_6$*, which plays a role in nervous system function, can cause some of the emotional disturbances seen during premenstrual tension.[26]

Various psychological factors, along with the hormonal changes, have been suggested as causing or at least influencing premenstrual tension. These include a conflict about motherhood, feelings about being a woman, and feelings that menstrual blood is related to urine and feces and is dirty. Negative attitudes of men about menstruation also can affect the feelings of women in the premenstrual phase of their cycle. Finally, some believe that if a woman was taught as a child that menstruation is dirty or is something to be ashamed of, she is more likely to experience premenstrual difficulties.

Cyclic Mood Changes

Numerous studies indicate that women may differ in mood depending on the phase of their menstrual cycle. When estrogen levels are high around the time of ovulation, they tend to have more self-esteem, self-confidence, and alertness. In the luteal phase, they tend to become more passive and self-directed. For the four days before menstruation, premenstrual anxiety, depression, irritability, hostility, and helplessness may occur. Then, during menstruation, some aspects of tension disappear, although depression often remains until estrogen begins to rise.[27] This is not to say that a woman is controlled by her physiology, because her degree of emotional maturity and life situation can influence and even eliminate these cyclic mood changes. Some earlier studies suggested that women in the premenstrual state showed an increased tendency to commit crimes, commit suicide, and to be committed to hospitals for mental or physical reasons.[28] It is now felt that conclusions from some of these studies were reached too hastily, and recent evidence indicates that women in the premenstrual condition are as capable physically and intellectually as they are in any other stage of the menstrual cycle.[29] We saw in Chapter 4 that men also can be subject to cyclic mood changes.

MENSTRUAL DIFFICULTIES

Dysmenorrhea

Many women (30 to 50 percent) experience some discomfort during menstruation, and a few experience pain that is severe enough to require lying in bed for a day or so. Pain during menstruation is called *dysmenorrhea*, and can be caused by severe contractions of the uterine smooth muscle, often termed *menstrual cramps*. It is common for muscle contraction to be painful and take the form of cramps when blood supply to that muscle is low, such as in the uterus during menstruation. The cause of these cramps is not clear, but prostaglandins present in the menstrual fluid may stimulate the uterine smooth muscles and cause their contraction. These prostaglandins, in combination with lowering progesterone levels, may actually initiate menstruation, and it is only when prostaglandin levels are unusually high that menstrual cramping occurs.[30] Evidence that prostaglandins cause menstrual cramps is that aspirin, which inhibits synthesis of prostaglandins, relieves these cramps and associated back pain in many women. Newer studies show that other antiprostaglandins (ibuprofen, naproxin, mefamic acid, indomethacin) are even more effective than aspirin.[31]

Additional factors suggested as being involved in dysmenorrhea are the changing levels of estradiol and progesterone at the end of the luteal phase, incomplete sloughing of the endometrium, and blockage of the cervical opening so that menstrual fluid does not readily pass to the vagina. However, the use of tampons does not increase the incidence or intensity of cramping; nor does exercise have any harmful effects on a menstruating woman—it can even relieve dysmenorrhea.

It is a myth that a woman must not carry on normal activities, such as exercise or taking a bath, when menstruating. In actuality, frequent exercise can relieve dysmenorrhea by increasing uterine blood flow. Another myth is that coitus during menstruation can lead to physical distress, whereas, in fact, the increase in uterine blood flow during orgasm (Chapter 17) can stop the uterus from cramping, at least temporarily.[32] Other myths about dysmenorrhea include far-fetched beliefs such as plucking hair at night will ensure a painless menstruation, or shaving armpits or legs causes dysmenorrhea, or if menstrual cramps go away while a woman is petting, the pain will be transmitted to her lover![33]

Toxic Shock Syndrome

Toxic shock syndrome is, in a way, a disease of menstruation, since the use of certain tampons increases its incidence. Let us assume that the use of tampons increases the risk of contracting this syndrome, even though there is a controversy about this association.[34] In the 1970s, tampon companies began to put new, more absorbent materials in their products, including foam sponge, carboxycellulose, and polyacrylate fibers. In 1978, toxic shock syndrome was first associated with the use of these highly absorbent tampons. Symptoms of this disease include high fever, a sunburnlike rash, vomiting, diarrhea, muscle aches, and a marked decrease in blood pressure. Toxic shock syndrome occurs in about 3 to 15 out of 100,000 tampon users, so the risk of contracting this disease is not high, even if a woman uses tampons. This disease, however, is serious; about 10 percent of those contracting it die. For reasons not clear, a few nonmenstruating women and even men and children have contracted this syndrome.

How do tampons increase the risk of toxic shock syndrome? About 10 percent of all women harbor a bacterium, *Staphylococcus aureus*, in their vagina (Chapter 3); and about 98 percent of women who have toxic shock harbor these bacteria in their reproductive tract. Menstrual flow accumulates in the tampon, and this serves as a place for *staphylococcus aureus* to multiply and produce a toxin, which enters the blood and causes the symptoms. A recent decrease in the incidence of toxic shock syndrome correlates with a reduction in the use of highly absorbent tampons.[35] If a woman does choose to wear tampons, it is suggested that the tampon be changed frequently and that there be a time each day when a tampon is left out.

ABSENCE OF MENSTRUATION

Amenorrhea is the absence of menstruation when it normally should occur. It is called *primary amenorrhea* if a female has not reached menarche by age 16. Anorexia nervosa, which occurs especially in young women who have had some sort of sexual trauma, can result in primary amenorrhea.[36] In this condition, a female will eat very little and become emaciated. Inasmuch as young females who are very thin for other reasons can fail to reach puberty, the primary amenorrhea of a person with anorexia may be due to the low amount of fat in her body. We discussed the role of body fat in the occurrence of primary amenorrhea in Chapter 6.

Oligomenorrhea is when an adult woman skips one or a few cycles. It is common for many women to exhibit oligomenorrhea because of emotional distress such as the strain of moving to a new apartment or house, losing a loved one, or even a difficult school examination. Oligomenorrhea could also be caused by disorders of the reproductive system as well as by chronic stress, and a woman should consult a physician if she has a pattern of missed menstrual cycles.

Secondary amenorrhea is when menstruation has failed to occur in an adult woman for at least six months. This condition normally occurs, for example, during pregnancy; a pregnant woman may menstruate for a month or two but eventually will stop cycling (Chapter 9). Another time when secondary amenorrhea occurs is during lactation, because suckling of the breasts tends to inhibit ovulation (Chapter 11). Secondary amenorrhea, of course, is also a characteristic of a woman after she has reached menopause.

Other situations can result in secondary amenorrhea. Women who exercise so much that their body contains a relatively low amount of fat can exhibit secondary amenorrhea. These include many ballet dancers and women who run several miles each day.[37] *Galactorrhea* (abnormal milk secretion from the mammary glands) can be associated with secondary amenorrhea in a few women after they stop taking the combination birth control pill (Chapter 13). Secondary amenorrhea also can be associated with more serious conditions, such as tumors of the pituitary gland or ovaries, so a woman who has secondary amenorrhea should discuss her condition with a physician.

MENSTRUAL TABOOS

For centuries, menstruating women were believed to exert many horrible effects on the world around them, thus the phrase "the curse." Menstruating women have been considered unclean, unfit for coitus, and sometimes have been beaten. It was also believed that they spoiled food, drink, and crops. The early Hebrews, for example, punished those who had intercourse while a woman was menstruating. Consider this Bible quotation (Leviticus 15:19): "And if a woman have an issue, and her issue in her flesh be blood, she shall be put apart seven days: and whoever touches her shall be unclean until the even." In medieval times, menstruating women were excluded from going to church and even from entering wine cellars (their presence supposedly spoiled the wine).

In the United States today, only about one-fifth as much coitus occurs in the five days of the menstrual phase as occurs in any other five days of the cycle—even though there is no physical reason for avoiding coitus during menstruation. Nevertheless, many couples avoid coitus at this time because of some sort of repulsion or belief system. Advertisements for tampons and sanitary napkins stress that their product hides the "abnormality" of the menstruating condition, and many women and men are still embarrassed by the subject of menstruation.

MENOPAUSE

Menstrual cycles of women near 50 years of age become shorter, oligomenorrhea is common, and fertility is lower. Finally, a woman will cease ovulating and menstruating altogether, a phenomenon called *menopause* (the *female climacteric*). The average age of menopause in the United States is 49 years, but women ranging in age from 35 to 55 may experience this "change of life." Women in developing countries tend to reach menopause at an earlier age, probably because of poor nutrition.[38] Mothers of twins reach menopause about a year earlier than those with single births.[39] Smoking also causes an earlier menopause.[40]

Symptoms

Before and after a woman reaches menopause, she can have many symptoms, the occurrence and intensity of which vary from woman to woman. These symptoms can include dizziness, numbness, heart palpitations, insomnia, skin spots, backaches, and a dry mouth. Another symptom is *hot flashes*, which are intense feelings of warmth in the skin (especially the face) and profuse sweating. About one-fourth of women undergoing menopause experience them. These hot flashes are caused by widening (dilatation) of blood vessels in the skin, which results in warm blood being carried to these regions. Hot flashes can occur as often as every ten minutes day and night, or can be less frequent. Other physical symptoms of menopause include slight shrinking of the external genitalia, breasts, and uterus. The vagina can decrease in size and its lining may become drier. Also chronic inflammation of the urethra (*urethritis*) is common. A woman may gain weight, and fat may accumulate in her abdomen. Her skin may become wrinkled, her voice may deepen, and hair may appear on her chin and upper lip. Calcium and phosphorus are lost at a greater rate from the bones, and a condition called *osteoporosis* can develop. In this condition, the bones become weak and prone to fracture, and a woman may lose height because her vertebrae compress.

Menopause can be a period of psychological adjustment for some women. Emotional symptoms that can occur include irritability and depression. Some women can sleep more than before, whereas others develop insomnia. Because postmenopausal women no longer have menstrual cycles, and because some of their physical appearance drifts away from what is considered feminine, they may begin to feel like they have lost their womanhood. The degree to which a woman's identity is based on her femininity can relate to the intensity and kind of emotional difficulty she has during the climacteric. However, only about 25 percent of women have exaggerated distressful emotional symptoms such as mood swings during menopause, and most adjust to this change in a healthy and comfortable way. It must be emphasized that menopause is a normal part of the human life cycle; it is *not* a "disease." We saw in Chapter 3 that older men can also suffer from a decline in sex hormones with associated symptoms (the "male climacteric").

Danger of Pregnancy

When menstrual cycles begin to decline as menopause approaches, and even after menstruation has ceased, a woman may still ovulate. Therefore, some method of birth control should be observed for at least a year after menopause has occurred if possible pregnancy is to be avoided. Not only does a woman of this age usually not want a baby, but older women tend to have more miscarriages and stillbirths, and babies born to older women have a slightly increased tendency to have birth defects (Chapter 12).

Causes of Menopause

What causes menopause? Basically, the ovaries cease to respond to FSH and LH, possibly because their blood supply decreases. In addition, most of a woman's stock of germ cells in her ovaries is gone by this age. Because the ovaries degenerate, the production of estradiol-17β decreases to very low levels, and this results in some of the symptoms described above such as the shrinking of the vagina, external genitalia, breasts, and uterus. The adrenal glands and ovaries of postmenopausal women also secrete increased androgens;[41] and these hormones, in the absence of estrogens, cause some menopausal symptoms such as voice deepening, enlargement of the clitoris, and appearance of facial hair. It had long been thought that hot flashes were caused by the abrupt lowering of estradiol levels, but now we know another factor is that a woman's sympathetic nervous system is more active after menopause, and this causes the dilation of skin vessels and sweating, as well as a 1°C rise in body temperature and an increase in heart rate during the flashes.[42] Hour-to-hour changes in secretion of LH from the pituitary gland of postmenopausal women have also been associated with hot flashes.[43] Because estradiol levels are so low, there is little if any negative feedback on GnRH secretion, and the pituitary gland secretes ten times as much FSH and four times as much LH as in younger women. When these gonadotropins are extracted from the urine of postmenopausal women, the extract is called *human menopausal gonadotropin* (hMG).

Estrogen Therapy—Risks and Benefits

Most women have mild menopausal symptoms that do not require treatment. Those who do have difficult symptoms have frequently been given an estrogen orally or by injection. Today, about one-third of the more than 30 million postmenopausal women in the United States receive estrogen. The estrogens used (e.g., estrone sulfate) are more natural and not as potent as those used in the combination birth control pill (Chapter 13). Many women receiving estrogen do show remarkable disappearance of many physiological and psychological symptoms, although some of these effects of estrogen may be caused by psychological expectations that the treatment will work.[44]

There is increasing evidence, however, that this estrogen treatment could increase the chance of women developing endometrial cancer. In one study, 94 postmenopausal women with endometrial cancer were compared to twice as many women of the same age not having cancer.[45] Fifty-seven percent of the women with cancer had been given estrogen to relieve menopausal difficulties, whereas only 15 percent of those not having cancer had received estrogen. The risk of endometrial cancer appears to be 5.6 times greater if estrogen has been given for one to five years, and 13.9 times greater if given for greater than seven years. Remember, however, that the incidence of endometrial cancer, even in women receiving estrogen, is still only less than 1:10,000 women, and the increased risk of cancer is gone within six months after stopping estrogen treatment.[46] There are also distinct benefits of the estrogen treatment. Estrogen given to postmenopausal women virtually eliminates osteoporosis, whereas postmenopausal women suffer 750,000 fractures a year, and these bone fractures kill 58,000 a year.[47] Also, estrogen treatment lowers the risk of heart disease in postmenopausal women.[48] In the final analysis, a woman should discuss the potential effects associated with estrogen therapy with her doctor before considering such treatment.

CHAPTER SUMMARY

Many mammals exhibit seasonal reproduction with estrous behavior during the breeding season, but little evidence exists for this phenomenon in human females. Human females have spontaneous ovulation, but they also may exhibit induced ovulation in some situations. Daily light cycles have a great influence on estrous cycles of some mammals but not on the menstrual cycle of humans. Social interaction, however, can influence human menstrual cycles.

The human menstrual cycle, which usually is 25 to 30 days long, consists of menstrual, follicular, and luteal phases. Cyclic secretion of hormones from the hypothalamus (gonadotropin-releasing hormone), anterior pituitary gland (FSH, LH), and ovaries (estradiol-17β, progesterone) control the cycle, and negative and positive feedback play important roles. The mechanisms causing regression of the corpus luteum at the end of each cycle are not yet understood. Methods used to detect ovulation in the home or laboratory are not reliable in predicting when ovulation will occur.

Many women experience a premenstrual syndrome before menstruation, which is caused by an interaction of biological and psychological factors. Similarly, pain of menstruation (dysmenorrhea) is influenced by both biological and psychological factors. Toxic shock syndrome is associated with tampon use during menstruation. Missing menstruation (amenorrhea) can be caused by emotional stress or normal reproductive events during pregnancy and lactation. Past taboos about menstruating women still influence present feelings and behavior.

Menopause occurs as menstrual cycles cease when women reach 35 to 55 years of age. At this time, the ovaries stop producing estradiol-17β, the result being a syndrome of physical changes and possible emotional problems. Estrogen treatment relieves symptoms of menopause but may increase slightly the risk of endometrial cancer. This treatment, however, lowers the risk of bone fracture and heart disease in postmenopausal women.

NOTES

1. J. H. Clark and M. X. Zarrow, "Influence of Copulation on Time of Ovulation in Women," *American Journal of Obstetrics and Gynecology*, 109 (1971), 1083–85; and W. Jockle, "Current Research in Coitus-Induced Ovulation: A Review," *Journal of Reproduction and Fertility*, suppl. 22 (1975), pp. 165–207.

2. C. E. McCormack, "Acute Effects of Altered Photoperiod on the Onset of Ovulation in Gonadotropin-Treated Immature Rats," *Endocrinology*, 93 (1973), 403–10.

3. Anonymous, "Lunar Phases and Menstrual Cycles," *Research in Reproduction*, 12, no. 4 (1980).

4. M. K. McClintock, "Menstrual Synchrony and Suppression," *Nature,* 229 (1971), 244–46.

5. W. B. Cutler et al., "Sexual Behavior and Frequency and Menstrual Cycle Length in Mature Premenopausal Women," *Psychoneuroendocrinology*, 4 (1979), 297–309.

6. J. D. Palmer, "Human Rhythms," *Bioscience*, 27, no. 2 (1977), 93–99.

7. A. E. Treloar et al., "Variation of the Human Menstrual Cycle through Reproductive Life, "*International Journal of Fertility*, 12 (1967), 77.

8. H. Pepper and S. Lindsay, "Levels of Eosinophils, Platelets, Leukocytes, and 17-Hydroxycorticosteroids during Normal Menstrual Cycle," *Proceedings of the Society of Experimental Biology and Medicine*, 104 (1960), 145–47.

9. C. J. Pauerstein et al., "Temporal Relationships of Estrogen, Progesterone, and LH Levels to Ovulation in Women and Infrahuman Primates," *American Journal of Obstetrics and Gynecology*, 130 (1978), 876–86.

10. M. L. Marder et al., "Suppression of Serum Follicle Stimulating Hormone in Intact and Acutely Ovariectomized Rats by Porcine Follicular Fluid," *Endocrinology*, 101 (1977), 1639–47; J. C. Hoffman et al., "Selective Suppression of the Primary Surge of Follicle-Stimulating Hormone in the Rat: Further Evidence for Folliculostatin in Porcine Follicular Fluid," *Endocrinology*, 105 (1979), 200–203; and C. P. Channing et al., "Relationship between Human Follicular Fluid, Inhibin F Activity and Steroid Content," *Journal of Clinical Endocrinology and Metabolism,* 52, (1981), 1193.

11. R. B. Jaffee and W. R. Keye, Jr., "Estradiol Augmentation of Pituitary Responsiveness to Gonadotropin-Releasing Hormone in Women," *Journal of Clinical Endocrinology and Metabolism*, 39 (1974), 850–55.

12. G. Leyendecker et al., "Experimental Studies on the Endocrine Regulations during the Periovulatory Phase of the Human Menstrual Cycle," *Acta Endocrinologica, Copenhagen,* 71 (1972), 160–78.

13. M. Ferin et al., "Estrogen-Induced Gonadotropin Surges in Female Rhesus Monkeys after Pituitary Stalk Section," *Endocrinology*, 104 (1979), 50–52.

14. E. E. Wallach et al., "Serum Gonadotropin Responses to Estrogen and Progesterone in Recently Castrated Human Females," *Journal of Clinical Endocrinology and Metabolism*, 31 (1970), 376–81.

15. H. Friesen and P. Hwang, "Human Prolactin," *Annual Review of Medicine*, 24 (1973), 251.

16. E. W. Horton and N. L. Poyser, "Uterine Luteolytic Hormone: A Physiological Role for Prostaglandin $F_{2\alpha}$" *Physiological Reviews*, 56 (1976), 595–651.

17. Anonymous, "Prostaglandin Receptors in Human Corpora Lutea," *Research in Reproduction*, 9, no. 4 (1977), 4.

18. C. G. Beling et al., "Functional Activity of the Corpus Luteum Following Hysterectomy," *Journal of Clinical Endocrinology and Metabolism*, 30 (1970), 30–39.

19. F. W. Hanson et al., "Effects of HCG and Human Pituitary LH on Steroid Secretion and

Functional Life of the Human Corpus Luteum," *Journal of Clinical Endocrinology and Metabolism*, 32 (1971), 211–15.

20. S. Wardlaw et al., "The LH-hCG Receptor of the Human Ovary at Various Stages of the Menstrual Cycle," *Acta Endocrinologica*, 79 (1975), 568–76.

21. F. Hoffmann, "Untersuchungen uber die hormonale Beeinflussung der Lebensdauer des Corpus Luteum in Zyplus der Frau," *Geburtshilfe Frauenheilkd*, 20 (1960), 1153–59.

22. R. F. Casper and S. S. C. Yen, "Induction of Luteolysis in the Human with a Long-Lasting Analog of Luteinizing Hormone-Releasing Factor," *Science*, 205 (1979), 408–10.

23. B. M. Sherman and S. G. Korenman, "Hormonal Characteristics of the Human Menstrual Cycle throughout Reproductive Life," *Journal of Clinical Investigation*, 55 (1975), 699–706.

24. E. Odeblad, "The Functional Structure of Human Cervical Mucus," *Acta Obstetrics and Gynecology, Scandanavia*, 47, suppl. 1 (1968), 57–79.

25. G. Oster, "Protease Activity and Impending Ovulation," *Research in Reproduction*, 10 (1978), 3–4.

26. R. A. H. Kinch, "Help for Patients with Premenstrual Tension," *Consultant* (April 1979), pp. 187–91; and W. Herbert, "Premenstrual Changes," *Science News*, 122, no. 24 (1982), 380–81.

27. K. Dalton, *The Premenstrual Syndrome* (Springfield, Ill.: Charles C Thomas, Publisher, 1964).

28. J. Bardwick, "Her Body, the Battleground," *Psychology Today*, 5 (February 1972), 50–54, 76, 82.

29. M. P. Parlee, "The Premenstrual Syndrome," *Psychological Bulletin*, 80 (1973), 454–65; S. J. Hutt, "Perceptual-Motor Performance during the Menstrual Cycle," *Hormones and Behavior*, 14 (1980), 116–25; and Anonymous, "The Premenstrual Syndrome," *Research in Reproduction*, 14, no. 1 (1982), 1.

30. J. L. Marx, "Dysmenorrhea: Basic Research Leads to a Rational Therapy," *Science*, 205 (1979), 175–76.

31. Ibid.

32. W. H. Masters and V. E. Johnson, *Human Sexual Response* (Boston: Little, Brown & Company, 1966).

33. J. L. McCary, *McCary's Human Sexuality*, 3rd ed. (New York: Van Nostrand Reinhold Company, 1978).

34. M. Gold, "Toxic Shock," *Science 80*, 1, no. 2 (1980), 10; Anonymous, "Toxic Shock and Tampon Content," *Science News*, 121, no. 13 (1982), 219; and M. Harvey et al., "Toxic Shock and Tampons: Evaluation of the Epidemiologic Evidence," *Journal of the American Medical Association*, 248 (1982), 840–46.

35. Anonymous, "Toxic Shock Declines as Does Tampon Use," *Science News*, 119, no. 6 (1981), 85.

36. M. P. Warren and R. M. Vandewiele, "Clinical Features of Anorexia Nervosa," *American Journal of Obstetrics and Gynecology*, 117 (1973), 935.

37. Anonymous, "Delayed Menarche and Amenorrhea in Ballet Dancers," *Research in Reproduction*, 13, no. 1 (1981), 2–3; and C. B. Feicht et al., "Secondary Amenorrhea in Athletes," *Lancet*, i (1978), 1145.

38. R. E. Frisch, "Population, Food Intake, and Fertility," *Science*, 199 (1978), 22–30.

39. G. Wyshah, "Menopause in Mothers of Multiple Births and Mothers of Singletons Only," *Social Biology*, 25 (1978), 52–61.

40. Anonymous, "Smoking and the Age of Menopause," *Research in Reproduction*, 9, no. 4 (1977), 2.

41. F. E. Purifoy et al., "Steroid Hormones and Aging: Free Testosterone, Testosterone, and Androstenedione in Normal Females Aged 20–87 Years," *Human Biology*, 52 (1980), 181–91.

42. D. J. Hendrick et al., *British Medical Journal*, 2 (1978), 81.

43. R. F. Casper et al., "Menopausal Flushes: A Neuroendocrine Link with Pulsatile Luteinizing Hormone Secretion," *Science* 205 (1979), 823–25.

44. R. C. Strickler et al., "The Role of Oestrogen Replacement in the Climacteric Syndrome," *Canadian Psychological Medicine*, 7 (1977), 631–39.

45. H. K. Ziel and W. D. Finkle, "Increased Risk of Endometrial Carcinoma among Users of Conjugated Estrogens," *New England Journal of Medicine*, 293 (1975), 1167–70.

46. H. Jick, "Replacement Estrogens and Endometrial Cancer," *New England Journal of Medicine*, 300 (1979), 218–22.

47. R. Lindsay, "Bone Response to Termination of Estrogen Treatment," *Lancet,* i (1978), 1325–28.

48. E. Barrett-Conner et al., "Heart Disease Risk Factors and Hormone Use in Postmenopausal Women," *Journal of the American Medical Association*, 241, no. 20 (1979), 2167–69.

FURTHER READING

BARDWICK, J., "Her Body, the Battleground," *Psychology Today,* 5 (February 1972), 50–54, 76, 82.

DALTON, K., *The Premenstrual Syndrome.* Springfield, Ill.: Charles C Thomas, Publisher, 1964.

DELANEY, J., et al., *The Curse.* New York: E. P. Dutton & Co., Inc., 1976.

HERBERT, W., "Premenstrual Changes," *Science News*, 122, no. 24 (1982), 380–81.

LIEN, A., *The Cycling Female.* San Francisco: W. H. Freeman & Company, Publishers, 1979.

SEGAL, S. J., "The Physiology of Human Reproduction," *Scientific American*, 231 (1974), 53–62.

VINCENT, L. M., *Competing with the Sylph. Dancers and the Pursuit of the Ideal Body Form.* New York: Andrews and McMeel, Inc., 1979.

WEIDEGER, P., *Menstruation and Menopause.* New York: Alfred A. Knopf, Inc., 1976.

8 Gamete Transport and Fertilization

LEARNING GOALS

Having read this chapter, you should be able to:

1. Describe the three stages of semen release.
2. List the major constituents of seminal plasma, and discuss how each plays a role in sperm transport, maintenance, and maturation.
3. List the major structural regions of a sperm and the main function of each region.
4. Discuss sperm movement through the vagina, and factors in the vagina that influence sperm movement and survival.
5. Describe how the cervix serves as a barrier to sperm movement such that abnormal sperm are eliminated.
6. Describe how sperm move up the uterus and into the oviduct.
7. List factors that move the ovum from the ovary to the point of fertilization in the oviduct.
8. List the mechanisms that result in sperm moving up the oviduct while the ovum is moving down.
9. Describe how oviductal fluid is important in gamete transport and sperm maturation.
10. Define sperm capacitation and activation, and factors that may cause these phenomena.
11. Describe the acrosome reaction, and discuss how acrosomal enzymes are important in penetration of the ovum by a sperm.
12. Define ovum cortical granules, and discuss their importance in prevention of polyspermy.

13. Describe the processes occurring in the ovum and sperm after sperm penetration that result in formation of a zygote.
14. List chemicals that can block fertilization and their mechanism of action.
15. Describe how chromosomal sex is determined at fertilization.
16. Describe two possible ways a couple may be able to choose the sex of their baby.
17. Describe how identical and fraternal twins are formed.
18. Define parthenogenesis, and discuss the probability of its occurrence in humans.
19. Describe how a new individual could be formed by a procedure called cloning.

INTRODUCTION

The process of *fertilization* (or *conception*) involves fusion of the nucleus of a male gamete (sperm) and a female gamete (ovum) to form a new individual. Since each gamete is haploid (N), fertilization restores the normal diploid (2N) chromosomal complement. Fertilization, however, is more than simple fusion of gametes in that it is preceded by and requires a series of precisely timed events. Once sperm are deposited in the female reproductive tract, they travel a long distance and overcome several obstacles before reaching the ovum. Similarly, the ovum travels through a portion of the female reproductive tract before it is fertilized. Not only do the gametes move to the appropriate regions of the female tract, but they undergo important physical and biochemical maturations that are a prerequisite for fertilization. Abnormalities in these maturational and/or transport processes, as well as in fertilization itself, can lead to infertility, spontaneous abortion (miscarriage), or birth defects.

SEMEN RELEASE

After leaving the epididymides, sperm enter the vasa deferentia, which are long paired ducts serving as sperm storage and transport organs (Chapter 4). Secretions of the male sex accessory glands (seminal plasma) mix with the sperm during ejaculation to form semen or seminal fluid. It has been theorized that depletion of the entire reserve of sperm in the epididymides and vasa deferentia would occur if an adult male had 2.4 ejaculations per day for ten consecutive days. However, this normally does not occur, even with such herculean ejaculation frequency, because new sperm are continuously produced by the testes—about 200 million per day! Thus, frequent ejaculation is not an effective method of contraception.

Semen is released in three stages. Before male orgasm, a small amount of semen comes from the bulbourethral glands. In the second stage, the majority of semen is released; most of the seminal plasma of this stage comes from the seminal vesicles and prostate gland. In the third stage, another small amount of fluid, produced by the seminal vesicles, is exuded. Most of the sperm are expelled in the second stage. It is very important, however, to know that some sperm are present in the semen of the first and third stages. That sperm are present in the first stage means that pregnancy can occur without male orgasm, and this is one reason why *coitus interruptus* (withdrawal of the penis before ejaculation) is not as efficient a birth control method as some couples would hope (Chapter 13).

CONTENTS OF SEMINAL PLASMA

Seminal plasma contains several substances, but the precise function of many of these components is not known. We do know, however, that some of them have roles in the maintenance, maturation, and transport of sperm.[1] Water is present, which serves as a liquid vehicle for the sperm and seminal plasma constituents. There is also mucus from the bulbourethral glands, which serves as a lubricant for passage of semen through the male reproductive tract. The prostate gland and the bulbourethral glands both secrete buffers, which neutralize the acidity in the male urethra and in the vagina. Some nutrients for sperm are present in the seminal plasma deposited in the vagina, the major ones being the sugar fructose and citric acid (from the seminal vesicles). *Carnitine,* concentrated from the blood by the epididymis, is also in the seminal plasma. This chemical is involved in the metabolism of fatty acids, with the metabolites being used as another nutrient source for the sperm. Another constituent of seminal plasma secreted by the epididymis is *glycerylphosphocholine*. The enzyme *diesterase* in the uterus hydrolyzes (breaks down) this molecule, and the products of this digestion are used by the sperm as nutrients. Other enzymes secreted by the prostate gland and seminal vesicles are involved in the clotting and subsequent liquification of semen in the vagina. Finally, some kinds of prostaglandins are secreted into the seminal plasma, mostly by the seminal vesicles. Prostaglandins in seminal plasma may be involved in sperm transport. Table 8–1 summarizes the sources of the major components of seminal plasma.

SPERM NUMBER AND STRUCTURE

From 200 to 500 million sperm (about 120 million per ml of 2.5–3.5 ml of semen) are deposited at each ejaculation (Fig. 8–1). A male produces about 1 billion sperm for every ovum ovulated in a woman. Many ejaculated sperm (about 30 percent), however, are structurally or biochemically abnormal and are either dead or are incapable of fertilizing;[2] these are resorbed by the female reproductive tract or are lost through the vagina. For a male to be minimally fertile, his sperm count should be at least 20 million sperm/ml of semen; 40 percent of these sperm must swim, and 60 percent should be of normal shape and size (Table 8–1).

TABLE 8-1. SOME CHARACTERISTICS OF HUMAN SEMEN

General Properties

Creamy texture: gray to yellow color

Average volume: 2.5–3.5 ml after three days of abstinence (range = 2–6 ml)

Fertility index (minimum qualifications for male fertility):
1. At least 20 million sperm/ml.
2. At least 40% sperm must show vigorous swimming.
3. At least 60% sperm must have normal shape and size.

pH: 7.35–7.50 (slightly basic)

Major Components of Seminal Plasma			
Epididymis (a slight amount)	Seminal Vesicles (about 2/3 of total volume)	Prostate Gland (about 1/3 of total volume)	Bulbourethral Glands (a few drops)
Water	Water	Water	Water
Carnitine (a nutrient)	Fructose (a nutrient)	Bicarbonate buffers (neutralize vaginal pH)	Phosphate and bicarbonate buffers (neutralize vaginal pH)
Glycerylphosphor- choline (a nutrient)			
	Fibrinogen (clots semen)	Fibrinogenase (clots semen)	
	Ascorbic acid (a nutrient)	Fibrinolytic enzyme (liquifies semen)	Mucus (lubrication)
	Citric acid (a nutrient)		
	Most of the prosta- glandins (contract the vas deferens)	A little prosta- glandin	

Source: Adapted from B. Goldstein, *Human Sexuality* (New York: McGraw-Hill Book Company, 1976), p. 20. © 1976; used with permission.

A healthy human sperm is 40 to 250 μm long and is composed of the following structures (Fig. 8–2): First is the sperm head, which consists of an elongated haploid nucleus surrounded by a nuclear membrane, an *inner acrosomal membrane,* and an *outer acrosomal membrane.* The gap between the latter two membranes (*acrosomal space*) is filled with enzymes important in penetration of the ovum by a sperm. An external *sperm plasma membrane* covers the sperm head. Everything in the sperm head outside of the nuclear membrane is called the *acrosome* (Greek for "sharp body"). Next is the *sperm neck,* then the *sperm midpiece,* which contains mitochondria that produce energy for tail movement, and finally, the sperm tail.

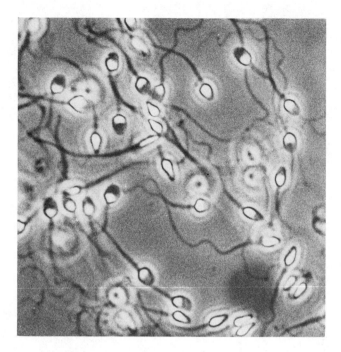

Figure 8–1. Photomicrograph of human sperm swimming in seminal fluid. Courtesy of Dr. R. Yanagimachi.

SPERM TRANSPORT AND MATURATION IN THE FEMALE REPRODUCTIVE TRACT

Let us now follow the sperm on their journey through the female reproductive tract to the point of fertilization in the oviduct, a distance of about 15 cm (6 in.). The sperm are first deposited in the vagina; they then pass up this cavity to and through the cervix into the uterus, up the uterus, through the junction between the uterus and oviduct (uterotubal junction), and up the isthmus of the oviduct to the usual area of fertilization in the oviduct—the ampullary-isthmic junction. Many of the millions of deposited sperm are lost during this journey, and only about 100 to 1000 of them reach the oviduct. In addition, the sperm must undergo maturational processes during their journey, which give them the capacity to move and to fertilize an ovum.

Vaginal Sperm

About one minute after deposition in the vagina, the semen becomes thicker and less liquid. This *semen coagulation* is brought about by the enzyme *fibrinogenase* in the seminal plasma that converts the protein *fibrinogen* to *fibrin,* another protein. The

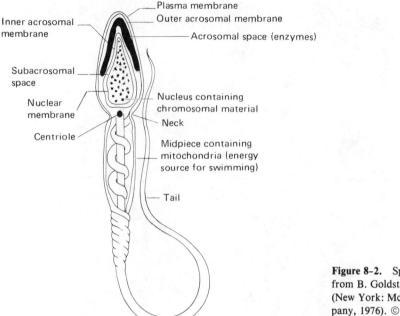

Inner acrosomal membrane

Plasma membrane

Outer acrosomal membrane

Acrosomal space (enzymes)

Subacrosomal space

Nuclear membrane

Centriole

Nucleus containing chromosomal material

Neck

Midpiece containing mitochondria (energy source for swimming)

Tail

Figure 8-2. Sperm structure. Adapted from B. Goldstein, *Human Sexuality* (New York: McGraw-Hill Book Company, 1976). © 1976; used with permission.

major function of this coagulation may be prevention of sperm loss from the vagina. After about 20 minutes, however, the semen again liquifies. This *semen liquification,* which is caused by a *fibrinolytic enzyme* in the seminal plasma, allows the sperm to move and begin their journey to the oviduct.

The environment in the vagina is usually acidic (about pH 4.2), and this level of acidity inhibits sperm motility. The presence of semen in the vagina, however, increases the vaginal pH to a basic 7.2, which in turn increases sperm motility.

During coitus, female orgasm is accompanied by muscular contraction of the vaginal walls (Chapter 17), and this creates a pressure in the vagina that is higher than that in the uterus. Sperm movement through the cervix may be aided by this pressure differential.[3] Sperm, however, can move up the female tract without female orgasm. Thus, the idea that pregnancy cannot occur unless a woman has an orgasm is a myth.

Cervical Sperm

The cervical canal is lined by a complicated series of narrow folds and crypts and is blocked by a sticky mass of cervical mucus and tiny cervical fibers (Chapter 3). In most stages of the menstrual cycle, the mucus is thick and the fibers within it are densely packed. Shortly before ovulation, however, the rise in circulating estrogen levels causes the mucus to become more liquid and the gaps between the cervical fibers to widen.[4] These gaps orient so that channels are formed. When the sperm enter the mucus, they line up in these channels almost in single file and pass through the cervix at a speed of about 1.2–3.0 mm per minute.

The cervical fibers may serve as a network upon which the sperm tails exert force, thus propelling the sperm upward. Also, these fibers may be of such dimension and length that they vibrate in rhythm with the tail beat frequency of normal sperm; this may allow normal sperm to move through the cervix, whereas sperm with abnormal or absent tail beats are detained.[5] These latter sperm then die and are resorbed or lost from the body. Other sperm enter *cervical crypts* (deep recesses in the cervical wall), where they die or are lost, or they may remain alive as a reservoir of sperm that later may enter the uterus. Only about 1 million of the original 200 to 500 million sperm make it through the cervix.

Uterine Sperm

Upon leaving the cervix, the sperm travel up the uterus to the uterotubal junction. The uterine fluid is watery but sparse in humans, and the sperm essentially "climb" up the uterine lumen by beating their tails. The swimming rate of sperm (about 3 mm/min), however, cannot account for their traveling a distance of about 15 cm in 30 minutes after ejaculation. In a similar manner, inert particles placed into the vagina of sheep arrive in the oviduct in less than 15 minutes. Also, dead sperm reach the oviduct at about the same time as do live sperm in at least some laboratory animals.[6] Thus, sperm tail-beating probably is not important during sperm transport through the uterus; so it must be the muscle contraction in the female reproductive tract that facilitates sperm transport.

Mechanical stimulation of the cervix by the penis causes release of the hormone oxytocin from the posterior pituitary gland.[7] This hormone quickly travels via the blood to the uterus and increases the force of rhythmic uterine muscle contractions. These contractions act as waves to help the sperm move to the uterotubal junction. Prostaglandins in the seminal fluid also may cause uterine muscles to contract, but this is unlikely since very little if any seminal fluid enters the uterus through the cervix. The main function of the prostaglandins in seminal fluid is probably to contract the muscles of the vasa deferentia, thus aiding in sperm passage during ejaculation.

The presence of sperm in the uterus initiates a massive invasion of white blood cells (*leucocytes*) into the uterine lumen. These cells then begin to engulf the dead or dying sperm that have not as yet moved up to the uterotubal junction. No more than a few thousand sperm reach this junction.

The uterotubal junction is a muscular, tightly constricted region separating the uterus from the oviduct (Chapter 3). Sperm enter the narrow opening of this junction and move through it at a relatively slow rate. Thus, the uterotubal junction allows gradual entrance of sperm into the isthmus of the oviduct.

TRANSPORT OF THE SPERM AND OVUM
IN THE OVIDUCT

Once entering the isthmus, several sperm meet the ovum, and fertilization by a single sperm usually occurs at the point where the isthmus joins the wider oviductal ampulla (ampullary-isthmic junction). Other sperm swim up the ampulla, through the infundibulum, and are lost in the body cavity.

Once ovulation has occurred, the infundibulum (funnel-shaped free end) of the oviduct moves to the ovary and envelopes the ovulated ovum along with fluid derived from the ovulated follicle. Movement of the infundibulum in some mammals is accomplished by contraction of muscles in the membrane supporting the oviduct. Cilia are present in the wall of the fimbria (the edge of the infundibulum), and these beat toward the uterus. Thus, when the infundibulum envelops the ovary, the beating of the cilia moves the ovum into the ampulla of the oviduct. Cilia in the ampulla and isthmus of the oviduct also beat in a uterine direction, and this sets up a flow of fluid toward the uterus.

The muscles of the rabbit oviduct also exhibit waves of muscular contraction after ovulation. These waves travel in the direction of the uterus, and along with the cilia help the oocyte move down the oviduct.[8] Both ciliary beating and muscular contraction in the rabbit oviduct are influenced by ovarian sex hormones. Estrogens increase cilia number, and progesterone increases ciliary beating and egg transport.[9]

The human oviduct also has cilia and muscles that contract, but their precise role in ovum transport is not clear.[10] Movement of the human ovum through the ampulla is rapid, but it takes 2 to 2.5 days for the ovum to move through the ampullary-isthmic junction (Fig. 8–3).

In order to be fertilized, an ovum must be penetrated by a sperm within 24 to 48 hours after ovulation. A sperm can be up to 72 hours old before its fertilizing ability declines. However, it is possible that sperm stored in the cervical crypts remain viable a good deal longer than 72 hours. This, in fact, could explain why a female could conceive after coitus during menstruation, especially if her follicular phase is short.

An interesting transport problem exists in the oviduct in that the ovum moves in an direction opposite to sperm movement. One theory suggests that oviductal fluid moves in a primary current toward the uterotubal junction because of the beating of oviductal cilia and muscular contraction. When this current reaches the tight uterotubal junction, however, little fluid gets through, and a back current is created in the isthmus. It is this back current that helps sperm swim up the isthmus. Once the fluid reaches the wider oviductal ampulla, the force of the back current is reduced enough so that the primary current dominates and carries the ovum to the region of the ampullary-isthmic junction. Equalization of current flowing in opposite directions at the ampullary-isthmic junction may explain why the oocyte and sperm remain at this position for a day or two.[11]

Another factor involved in the opposite movement of egg and sperm may be the direction of ciliary beating in the oviduct. Oviductal cilia exist in deep recesses in which cilia beat toward the ovary and on ridges where these cilia beat toward the

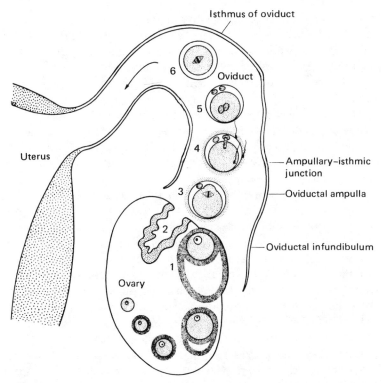

Figure 8–3. Diagram of the human ovary, oviduct, and part of uterus, showing fertilization: (1) Follicle in ovary ready to ovulate; (2) new corpus luteum; (3) ovulated ovum arrested in second meiotic division (note the first polar body); (4) formation of second polar body after fertilization; (5) fusion of ootid and sperm pronuclei; (6) beginning of first mitotic division of zygote. Adapted from C. R. Austin, "Fertilization," in *Reproduction in Mammals, Book 1, Germ Cells and Fertilization,* eds. C. R. Austin and R. V. Short (Cambridge, England: Cambridge University Press, 1972). © 1972; used with permission.

uterus.[12] Sperm may travel in these recesses, whereas the ovum may be propelled along the ridges. The presence of considerable amounts of mucus in the oviducts for three to four days after ovulation may serve as a medium for sperm transport. This mucus is gone when the fertilized ovum (embryo) travels down the oviduct to the uterus, as we discuss in the next chapter.

SPERM CAPACITATION AND ACTIVATION

Freshly ejaculated human sperm are not capable of fertilization. A period of an hour or more in the female reproductive tract is necessary before sperm can fertilize an oocyte. Thus, sometime during their journey and usually prior to arrival at the oviduct, sperm gain the ability to move rapidly (a process called *sperm activation*) and to fertilize an egg (a process called *sperm capacitation*). Little is known about sperm activation and capacitation in humans, and the discussion that follows is based mainly on research using laboratory mammals.[13]

In general, the present scientific opinion is that capacitation involves removal or modification of molecules (glycoproteins) associated with the sperm head that stabilize the sperm plasma membrane. Capacitation probably occurs in the uterus, and takes less than five hours in humans.[14] One aspect of capacitation may involve removal of a "decapacitation factor" that adheres to the sperm head. This decapacitation factor suppresses the ability of sperm to fertilize. Removal of this factor from the sperm head probably occurs in the uterus.

Certain substances in the oviductal fluid as well as in the uterus may render sperm capable of fertilization.[15] For example, antral fluid escaping from the ovulating follicle makes up a small part of the oviductal fluid (Chapter 3). This antral fluid contains the protein *albumin,* which makes the sperm capable of undergoing the acrosome reaction, an important step in the process of fertilization as we see next. Another substance in follicular fluid causes sperm activation and increases the vigor of tail-beating. It is not clear if these substances are present in humans, although follicular fluid does activate human sperm.[16] If operative in humans, they may have their effect when the sperm penetrates the cumulus oophorus (see below), which surrounds the ovum and is bathed in follicular fluid. *Calmodulin,* a protein in seminal plasma, may also play a role in sperm capacitation.[17] This protein (or another epididymidal secretion) may give the sperm the ability to be capacitated while they are still in the epididymis.[18]

THE PROCESS OF FERTILIZATION

Once a sperm and ovum are in the region of the ampullary-isthmic junction of the oviduct (Fig. 8-3), fertilization occurs. There is no evidence in humans that a chemical produced by the ovum attracts sperm. Instead, many sperm reach the ovum by bumping into it by chance. In the fertilization process, a sperm first penetrates between the cells constituting the cumulus oophorus, then between the cells of the corona radiata, and finally through the zona pellucida and into the pervitelline space. The sperm then enters the oocyte through its cell membrane (the vitelline membrane). The following is a discussion of what happens during each of these processes, and Figs. 8-4, 8-5, and 8-6 depict these processes. This information is based mainly on studies of ova and sperm of sea urchins and laboratory mammals, but the mechanisms are probably similar in humans.[19] The entire process of fertilization takes about 24 hours.

Sperm Passage through the Cumulus Oophorus

The ovulated ovum is surrounded by the cumulus oophorus, which is a sphere of loosely packed follicle cells (Fig. 8-4). Appropriately, cumulus oophorus means

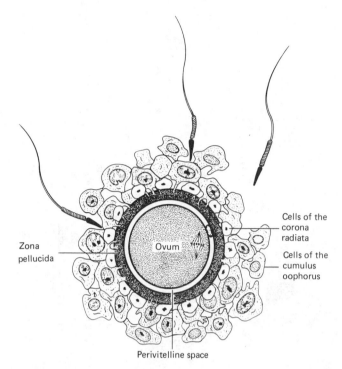

Zona
pellucida

Ovum

Cells of the
corona
radiata

Cells of the
cumulus
oophorus

Perivitelline space

Figure 8-4. Illustration of the barriers around the recently ovulated ovum through which the capacitated sperm must pass to reach the perivitelline space and achieve activation and fertilization of the ovum. Adapted from *Reproduction and Human Welfare: A Challenge to Research* by R. O. Greep, M. A. Koblinsky, and F. S. Jaffe (Cambridge, Mass.: M.I.T. Press, 1976). Used by permission of the M.I.T. Press. © 1976 by the Ford Foundation.

"egg-bearing little cloud." As a sperm enters the cumulus oophorus, or perhaps before,[20] it undergoes the *acrosome reaction* (Fig. 8-6). Its outer acrosomal membrane and plasma membrane fuse at discrete points to form a *composite membrane* containing many small openings. Several enzymes contained in the acrosomal space are released through these openings, and the enzymes help the sperm penetrate the layers surrounding the ovum. One of these enzymes is *hyaluronidase*. Hyaluronic acid is a major component of the cementing material found between the cells of the cumulus oophorus as well as between other cells in the body. Enzymatic dissolution of hyaluronic acid allows the swimming sperm to penetrate the cumulus oophorus and to reach the corona radiata. Electron micrographs of sperm penetration suggest that the cells of the cumulus oophorus actually may attach to sperm and pull them through.

Sperm Passage through the Corona Radiata

The corona radiata ("circular crown") is a sphere of tightly packed follicle cells attached to the zona pellucida of the ovum. The acrosome releases a *corona-penetrating enzyme,* which dissolves the material between the adjacent corona cells. As the sperm passes through the corona, the composite membrane of the sperm head is discarded. The inner acrosomal membrane of the sperm head is now ready to contact the zona pellucida.

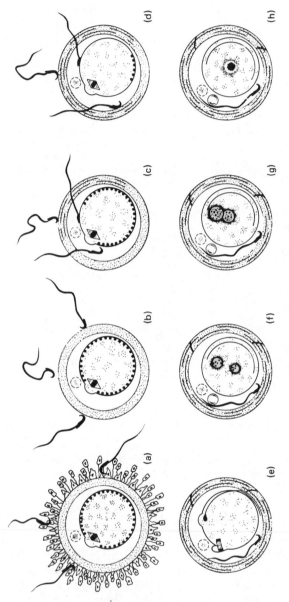

Figure 8–5. Stages of fertilization: (a) Sperm pass through the cumulus oophorus and corona radiata. (b) Sperm reach the zona pellucida and penetrate this layer to reach the perivitelline space. (c) Note that, for the purposes of clarity, the cumulus oophorus and corona radiata cells are not included in diagrams (b)–(h). (d) The sperm head attaches sideways to the vitelline membrane, and the cortical granules in the ovum (black dots) begin to release their secretions into the perivitelline space. The latter event occurs all around the ovum and prevents polyspermy. (e) The sperm nucleus sinks into the ooplasm, and this activates the ovum to complete the second meiotic division, resulting in (f), extrusion of the second polar body. (f) Haploid sperm and ovum pronuclei then form (g). Fusion of pronuclei occurs (h); and a zygote (2N) is now ready to divide to form a two-celled embryo. Adapted from M. C. Chang and C. R. Austin, "Fertilization," *Research in Reproduction,* vol. 8, no. 4 (1976), map. Used by permission of the International Planned Parenthood Federation and Drs. M. C. Chang and C. R. Austin.

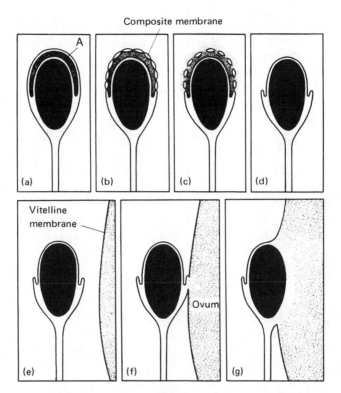

Figure 8-6. Diagram showing the acrosome reaction (above) and the first steps in sperm-ovum fusion (below). (a) The outline of the sperm is the plasma membrane; the nucleus is solid black; A = acrosome contents. (b) Outer acrosomal membrane and plasma membrane fuse at discrete points, which results in openings at these points of fusion. (c) Acrosomal enzymes are released through these openings. (d) The composite membrane (outer acrosomal *plus* plasma membrane) is discarded. (e) Sperm head approaches ovum vitelline membrane. (f) Sperm head attaches sideways to vitelline membrane. (g) Inner acrosomal membrane and vitelline membrane fuse, a large opening develops, and the sperm nucleus sinks into the ooplasm. Adapted from C. R. Austin, "Fertilization," in *Reproduction in Mammals, Book 1, Germ Cells and Fertilization,* eds. C. R. Austin and R. V. Short (Cambridge, England: Cambridge University Press, 1972). © 1972; used with permission.

Sperm Passage through the Zona Pellucida

The zona pellucida has specific receptor sites that attach to the inner acrosomal membrane of the sperm head. At contact, the acrosome of the sperm head releases two enzymes that dissolve constituents of the zona. One of these is *acrosin,* which is a *protease* that hydrolyzes (breaks down) the protein of the zona.[21] The other enzyme, *neuramidase,* breaks down neuraminic acid, a complex carbohydrate present in the zona pellucida. There is a paradox here. These two enzymes are present in the acrosome, which has been discarded by the time these enzymes are used. Then how is it that they still are present? Perhaps the enzymes adhere to the inner acrosomal membrane. Some scientists believe that the sperm penetrates the zona mainly due to its tail movement, and that acrosomal enzymes play only a minor role. Regardless of the mechanisms involved, once the sperm has penetrated the zona pellucida, it moves through a narrow, oblique path into the *perivitelline space* (the area between the zona pellucida and vitelline membrane); see Fig. 8–4. Penetration of the human zona pellucida by a sperm takes less than 10 minutes under experimental conditions.

Sperm Attachment to the Vitelline Membrane

The inner acrosomal membrane of the sperm now attaches to specific receptor sites on the surface of the vitelline (ovum) membrane. The sperm attaches sideways instead of head-on, and part of the inner acrosomal membrane fuses with the ovum. An opening then appears, and the sperm nucleus sinks into the ovum cytoplasm (Fig. 8–6). The sperm neck and midpiece enter with the sperm nucleus, whereas the tail remains in the perivitelline space.

Dissolution of Ovum Cortical Granules

Cortical granules are contained in small, membrane-bound *cortical vesicles* that lie along the outer margins of the ovum (Fig. 8–5). When the sperm head attaches to the ovum, the membrane of each cortical vesicle fuses with the vitelline membrane, an opening develops to the perivitelline space, and the contents of the cortical granules are expelled into this space. The cortical granules contain substances that prevent more than one sperm from penetrating the egg.

Completion of Second Meiotic Division of Ovum

Once the cortical granules are released, *ovum activation* has commenced. The single set of chromosomes of the ovulated ovum (secondary oocyte) is arrested in the second meiotic division (Chapter 3). Penetration of the sperm head into the ovum cytoplasm activates the ovum nucleus so that it completes the second meiotic division, and a second polar body is produced. The haploid ovum is now called an ootid (Fig. 8–7). The second polar body degenerates, sometimes dividing before it dies.

Formation and Fusion of Sperm and Ootid Pronuclei

Soon after the sperm nucleus enters the ovum, its chromosomes collect to form a *sperm pronucleus,* and the chromosomes of the ootid nucleus collect to form an *ootid pronucleus.* The two pronuclei now become positioned close to each other near the center of the ovum. Their nucleoli disappear, and the chromosomes condense and aggregate. Next, the chromosomes mingle and begin a mitotic division (Fig. 8–7).

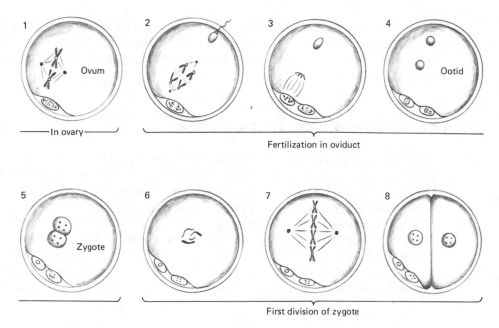

Figure 8–7. The nucleus of the ovulated ovum is haploid, and its chromosomes are arrested in the second meiotic division (1). The first polar body divides into two small cells (1), one of which is pictured in further figures. Sperm penetration activates the ovum so that the second meiotic division is completed (2, 3), and an ootid and second polar body are formed (4). The ootid and sperm pronuclei then fuse (5), and the resultant diploid zygote now divides mitotically (6, 7) to form a two-cell embryo (8) consisting of two blastomeres.

The fertilized cell (*zygote*) is now diploid (46 chromosomes), and conception of a new individual has occurred. In mammals, it takes about 12 hours from the beginning of ovum activation to pronuclear fusion. The zygote next divides mitotically, and two identical daughter cells, called *blastomeres,* are formed (Fig. 8–7); embryonic development has commenced.

PREVENTION OF POLYSPERMY

Many sperm reach the ovum, but usually only one fertilizes it. It is fatal to the embryo if more than one sperm penetrates the oocyte, an event called *polyspermy*. How is polyspermy prevented? There are several answers to this question.[22] After the cortical granules release their products, the structural appearance and chemical properties of the zona pellucida are altered (the *zona reaction*). In humans, it appears that cortical granules contain an *acrosin inhibitor.* Acrosin is one enzyme contained in the sperm acrosome that dissolves the zona pellucida. Also, the granules may produce a protease that destroys zona pellucida receptor sites for the sperm head. In rare cases, the mechanisms that usually prevent polyspermy fail, and embryos are formed with more than the normal number of chromosomes in their cells. The various kinds of fertilization errors and chromosomal disorders, and their effects on development, are discussed in Chapter 12.

CHEMICAL INHIBITION OF FERTILIZATION

In the future, it may be possible to block fertilization by interfering with steps in the fertilization reaction. In some laboratory mammals, for example, antibodies made against hyaluronidase block fertilization when administered to the female. Antibodies made against ovarian or testicular tissue also can block fertilization in some species.[23] In hamsters, *trypsin* (an enzyme from the pancreas) blocks fertilization, probably by mimicking the action of the protease secreted by the cortical granules of the oocyte. Finally, plant proteins called *lectins* bind to sugar molecules in the zona pellucida and prevent sperm penetration. Whether or not these methods can be used as effective contraceptive methods in humans (Chapter 13) remains to be determined.

SEX RATIOS

As discussed in Chapter 5, the normal chromosome number in humans is 46 (2N, diploid). Females have 22 pairs of autosomes and two X chromosomes. Males have 22 pairs of autosomes and an X and Y chromosome. The genes for male sex determination are carried on the Y chromosome. Thus, embryos without a Y chromosome are female.

As a result of meiosis in the adult testis, one diploid male germ cell (spermatogonium) gives rise to four haploid spermatozoa (Chapter 4). Two of these spermatozoa will have 22 autosomes and a Y chromosome, whereas the other two will have 22 autosomes and an X chromosome. If one of the former fertilizes an ovum (22 autosomes and an X chromosome), the embryo will be male; if one of the latter fertilizes an ovum, the offspring will be female. Thus, given an equal chance of X and Y sperm to fertilize, the sex ratio of embryos should be 100:100 (Fig. 8–8). However, the ratio of male to female embryos at conception (the *primary sex ratio*) usually is stated to be about 160:100, although new evidence suggests that the primary sex ratio is closer to 120:100.[24] This latter ratio is based on sexes of early aborted embryos. It is assumed that this means a greater fertilization rate by Y sperm than X sperm, perhaps because Y sperm are lighter and are faster swimmers than X sperm. However, female embryos may die more frequently at an earlier age than male embryos, or more X sperm may die in the female reproductive tract than Y sperm. The sex ratio of male births to female births (the *secondary sex ratio*) is 105:100. Thus, for reasons not yet understood, male fetuses suffer a greater mortality than female fetuses in the uterus.

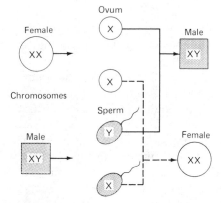

Figure 8–8. The chromosomal basis for the existence of a theoretically equal number of X and Y sperm, and thus a theoretical primary sex ratio of 100:100. As discussed in the text, this theoretical ratio is not borne out, and more embryos are male than female.

Myths about Sex Ratios

Several beliefs have existed about the numbers of male and female babies born in different circumstances. Examples of these are: (1) more males are born during or after a war; (2) more females are born to older parents; (3) first children of couples are more likely to be female, whereas subsequent children are more frequently male; (4) more males are born to parents of higher socio-economic class; (5) black couples tend to have more male babies; (6) more males are born in the spring; and (7) more males occur after disasters such as floods or plague. There is no sound scientific evidence, however, to support any of these ideas.[25]

SEX PRESELECTION

It now appears that couples who desire to do so may be able to choose the sex of their baby. This has been done for a few couples by collecting semen from the father, separating Y sperm and X sperm in the laboratory, and then selectively implanting a certain sperm type in the female's vagina or uterus to fertilize her oocyte at the proper time.

At present, methods for effectively separating human X and Y sperm are in the preliminary stages of success. In one method, sperm are placed in a test tube containing an appropriate medium (albumin). The Y sperm, being lighter and better swimmers, reach the bottom of the tube faster and thus can be separated to some degree from the X sperm. This method has been successful in some laboratories but not in others. Even when successful, however, the Y sperm are not totally separated from the X sperm.

Recently, human serum albumin density gradients have been used to recover Y sperm, successful fertilization followed, and 70 percent of the babies born were male. This percentage was similar to that of Y sperm in the bottom of the test tube. Present efforts are directed toward obtaining better separation of sperm types and isolating X as well as Y sperm.[26] A staining method (using the dye quinacrine) for living cells detects presence of the Y chromosome (Chapter 5), and this may help in sperm separation techniques.

L. B. Shettles has advocated a technique for choosing offspring sex that has been widely publicized.[27] The rationale behind this method is that the Y sperm are faster swimmers, but their fertilizing capability does not last as long as that of X sperm. According to Shettles, to have a boy 80 percent of the time, a woman should perform an alkaline douche (using two tablespoons of baking soda in one quart of water) prior to coitus. This reduces vaginal acidity and favors survival of male sperm. Deep penile penetration should be achieved during ejaculation, and the woman should have orgasm. These factors favor migration of the shorter-lived male sperm. Finally, the couple should abstain from coitus until just after her time of ovulation, abstinence increasing the number of male sperm in the ejaculate. In contrast, to have a girl 80 percent of the time, the douche should be acidic (two tablespoons of distilled white vinegar in one quart of water). There should be only shallow penile penetration at ejaculation, the woman should not have orgasm, and coitus can occur any time except a day or so before ovulation, and then cease. The reliability of Shettles' method has been contested by some, and there also is some evidence that fertilization early in the menstrual cycle produces more male than female offspring.[28]

MULTIPLE EMBRYOS

Twins occur in about 1 of every 80 or 90 pregnancies. When two ova are released and each is fertilized by a different sperm, *fraternal twins* are produced. These twins are *dizygotic* (the products of two different zygotes) and can be of the same or different sex. Fraternal twins, which are *nonidentical* and are as different from each other as are nontwin brothers and sisters, account for two-thirds of all twins. The incidence of dizygotic twins is influenced by race and inherited factors from the mother (not the father). Fraternal twins are more common in older mothers.

Identical twins usually occur when an early embryo (inner cell mass; Chapter 9) divides into two. These twins are *monozygotic* (derived from one zygote) and are genetically identical. The incidence of identical twins is not related to race, inheritance, or age of the mother. Rarely, identical twins are *conjoined* (fail to separate completely during embryonic development). These are called *Siamese twins*. The first publicized Siamese twins were "Chang" and "Eng" (1811-1874), born in Siam but of Chinese extraction. They were united at the chest by a thick mass of flesh. Some Siamese twins have been surgically separated after birth.

When the number of embryos is greater than two (i.e., triplets, quadruplets, etc.), all are usually of multizygotic origin; in a few cases, some are multizygotic and some are monozygotic.

PARTHENOGENESIS

Is it possible that an embryo can develop in a human female without previous fertilization? Embryonic development from an ovum not previously stimulated or penetrated by a sperm is called *parthenogenesis*. Such "virgin birth" is common in many insects, in some fish, amphibians, and reptiles, and in a strain of domestic turkeys. In addition, parthenogenetic mouse embryos can be produced in the laboratory, but they do not develop to term.[29] If parthenogenesis could occur in humans, reduction division in the oocyte must not occur, the offspring would always be female, and the child would be genetically identical to the mother.

CLONING

There is speculation that in the future new human beings could be produced by a process of *cloning*. Scientists have succeeded in removing the male and female pronuclei from a fertilized mouse ovum. These are replaced with a nucleus obtained from a cell

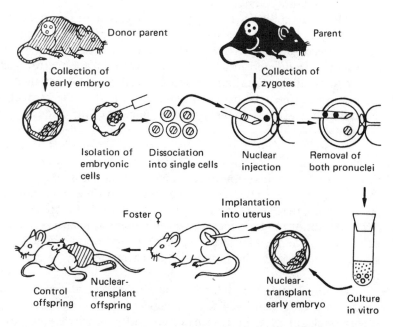

Figure 8–9. Diagram of the process by which mice have been "cloned" using a nucleus from a mouse embryo. From K. Illmensee and P. C. Hoppe, "Nuclear Transplantation in *Mus musculus:* Developmental Potential of Nuclei from Preimplantation Embryos," *Cell,* 23 (1981), 9–18. Adapted with permission of the M.I.T. Press, © 1981, and Dr. K. Illmensee.

of a mouse embryo. The "new" cell then is implanted in the uterus of an adult mouse, where it develops to term as a copy of the donor embryo mouse (Fig. 8–9).[30] This is not "true" cloning, however, which would involve transplantation of a nucleus from an *adult* mouse cell into an enucleated ovum, which would then develop into an adult. Complete cloned copies of frogs and carrots have been obtained using the latter technique, and only the future will reveal if a "human copy" can be obtained in this manner. Thus far, all claims that certain individuals were the products of human cloning have been revealed to be hoaxes.[31] If human cloning becomes possible, research in this area would best be preceded and accompanied by thorough discussions about its moral and ethical repercussions.

CHAPTER SUMMARY

After sperm mature in the epididymides, they move down the vasa deferentia. Seminal plasma consists of secretions from male sex accessory glands. These secretions are added to the sperm to form semen or seminal fluid, which leaves the male urethra during ejaculation. Seminal plasma contains substances necessary for sperm movement, maturation, and maintenance.

About 120 million sperm are present in each ml of semen. Some of these sperm are abnormal and die. A healthy sperm is made up of a head (nucleus plus acrosome), neck, midpiece, and tail. After insemination of the female, the sperm move through the vagina, uterus, and into the oviduct. While in the uterus, sperm acquire the ability to fertilize (capacitation) and are activated so that their tails beat more rapidly. Meanwhile, the ovulated ovum moves down the oviduct, and the sperm and ovum meet at the ampullary-isthmic junction of the oviduct, where fertilization occurs.

Before penetrating the ovum, a sperm moves first through the cumulus oophorus, then the corona radiata, zona pellucida, and vitelline membrane. As it penetrates the cumulus oophorus, it undergoes the acrosome reaction, during which the sperm acrosome releases enzymes that help dissolve chemical barriers to its passage. Once the sperm enters the ovum, it causes the completion of oocyte meiosis (ovum activation) and stimulates changes in the zona pellucida that act as a barrier to polyspermy. The haploid sperm pronucleus and ootid pronucleus then fuse, and a zygote is formed. In the future, certain chemicals may be used to block fertilization.

Chromosomal sex is determined at fertilization, and couples may in the future be able to choose their baby's sex. Identical twins (monozygotic twins) are formed when a single sperm fertilizes a single ovum, after which the embryo divides into two. Fraternal twins (dizygotic twins) are formed by fertilization of two separate eggs and sperm. Although several lower animals have offspring without fertilization (parthenogenesis), this has not occurred in humans. Attempts to produce new individuals by nuclear transplantation (cloning) have been successful in some organisms, but not in humans.

NOTES

1. T. Mann, "Semen: Metabolism, Antigenicity, Storage and Artificial Insemination," in *Frontiers in Reproduction and Fertility Control,* eds. R. O. Greep and M. A. Koblinsky, pp. 427-33 (Cambridge, Mass.: M.I.T. Press, 1977).

2. W. W. Williams, "Semen Variability," *Medical Aspects of Human Sexuality* (September 1974), pp. 67-73.

3. E. S. E. Hafez, "Gamete Transport," in *Human Reproduction: Conception and Contraception,* eds. E. S. E. Hafez and T. M. Evans, pp. 85-118 (New York: Harper & Row, Publishers, Inc., 1973).

4. E. Odeblad, "The Functional Structure of Human Cervical Mucus," *Acta Obstetrics and Gynecology, Scandinavia,* 47, suppl. 1 (1968), 57-79.

5. M. Dubois et al., "Spermatozoa Motility in Human Cervical Mucus," *Nature,* 252 (1974), 711-13.

6. D. W. Bishop, "Sperm Physiology in Relation to the Oviduct," in *The Mammalian Oviduct,* eds. E. S. E. Hafez and R. J. Blandau, pp. 231-50 (Chicago: University of Chicago Press, 1969).

7. C. A. Fox and G. S. Knaggs, "Milk Ejection Activity (Oxytocin) in Peripheral Venous Blood in Man during Lactation and in Association with Coitus," *Journal of Endocrinology and Metabolism,* 45 (1969), 145-46.

8. S. A. Halbert et al., "Egg Transport in the Rabbit Oviduct: The Roles of Cilia and Muscle," *Science,* 191 (1976), 1052-53.

9. M. C. Chang, "Effects of Oral Administration of Medoxyprogesterone Acetate and Ethinyl Estradiol on the Transportation and Development of Rabbit Eggs," *Endocrinology,* 79 (1966), 939-48.

10. A. Ferenczy et al., "Scanning Electron Microscopy of the Human Fallopian Tube," *Science,* 175 (1972), 783-84.

11. Hafez (1973), op. cit.

12. Ferenczy et al. (1972), op. cit.

13. M. C. Chang et al., "Capacitation of Spermatozoa and Fertilization," in *Frontiers in Reproduction and Fertility Control,* eds. R. O. Greep and M. A. Koblinsky, pp. 434-51 (Cambridge, Mass.: M.I.T. Press, 1977).

14. A. Lopata et al., *"In Vitro* Fertilization of Preovulatory Human Eggs," *Journal of Reproduction and Fertility,* 52 (1978), 339-42.

15. R. Yanagimachi, *"In Vitro* Acrosome Reaction and Capacitation of Golden Hamster Spermatozoa by Bovine Follicular Fluid and Its Fractions," *Journal of Experimental Zoology,* 170 (1969), 269-80.

16. R. G. Edwards et al., "Fertilization and Cleavage In Vitro of Preovulatory Human Oocytes," *Nature, London,* 227 (1970), 1307-9.

17. Anonymous, "More Studies in Capacitation," *Research in Reproduction,* 12, no. 4 (1980), 4.

18. M. J. Hinrichsen and J. A. Blaquier, "Evidence Supporting the Existence of Sperm Maturation in the Human Epididymis," *Journal of Reproduction and Fertility,* 60 (1980), 291-94.

19. Chang et al. (1977), op. cit.

20. J. W. Overstreet and G. W. Cooper, "The Time and Location of the Acrosome Reaction during Sperm Transport in the Female Rabbit," *Journal of Experimental Zoology,* 209 (1979), 97–104.

21. R. Stambaugh and M. Smith, "Amino Acid Content of Rabbit Acrosomal Proteinase and Its Similarity to Human Trypsin," *Science,* 186 (1974), 743–46.

22. A. W. H. Braden et al., "The Reaction of the Zona Pellucida to Sperm Penetration," *Australian Journal of Biological Science,* 7 (1954), 391–409; C. Barros and R. Yanagimachi, "Polyspermy-Preventing Mechanisms in the Hamster Egg," *Journal of Experimental Zoology,* 180 (1972), 251–66; and D. Epel and E. J. Carroll, Jr., "Molecular Mechanisms for Prevention of Polyspermy," *Research in Reproduction* 7, no. 2 (1975), 2–3.

23. M. C. Chang, "Effects of Antisera on Mammalian Fertilization," *Research in Reproduction,* 10, no. 1 (1978), 2.

24. M. M. McMillen, "Differential Mortality by Sex in Fetal and Neonatal Deaths," *Science,* 204 (1979), 89–91.

25. W. Rinehart, "Sex Preselection—Not Yet Practical," *Population Reports,* ser. I, no. 2 (1975), pp. 21–31.

26. Ibid.

27. D. M. Rorvick and L. B. Shettles, *Your Baby's Sex: Now You Can Choose* (New York: Dodd, Mead & Company, 1970); and L. B. Shettles, "Predetermining Children's Sex," *Medical Aspects of Human Sexuality* (June 1972), p. 172.

28. W. H. James, "Timing of Fertilization and Sex Ratio of Offspring—A Review," *Annals of Human Biology,* 3 (1976), 549–56.

29. Anonymous, "Two Mothers, No Father = One Embryo," *Science News,* 116, no. 7 (1979), 116.

30. J. L. Marx, "Three Mice 'Cloned' in Switzerland," *Science,* 211 (1981), 375–76.

31. W. J. Broad, "Saga of Boy Clone Ruled a Hoax," *Science,* 211 (1981), 902.

FURTHER READING

AUSTIN, C. R., "Fertilization," in *Reproduction in Mammals, Book 1, Germ Cells and Fertilization,* eds. C. R. Austin and R. V. Short. Cambridge, England: Cambridge University Press, 1972. Pp. 103–34.

BLOCK, I., "Sperm Meets Egg," *Science Digest,* 89, no. 3 (1981), 96–99.

CHANG, M. C., AND C. R. AUSTIN, "Mammalian Fertilization," *Research in Reproduction,* vol. 8, no. 4 (1976).

EDWARDS, R. G., AND R. E. FOWLER, "Human Embryos in the Laboratory," *Scientific American,* 223 (1970), 44–54.

EPEL, D., "The Program of Fertilization," *Scientific American,* 237 (1977), 128–38.

GROBSTEIN, C., "External Human Fertilization," *Scientific American,* 240 (1979), 57–67.

GWATKIN, R. B. L., *Fertilization Mechanisms in Man and Mammals.* New York: Plenum Publishing Corporation, 1977.

KARP, L. E., "Genetic Crossroads," *Natural History Magazine,* 87 (1978), 8–22.

LONGO, F. J., "Fertilization: A Comparative Structural Review," *Biology of Reproduction,* 9 (1973), 149–215.

MITTWOCH, U., "Sex Differences in Cells," *Scientific American,* 209 (1963), 54–69.

THIBAULT, C., "Sperm Storage and Transport in Vertebrates," *Journal of Reproduction and Fertility,* suppl. 18 (1973), pp. 39–53.

9 *Pregnancy*

LEARNING GOALS

Having read this chapter, you should be able to:

1. List some presumptive, probable, and positive signs of pregnancy.
2. Describe the procedure used in the latex agglutination inhibition pregnancy test.
3. Discuss the procedures to be followed by a woman if she is pregnant and desires to carry the fetus to term.
4. Describe the structure of the blastocyst, and name the future products of each blastocyst component.
5. Describe, in detail, the process of implantation.
6. Discuss theories that explain why the embryo is not immunologically rejected by the mother.
7. List the three primary germ layers, and name the major derivatives of each layer in the adult.
8. List the four extraembryonic membranes, and discuss the fate and function of each.
9. Describe how the placenta is formed, and how it functions as a respiratory, excretory, and nutritive organ for the fetus.
10. List major characteristics of a fetus after 9, 12, and 20 weeks of pregnancy.
11. Describe adaptations of the fetal circulatory system to its existence within the uterus.
12. Describe how rhesus disease develops and how it can be treated.
13. Give three examples of a teratogen.

14. Describe how ingestion of aspirin can harm the fetus.
15. Describe how ingestion of alcohol can harm the fetus.
16. Describe how smoking harms the fetus.
17. Describe three different methods used to assess fetal condition.
18. Discuss the importance of a pregnant woman's nutritional state to the well-being of the fetus.
19. List the hormones secreted by the placenta and at least one major function of each hormone.
20. Describe the feto-placental unit and its role in secretion of estrogen by the placenta.
21. Discuss the causes and dangers of ectopic pregnancies.
22. Describe the role of the placenta in maintaining function of the corpus luteum of pregnancy.
23. List some major causes of miscarriage.

INTRODUCTION

As discussed in the previous chapter, conception occurs when a sperm and ovum fuse to become a zygote. Then, the zygote divides mitotically to form two blastomeres. These cells, in turn, divide to produce four smaller blastomeres, and so on. It has been estimated that it takes about 42 sets of mitotic cell divisions to produce a new-born baby! Thus, the number of cells in the developing human increases exponentially. Not only that, but there is cell differentiation, so that cells become disparate in form and function; some become liver cells, some nerve cells, some muscle cells, and so on. Not only does the zygote give rise to the fetus, but the same single cell produces the placenta as well. The purpose of this chapter is to describe the process of *pregnancy,* or *gestation,* during which the mother supports a developing human to a stage when it can exist in the outside world.

SIGNS OF PREGNANCY

How does a woman know she is pregnant? *Presumptive signs of pregnancy* are possible indications of pregnancy. One presumptive sign is a missed menstrual period associated with coitus the previous month. As we discuss later, secondary amenorrhea is associated with pregnancy. Some women, however, continue to menstruate (albeit, to a lesser degree) in the first two or three months of pregnancy, so secondary amenorrhea is not a reliable sign of pregnancy. Another presumptive sign of pregnancy is nausea, often after awakening. This is called *morning sickness* and is due to a decrease in stomach function at this time. Morning sickness usually but not always goes away in a few weeks. Another presumptive sign that a woman is pregnant is an increase in size and tenderness of the breasts as well as a darkening of the areola surrounding the nipples. If a woman experiences any of these presumptive signs, she should consult a physician or clinic.

Probable signs of pregnancy indicate that, in all likelihood, a woman is pregnant. These include an increase in frequency of urination (because the growing uterus presses on the urinary bladder) and an increase in size of the abdomen. Also, the uterine cervix becomes softer by the sixth week of pregnancy; this condition, called *Hegar's sign,* is detected by a physician during a pelvic exam. Another probable sign of pregnancy is a positive pregnancy test, as we discuss later.

Positive signs of pregnancy include detection of a fetal heartbeat, feeling the fetus moving, and visualization of the fetus by ultrasound or fetoscopy, methods we review later in this chapter.

Sometimes, a woman with either a great desire or fear of pregnancy can develop some of these presumptive or even probable signs of pregnancy. This *false pregnancy (pseudocyesis)* is a good example of how our brain can influence our physiology. In some cases, the false pregnancy can last nine months!

PREGNANCY TESTS

Pregnancy tests detect a hormone that is present in the blood or urine of a pregnant woman.[1] This hormone is human chorionic gonadotropin (hCG), which is secreted by the placenta very soon after pregnancy is established (see below). In the past, bioassays were used to detect the presence of hCG. In these tests, a woman's urine was administered to test animals such as mice, rabbits, frogs, or toads. Since hCG acts like luteinizing hormone (LH), when present in the urine it causes ovulation in female animals or spermiation in male animals. One disadvantage of this test is that most of these bioassays only detect presence of hCG two to four weeks after the last missed menstrual period, and another disadvantage is that many animals are killed. Therefore, newer, more effective pregnancy tests have been developed.

A current method of detecting pregnancy utilizes the fact that an antibody to hCG (anti-hCG) can be obtained. When hCG is present in a solution, anti-hCG combines with it to produce a visible brown reaction. In such an *immunoassay pregnancy test,* anti-hCG and urine are mixed in a test tube or on a glass slide, and the presence or absence of a brown ring is noted. This kind of test takes about two hours. A serious deficiency of this method is that, in 3 percent of cases, a brown ring forms when hCG is not present; this is due to the interaction of anti-hCG with other proteins in the urine and not to hCG. Also, in 20 percent of cases, a negative response can be obtained in a newly pregnant woman! This can be because the test is done before hCG is secreted in high enough levels to be detected. Several home pregnancy test kits, available without prescription, use this immunoassay.[2] Therefore, if one uses one of these kits, one should be aware of the above inaccuracies.

A more accurate and effective method used in at least one home kit and in clinics and hospitals is the *latex agglutination inhibition test,* introduced in 1962. This test takes two minutes to two hours to finish, and its accuracy increases the longer the test is run. In this test, small latex rubber particles are coated with hCG. If these particles are added to a solution containing anti-hCG, the particles will agglutinate (the solution looks like it curdles). If, however, the solution of anti-hCG has previously been exposed to hCG in urine (by mixing), the hCG on the latex particles will have no anti-hCG left to react with, and so the particles will not agglutinate (the solution will remain milky white). To do this test, a solution of anti-hCG is first added to the urine. Then the latex particles are added to the mixture of anti-hCG and urine. Figure 9–1 summarizes this procedure.

An extremely accurate pregnancy test uses radioimmunoassay of hCG in a woman's blood. With this method, hCG can be detected a few days after conception.[3] However, only a limited number of hospitals have the facilities to do this procedure, and it is more costly than the previously described methods.

One problem with all of these pregnancy tests is that they can be "tricked" (especially if done in the first two weeks of pregnancy). Not only do the immunoassay and (less so) the latex agglutination inhibition tests sometimes give false positives and negatives, but also some abnormal kinds of embryonic tissue, such as hydaditiform moles as we discuss later, can secrete large amounts of hCG when there is no embryo present. Also, a woman with an ectopic pregnancy may not have detectable levels of hCG in her blood. Failure to detect this kind of pregnancy, as we see later, can be dangerous. Thus, a woman should consult with a physician in conjunction with any pregnancy test, positive or negative.

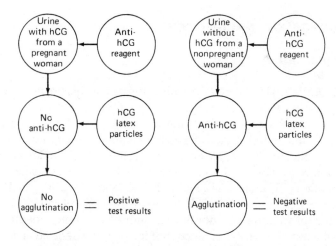

Figure 9–1. Diagram of the "latex agglutination inhibition" pregnancy test. If the level of urinary hCG is high enough to be detected, the anti-hCG reagent is neutralized, and no agglutination will occur when hCG-coated latex particles are added. If hCG is not present or is below detectable levels, the anti-hCG is not neutralized and will agglutinate the hCG-coated particles. Adapted from W. B. Hunt II, "Pregnancy Tests—The Current Status," *Population Reports,* ser. J, no. 7 (1975), pp. 109–24. Used with permission of the Population Information Program, Johns Hopkins University, Baltimore, Md.

WHAT TO DO IF YOU ARE PREGNANT

Pregnancy can be wanted or unwanted (or sometimes both). If a woman is pregnant but does not want to have a child, she has the options of induced abortion or adoption (Chapter 14). If she wants her pregnancy to go to term (or even if she does not), she should go to a physician or clinic within two or three weeks after her missed menstrual period. Once there, she will have her medical history taken, including information on the patterns of her menstrual cycles, previous pregnancies or abortions, miscarriages, surgery (especially of the pelvic region), childhood diseases, past or present venereal disease infections or other pelvic infections, and general health. Any genetically related disorder in her or in her partner's past should be communicated. The woman should also have a complete pelvic examination for evidence of infection and structure of her pelvis. A Pap smear should be done to check for uterine precancerous or cancerous growth. Her blood should be tested for syphilis, blood type, and diabetes. She should find out about the physician's and hospital fees, and should discuss with the physician his or her opinions about anesthetics, role of the male partner in delivery, and availability of alternative birthing experiences (Chapter 10).

After the initial visit, most physicians advise a pregnant woman to be checked monthly for the first five or six months of pregnancy. Then, she is examined at least twice a month, until the last months, when she should be seen once a week.

THE PROCESS OF PREGNANCY

Implantation

As mentioned in Chapter 8, fertilization usually occurs in the ampullary-isthmic portion of the oviduct. Then, the dividing ball of cells moves down the oviduct and into the uterus, where it implants in the uterine wall. We now describe this process.

After fertilization, the potential embryo undergoes cleavage (mitotic division) to become first 2, 4, 8, 16, and then 32 cells by the third day after fertilization. After it contains 8 cells or more, the solid ball of cells is called a *morula* (Latin for "mulberry"), as shown in Fig. 9-2. The zona pellucida, which originally surrounded the oocyte within the follicle before ovulation, still remains as a translucent membrane surrounding the morula. The morula continues dividing as it passes down the oviduct to the uterotubal junction (Fig. 9-3). It is assisted in its journey by the relative absence of oviductal mucus at this stage and by the beating of oviductal cilia in a uterine direction.[4]

At about three to four days after fertilization, the potential embryo enters the uterus (Fig. 9-3). Now, it is a larger mass of cells called a *blastocyst* (Fig. 9-2). The blastocyst looks from the outside like a solid ball of cells. If, however, it is cut in half, a fluid-filled cavity, the *blastocoel* or *blastocyst cavity,* is revealed. A single layer of cells, the *trophoblast,* forms the outer layer of the blastocyst just inside the zona pellucida (Fig. 9-2). A clump of cells at one end of the blastocyst under the trophoblast

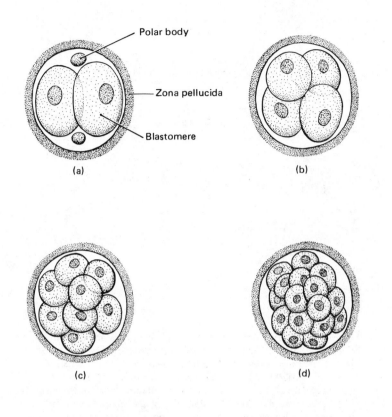

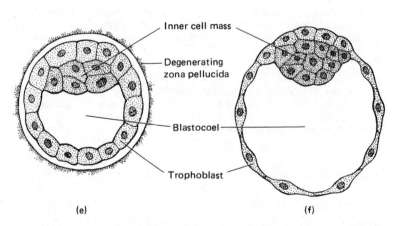

Figure 9-2. Cleavage of the zygote and formation of the blastocyst. (a), (b), and (c) show various stages of cell division (cleavage). (d) is a solid ball of cells called the morula. Note that as cleavage proceeds, cells become smaller. (e) and (f) represent stages of the blastocyst. Adapted from K. L. Moore, *The Developing Human. Clinically Oriented Embryology,* 3rd ed. (Philadelphia: W. B. Saunders Company, 1982). © 1982; used with permission.

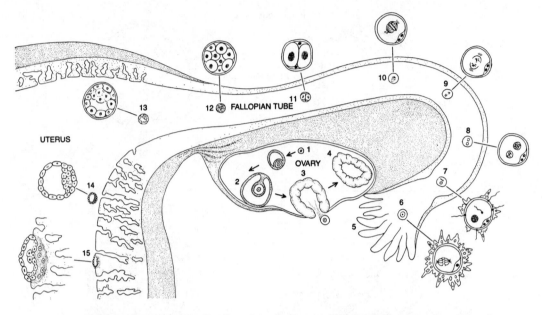

Figure 9–3. Internal fertilization of the human egg is one event in a complex process that begins when a primary follicle in the ovary (1) develops to become a mature follicle (2), which ruptures, releasing the egg (3). The follicle becomes a corpus luteum (4). (Follicular development is arbitrarily shown at different sites in the ovary for clarity.) The ovulated egg is swept from the surface of the ovary by the open end of the oviduct, or Fallopian tube (5). The egg has just completed the first meiotic division, a division whereby its chromosomes are reduced to half the normal complement and a polar body is extruded to the margin of the egg; a second meiosis has begun (6). A sperm penetrates the egg cytoplasm, activating the egg to complete the second meiotic division, extrude a second polar body, and form the female pronucleus (7). The sperm head swells to form a male pronucleus (8), and the two pronuclei fuse, mingling male and female chromosomes (9). The chromosomes replicate and divide (10), and the fertilized egg, or zygote, undergoes cleavage (11). Successive cleavages as the zygote moves along produce a morula (12), an early blastocyst (13), and after some 4½ days, a late blastocyst with an inner cell mass that gives rise to the embryo and an outer trophoblast (14). On the sixth or seventh day, the blastocyst is implanted in the uterine wall (15). Reproduced from C. Grobstein, "External Human Fertilization," *Scientific American,* 240, no. 6 (1979), 57–67. Copyright © 1979 by Scientific American, Inc.; all rights reserved.

layer is called the *inner cell mass* (Fig. 9–2), a structure that we see later gives rise to the embryo.

The blastocyst rests freely in the uterine cavity for about two or three days, during which time it derives nutrients secreted by the uterine glands and increases slightly in size. On about the sixth day after fertilization, the uterus secretes an enzyme (protease) that dissolves the zona pellucida surrounding the blastocyst. Once the zona pellucida disappears, the inner cell mass end of the blastocyst attaches to the uterine wall. Then, the blastocyst begins to invade the endometrium, a process called *implantation* or *nidation*. Implantation occurs five to eight days after fertilization.

During the early phases of implantation, the trophoblast differentiates into an outer *syncytiotrophoblast* and an inner *cytotrophoblast* (Fig. 9–4). The syncytiotrophoblast secretes proteases that break down cells of the uterine endometrium, thus allowing the blastocyst to penetrate into the uterine stroma.[5] The syncytiotrophoblast acquires its name from the fact that it consists of a mass of cells that have lost their cell membranes and have communicating cytoplasm; such a structure is called a *syncytium*. Meanwhile, the cells of the *uterine stroma* (the connective tissue framework of the uterus) multiply rapidly and form a cup that grows over the blastocyst. This growth of uterine stromal cells is called the *deciduoma response*. Implantation is now complete. Figures 9–3 and 9–4 summarize the process of implantation.

Implantation is a very important event. If it does not occur, the blastocyst will age, degenerate, and the potential pregnancy will be terminated. Implantation requires a uterus that has been exposed to just the right amounts of estrogen and progesterone at the right time. Remember from Chapter 7 that the corpus luteum that is formed after ovulation secretes moderate levels of estradiol-17β and higher amounts of progesterone. This ratio of steroid hormones primes the uterus, making the endometrium vascular, secretory, and ready for implantation. Progesterone also causes the uterus to secrete the protease that dissolves the zona pellucida surrounding the blastocyst, at least in rabbits.[6] Recent evidence in several laboratory mammals indicates that the blastocyst itself may secrete hormones that play a role in implantation. More specifically, the blastocyst secretes estradiol and progesterone, and it is thought that these steroid hormones diffuse from the blastocyst and interact with a properly primed uterus to induce implantation.[7] These steroids from the blastocyst may induce formation of steroid receptors on endometrial cells.[8] The morula and blastocyst of some laboratory mammals contain a molecule like hCG but about one-half as large.[9] This hCG-like molecule may control steroid hormone production in the blastocyst. Prostaglandins secreted by the endometrium may also play a role in implantation because suppression of prostaglandin synthesis blocks this process.[10]

The fetal cells are genetically different from those of the mother, so why doesn't the mother's immune system reject the fetus? The answer is not clear. One theory, however, is that the zona pellucida protects the blastocyst from immunologic rejection. After this structure is shed, however, the embryo implants, and the human trophoblast has antigens to which antibodies are formed by the mother; these antibodies then would suppress further immunological rejection.[11] A coating of hCG on the trophoblast could also protect against rejection. In addition, there may be factors in the blood of the mother and fetus that block immunological rejection.[12] Changes in the number of circulating white blood cells in the mother's blood, however, are not likely to be important in fetal protection.[13]

Early Embryonic Development

Soon after implantation, the inner cell mass differentiates into two layers of cells, the *ectoderm* and *endoderm* (Fig. 9–4). This two-layered structure now is called the *embryonic disc* because it is the early embryo. Then, a middle cell layer, the *mesoderm*,

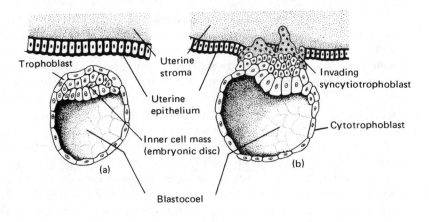

Trophoblast

Uterine stroma

Uterine epithelium

Inner cell mass (embryonic disc)

Invading syncytiotrophoblast

Cytotrophoblast

(a)

(b)

Blastocoel

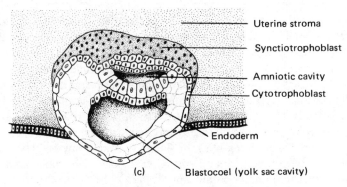

Uterine stroma

Synctiotrophoblast

Amniotic cavity

Cytotrophoblast

Endoderm

(c)

Blastocoel (yolk sac cavity)

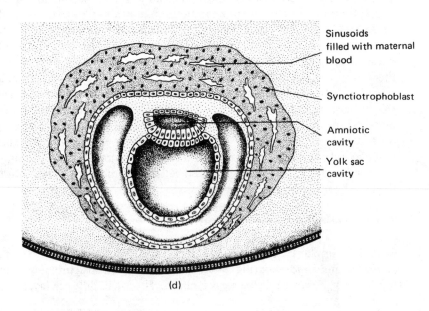

Sinusoids filled with maternal blood

Synctiotrophoblast

Amniotic cavity

Yolk sac cavity

(d)

appears between the ectoderm and endoderm. The three layers, which are called the *primary germ layers,* give rise to all adult tissues. Some derivatives of each germ layer are listed in Table 9–1.

The Extraembryonic Membranes

The inner cell mass gives rise to three of the four *extraembryonic membranes.*[14] The first of these to be formed is the *yolk sac,* which is an endoderm-lined membrane that surrounds the blastocoel; the blastocoel now is called the *yolk sac cavity* (Figs. 9–4 and 9–5). Another extraembryonic membrane formed from the inner cell mass, the *amnion,* then grows over the forming embryo (Fig. 9–5). The amniotic cavity becomes filled with *amniotic fluid.* The amount of amniotic fluid is about 5 to 10 ml after 8 weeks of development, about 250 ml at 20 weeks, and then increases to a maximum of 1000 to 1500 ml by the thirty-eighth week of pregnancy. The amount of amniotic fluid then declines to 500 to 1000 ml near the time of birth. Amniotic fluid is secreted and absorbed rapidly, at a maximum rate of about 300 to 600 ml per hour. The third membrane is the *allantois,* which forms as a small pouch. A fourth membrane, the *chorion,* is derived from the trophoblast and surrounds the embryo after about one month of development; the chorion eventually fuses with the amnion (Fig. 9–5).

In lower animals, the yolk sac contains a large amount of yolk, which is used as a source of energy for the developing embryo. An example is the yolk of a chicken egg. In the ova of humans and most other mammals, however, little yolk is present, and the yolk sac degenerates very early in development. Before it degenerates, however, it supplies the embryo with red blood cells as well as primordial germ cells (Chapter 5). The amnion is an important extraembryonic membrane throughout development. The fluid in this sac supports and protects the fetus against mechanical shock and supplies water and other materials to the fetus. The chorion and allantois, as we discuss below, form important components of the placenta.

The Placenta

The *placenta* is an organ that is vital to the developing fetus.[15] Through this organ, the fetus receives important substances such as oxygen and glucose, and it eliminates toxic substances such as carbon dioxide and other wastes. The placenta is formed in

Figure 9–4. Implantation of the human embryo: (a) The blastocyst is not yet attached to the uterine epithelium; (b) the trophoblast has penetrated the epithelium and is beginning to invade the stroma; (c) the blastocyst sinks further into the stroma and the amniotic cavity has appeared; (d) the uterine tissue has grown over the implantation site (deciduoma response), and irregular spaces, the blood sinusoids, have appeared in the syncytiotrophoblast. Adapted from A. McLaren, "The Embryo," in *Reproduction in Mammals, Book 2, Embryonic and Fetal Development,* eds. C. R. Austin and R. V. Short, pp. 1–43 (Cambridge, England: Cambridge University Press, 1972). © 1972; used with permission.

TABLE 9-1. STRUCTURES PRODUCED BY THE THREE PRIMARY GERM LAYERS

Endoderm	Mesoderm	Ectoderm
Epithelium and glands of digestive tract	Skeletal, smooth, and cardiac muscle	Epidermis of skin
Epithelium of urinary bladder and gallbladder	Cartilage, bone, and other connective tissues	Hair, nails
Epithelium of pharynx, auditory tube, tonsils, larynx, trachea, bronchi, lungs	Blood, bone marrow, and lymphoid tissue	Receptor cells of sense organs
Epithelium of thyroid, parathyroid, thymus glands	Endothelium of blood vessels and lymphatics	Epithelium of mouth, nostrils, sinuses, oral glands, anal canal
Epithelium of vagina, vestibule, urethra, associated glands	Mesothelium of coelomic and joint cavities	Enamel of teeth
	Epithelium of kidneys and ureters	Entire nervous tissue
	Epithelium of gonads and associated ducts	Adenohypophysis
	Epithelium of adrenal cortex	
	Stroma of most soft organs, except those of the central nervous system	

Source: Adapted from *Principles of Anatomy and Physiology,* 3rd ed., by Gerard J. Tortora and Nicholas P. Anagnostakos (New York: Harper & Row, Publishers, Inc., 1981), Exh. 29–1, p. 750. © 1981 by Gerard J. Tortora and Nicholas P. Anagnostakos. Reprinted by permission of Harper & Row, Publishers, Inc., and Dr. Gerard J. Tortora.

the following way (Figs. 9–4 and 9–5): about the fourteenth day after fertilization, fingerlike projections of the cytotrophoblast extend through the syncytiotrophoblast and into the vascular uterine stroma. These villi are called *chorionic villi* because the cytotrophoblast grows to surround the embryo and thus forms the chorion. The chorionic villi secrete enzymes that dissolve the walls of small uterine blood vessels present in the stroma so that the mother's blood actually bathes the villi. The human placenta is called a *hemochorial placenta* because the uterine blood directly bathes The chorionic villi. It must be emphasized, however, that at no time does the mother's blood mix with that of the fetus. Each chorionic villus contains small blood vessels, derived from the allantois, that are fed by blood coming from the fetus via the umbilical arteries present in the umbilical cord. Materials from the mother's blood such as oxygen and glucose then diffuse from the uterine blood spaces (sinusoids) through the wall of each chorionic villus and into the fetal vessel within each villus. These materials travel to the fetus via the umbilical vein. In an opposite way, fetal waste products such as carbon dioxide leave the fetal blood and diffuse into the mother's blood to be excreted. The detailed structure of the placenta is shown in Fig. 9–6. The placenta serves as a nutrient, respiratory, and excretory organ for the fetus. As we discuss later, the placenta also secretes hormones that are vital to pregnancy.

Molecules greater than about 500 molecular weight will not pass from the mother's blood into the fetal circulation because they are too large to diffuse through

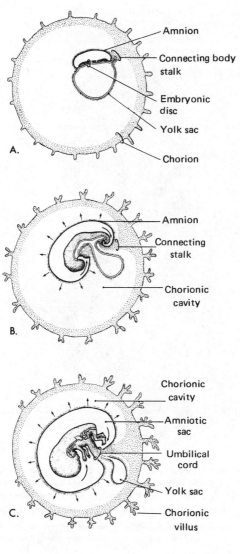

A.
- Amnion
- Connecting body stalk
- Embryonic disc
- Yolk sac
- Chorion

B.
- Amnion
- Connecting stalk
- Chorionic cavity

C.
- Chorionic cavity
- Amniotic sac
- Umbilical cord
- Yolk sac
- Chorionic villus

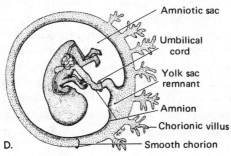

D.
- Amniotic sac
- Umbilical cord
- Yolk sac remnant
- Amnion
- Chorionic villus
- Smooth chorion

Figure 9-5. These drawings show the formation of the amnion, chorion, and yolk sac. Note that the chorion eventually fuses with the amnion, and that the yolk sac eventually degenerates. The allantois is not shown. In the bottom figure, the region of the extended chorionic villi is the placenta. Adapted from K. L. Moore, *The Developing Human. Clinically Oriented Embryology,* 3rd ed. (Philadelphia: W. B. Saunders Company, 1982). © 1982; used with permission.

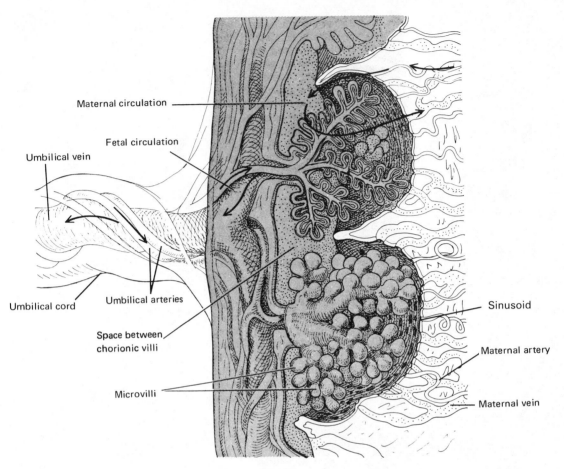

Maternal circulation

Fetal circulation

Umbilical vein

Umbilical cord

Umbilical arteries

Space between chorionic villi

Microvilli

Sinusoid

Maternal artery

Maternal vein

Figure 9–6. Structure of the human placenta in the vicinity of the umbilical cord. The human placenta is "hemochorial" because the chorionic villi are directly bathed by maternal blood. Adapted from P. Beaconsfield, G. Birdwell, and R. Beaconsfield, "The Placenta," *Scientific American,* 243, no. 2 (1980), 94–102. Copyright © 1980 by Scientific American, Inc.; all rights reserved.

the chorionic villi and into fetal blood vessels. This means that most protein hormones of the mother do not reach the fetus; this is important because maternal pituitary hormones reaching the fetus could adversely alter fetal development. Actually, some large proteins from the mother do reach the fetus near the end of pregnancy. These maternal antibodies are actively pumped into the fetal circulation by placental cells, and this is how the fetus is born with its mother's immunological protection against disease. Steroid hormones in the mother's blood, however, are small enough to pass, but most do not because they are degraded by placental enzymes. Other maternal hormones, such as thyroxin, can enter the fetus. In fact, hypersecretion or hyposecretion of a woman's thyroid gland during pregnancy can harm fetal development.

The placenta continues to grow, like an expanding disk, throughout pregnancy. At the fourth week of pregnancy, this organ covers about 20 percent of the inner wall of the uterus, and by week 20 it covers about half of the uterine wall. At this time, the placenta weighs about 200 g, and the fetus weighs 500 g. At term, the placenta is a disklike structure that weighs about 500 g and has a diameter of 20 cm. Actually, the size of the term placenta varies among individuals but usually is one-sixth of the weight of the term fetus. It is obvious from the above discussion that the placenta is a very vascular organ. Near the end of pregnancy, about 75 gal (285 l) of blood pass through the placenta daily; this is about 10 percent of the total blood flow in the mother. This large organ, although supporting the fetus, has a life of its own; if the fetus dies or is removed, the placenta continues to flourish!

The *umbilical cord* connects the fetus with the placenta (Figs. 9–5 and 9–6) and is the lifeline of the fetus. It is derived from a structure connecting the embryo and chorion called the *body stalk* (Fig. 9–4). At birth, this cord is 0.3 to 1.0 in. (1 to 2 cm) in diameter and 20 to 22 in. (50 to 55 cm) long. It is covered by the amniotic membrane and contains two *umbilical arteries* (which carry deoxygenated fetal blood to the placenta) and one *umbilical vein* (which carries oxygenated blood back to the fetus). The vessels within the cord are cushioned by a gelatinous substance, *Wharton's jelly*. If one views a living umbilical cord, its pale color with spiraling red and blue vessels explains its symbolism in the barber's pole, handed down by medieval barber surgeons.[16]

Twin Pregnancies

As mentioned in the previous chapter, dizygotic (fraternal) twins develop from two separate zygotes. The two embryos can implant separately, which results in two separate placentas, chorions, and amnions (Fig. 9–7). In some cases, however, the embryos implant close to one another, which results in a common placenta, fused chorions, and fused amnions (Fig. 9–7). In most cases, monozygotic (identical) twins result from separation of a single inner cell mass into two. This results in the two embryos sharing a common placenta and chorion, but each being enclosed in its own amniotic sac (Fig. 9–8).

Embryonic and Fetal Development

Pregnancy, which lasts about nine calendar months, can be divided into three *trimesters,* each of which lasts three months. Weeks 2 through 8 of the first trimester are the *embryonic period,* and the developing human is called an *embryo* during this time (Fig. 9–9). After the eighth week, it is called a *fetus.* During pregnancy, the developing person grows from about the size of this period (.) to an average birth weight of 7.0 to 7.5 lb.

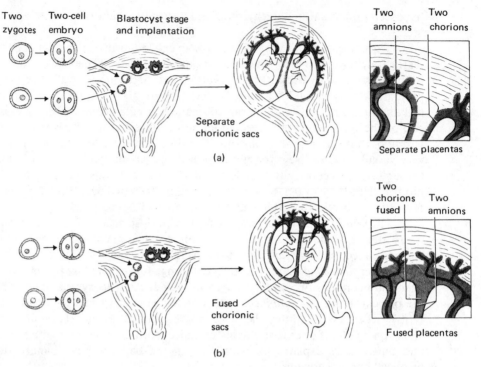

Two zygotes · Two-cell embryo · Blastocyst stage and implantation · Separate chorionic sacs · (a) · Two amnions · Two chorions · Separate placentas

Fused chorionic sacs · (b) · Two chorions fused · Two amnions · Fused placentas

Figure 9-7. Pregnancy with dizygotic (fraternal) twins. Two zygotes are formed, and two embryos implant in the uterus. When they implant separately, two separate placentas, chorions, and amnions appear (a). If they implant together, a single placenta, fused chorions, and two amnions are present (b). Adapted from K. L. Moore, *The Developing Human. Clinically Oriented Embryology,* 3rd ed. (Philadelphia: W. B. Saunders Company, 1982). © 1982; used with permission.

By the end of the first month, the heart is formed and begins to beat, and most major organ systems are developing. By the end of the embryonic period, the early fetus has distinct limbs and digits (Fig. 9-9). By the end of the first trimester, most major organ systems are present, and the fetus has a large head, a well-formed face, and the heartbeat can be detected with a stethoscope. By the end of the second trimester, an oily layer of cells, the *vernix caseosa,* covers the surface of the skin. By the fourth or fifth month, the mother can usually feel fetal movement through her abdomen, and the fetal skin is covered with downy hair *(lanugo).* This hair usually is shed before birth. By the end of six months, about 10 percent of fetuses can survive with special care if born; and by seven months, many premature newborns can survive.

Some of the major events during fetal development are presented in Table 9-2, and embryos and fetuses of different ages are depicted in Figs. 9-9 and 9-10. The position of the fetus in relation to the woman's body is depicted in Fig. 9-11.

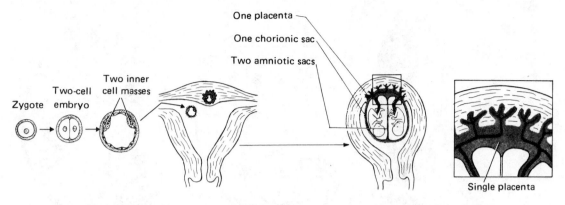

One placenta
One chorionic sac
Two amniotic sacs

Zygote   Two-cell   Two inner
         embryo   cell masses

Single placenta

Figure 9–8. Pregnancy with monozygotic (identical) twins usually occurs when the inner cell mass of the blastocyst divides, producing two embryos with a single placenta and chorion but two amniotic sacs. Adapted from K. L. Moore, *The Developing Human. Clinically Oriented Embryology,* 3rd ed. (Philadelphia: W. B. Saunders Company, 1982). © 1982; used with permission.

Digestive/Urinary Systems. As mentioned earlier, the fetus derives nutrients from the mother's blood in the form of glucose, amino acids, fatty acids, vitamins, salts, and minerals. These nutrients pass to the fetus in the umbilical vein and are utilized by fetal tissues. Carbon dioxide and other wastes produced by fetal reactions then pass back to the placenta via the umbilical arteries and are excreted by the mother. In late pregnancy, the fetus swallows about 500 ml of amniotic fluid each day, and this fluid contains water, salts, glucose, urea, and cell debris from the amnion and fetal skin. These ingested materials provide some nourishment for the fetus, and the waste products from this digestion combine with bile pigments to form green feces in the large intestines. This green fecal material, called *meconium,* is the first to be defecated by the neonate. The fetal kidneys are functional throughout pregnancy and produce about 450 ml of urine per day, which is excreted into the amniotic fluid.

Circulatory System. In the adult, deoxygenated blood (poor in oxygen and rich in carbon dioxide) enters the right side of the heart and then is pumped to the lungs via the pulmonary trunk and arteries. In the lungs, carbon dioxide is removed and oxygen is picked up. The oxygenated blood then returns to the left side of the heart via the pulmonary veins and then is pumped to the dorsal aorta, which carries arterial blood to all other tissues.

The placenta, not the lungs, is the respiratory organ of the fetus, so the fetal circulatory system is different from that of the adult (Fig. 9–12). More specifically, after the oxygenated blood enters the right side of the fetal heart and is pumped into the pulmonary trunk, it is prevented from going to the collapsed lungs by a blood vessel shunt that goes from the pulmonary trunk to the dorsal aorta, the *ductus ateriosus.* The deoxygenated blood then reaches the placenta via the umbilical arteries, where it

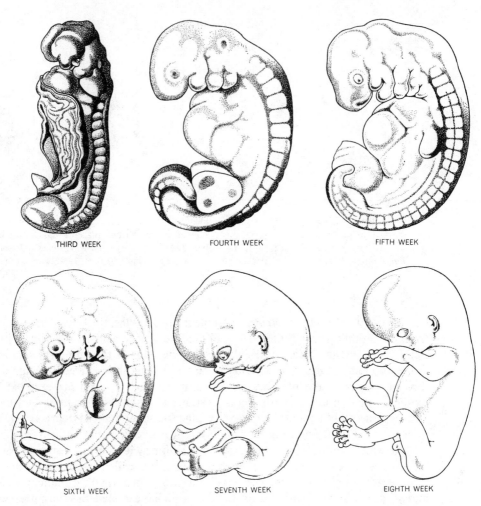

| THIRD WEEK | FOURTH WEEK | FIFTH WEEK |
| SIXTH WEEK | SEVENTH WEEK | EIGHTH WEEK |

Figure 9–9. Development of human embryo from the third week through the eighth week after conception. The embryo grows from about one in. in length after the third week to 1¼ in. at the end of the eighth week. Reproduced from H. B. Taussig, "The Thalidomide Syndrome," *Scientific American,* 207, no. 2 (1962), 29–35. Copyright © 1962 by Scientific American, Inc.; all rights reserved.

loses carbon dioxide and picks up oxygen. The oxygenated blood then travels via the umbilical vein back to the fetus, eventually reaching the right side of the fetal heart. Another shunt, the *ductus venosus,* causes blood to bypass the fetal liver (Fig. 9–12). The ductus arteriosus and ductus venosus close after birth.

The above-described fetal system, however, presents a problem. The volume of blood flowing through the left and right side of the heart must be equal or the system would become imbalanced. In the fetus, no blood is flowing from the lungs into the left side of the heart as in the adult, so how are the left and right sides of the heart kept

198   Part II   Procreation

TABLE 9-2. CHANGES ASSOCIATED WITH FETAL GROWTH

End of Month	Approximate Size and Weight	Representative Changes
1	0.6 cm ($\frac{3}{18}$ in.)	Eyes, nose, and ears not yet visible. Backbone and vertebral canal form. Small buds that will develop into arms and legs form. Heart forms and starts beating. Body systems begin to form.
2	3 cm ($1\frac{1}{4}$ in.) 1 g ($\frac{1}{38}$ oz)	Eyes far apart, eyelids fused, nose flat. Ossification begins. Limbs become distinct as arms and legs. Digits are well formed. Major blood vessels form. Many internal organs continue to develop.
3	7.5 cm (3 in.) 28 g (1 oz)	Eyes almost fully developed but eyelids still fused; nose develops bridge; and external ears are present. Ossification continues. Appendages are fully formed, and nails develop. Heartbeat can be detected. Body systems continue to develop.
4	18 cm ($6\frac{1}{2}$–7 in.) 113 g (4 oz)	Head large in proportion to rest of body. Face takes on human features, and hair appears on head. Skin bright pink. Many bones ossified, and joints begin to form. Continued development of body systems.
5	25–30 cm (10–12 in.) 227–454 g ($\frac{1}{2}$–1 lb)	Head is less disproportionate to rest of body. Fine hair (lanugo hair) covers body. Skin still bright pink. Rapid development of body systems.
6	27–35 cm (11–14 in.) 567–681 g ($1\frac{1}{4}$–$1\frac{1}{2}$ lb)	Head becomes less disproportionate to rest of body. Eyelids separate and eyelashes form. Skin wrinkled and pink.
7	325–425 cm (13–17 in.) 1135–1362 g ($2\frac{1}{2}$–3 lb)	Head and body become more proportionate. Skin wrinkled and pink. Seven-month fetus (premature baby) is capable of survival.
8	40–45 cm ($16\frac{1}{2}$–18 in.)	Subcutaneous fat deposited. Skin less wrinkled. Testes descend into scrotum. Bones of head are soft. Chances of survival much greater at end of eighth month.
9	50 cm (20 in.) 3178–3405 g (7–$7\frac{1}{2}$ lb)	Additional subcutaneous fat accumulates. Lanugo hair shed. Nails extend to tips of fingers and maybe even beyond.

Source: Adapted from *Principles of Anatomy and Physiology,* 3rd ed., by Gerard J. Tortora and Nicholas P. Anagnostakos (New York: Harper & Row, Publishers, Inc., 1981), Exh. 29-2, p. 753. © 1981 by Gerard J. Tortora and Nicholas P. Anagnostakos. Reprinted by permission of Harper & Row, Publishers, Inc., and Dr. Gerard J. Tortora.

in balance? The solution to this dilemma is a hole (covered by a flap) in the wall separating the left and right atria of the fetal heart, allowing blood to mix between the sides and to balance the heart. This hole, the *foramen ovale* (Fig. 9–12), closes after birth (Chapter 11).

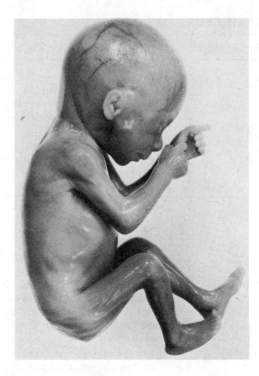

Figure 9-10. Photograph of a 17-week fetus. Fetuses of this age are unable to survive if born prematurely, mainly because their respiratory system is immature. From K. L. Moore, *The Developing Human. Clinically Oriented Embryology,* 3rd ed. (Philadelphia: W. B. Saunders Company, 1982). © 1982; used with permission.

Nervous System.   The nervous system of the fetus is formed very early in development. The central nervous system (brain and spinal cord) and peripheral nerves develop and by the eighth week influence fetal development and function, so that muscle sense and coordinated movements of the fetus appear very early. After the fetus begins to move on its own, some of its movements are very well coordinated, as evidenced by actual pictures of the fetus sucking its thumb (Fig. 9-13). The fetal sensory nervous system also is functional, and the environment in the uterus is not totally devoid of stimuli. For example, the noise level within the amniotic sac is similar to that of a quiet room (about 50 decibels), and the light is like a dark room. The temperature of the amniotic fluid is about 0.5°C higher than the mother's body temperature because the rapidly growing fetal tissues are producing a lot of heat. It has been shown that a loud noise outside but near the woman's abdomen, or a flash of light within the amniotic fluid, or pricking the fetal skin with a small instrument can evoke vigorous fetal reactions. Thus, the fetus, even in early stages, is not simply an inert, growing lump of tissue but a moving, sensing being that is capable of responding to changes in its environment.

Endocrine System.   The endocrine system of the fetus is functional during most of pregnancy. The fetal anterior pituitary gland secretes the gonadotropins, FSH and LH,[17] and these hormones may influence development of the fetal gonads

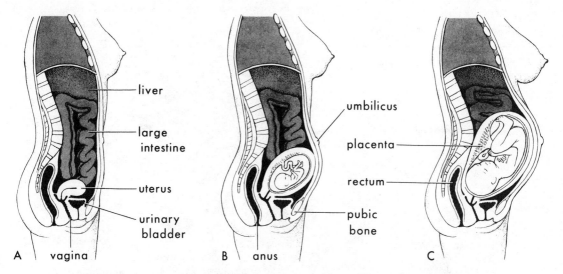

Figure 9–11. Diagrams of sections through a female: (A) Not pregnant; (B) 20 weeks of pregnancy. Note that as the fetus enlarges, the uterus increases in size. (C) 30 weeks of pregnancy. Note that the uterus and fetus now extend above the umbilicus (belly button). The mother's abdominal organs are displaced, and the skin and muscle of her anterior abdominal wall are greatly stretched. Reproduced from K. L. Moore, *The Developing Human. Clinically Oriented Embryology,* 3rd ed. (Philadelphia: W. B. Saunders Company, 1982). © 1982; used with permission.

(Chapter 5). In addition, some hCG reaches the fetus and may be involved in fetal gonadal function. The fetal pancreas secretes insulin, which allows the fetal cells to use glucose supplied by the mother. As we discuss later in this chapter, the fetal adrenal glands contain a special region that secretes steroid hormones that may play a role in the initiation of labor.

As mentioned, the fetus is surrounded by amniotic fluid, and there is a constant turnover of this fluid throughout pregnancy. There is some evidence that two hormones are involved in regulation of amniotic fluid secretion and absorption. One is prolactin secreted by the uterus, which is found in high levels in amniotic fluid and is known to play a role in pumping sodium and therefore water across membranes.[18] The second is a hormone secreted by the fetal neurohypophysis, arginine vasotocin (AVT).[19] As we discussed in Chapter 2, the adult neurohypophysis secretes oxytocin and vasopressin, not AVT. In some aquatic animals, AVT is an adult hormone that is involved in water transport across membranes. It is thought by some that ontogeny (development of an individual) tends to recapitulate phylogeny (evolutionary history of an animal). Since AVT is secreted by the fetus and is involved in regulation of the amniotic "pond," fetal AVT secretion could be a recapitulation of the role of this hormone in distant aquatic ancestors of *Homo sapiens.*

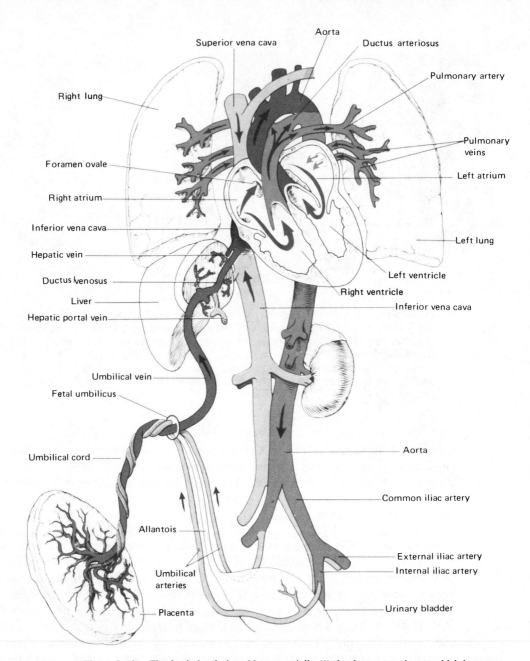

Figure 9-12. The fetal circulation. Note especially (1) the ductus arteriosus, which bypasses the fetal lungs; (2) the foramen ovale, which allows mixing of blood between the right and left heart chambers; and (3) the ductus venosus, which shunts blood away from the fetal liver. Adapted by permission from Alexander P. Spence and Elliott B. Mason, *Human Anatomy and Physiology,* 1st ed., Fig. 29.18, p. 809 (Menlo Park, Calif.: Benjamin/Cummings Publishing Co., 1979). © 1979 by the Benjamin/Cummings Publishing Co.

Figure 9-13. A human fetus sucking on its thumb. Reproduced from G. C. Liggins, "The Fetus and Birth," in *Reproduction in Mammals, Book 2, Embryonic and Fetal Development,* eds. C. R. Austin and R. V. Short, pp. 72–109 (Cambridge, England: Cambridge University Press, 1972). © 1972 used with permission.

FETAL DISORDERS

As you have seen, development of the fetus is a complex process beginning with conception and ending with birth. Although many healthy babies are born, some situations can result in abnormal fetal development resulting in fetal death or the birth of an infant with congenital ("at birth") disorders. We now discuss some causes of abnormal fetal development.

Genetic and Chromosomal Disorders

The fetus can inherit some genetic disorders (Chapter 12). In fact, about 1600 human diseases are of genetic origin.[20] Some of these disorders are mild, such as colorblindness. Others, however, can cause troublesome handicaps, such as harelip, cleft palate, and club foot. Unfortunately, many others can kill the embryo or fetus. Of the many possible disorders having a genetic basis, I will discuss only Rhesus disease here because this disease involves an association between a woman and her fetus. Others, such as cystic fibrosis, Tay-Sachs disease, phenylketonuria, sickle-cell anemia, and ABO incompatibility, are discussed in Chapter 12. In addition, chromosomal abnormalities account for 42 percent of aborted fetus and are present in 1 out of 200 newborns. These disorders also are discussed in Chapter 12.

Rhesus Disease. *Rhesus disease,* commonly called "Rh incompatibility," is an inherited phenomenon that damages not the present fetus but the fetus of a future pregnancy. This disease involves a gene with a dominant allele (R) and a recessive allele (r) for the Rhesus factor. (For a discussion of alleles, see Chapter 12.) Thus, cells of an individual can be Rh+ (RR or Rr) or Rh- (rr).

The cause for concern is when the pregnant woman is Rh- and the father is Rh+, which occurs in about 10 percent of marriages. In this situation, the Rh- woman could be carrying an Rh+ fetus. During labor and delivery, fetal blood cells could enter the maternal tissues because of broken blood vessels as the placenta detaches (Chapter 10). Since the mother's tissues are Rh- and the fetal blood cells are Rh+, she will form antibodies to the fetal Rh+ cells. If she then becomes pregnant again with an Rh+ baby, these antibodies will enter the second fetus and destroy its mature red blood cells. Since the mature fetal red blood cells are destroyed, the fetus will develop *jaundice* (yellowish skin) due to the accumulation of *bilirubin* (a breakdown product of red blood cells) in its tissues. Bilirubin is toxic and can cause brain damage. Also, because the mature fetal red blood cells are being destroyed, there will be many new immature red blood cells (erythroblasts) in the fetal blood, a condition called *erythroblastosis fetalis*. These immature red blood cells are not efficient in carrying oxygen, so the fetus is anemic and its tissues are unable to grow properly. It should be noted, however, that certain combinations of ABO blood type (Chapter 12) in the mother and fetus can prevent rhesus disease.

In the past, it was necessary to give a complete blood transfusion to a newborn with rhesus disease. A new treatment is, however, much safer. An injection of Rhogam or Rho Immune (antibodies to Rh factor) is given to the mother within two or three days after delivery of the first infant or after a miscarriage or induced abortion (Chapter 14). This drug destroys all Rh+ fetal red blood cells that may have entered her blood, and thus she does not form antibodies that could harm her future fetus.

Teratogens, Mutagens, and Other Agents That Damage the Fetus

Exposure of the fetus to certain drugs, chemicals, or radiation can be *mutagenic* (damaging the genes or chromosomes of fetal cells) or *teratogenic* (damaging fetal growth). *Mutagen* means "mutant producing," and *teratogen* means "monster producing." There are many factors that have been shown in laboratory mammals and humans to be teratogenic or mutagenic; we discuss a few of them here.

Viruses and Bacteria. Some viruses can severely damage or kill the fetus. Examples are smallpox, chicken pox, mumps, and herpes. German measles virus *(rubella)* can produce heart defects, blindness (due to cataracts), deafness, microcephaly (small brain), mental deficiency, cleft palate, harelip, and spina bifida (exposed spinal cord). Exposure of the fetus to the rubella virus is most damaging from the third to the twelfth week of pregnancy. Before this time, little or no damage occurs. To prevent fetal exposure to rubella or other damaging viruses, a female child should be exposed to the virus to form antibodies, either by naturally contracting the disease or by vaccination. Bacterial infections such as syphilis, pneumonia, tuberculosis, and typhoid can cause spontaneous abortion.

Environmental Pollutants. Many environmental pollutants can be teratogenic or mutagenic. Most of these agents have the most damaging effects during the fourth to the seventh week of pregnancy (Fig. 9–14). Before this time, they kill the embryo; and after this time, they have less chance of harming the fetus. Examples of mutagens are mercury, lead, cadmium, arsenic, PCBs, DDT, benzene, and carbon tetrachloride. In Japan, mercury from fertilizers in factory effluent got into the fish population and caused *Minimata's disease,* which is characterized by damage to the fetal brain, resulting in abnormal muscle movement. *Cerebral palsy* (spastic muscle paralysis due to brain damage), is another name for this type of damage, and can be caused by fetal exposure to bacterial infection, oxygen deficiency, anemia, jaundice, and low blood sugar.

Drugs, Alcohol, and Tobacco. *Thalidomide* is a mild tranquilizer that was used to treat morning sickness and to stop bleeding in pregnant women in the late 1950s and early 1960s.[21] In 1958 through 1961, however, several thousand cases of severely deformed infants were related to the use of this drug by pregnant women, especially during the fourth through the seventh week of pregnancy. Thalidomide causes the fetus to develop hands and feet but not arms or legs, a condition called *phocomelia* (Greek *phoke* = ''seal''; *melos* = ''limb'') or other abnormalities such as total absence of limbs *(ectromelia).*

The drug *bendictin* may be involved in a drama similar to that of thalidomide. About 25 percent of women in the United States have been using this drug for morning sickness. Altogether, about 1.5 million pregnant women in 31 countries are now taking this drug, in spite of the fact that recent publicity has emphasized that bendictin may cause fetal abnormalities such as heart defects, hernia of the diaphragm, abnormal limbs, cleft palate, and stomach defects.[22] More studies are needed, however, to confirm these reports.

From the 1940s until the early 1970s, a synthetic estrogen, *diethylstilbestrol* (DES), was given to pregnant women to prevent miscarriage. About 2 million women in the United States were exposed to the drug during the first trimester of pregnancy. In the early 1970s, it was found that daughters born to these women exhibited an increased incidence of vaginal and cervical cancer, as well as an increase in miscarriage and premature births. Some of the sons born to these women also developed abnormalities in the male reproductive tract, including undescended testes and a low sperm count in their semen.[23] Therefore, DES usually is no longer given to pregnant women, although it is still an approved drug in the United States and is used as a morning-after pill contraceptive (Chapter 13).

Alcohol ingested by a pregnant woman can cross the placenta and adversely affect fetal development. Having at least two alcoholic drinks a week can increase the risk of miscarriage. Chronic use of alcohol (six or more cocktails, or more than 3 oz. of alcohol, daily) can produce the *fetal alcohol syndrome* 30 to 45 percent of the time. This syndrome is the third most common cause of mental retardation in infants in the

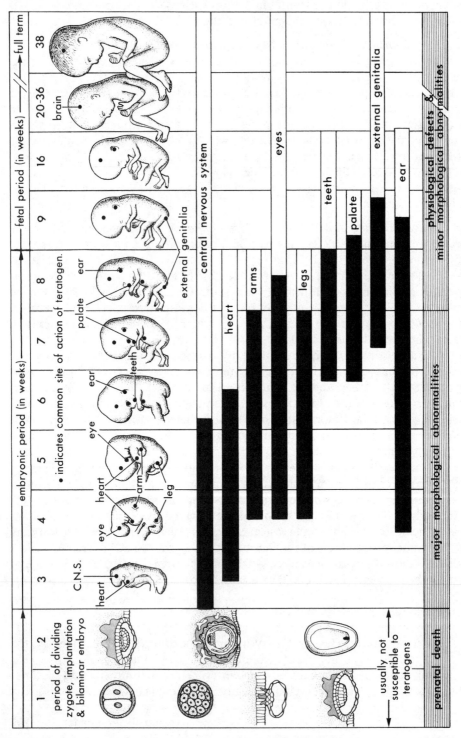

Figure 9–14. Critical periods of human development during which time teratogens are most effective. Black areas show the most critical spans of time and denote the major morphological abnormalities most likely to occur. Reproduced from K. L. Moore, *The Developing Human. Clinically Oriented Embryology*, 3rd ed. (Philadelphia: W. B. Saunders Company, 1982). © 1982; used with permission.

United States today. It is characterized by the birth of relatively small infants with small heads. These children usually are retarded or have learning disabilities. More mild alcohol intake (1 to 2 oz. daily) can constrict umbilical blood vessels and cause spontaneous abortion or birth of an abnormally small infant.

Narcotics such as heroin or methadone will cross the placenta and can cause addiction of the newborn.[24] *Lysergic acid diethylamide* (LSD) also can cross the placenta and damage the fetal chromosomes, leading to deformities.[25] Marijuana smoking (exposure to *tetrahydrocannabinol*), besides reducing fertility (Chapter 17), can decrease estrogen secretion from the placenta and can cause spontaneous abortion.[26]

Tobacco smoking can have adverse affects on the fetus in two ways.[27] First, the nicotine in tobacco smoke constricts blood vessels in the placenta and fetus, resulting in poor delivery of blood-borne substances such as oxygen and glucose to the fetal tissues. Second, the carbon monoxide in tobacco smoke can bind to the hemoglobin of fetal red blood cells, thus preventing the oxygen from binding. Tobacco smoking during pregnancy can impair fetal growth, damage the fetus and placenta, and lead to miscarriage or stillbirth. In fact, about 35 percent of stillbirths can be attributed to smoking during pregnancy. Smoking also can damage the part of the brain that will control respiration in the infant, and it has been estimated that 50 percent of crib deaths (Chapter 11) are related to smoking during pregnancy. Obviously, then, it is recommended that a woman not smoke while pregnant.

Some evidence in laboratory mammals suggests that ingesting high levels of caffeine also may harm the fetus. A recent thorough study, however, indicates no adverse effects of caffeine on fetal development in humans.[28]

Ingestion of aspirin or *indomethacin* by a pregnant woman can harm the fetal heart.[29] The ductus arteriosus of the fetal heart is kept open by secretion of prostaglandins, and aspirin and indomethacin are antiprostaglandins. Thus, exposure of the fetus to these drugs can partially close the ductus arteriosus, resulting in babies born with poorly oxygenated blood and bluish skin *(cyanosis);* see Chapter 11. Also, because the ductus arteriosus is partially closed in these fetuses, too much blood is pumped into the vessels of the still-collapsed fetal lungs. This thickens the walls of the fetal lung blood vessels and can lead to *persistent pulmonary hypertension,* a condition in the newborn in which the arteries in the lungs have thick walls and blood cannot pass through the lungs as well as it should. What is too much aspirin or indomethacin during pregnancy? Not much, about two tablets a day for four or five days in a row. *Acetaminophen,* the nonprescription aspirin substitute, only has slight antiprostaglandin activity, and probably is safer than aspirin during pregnancy.

Radiation. Radiation also can harm the fetus. X-rays and other radiation can be mutagenic in that they damage fetal genes or chromosomes. For example, infants born in Southern Utah whose mothers were exposed to radiation fallout from atomic bomb tests in Nevada had twice the rate of birth defects as did a control population. Also, prior to 1955, X-rays were used to test for pregnancy. Because of possible damage to the fetus, however, X-rays are now used only sparingly and in low dosages during pregnancy.

For several reasons, it may be desirable to examine the condition of the fetus during its existence within the mother. For example, parents may be worried about Down's syndrome (Chapter 12), or there may be fear that the fetus has been exposed to teratogenic or mutagenic substances. At present, there are several ways that the condition of the fetus can be scanned. In the future, there may be more attention paid to this matter, not only for health reasons but because physicians in the United States could be held liable if a deformed infant is born without previous fetal scanning.[30]

Amniocentesis is best done during the sixteenth week of pregnancy. In this method (Fig. 9–15), a needle is inserted through the abdominal and uterine wall and into the amniotic fluid, using an ultrasound image as a guide. Then, a small sample of amniotic fluid is withdrawn for analysis. Since fetal cells are present in the fluid, both these and the fluid itself can be examined. The cells can be cultured in dishes and their structure and function studied. Over 40 genetic abnormalities can be detected by amniocentesis.[31] Sex of the fetus also can be checked by examining the sex chromosomes of the fetal cells. Although this procedure is relatively safe if done cautiously, some studies show an increase in uterine hemorrhage and bone deformities in babies born to women who have had an amniocentesis. In fact, one study from England suggests

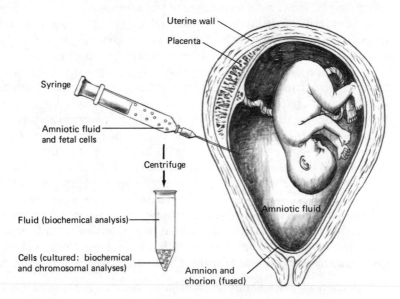

Figure 9–15. The process of amniocentesis is depicted here. A sample of amniotic fluid is taken by inserting a sterile needle through the abdominal and uterine wall and into the amniotic cavity and withdrawing a small amount of fluid that contains fetal cells (not drawn to scale). The sample is centrifuged to separate the cells from the fluid. Then, a variety of tests can be made. Optimum time for this procedure is about the 16th week of pregnancy.

that amniocentesis kills 1.5 percent of the fetuses.[32] Another disadvantage of amniocentesis is that it may take several weeks for the results to be known, and by that time an induced abortion, if desired, is more complex and dangerous (Chapter 14).

Ultrasound can also be used to assess the fetal condition. In this method, a high-frequency sound source is applied to a pregnant woman's abdomen, and sound waves penetrate to the fetus. Dense fetal tissues, such as bone, reflect the waves, and these are detected by a receiver. In this way, fine measurements of the size and dimensions of the fetus can be made. Fetal heart rate also can be detected as early as the tenth week of pregnancy using ultrasound. There is some concern, however, that ultrasound may affect fetal cells.[33]

Fetoscopy is a newer and more involved and expensive method of fetal scanning. In this method, a small incision is made in the abdomen and uterus after injection of a local anesthetic. Then, an optical viewer is inserted into the uterus, and the fetus is viewed directly. Fetoscopy usually is done from the fifteenth to the twentieth week of pregnancy. There is some risk, however, because fetoscopy causes miscarriage about 5 percent of the time.

THE PREGNANT WOMAN

Maternal Nutrition

During pregnancy, a woman not only must maintain her own health and well-being, but she faces the additional demands of a rapidly growing fetus and placenta. This requires her careful attention to diet, weight gain, and general health. Because the woman is supporting both herself and the fetus, her caloric intake must increase during pregnancy as well as during lactation (Table 9–3). Also, special dietary supplements usually are prescribed by her physician. Specific nutritional requirements for pregnancy and lactation include extra protein, iron, calcium, folic acid, and vitamin B_6. Undernutrition can harm the fetus, resulting in a low birth weight or even spontaneous abortion.

In the past, it was often recommended that a pregnant woman should limit her weight gain. It is now felt, however, that she should gain about 25 lb during pregnancy.[34] Of these 25 lb, about 11 lb should be fat. The increase in breast and uterine size adds about 3 lb, and the growing placenta another 2 lb. The amniotic fluid eventually weighs 1 lb, and the increase in maternal blood volume adds another 1 lb. The fetus itself weighs about 7 lb at term, bringing the total to 25 lb.

Exercise during pregnancy is recommended as long as the woman maintains her weight and consumes the specific nutritional requirements of the fetus. The amount of exercise recommended usually varies from physician to physician, and it depends a lot on how the woman had exercised previously. In general, however, most forms of exercise are not harmful and are good for the mother's cardiovascular demands in supporting her growing fetus and in controlling excessive weight gain.

TABLE 9-3. COMPARISON OF NUTRITIONAL NEEDS OF A NONPREGNANT, NONLACTATING WOMAN WITH THOSE OF A WOMAN DURING A NORMAL 280-DAY PREGNANCY FOLLOWED BY 180 DAYS OF LACTATION

Nutrient	Total Requirements for a 460-Day Period[a]		
	Nonpregnant, Nonlactating Woman[b]	Total Needs for a Woman during Normal Pregnancy and 6 Months' Lactation	Increased Needs during Pregnancy and Lactation
Calories	920,000	1,156,000	236,000
Protein (g)	25,300	31,700	6,400
Calcium (g)	368	570	202
Iodine (g)	46	62	16
Vitamin A (I.U.)	2,300,000	3,120,000	820,000
Ascorbic acid	25,300	27,600	2,300
Folacin (mg)	184	314	130
Niacin equivalent (mg)	5,980	7,800	1,820
Riboflavin (mg)	690	864	174
Thiamin (mg)	460	578	118

Source: R. Buchanan, "Effects of Childbearing on Maternal Health," *Population Reports,* ser. J, no. 8 (1975), pp. 125–39. Used with permission of the Population Information Program, Johns Hopkins University, Baltimore, Md. Originally from the Pan American Health Organization Technical Group, "Maternal Nutrition and Family Planning in the Americas"; *Report of a PAHO Technical Group Meeting,* Washington, D.C., 1970, p. J-130.

[a] 280 days of pregnancy plus 180 days of lactation.

[b] Based on the needs of a woman 22 years old, weighing 58 kg (128 lb), engaged in light activity.

The Endocrinology of Pregnancy

The endocrine system of a pregnant woman operates differently from that of a nonpregnant woman. Many of the changes in hormone secretion are adaptations to maintain the fetus and to adapt the woman's body to her new nurturing role.

As mentioned in Chapter 7, the corpus luteum formed after ovulation in the menstrual cycle secretes estradiol-17β and progesterone during the luteal phase, but this organ degenerates before menstruation in the nonpregnant menstrual cycle. If implantation occurs, however, the corpus luteum does not die but continues to secrete high amounts of progesterone and low amounts of estradiol during the first trimester of pregnancy. This continued activity of the corpus luteum of a pregnant woman maintains the placenta in a functional condition. Steroids secreted by the corpus luteum of pregnancy also initiate the development of the mammary glands and inhibit ovulation by exerting negative feedback on pituitary gonadotropin secretion (Chapter 7). Progesterone also increases fat deposition in early pregnancy by stimulating the appetite and diverting energy stores from sugar to fat.

What extends the life of the corpus luteum in a pregnant woman? The probable answer is that hCG prevents the corpus luteum from regressing and causes it to continue to secrete estradiol-17β and progesterone. Secretion of hCG by cells of the cytotrophoblast begins as soon as 48 hours after implantation.[35] The human placenta produces GnRH, which may regulate secretion of hCG as it regulates gonadotropin secretion from the anterior pituitary gland.[36] Levels of hCG in the pregnant woman's blood then steadily rise, reaching a peak in the third month of pregnancy (Fig. 9–16). Then, hCG secretion declines, exhibiting another, smaller rise in late pregnancy.

Even though hCG levels peak in the third month, secretion of steroid hormones from the corpus luteum begins to decline after the second month.[37] What causes this slow decline of corpus luteum function in the second month, even while hCG levels are still high, is not known. In sheep, the fetus secretes a luteotropic factor that maintains the corpus luteum, and decline of this factor may be involved in regression of the corpus luteum during the human pregnancy.[38]

After about five weeks of pregnancy, the placenta begins secreting three estrogens—estradiol-17β, estrone, and (predominantly) estriol. The placenta also begins secreting progesterone. Secretion of estrogens and progesterone by the placenta is stimulated by hCG. Levels of these steroid hormones from the placenta increase throughout pregnancy (Fig. 9–16), and they continue to support the placenta and mammary glands and to inhibit ovulation. Can you now explain why removal of the corpus luteum before the seventh week of pregnancy causes miscarriage but not after this time?[39]

The fetus and placenta both are involved in estrogen and progesterone secretion from the placenta. This cooperative arrangement, the *feto-placental unit,* operates as follows (Fig. 9–17): First, the placenta converts cholesterol to progesterone, a con-

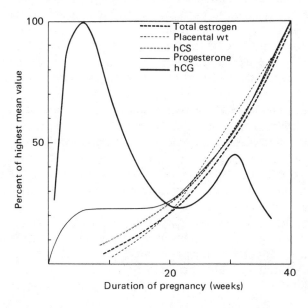

Figure 9-16. Hormone concentrations in a pregnant woman's blood and their relation to placental weight. Total estrogen includes estrone, estradiol-17β, and estriol. Adapted from R. B. Heap, "Role of Hormones in Pregnancy," in *Reproduction in Mammals, Book 3, Hormones in Reproduction,* eds. C. R. Austin and R. V. Short, pp. 73–105 (Cambridge, England: Cambridge University Press, 1972). © 1972; used with permission.

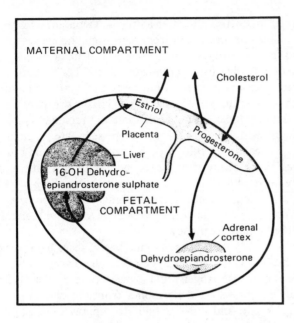

MATERNAL COMPARTMENT

Cholesterol

Estriol

Placenta

Liver

16-OH Dehydro-
epiandrosterone sulphate

Progesterone

FETAL
COMPARTMENT

Adrenal
cortex

Dehydroepiandrosterone

Figure 9–17. The human feto-placental unit, showing how the mother provides cholesterol to the placenta, which converts it to progesterone for release into the maternal and fetal circulations. In the fetus, progesterone is converted to dehydroepiandrosterone (DHEA) by the fetal zone of the adrenal glands. DHEA then is converted to 16-OH DHEA sulphate in the fetal liver. This steroid then goes to the placenta and is converted to estriol, the major estrogen secreted by the placenta. Adapted from D. J. Baird, "Reproductive Hormones," in *Reproduction in Mammals, Book 3, Hormones in Reproduction,* eds. C. R. Austin and R. V. Short, pp. 1–41 (Cambridge, England: Cambridge University Press, 1972). © 1972; used with permission.

version that the fetus is not capable of performing. Then, progesterone passes from the placenta to the fetus and reaches the fetal adrenal glands. These glands of the fetus contain a region called the *fetal zone.* This zone is very large but disappears soon after birth. The fetal zone converts progesterone to large amounts of the weak androgen dihydroepiandrosterone (DHEA), which is changed by the fetal liver to 16-OH DHEA-sulphate. Then, 16-OH DHEA-sulphate is carried in the fetal blood back to the placenta to be converted to the estrogen estriol. Thus, secretion of estrogens by the placenta requires the fetal adrenal glands. The fetal adrenals also secrete cortisol (a corticosteroid hormone), and this hormone may go to the placenta, where it influences estrogen and progesterone secretion. In fact, as we see in the next chapter, cortisol secretion by the fetal adrenal near term may initiate labor in this manner.

The placenta, besides secreting hCG, estrogens, and progesterone, also secretes a protein hormone that is similar in biological effect to pituitary growth hormone and prolactin. This hormone is called *human chorionic somatomammotropin* (hCS), and it rises in the female's blood in late pregnancy (Fig. 9–16) and causes an increase in sugar in the mother's blood.[40] Thus, hCS provides the fetus with additional glucose for its growth. This hormone (along with estrogens and progesterone) also helps prime the mammary glands for later milk secretion.

As mentioned above, the placenta secretes prolactin, which enters the amniotic fluid. The human placenta may also secrete chorionic corticotropin and thyrotropin, both similar to the pituitary hormones of the same name. In addition, prolactin secretion from the mother's pituitary gland increases during pregnancy.[41] Another hormone, *relaxin,* is secreted by the corpus luteum and placenta.[42] Levels of this polypeptide rise during pregnancy, and this substance relaxes the connective tissue connecting the two pubic bones (pubic symphysis) so that the fetus can pass through the birth canal with more ease during labor. Relaxin also helps to efface the cervix and to inhibit premature uterine contractions. As we see in the next chapter, the cervix also dilates during early labor, and relaxin helps prepare the cervix for this event. And it recently has been discovered that the placenta produces endorphins (opiatelike natural painkillers).[43] This may mean that a woman in late pregnancy is less sensitive to pain.

Maternal Complications of Pregnancy

Some ailments of pregnancy are relatively common. These include constipation, headaches, and the development of enlarged veins that bulge under the skin *(varicose veins)*. Many women also report changes in sleep patterns during pregnancy. A woman in the first trimester often requires more sleep than usual, but during late pregnancy, she may have difficulty sleeping.

Toxemia. There are other ailments of pregnancy that can be dangerous to the fetus and mother. For example, *toxemia* is a condition that develops in the last one or two months of pregnancy in 6 to 7 percent of all pregnancies in the United States.[44] It is more common in primiparous women and in multiparous women over 35 years of age than in other females. Consumption of fat and cholesterol during pregnancy increases the risk of developing toxemia. The symptoms of toxemia are rapid weight gain, fluid accumulation in tissues (edema), high blood pressure (hypertension), and an increased excretion of proteins in the urine (proteinuria). Early toxemia is called *preeclampsia.* If this early condition is not controlled by diet, more severe *eclampsia* can develop, characterized by convulsions, coma, and death in about 15 percent of cases. The cause of toxemia is not clear.

Diabetes Mellitus. About 1 out of 350 pregnant women in the United States develops *diabetes mellitus,* a condition in which the pancreas is not producing enough insulin so that the maternal tissues can utilize glucose as energy.[45] As a result, copious amounts of urine, containing glucose, are produced. This condition is not only damaging to the mother if not controlled, but also kills the fetus in 30 percent of cases.

Ectopic Pregnancies. Normally, implantation occurs on the posterior wall of the uterus. If, however, implantation occurs outside the uterus, an *ectopic pregnancy* develops.[46] In the United States about 1 percent of pregnancies are ectopic. About 96 percent of these are in the oviduct; they are called *tubal pregnancies.* The remaining 4 percent are in the abdomen (*abdominal pregnancies*), implantation occurring on the gut, mesenteries, or ovaries. Tubal pregnancies are very dangerous to the mother because the embryo and placenta are growing in a restricted area and the oviduct walls are thin and vascular. These pregnancies are accompanied by pain and serious hemorrhage, and they require surgical removal of the embryo and placenta. Tubal pregnancies account for 10 percent of all maternal deaths in the United States; this is about 1 death per 1000 ectopic pregnancies. When ectopic pregnancies occur in the abdomen, the fetus dies and often is surrounded by calcium. These calcium deposits (*lithopedions,* or "stone babies") often are not discovered until later abdominal surgery for some other reason. In very rare cases, however, abdominal pregnancies can result in the birth of a healthy infant by cesarean section.

Ectopic pregnancies are more common in older, multiparous women, in non-Caucasians, and in women who have had a previous abortion, pelvic infection, or endometriosis than in other women. Birth control devices can have some influence on the incidence of ectopic pregnancies.[47] Barrier contraceptive methods (such as condoms and diaphragms) and use of the combination pill slightly decrease the risk of ectopic pregnancy. On the other hand, use of the minipill, intrauterine devices, and tubal sterilization can lead to an increased risk of ectopic pregnancy (Chapter 13).

Hydatidiform Moles.   In rare cases, although implantation occurs, the implanted body contains swollen chorionic villi and no embryo. Some of these *hydatidiform moles* become malignant and secrete large amounts of hCG. Therefore, they can be mistaken for normal pregnancy. The cells of these moles are diploid, but the chromosomes are all from the father.[48] A *partial hydatidiform mole* is one that contains a dead embryo; these moles are triploid (3N). Hydatidiform moles are removed surgically.

Septic Pregnancy.   A *septic pregnancy* occurs when bacteria enter the uterus and cause infection.[49] This condition is dangerous to both the mother and fetus.

Hemorrhage.   Excessive uterine bleeding during or immediately after labor can be dangerous to the mother if not controlled. One cause of excessive uterine hemorrhage during labor is *placenta previa,* a condition in which the placenta adjoins or covers the cervix so that the fetus is unable to be born properly.[50] Besides resulting in excessive bleeding, this condition sometimes requires a cesarean delivery (Chapter 10). Excessive hemorrhage also can occur if the placenta detaches prematurely, a condition termed *abruptio placenta.* This is not only dangerous to the mother, but in addition, the fetus can no longer receive life-giving materials from the mother.

Spontaneous Abortion.   About 30 percent of fertile zygotes undergo *spontaneous abortion (miscarriage)* before or right after implantation. This occurs because of genetic or chromosomal damage to the cells, immune responses, improper hormonal priming of the uterus, or exposure to drugs or pollutants.

After pregnancy is established, chromosomal or genetic errors account for 42 percent of spontaneous abortions. About 20 percent of established pregnancies terminate by spontaneous abortion, usually in the first trimester.[51] Other factors that can induce spontaneous abortion after pregnancy is established include exposure to teratogens or mutagens, maternal stress, under- or overnutrition, or vitamin deficiencies or excessive amounts of vitamins (especially vitamin A). When pelvic infection causes spontaneous abortion, it is called a *septic abortion.*

As we discussed, a critical stage in pregnancy is just when the corpus luteum is regressing and before the placenta is fully functional and is secreting steroid hormones. Thus, weeks 10 to 14 are marked by a low level of estrogens and progesterone in the mother's blood, and in some women this drop may be low enough to cause placental detachment and spontaneous abortion.

Many women can experience a miscarriage and then have one or more successful pregnancies. (A woman's menstrual cycle usually resumes within three months after a spontaneous abortion.) Other women, however, are functionally infertile because of continually losing pregnancies through miscarriage. This often is because the cervix is not capable of "holding" a fetus to term (an *incompetent cervix*).

In the future, it may be possible to predict the danger of a miscarriage by detecting levels of certain chemicals in a woman's blood. One such chemical is B_1-*glycoprotein,* which is produced by the trophoblast. Abnormally low levels of this chemical after the ninth week of pregnancy have been shown to be associated with future spontaneous abortion.[52]

Spontaneous abortion is an emotionally draining experience for most couples. A woman may feel guilty over her inability to carry a fetus and may be fearful of future pregnancies. Many couples experience a profound degree of grief and sadness after a miscarriage similar to that experienced by those who have lost a child. Counseling is helpful to these people, and women who continually experience miscarriage should seek medical help at infertility clinics.

Even though the above conditions are potentially serious, they seldom result in death of a pregnant woman. In developed countries such as the United States, only about 14 per 100,000 pregnant women die of problems related to pregnancy. In underdeveloped countries, however, this number is about 740 per 100,000 because of poorer health care and nutrition.[53]

Sex during Pregnancy

Coitus presents little danger to the fetus or mother. Coitus, however, should be avoided if there is any uterine bleeding during pregnancy, or if the extraembryonic membranes have ruptured. For this reason, couples are often counseled to avoid coitus during the last few weeks of pregnancy. Oral-genital contact that involves air blown into the vagina could be dangerous in the last few weeks because of possible introduction of air into the woman's bloodstream via the placenta. In general, the period of pregnancy requires sexual adjustments by both parents. A physician or counselor may suggest certain coital positions to be used during this time (Chapter 17).

CHANCES FOR A SUCCESSFUL PREGNANCY

Given the complexity of pregnancy and its possible complications, you may be wondering how on earth a healthy baby is ever born! Well, the human body contains remarkable defenses against malfunction and disease. Of all newborns, about 87 percent will go home as healthy babies. Of the remaining 13 percent, about 2 percent have congenital defects that kill them or result in a major handicap;[54] recently, however, a few successful operations on fetuses with defects, while they are still in the womb, have been performed. There is also a 1 percent chance of a baby dying in the first neonatal month. The other 11 percent are born with minor congenital defects, many of which are barely noticeable. The great chance that a healthy baby will be born and continue to be a healthy child should help alleviate the fear that some couples have about pregnancy.

CHAPTER SUMMARY

Pregnancy is the condition in which a developing human is nurtured within the uterus. There are presumptive, probable, and positive signs of pregnancy. Pregnancy tests, which detect the presence of human chorionic gonadotropin (hCG) in a woman's urine or blood, can utilize bioassay, immunoassay, latex agglutination inhibition, or radioimmunoassay. If a woman believes she is pregnant, she should consult a physician or clinic within two or three weeks after her first missed menses.

After fertilization, cleavage occurs to form a morula, which enters the uterus and develops into a blastocyst. Then, the blastocyst implants in the uterine lining. Implantation requires a uterus primed with estrogen and progesterone.

The embryo forms three primary germ layers (ectoderm, mesoderm, endoderm), which give rise to all adult tissues. Then, the four extraembryonic membranes (yolk sac, amnion, allantois, chorion) form. The allantois and chorion contribute to the fetal portion of the placenta, which is a respiratory, excretory, and nutritive organ for the fetus.

Embryonic and fetal development is complex and is accompanied by growth and development of all adult organ systems. Special adaptations of the fetal circulatory system reflect the use of the placenta (and not the lungs) for fetal respiration. Many fetal organ systems are functional early in development.

Although most fetuses are healthy, fetal disorders sometimes occur, leading to miscarriage or the birth of children with congenital defects. Some of these abnormalities have a genetic or chromosomal basis. Others, however, are caused by exposure of the fetus to teratogens or mutagens. Environmental pollutants and drugs such as aspirin, indomethacin, diethylstilbestrol, alcohol, and nicotine can adversely affect fetal health, as can exposure to radiation. Because of possible fetal abnormality, some women choose to have their fetus examined by amniocentesis, ultrasound, or fetoscopy.

The body of a pregnant woman must support not only itself but the developing fetus. Under the influence of hCG secreted by the placenta, the corpus luteum life span is prolonged to last into the first trimester of pregnancy. After this time, the corpus luteum regresses, and the placenta secretes not only hCG but human chorionic somatomammotropin (hCS) as well as estrogens and progesterone. The fetus interacts with the mother in controlling estrogen secretion from the placenta. Some maternal complications of pregnancy include toxemia, diabetes mellitus, placenta previa, ectopic pregnancy, hydatidiform moles, septic pregnancy, and maternal hemorrhage. Nevertheless, the maternal death rate associated with pregnancy is low, at least in developed countries. About 20 percent of established pregnancies end in miscarriage. Of newborns, 2 percent have serious birth defects, and 11 percent have minor problems. The remaining 87 percent are perfectly healthy.

NOTES

1. W. B. Hunt, II, "Pregnancy Tests—The Current Status," *Population Reports,* ser. J, no. 7 (1975), pp. 109-24; and Anonymous, "Simplified Methods of Assaying Urinary HCG Early in Pregnancy," *Research in Reproduction,* 11, no. 1 (1979), 3.

2. Anonymous, "Test Yourself for Pregnancy," *Consumer Reports* (November 1978), pp. 644-45; and M. Clark et al., "Home Tests for Pregnancy," *Newsweek* (September 1979), p. 69.

3. B. B. Saxena, "Criteria for Clinically Valid Measurement of Human Chorionic Gonadotropin," *Research in Reproduction,* 12, no. 4 (1980), 1-2.

4. R. P. S. Jansen, "Fallopian Tube Isthmic Mucus and Ovum Transport," *Science,* 201 (1980), 349-51.

5. Anonymous, "Progesterone and Enzymes in the Blastocyst at Implantation," *Research in Reproduction,* 4, no. 4 (1972), 2.

6. Ibid.

7. Z. Dickman, "Steroidogenesis in Preimplantation Embryos," *Research in Reproduction,* 7, no. 1 (1975), 3; and Z. Dickman, "Systemic versus Local Hormonal Requirements for Blastocyst Implantation: A Hypothesis," *Perspectives in Biology and Medicine* (Spring 1979), pp. 390-93.

8. F. Logeat et al., "Local Effect of the Blastocyst on Estrogen and Progesterone Receptors in the Rat Endometrium," *Science,* 207 (1980), 1083-85.

9. L. D. Wiley, "Presence of a Gonadotropin on the Surface of Preimplanted Mouse Embryos," *Nature, London,* 252 (1974), 715-16.

10. Anonymous, "Are Prostaglandins Involved in Implantation in Women?" *Research in Reproduction,* 10, no. 4 (1978), 2.

11. Anonymous, "Antigens of Human Trophoblast," *Research in Reproduction,* 11, no. 4 (1979), 3.

12. Anonymous, "The Immunobiology of Pregnancy," *Research in Reproduction,* 5, no. 1 (1973), 4.

13. Anonymous (1979), op. cit.

14. Anonymous, "The Differentiation of the Blastocyst," *Research in Reproduction,* 11, no. 4 (1979), 1-2.

15. P. Beaconsfield et al., "The Placenta," *Scientific American,* 243, no. 2 (1980), 94-102.

16. Ibid.

17. S. L. Kaplan and M. M. Griembach, "The Ontogenesis of Human Fetal Hormones. II. Luteinizing Hormone (LH) and Follicle Stimulating Hormone (FSH)," *Acta Endocrinologica, Copenhagen,* 81 (1976), 808-29.

18. Anonymous, "Prolactin in Amniotic Fluid," *Research in Reproduction,* 11, no. 3 (1979), 4.

19. A. M. Perks, "Developmental and Evolutionary Aspects of the Neurohypophysis," *American Zoologist,* 17 (1977), 833–50.

20. T. Friedmann, "Prenatal Diagnosis of Genetic Disease," *Scientific American,* 225, no. 5 (1971), 34–42.

21. H. B. Taussig, "The Thalidomide Syndrome," *Scientific American,* 207, no. 2 (1962), 29–35.

22. G. B. Kolata, "How Safe Is Bendictin?," *Science,* 210 (1980), 518–19; Anonymous, "Bendictin Linked to Birth Defects," *Science News,* 122, no. 1 (1982), 7; and L. Garmon, "Anti-Nausea Drug Linked to Stomach Birth Defects," *Science News,* 123, no. 1 (1983), 5.

23. Anonymous, "Effects of DES Exposure Reevaluated," *Drug Therapy* (May 1979), p. 26; and H. A. Bern and A. L. Herbst, "Present Problems and Future Concerns," in *Developmental Effects of Diethylstilbestrol (DES) in Pregnancy,* eds. A. L. Herbst and H. A. Bern, pp. 194–98 (New York: Threme-Stratton, Inc., 1981).

24. D. F. Dinges et al., "Fetal Exposure to Narcotics: Neonatal Sleep Disturbance as a Measure of Nervous System Disturbance," *Science,* 209 (1980), 619–21.

25. N. Dishotsky et al., "LSD and Genetic Damage," *Science,* 172 (1971), 431–40.

26. Anonymous, "THC, Placentas: The Estrogen Connection," *Science News,* 119, no. 1 (1981), 9.

27. S. Coleman et al., "Tobacco-Hazards to Health and Human Reproduction," *Population Reports,* ser. L, no. 1 (1979), pp. 1–37.

28. S. Linn et al., "No Association between Coffee Consumption and Adverse Outcomes of Pregnancy," *New England Journal of Medicine,* 302 (1982), 141–45.

29. Anonymous, "Maternal Aspirin, Indomethacin Are Bad for Fetal Lungs," *Medical World News* (February 19, 1979), p. 21.

30. Anonymous, "Doctor May Get Bill for Lifelong Care," *Medical World News* (February 19, 1979), p. 29.

31. Friedmann (1971), op. cit.

32. Anonymous, "Amniocentesis Kills 1.5% of Fetuses in British Study," *Medical World News* (February 19, 1979), p. 25.

33. S. Campbell et al., "Anencephaly: Early Ultrasonic Diagnosis and Active Management," *Lancet,* ii (1972), 1226–27; and J. Arehart-Treichel, "Fetal Ultrasound: How Safe?" *Science News,* 121, no. 24 (1982), 396–97.

34. M. Winick, "Food for the Fetus," *Natural History Magazine,* 90, no. 1 (1981), 79–81.

35. Hunt (1975), op. cit.

36. Anonymous, "Polypeptides and Their Function in the Placenta," *Research in Reproduction,* 12, no. 3 (1980), 2.

37. Anonymous, "Endocrinology of Human Pregnancy," *Research in Reproduction,* 7, no. 2 (1975), p. 1; and R. de Hertogh et al., "Plasma Levels of Unconjugated Estrone, Estradiol and Estriol and of HCS throughout Pregnancy in Normal Women," *Journal of Clinical Endocrinology and Metabolism,* 40 (1975), 93–101.

38. Anonymous, "Luteotropic Actions of Mammalian Embryos," *Research in Reproduction,* vol. 11, no. 2 (1979).

39. Anonymous (1975), op. cit.

40. Ibid.

41. H. G. Friesen, "Prolactin and Human Reproduction," *Research in Reproduction,* 8, no. 3 (1976), 3–4.

42. M. X. Zarrow et al., "The Concentration of Relaxin in the Blood Serum and Other Tissues of Women during Pregnancy," *Journal of Clinical Endocrinology and Metabolism,* 15 (1955), 22–27.

43. Anonymous (1980), op. cit.

44. R. Buchanan, "Effects of Childbearing on Maternal Health," *Population Reports,* ser. J, no. 8 (1975), pp. 125–39.

45. Ibid.

46. D. A. Edelman, "Contraceptive Practice and Ectopic Pregnancy," *IPPF Medical Bulletin,* 14, no. 3 (1980), 1–3.

47. Ibid.

48. Anonymous, "The Genetics of Hydatidiform Moles," *Research in Reproduction,* 11, no. 4 (1979); and Anonymous, "Androgenetic Origin of Hydatidiform Moles," *Research in Reproduction,* 12, no. 1 (1980), 4.

49. Buchanan (1975), op. cit.

50. Ibid.

51. J. L. Marx, "Drugs during Pregnancy: Do They Affect the Unborn Child?" *Science,* 180 (1973), 174–75.

52. Anonymous, "Pregnancy Specific B_1-Glycoprotein and Early Pregnancy," *Research in Reproduction,* 13, no. 1 (1981), 2–3.

53. Buchanan (1975), op. cit.

54. Marx (1973), op. cit.

FURTHER READING

ANONYMOUS, "Test Yourself for Pregnancy," *Consumer Reports,* 43, no. 11 (1978), 644–45.

AUSTIN, C. R., "Pregnancy Losses and Birth Defects," in *Reproduction in Mammals, Book 2, Embryonic and Fetal Development,* eds. C. R. Austin and R. V. Short. Cambridge, England: Cambridge University Press, 1972. Pp. 134–52.

BEACONSFIELD, P. et al., "The Placenta," *Scientific American,* 243, no. 2 (1980), 94–102.

CEKAN, Z., "Steroid Biosynthesis in the Human Foeto-Placental Unit," *Research in Reproduction,* vol. 4, no. 3 (1972), map.

FREIDMANN, T., " Prenatal Diagnosis of Genetic Disease," *Scientific American,* 225, no. 5 (1971), 34–42.

GROBSTEIN, C., "External Human Fertilization," *Scientific American,* 240, no. 6 (1979), 57–67.

HEAP, R. B., "Role of Hormones in Pregnancy," in *Reproduction in Mammals, Book 3, Hormones in Reproduction,* eds. C. R. Austin and R. V. Short. Cambridge, England: Cambridge University Press, 1972. Pp. 73–105.

LIGGINS, G. C., "The Fetus and Birth," in *Reproduction in Mammals, Book 2, Embryonic and Fetal Development,* eds. C. R. Austin and R. V. Short. Cambridge, England: Cambridge University Press, 1972. Pp. 72–109.

MARX, J. L., "Drugs and Pregnancy: Do They Affect the Unborn Child?" *Science,* 180 (1973), 174–75.

MCLAREN, A., "The Embryo," in *Reproduction in Mammals, Book 2, Embryonic and Fetal Development,* eds. C. R. Austin and R. V. Short. Cambridge, England: Cambridge University Press, 1972. Pp. 1–42.

TANNER, J. M., *Foetus into Man: Physical Growth from Conception to Maturity.* London: Open Books, 1978.

TAUSSIG, H. B., "The Thalidomide Syndrome," *Scientific American,* 207, no. 2 (1962), 29–35.

10   *Labor and Birth*

LEARNING GOALS

Having read this chapter, you should be able to:

1. List two ways in which date of birth (due date) can be calculated.
2. Discuss the evidence for cycles in human births, and possible adaptive reasons for these cycles.
3. List evidence that the fetus controls the onset of labor in sheep, and discuss how this control occurs.
4. Describe evidence that sheep and humans are similar or dissimilar in the hormonal control of parturition.
5. Describe the role of oxytocin in birth, and the factors that may increase the secretion of oxytocin.
6. List evidence that prostaglandins are involved in human parturition.
7. List the three stages of labor, and describe what happens to the mother and fetus during these stages.
8. List methods and reasons for artificially inducing labor.
9. Define fetal distress, and list methods for determining fetal condition during labor.
10. List methods for stopping uterine hemorrhage after delivery of the placenta.
11. List the differences between immature, premature, and postmature births.
12. List some causes of premature births.
13. Describe the birth of twins and how it differs from single births.

14. Describe causes of difficult births, and reasons for performing cesarean or forceps deliveries.
15. List kinds of medication used to ease the discomforts of labor, and discuss the pros and cons of the use of such medication during childbirth.
16. Describe how prepared childbirth methods differ from traditional hospital births.
17. List some of the benefits and dangers of home births, and how the dangers can be avoided.

INTRODUCTION

In the last chapter, we saw that pregnancy involves physiological changes and psychological adjustments in the future mother as well as remarkable development and growth of the fetus. The internal relationship between mother and fetus terminates in childbirth, or *parturition*. In this chapter we first discuss the timing of birth and the role of hormones in the birth process. Then we see how knowledge about hormonal control of birth has allowed us to induce labor artificially. Next, we describe the stages of labor and birth and some aspects of premature, multiple, and difficult births. Finally, use of medications during labor and birth and new birthing methods are reviewed.

TIME OF BIRTH

The average length of pregnancy in humans is about nine calendar months, but labor can begin more than 2 weeks before or after the expected date of birth (see Table 10-1). The *due date* (or *term)* can be calculated by counting 280 days (40 weeks) from the first day of the last menstruation. This can be done quickly by adding 7 days to the first day of the last menstrual period and then counting back three months. The due date also can be determined by counting 265 days (38 weeks) from conception. Usually, ovulation and conception occur 13 to 15 days after the first day of the last menstruation, but conception also can happen at other lengths of time after day 1 of the cycle (Chapter 7). This is one reason why babies often are not born on the due date if it is calculated by date of conception. Female infants tend to be born a few days earlier than males, and women who exercise tend to give birth sooner. Also, women who have short menstrual cycles tend to have shorter pregnancies. The reasons for these differences are unknown.

There are subtle seasonal and daily cycles in the number of births. In the Northern Hemisphere, more babies are born in the last six months of the year, with more births occurring in July through October than during the rest of the year. Thus, slightly more conceptions occur in late fall and early winter. This slight seasonality in human birth may be an evolutionary remnant of a more pronounced seasonal cycle that occurs in some other primates as an adaptation to seasonal food supplies for their young. Also, more babies are born at night than during the day, with a small birth peak between 1:00 A.M. and 7:00 A.M.[1] No one knows why, but this nocturnal

TABLE 10-1. DAY OF DELIVERY, MEASURED FROM THE FIRST DAY OF THE LAST MENSTRUAL PERIOD

Day of Delivery	Percent
Before the 226th day	12.7
Second week before due date (days 266–272)	12.3
First week before due date (days 273–297)	22.1
On 280th day (the due date)	2.7
First week after due date (days 281–287)	24.2
Second week after due date (days 288–294)	15.6
After 294th day	9.4

Source: Adapted from *The Biology of People* by Sam Singer and Henry R. Hilgard (San Francisco: W. H. Freeman & Company, Publishers, 1978), p. 382. © 1978; used with permission.

peak is more pronounced in home births than in hospital births.[2] This daily birth cycle may be an evolutionary vestige in that nocturnal birth in our ancestors may have offered protection against predators active during the day.

HORMONES AND BIRTH

For nine months following conception, the human fetus has undergone marked development and has increased greatly in size (Chapter 9). Suddenly, it is expelled into the outer world. What factors determine when birth occurs? It is remarkable that much of our present answer to this question comes from observations of the food habits of sheep!

Delayed Birth in Sheep

Sheepherders in Idaho noticed that pregnant ewes grazing in certain pastures at specific times of the year failed to give birth to their lambs on time. Instead, extra-large lambs were born many days late, and the mother often died in the process. In a search for the cause of this delayed birth, investigators discovered that although the tissues of the mother appeared normal, the lambs had underdeveloped adrenal glands and malformed or absent pituitary glands.[3] These fetal abnormalities occurred because the pregnant ewes ate a plant called *Veratrum californicum*. This plant contains a chemical (an alkaloid) that passes across the placenta and harms the pituitary and adrenal glands of the fetus.

Further research has shown that birth was delayed in these sheep because the plant alkaloid reduced secretion of corticotropin (ACTH) from the fetal pituitary gland. ACTH is needed for the fetal adrenal glands to develop and secrete adrenal steroid hormones (Chapter 2). In the Idaho sheep, therefore, the fetal adrenal glands did not secrete steroid hormones, which play a major role in initiating birth. If the pituitary gland of a normal sheep fetus is removed (hypophysectomy), its adrenal glands are underdeveloped, and pregnancy is extended 37 days past the normal due date. Injection of ACTH into hypophysectomized sheep fetuses results in normal fetal adrenal glands, and delivery time is normal. Similarly, removal of both fetal adrenal glands *(adrenalectomy)* delays birth, and injection of adrenal steroid hormones into these fetuses reverses the effect of adrenalectomy. Finally, injection of either adrenal hormones or ACTH into the fetus any time after the second half of pregnancy results in early pregnancy. Normally, therefore, some as yet unknown factor causes release of *corticotropin-releasing hormone* (CRH) from the fetal hypothalamus. This, in turn, causes secretion of ACTH, secretion of fetal adrenal hormones (*cortisol,* for example), and ultimately initiation of birth.

How does fetal adrenal hormone secretion initiate birth? The apparent answer is that this hormone travels to the placenta and increases placental secretion of an estrogen.[4] Progesterone made by the placenta serves as a precursor for fetal cortisol (an adrenal steroid hormone) and an androgen (Chapter 9), both of which are carried in the blood to the placenta. The arrival of these steroid hormones at the placenta increases placental estrogen secretion and decreases placental progesterone secretion.[5] Thus, the ratio of estrogen to progesterone (E/P ratio) in the mother's blood increases. In many mammals, this ratio in the blood of pregnant females increases near the onset of birth. Estrogens stimulate contraction of the uterine muscles, whereas progesterone inhibits uterine contractions. Therefore, the increase in the E/P ratio in the blood initiates uterine contractions *(labor).*[6]

Hormonal Initiation of Human Birth

Is there evidence that a similar role of the fetal adrenal glands is present in humans as well as sheep? The answer is "maybe." Recent information shows that there is an increase in the E/P ratio in the blood of women in the last five weeks of pregnancy.[7] This elevated ratio may cause uterine contractions. Administration of progesterone to women who are undergoing premature labor does not delay labor initiation, but estrogen treatment does cause early labor.[8] Significantly, women exhibiting early delivery tend to have newborns with relatively large adrenal glands, whereas women who have delayed delivery have fetuses with relatively underdeveloped adrenal glands.[9] The latter women also have relatively low levels of estrogen in their blood.[10] *Postmature newborns* (babies born more than two weeks after the due date) have low levels of cortisol in their blood.[11] Thus, fetal adrenal glands may play a role in birth initiation in humans, but direct evidence is needed to substantiate this claim.[12]

Besides the ratio of estrogen to progesterone in the maternal blood, other hormonal factors appear to be important in initiating or maintaining human labor. Prostaglandins are a group of fatty acids widely distributed in various tissues (Chapter 2). Some prostaglandins stimulate contraction of uterine muscles and now are used by physicians to induce labor at term. But do prostaglandins play a role in normal human birth? We know that prostaglandin levels increase in the mother's blood a month before delivery and are very high during labor, [13] with bursts in secretion 15 to 45 sec after each uterine contraction. [14] We also know that levels of prostaglandin in the amniotic fluid increase during labor. [15] Certain drugs (such as aspirin) inhibit synthesis of prostaglandins, and these drugs may delay labor if given routinely near term. [16] In some mammals, secretion of fetal adrenal hormones increases synthesis of prostaglandins by the placenta. Prostaglandins also increase estrogen secretion from the human placenta. [17] In conclusion, prostaglandins do play a role in human labor.

It has long been known that a hormone secreted by the posterior lobe of the pituitary gland stimulates contraction of uterine muscle (Chapter 2). This hormone, oxytocin (in Greek, "quick birth"), appears also to be involved in human birth. In fact, a synthetic oxytocin, *pitocin,* often is used by physicians to hasten labor contractions. In humans, mechanical stimulation of the vagina, cervix, and/or uterus causes release of oxytocin, this reaction being called the *fetal ejection reflex.* [18] Once oxytocin is secreted, it increases the intensity of uterine contractions, which in turn causes more oxytocin release. A high E/P ratio as well as an increase in prostaglandins in the mother's blood causes oxytocin release in some mammals. [19]

In women, oxytocin levels in the blood are moderately low in the first stage of labor, and then higher and more variable in the second stage. [20] Therefore, this hormone may simply increase the intensity of uterine contractions during later stages of labor. A recent theory suggests, however, that even though oxytocin levels in the blood are low at the time of labor initiation, the number of oxytocin receptors in the uterus increases. Thus, oxytocin may actually initiate labor both directly and indirectly by causing release of prostaglandins from the uterus. [21] Ethyl alcohol inhibits oxytocin release, [22] and in the past it was not unusual for a physician to prescribe alcohol intravenously for women in premature labor. This practice is seldom done today. Instead, the woman may be asked to go home and have a glass of wine!

We have much to learn about the hormonal initiation of labor in humans. The sequence and relative importance of the processes described above are not well understood. Figure 10-1 presents a summary of how labor is initiated. You should realize, however, that mechanisms in other mammals may not be the same in humans. Indeed, some recent theories on control of parturition in man suggest that genetically programmed maturation of the placenta controls the onset of labor in women, and that the fetal adrenal glands may at best only help determine the exact day that this event occurs. [23]

INDUCED LABOR

Physicians can cause *induced labor* before term, near term, or after the due date, by administering either pitocin or prostaglandins, both of which are equally effective. [24] First, however, the physician will break the amniotic membrane (Chapter 9) because this often begins labor and hormones will not need to be administered. Pitocin is ad-

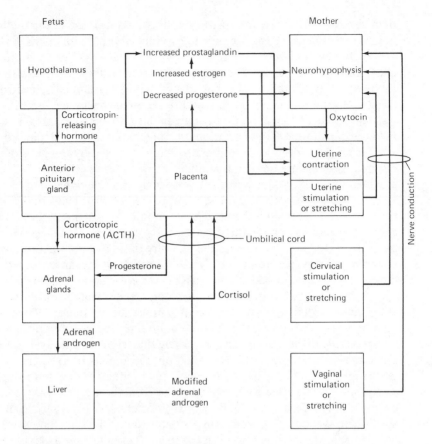

Figure 10–1. Diagram of the possible factors controlling the onset of labor in mammals. The fetal adrenal glands use progesterone supplied by the placenta to secrete steroid hormones (e.g., cortisol and the androgen dehydroepiandrosterone), which go to the placenta and increase the estrogen/progesterone ratio and prostaglandin secretion. These changes then increase uterine contraction directly, and indirectly by causing release of oxytocin. Stimulation of the uterus, cervix, and vagina also increases oxytocin release. This is a hypothesis mainly based on research using domestic and laboratory mammals, and much remains to be learned about the factors that initiate labor in women.

ministered intravenously (by inserting a needle into a vein) in continuous slow drops monitored by a pump. Prostaglandins also can be given intravenously or are injected into the amniotic sac, given orally, or administered as a vaginal suppository.

One reason for inducing labor is that the birth is two weeks overdue (i.e., the fetus is postmature). It should be noted, however, that many "late" pregnancies are miscalculations of the due date, and there may be danger of delivering a premature baby if labor is induced in such cases. The physician, however, usually can determine fetal maturity by physical and physiological characteristics. About 8 to 12 percent of babies are born postmaturely in the United States. Also, labor can be artificially induced after 28 to 36 weeks of pregnancy, and babies born this way are premature, as we discuss later. This can be done, for example, if the amniotic sac has burst 12 to 24 hours beforehand and labor has not yet begun. A broken amniotic sac increases the danger of neonatal and maternal infection.

PREPARATION FOR LABOR

About two or three weeks before labor, a woman may have a sensation of decreased abdominal distention produced by movement of the fetus down into the pelvic cavity. This is known as *lightening*. Or it is said that the baby has "dropped." Lightening occurs about two weeks before birth in a woman having her first baby, but may not occur until labor begins in her subsequent pregnancies. A woman literally breathes easier after lightening because of less pressure on her diaphragm (a dome-shaped muscle under the rib cage that assists in breathing). Also, she may urinate more frequently because the fetus is now pressing on her bladder. A few hours to a week before labor begins, the part of the fetus that will exit first (usually the head) moves down into the pelvic girdle. This is termed *engagement of the presenting part*.

THE BIRTH PROCESS

The birth process can be divided into three stages: (1) *cervical effacement and dilatation,* (2) *expulsion of the fetus,* and (3) *expulsion of the placenta.* The length of each stage varies among individuals and in the same individual between first and subsequent births. The entire process of labor and birth lasts from 8 to 14 hours in women giving birth for the first time (*primiparous* women), but is shorter (4 to 9 hours) in women who have previously had a child (multiparous women). Any length of labor up to 24 hours, however, is considered normal.

Stage 1: Cervical Effacement and Dilatation

Throughout pregnancy, and especially in the last few weeks, a woman can experience mild, irregular uterine contractions. These *Braxton-Hicks contractions* have little rhythmic pattern. They sometimes cause the abdominal wall to become hard to the touch, but usually are not felt and subside after a few minutes. Some women have *false labor* contractions during late pregnancy, which they may experience as being moderately intense. These contractions can be rhythmic in nature, but they do not cause much cervical effacement or dilatation. Also, they will go away after awhile.

Eventually, true labor commences; the uterine contractions become more intense, and they occur at regular intervals. These *effacement contractions* usually are felt in the back and then on the abdominal wall; they reach a peak and then relax. Some women may never be aware of them. The contractions can be felt and timed by placing a hand on the upper abdomen. Each contraction lasts about 30 to 60 seconds, with intervals between contractions of about 5 to 20 minutes (Fig. 10-2). The result of effacement contractions is *cervical effacement,* which means a thinning of the normally thick walls of the cervix and retraction of the cervix upward, making it easier for the fetus to pass into the *birth canal* (cervical and vaginal canals).

During pregnancy, the cervix is blocked by a *mucous plug.* At or immediately before the beginning of effacement contractions, mucus is dislodged along with a small amount of blood, and this *bloody show* (pinkish in color) exits through the vagina. Also at this time, or in the late first stage, a small tear appears in the amniotic sac and clear amniotic fluid trickles or gushes from the sac and is expelled from the vagina. This bursting of the amniotic sac *(breaking of the bag of waters)* and the bloody show are sure signs that true labor is commencing. In about 12 percent of pregnancies, the amniotic sac breaks before labor begins. These "dry labors" proceed normally but often are shorter than usual. It is also common for the sac to remain intact after labor has advanced considerably, and in these cases the physician will puncture the amnion with an instrument (this does not hurt the mother, since there are no pain receptors in the amnion).

A pregnant woman should notify her doctor of the advent of effacement labor, bloody show, or leaking of amniotic fluid. If she is planning a hospital birth, she may be advised to relax at home for a few hours, especially if this is her first baby. Leaving for the hospital before this time may not be necessary, even if the woman or her partner is nervous and anxious to get going. Meanwhile, they can time the contractions by feeling when the abdomen becomes hard and soft, and can report to their doctor the duration of contractions and the interval between the initiation of successive contractions. The woman should not eat during this time, since digestion is inhibited during labor, and vomiting should be avoided. An informed woman usually feels excited and confident during this stage of labor. Many physicians advise women to leave for the hospital when contractions are about 5 minutes apart—from the beginning of one contraction to the start of the next. Of course, women who are a long distance from the hospital will be advised to leave earlier!

After a variable amount of time, the woman will notice that her contractions begin to last longer (about 60 seconds), are more intense, and occur more frequently, with rest intervals of only 1–3 minutes (Fig. 10-2). She has now entered the early dilatation portion of stage 1; the result of these contractions is the beginning of *cervical dilatation.* The external cervical os increases from its normal 0.3-cm diameter to a diameter of about 7 cm. The entire early dilatation process lasts about 5 to 9 hours in primiparous women, but is shorter (2 to 5 hours) in multiparous women because the cervix is more pliable. The *dilatation contractions,* because of their frequency and intensity, may become uncomfortable. If the mother is still at home, she should enter the hospital soon after these contractions begin. When she arrives at the hospital, she will be admitted, prepared for labor, and given a bed in the labor or birthing room. A nurse or her physician then will check her for her degree of cervical dilatation and often will rupture her amniotic sac if this has not occurred naturally. After washing the genital area, all or some of the pubic hair may be shaved at this time to help pro-

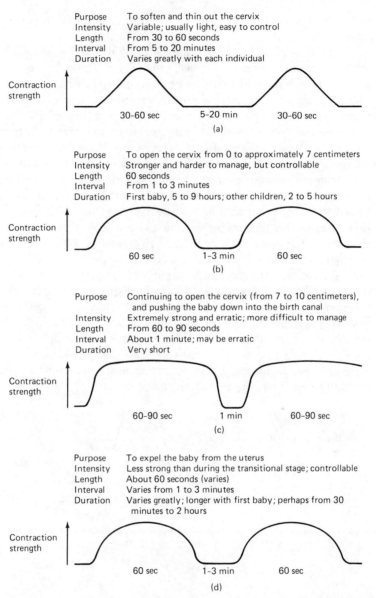

Purpose   To soften and thin out the cervix
Intensity   Variable; usually light, easy to control
Length   From 30 to 60 seconds
Interval   From 5 to 20 minutes
Duration   Varies greatly with each individual

Contraction strength

30-60 sec   5-20 min   30-60 sec

(a)

Purpose   To open the cervix from 0 to approximately 7 centimeters
Intensity   Stronger and harder to manage, but controllable
Length   60 seconds
Interval   From 1 to 3 minutes
Duration   First baby, 5 to 9 hours; other children, 2 to 5 hours

Contraction strength

60 sec   1-3 min   60 sec

(b)

Purpose   Continuing to open the cervix (from 7 to 10 centimeters),
            and pushing the baby down into the birth canal
Intensity   Extremely strong and erratic; more difficult to manage
Length   From 60 to 90 seconds
Interval   About 1 minute; may be erratic
Duration   Very short

Contraction strength

60-90 sec   1 min   60-90 sec

(c)

Purpose   To expel the baby from the uterus
Intensity   Less strong than during the transitional stage; controllable
Length   About 60 seconds (varies)
Interval   Varies from 1 to 3 minutes
Duration   Varies greatly; longer with first baby; perhaps from 30
            minutes to 2 hours

Contraction strength

60 sec   1-3 min   60 sec

(d)

Figure 10-2. Characteristics and purpose of stage 1 [(a) effacement, (b) early dilatation, and (c) transition dilatation] and stage 2 [(d) expulsion of fetus] contractions. (*Note:* Times given are approximate.) Stage 3 (expulsion of the placenta) contractions are not shown; these contractions are similar to those in early dilatation, but the duration of the stage is shorter. Adapted from D. Ewy and R. Ewy, *Preparation for Childbirth. A Lamaze Guide* (Boulder, Colo.: Pruett Publishing Co., 1972). © 1972; used with permission of Pruett Publishing Co. © 1982 D. Ewy and R. Ewy.

tect against infection. An enema (induction of defecation) often is given, except if labor is advanced.

The final phase of stage 1 labor, during which the cervix dilates from about 7 cm to 10 cm in diameter, is called *transition dilatation*. This phase lasts about 20 minutes to 1 hour and tends to be shorter in multiparous women. Transition is characterized by very intense *transition contractions* that last longer (60 to 90 seconds) than those in earlier stages of dilatation (Fig. 10-2). The interval between transition contractions is about 1 minute, but often is erratic. In primiparous women, the cervix usually dilates after it effaces, but these two events occur together in multiparous women (Fig. 10-3). This is one reason why stage 1 of labor happens more quickly in multiparous women. During transition, the fetus descends into the pelvic basin, which puts pressure on the pelvic floor (Fig. 10-4). The woman has an urge to push during these contractions, but she is advised not to do so. Pressure on the pelvic floor creates this pushing urge, which, I am told, feels like an urge to defecate. Pushing before complete dilatation to 10 cm will tire the mother and not move the fetus, and could cause edema of the cervix.

Transition is the most difficult part of labor, not only because of the severity of the contractions but because a woman may experience nausea, vomiting, trembling, leg cramps, discouragement, and restlessness. Normally, the face of the fetus points toward the sacrum of the back, but sometimes the hard back of its head is toward the sacrum. When the latter situation is present, back pain *(back labor)* is felt during contractions. Transition does not last very long, and medications to relieve discomfort often are used during this phase. A woman also can assume an upright or squatting position to allow gravity to help the labor process.

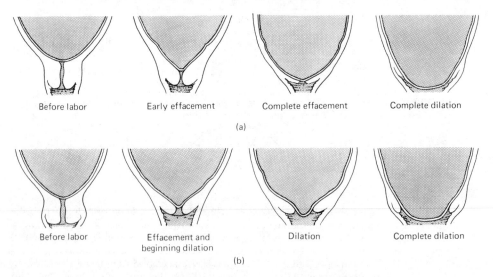

Before labor   Early effacement   Complete effacement   Complete dilation

(a)

Before labor   Effacement and   Dilation   Complete dilation
                beginning dilation

(b)

Figure 10-3. Degrees of cervical effacement and dilatation. Note that in primiparous women (a), effacement occurs before dilatation. In multiparous women (b), effacement and dilatation occur together.

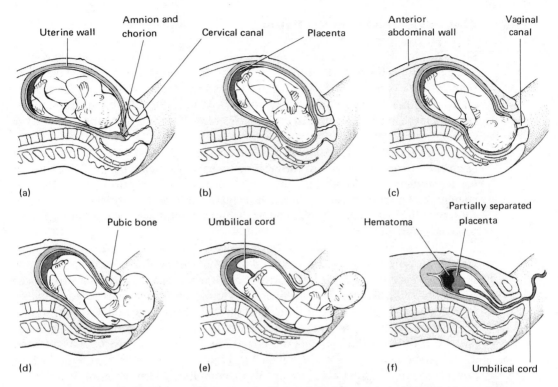

Uterine wall

Amnion and chorion

Cervical canal

Placenta

Anterior abdominal wall

Vaginal canal

(a)

(b)

(c)

Pubic bone

Umbilical cord

Hematoma

Partially separated placenta

(d)

(e)

(f)

Umbilical cord

Figure 10–4. Diagrams illustrating the birth process. The cervix is dilated during the first stage of labor (a and b). The fetus passes through the cervix and vagina during the second stage of labor (c to e). Note that the fetal head rotates and first bends forward (flexion) and then back (extension). Then, as the uterus contracts, the placenta folds up and pulls away from the uterine wall (f). Separation of the placenta results in bleeding, forming a large blood clot (hematoma). Later, the placenta is expelled. Adapted from K. L. Moore, *The Developing Human. Clinically Oriented Embryology,* 3rd ed. (Philadelphia: W. B. Saunders Company, 1982). © 1982; used with permission.

During transition, the condition of the fetus can be monitored for signs of *fetal distress.* Fetal heart function is monitored by placing wire leads either on the mother's abdomen or on the fetus's scalp through the cervical opening. For some women, *fetal monitoring* procedures are an unwanted intervention. These procedures, however, can have several benefits, including a reduction in unnecessary cesarean deliveries (see below), and early detection of fetal distress.

After full dilatation (10 cm) is reached, the mother is moved to the delivery room. (If she is in a "birthing room," both labor and delivery will occur there.) A multiparous woman may be moved before full cervical dilatation. In the delivery room, she moves onto a delivery table, where her legs are placed in stirrups with supports under her knees. Her inner thighs, abdomen, and genital regions then are cleansed with an antiseptic. The region then is partially covered with a sterile sheet.

Stage 2: Expulsion of the Fetus

Stage 2 of labor, expulsion of the fetus, begins when the cervix is maximally dilated and ends with delivery of the infant. The intensity of contractions in this stage is less than that in the transition portion of stage 1. Each contraction lasts about 60 seconds, with 1- to 3-minute rest intervals between contractions (Fig. 10-2). Expulsion of the fetus happens in a shorter time in multiparous women; the duration for all women is one-half to two hours.

Once the woman is settled on the delivery table, the physician may perform an *episiotomy* to prevent tearing of the perineal tissues as the baby emerges. To do this, a local anesthetic is injected into the *perineum* (the region between the anus and vagina), and a small incision is made in the perineal skin. This incision later will be sutured with absorbable material and heals quickly, although the mother may have some later discomfort.

The woman is encouraged to push during expulsion contractions, and the increased pressure due to this pushing and to the uterine contractions begins to move the fetus's head through the birth canal. Soon the top of the head begins to appear without receding between contractions. This show of the fetal head is called *crowning* (Fig. 10-4) and signifies that the baby is about to be born. The bones of the infant's head are not yet fused, and they overlap when the head is moving out. This facilitates passage of the fetus through the birth canal. The "pointed" appearance of the head disappears soon after delivery (Chapter 11). Usually, the head emerges with the face down. Once the head is out, any mucus or amniotic fluid in its nose or mouth is removed with a suction device. The infant rotates during labor so that the shoulders emerge in the up and down position that is the largest dimension of the birth canal (Fig. 10-4).

Once the shoulders have emerged, the physician determines if the umbilical cord is around the neck, a rather common occurrence with no special dangers if detected at this point. The infant then slides out (Fig. 10-5), takes his or her first breath, and emits an exhilarating cry. (The old tradition of slapping the back of the infant to stimulate its first breath usually is not necessary.) The physician then swabs or suctions the nose and mouth with a rubber bulb to remove any mucus. The umbilical cord then is clamped in two places about 3 in. from the baby's abdomen and cut between the clamps. There are no nerve endings in the cord, and neither the mother nor infant feels the procedure. A nurse puts drops of penicillin or silver nitrate in the infant's eyes to prevent bacterial infection. This is required by law in all states because without this treatment, the infant could be blinded by gonorrhea bacteria if the mother is infected (Chapter 16).

Stage 3: Expulsion of the Placenta

During expulsion of the placenta (stage 3), which lasts about 15–30 minutes, the placenta *(afterbirth)* is expelled. After the fetus is delivered, the next few contractions push the placenta, which has become detached from the uterine wall, through the

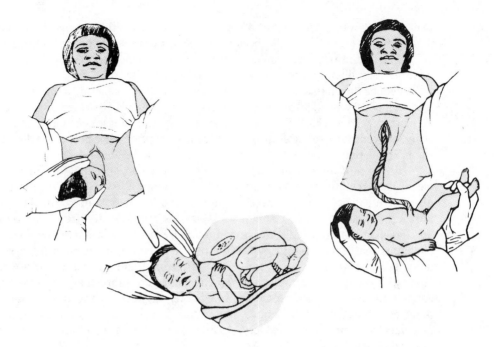

Figure 10-5. Stages in the final emergence of the infant. Adapted from W. G. Birch, *A Doctor Discusses Pregnancy* (Chicago: Budlong Press, 1963). © 1963; used with permission.

birth canal. If it does not come out, the physician will gently massage the abdomen or pull on the umbilical cord to help expulsion of the placenta. About 8 oz of blood normally are lost during delivery. Usually a nurse will massage the uterus through the abdominal wall to encourage contraction, which inhibits uterine blood flow. If uterine hemorrhage (excessive bleeding) persists, oxytocin or *Ergonamine* (derived from a fungus that grows on rye) is administered to contract the uterus and inhibit bleeding. Next, the episiotomy incision is sutured, and the mother is ready to attend to her infant.

PREMATURE BIRTHS

Prematurity is determined by the weight, not age, of the newborn. When a newborn weighs less than 2.49 kg (5.5 lb) but greater than 0.99 kg (2 lb, 3 oz), it is considered to be *premature*. If it weighs between the 0.99 kg (2 lb, 3 oz) and 0.5 kg (1 lb, 1.5 oz), it is *immature*. Below this lowest weight, the fetus usually is *stillborn* (born dead). A fetus normally weighs about 2.5 kg (5.5 lb) about one month before the due date, as compared to the average weight of 3.40 kg (7.5 lb) at birth. In the United States, about 7.7 percent of births are premature, and whites (for unknown reasons) tend to have fewer premature births than blacks do. Almost one-half of twins are born prematurely. The average birth weight of twins (2.49 kg; 5.5 lb) is less than the average birth weight of single infants (3.40 kg; 7.5 lb) because each twin must share maternal nutrients.

The organs of premature newborns are not fully developed, and this and other possible fetal disorders that may have precipitated the premature birth in the first place reduce the chance for survival of these infants. In general, when development has reached 40 weeks, 99 percent survive, but this figure is 50 percent at 31 weeks and 10 percent at 27 weeks. Of infants born with less than 23 weeks of development, less than 1 percent survive. Some aspects of the care of premature babies are discussed in the next chapter.

We do not understand fully what causes premature labor. The increased incidence of prematurity in twins may be due to greater distension of the uterus (fetal ejection reflex) or to a higher production of fetal adrenal hormones from two pairs of adrenals instead of one. Premature babies, as mentioned, often have oversized adrenal glands. Other evidence that premature labor is initiated by a hormonal change is that levels of estradiol in the mother's blood increase during premature labor.[25] Some women simply are prone to having premature babies; if a woman's first baby was premature, her subsequent babies are more than likely to be born early. Maternal disorders (discussed in Chapter 12) such as infection of the uterus, diabetes, hypertension, toxemia, and premature separation of the placenta are associated with premature delivery. Also fetuses suffering from birth defects tend to be born prematurely.

MULTIPLE BIRTHS

Considering all births, the odds for having twins are about 1 in 80; for triplets, 1 in 6400; and for quadruplets, 1 in 512,000. However, heredity can influence the odds of having fraternal (but not identical) twins. Fraternal twins occur more commonly in women who have a family history of twins (Chapter 9). Racial factors and age also appear to influence the incidence of fraternal twins; twinning is more likely to occur in blacks than in whites, and is least frequent in Orientals. Women in their thirties have more fraternal twins than those in their twenties; and women who have been receiving certain fertility drugs tend to have multiple births (Chapter 15). Delivery of multiple fetuses occurs about 22 days earlier, on the average, than do single births.

In about 70 percent of cases, the presence of twins can be diagnosed before they are born. This can be done by ultrasound (discussed in Chapter 9), ascertaining excessive fetal movement, or detection of two heartbeats. When twins are born, both can come out head first, or one can be head first and the other in a breech position (see below). One twin usually is expelled a few minutes to one hour before the other, but there are records of the second fetus being delivered up to 56 days after the first! A few of these cases where a twin is born many days after its sibling may be due to *superfetation*. That is, fertilization of a new ovulated egg occurs while a previous fetus is developing in the uterus. This would have to occur before the fourth month of pregnancy because after that time the amniotic sac obliterates the uterine lumen and would not allow sperm passage. There is, however, no direct proof of superfetation in humans.

DIFFICULT FETAL POSITIONS

In 95 percent of all births, the fetus presents in the normal, head-down position. In 3 to 4 percent of births, however, the fetus is in a *breech presentation* at the beginning of labor, which means that the feet or buttocks rest against the cervix (Fig. 10-6). Actually, 50 percent of all fetuses are normally in the breech presentation before the seventh month of pregnancy; most then naturally turn 180° to the normal head-down position before the ninth month. Breech deliveries often occur with no difficulties (although labor is longer), but they sometimes require cesarean delivery. In 1 out of 200 births, the fetus is in a *transverse presentation,* the shoulders and arms coming out first (Fig. 10-6). In these cases, a cesarean delivery is necessary.

HANDLING DIFFICULT BIRTHS

Forceps Delivery

If the fetus is not emerging easily, the physician can insert an instrument (forceps) into the birth canal and around the head, to affect a *forceps delivery*. The instrument then is gently pulled or twisted to assist in expulsion of the fetus. The forceps are inserted as two separate steel blades, the inner surfaces of which are curved to fit the

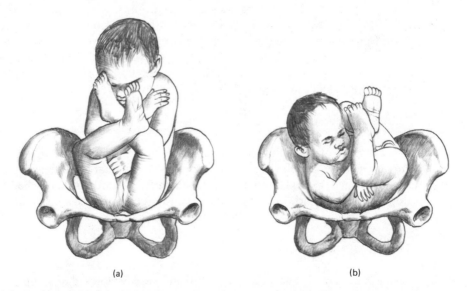

(a)   (b)

Figure 10–6. Breech (a) and transverse (b) fetal presentations. These relatively uncommon fetal positions may require forceps delivery or cesarean section.

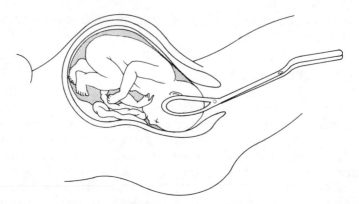

Figure 10–7. Use of forceps in delivery of the fetus. Some marks may appear on the infant's head where the forceps were applied, but these go away soon after birth. Forceps deliveries should be done only in some specific situations (see text). Adapted from W. G. Birch, *A Doctor Discusses Pregnancy* (Chicago: Budlong Press, 1963). © 1963; used with permission.

fetus's head (Fig. 10-7). After both blades are in place, the handles are joined. Forceps deliveries can be done if the head of the fetus is resting properly in the pelvis, the membranes are ruptured, and the cervix is dilated maximally (10 cm).

Some medical reasons for using forceps are: (1) acute distress of the fetus, such as irregular or weak heartbeat and lack of oxygen caused by premature separation of the placenta, compression of the umbilical cord, or excessive pressure on the fetal head; (2) illnesses of the mother, such as heart problems, tuberculosis, or toxemia; (3) a previous cesarean section, since the wall of the uterus might tear; (4) presentation of the fetus in a breech position; and (5) an abnormally slow labor. Some physicians, however, use forceps as a matter of routine, a practice not without controversy. The fetus rarely is damaged by forceps if they are used when the fetus is crowning and not sooner. Anesthetics are required during a forceps delivery.

Vacuum Extraction

A new method of extracting the fetus is used commonly in Europe. In this *vacuum extraction* method, a metal cup is placed on top of the fetal head, negative pressure is applied to this cup, and the cup is attached firmly. The fetus is then pulled out. There are some reports, however, of damage to the fetus using this instrument, and the method is not commonly used in the United States.

Cesarean Delivery

Cesarean deliveries (Latin *caedere* = "to cut") are performed in an operating room after a spinal or general anesthetic is given. (A few cesarean deliveries have been done under *acupuncture,* which involves pain relief by inserting steel needles in specific body regions.) A cesarean operation usually lasts 20 to 90 minutes. Apparently, the name of this operation had its origin in an order by Emperor Julius Caesar of such an operation to be done on dying pregnant women in hopes of saving the unborn children.

In this procedure, an abdominal incision is made below the navel in the midline and through the uterine wall, and the baby is removed (Fig. 10-8). In Caesar's time, few if any of these operations were successful, but with today's modern surgical and antiseptic techniques, these operations are relatively safe for the mother and infant. In fact, women can have four or five babies or more by this method. Cesarean delivery is sometimes performed not only when the fetus is in a transverse presentation or less commonly in a breech presentation but also when the pelvis of the woman is too small, the fetus is too large, or when the fetus shows signs of distress, such as abnormalities in heart function. Also, cesarean delivery is performed if the umbilical cord gets compressed between the head and the wall of the birth canal, if the placenta is coming out before the fetus (placenta previa), or if the placenta separates from the uterus prematurely. All of these situations are dangerous to the unborn child because the oxygen supply to the fetus is reduced, and prompt delivery is needed.

Some feel that too many Cesareans are being done. This type of delivery accounted for 5.5 percent of deliveries in 1970, and has increased to 18 percent today (about 500,000 annually in the United States). Reasons given for this increase include: (1) the assumption that once a woman has a cesarean, she can never deliver a future child vaginally (this is not true in many cases); (2) using a cesarean delivery for a breech birth (this often is not necessary); (3) the increased use of fetal monitoring to detect fetal problems during labor; and (4) the increasing pregnancies of older women. Many physicians argue that cesarean deliveries decrease fetal and maternal death. Others, however, are questioning the unnecessary use of this operation.[26]

After a cesarean delivery, a woman usually stays in the hospital two to four days longer than do those giving birth vaginally. This is because the stitched incision

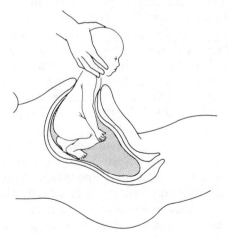

Figure 10-8. Removal of infant through incision in abdomen and uterus during a cesarean delivery. Adapted from W. G. Birch, *A Doctor Discusses Pregnancy* (Chicago: Budlong Press, 1963). © 1963; used with permission.

in the lower abdominal wall limits normal movement. Infants delivered with this operation can nurse normally, although the suckling response of some may be slow to develop. Milk production in mothers after this operation is normal once the infant's suckling pattern is established. Some women having this operation may feel inadequate in one way or another, but there appears to be no influence of cesarean delivery on the relationship between the mother and her child.

USE OF MEDICATIONS DURING LABOR

Anesthetics, which obliterate all sensations, can be used to relieve discomfort during childbirth. The anesthetic can be inhaled in the form of a gas. This procedure (general anesthesia), however, is usually not used during labor because of possible dangers to the baby and because the woman is unable to experience or participate in the birth process.

Conduction anesthetics are given by injection, and there are several procedures for giving anesthesia in this way during labor. In an *epidural,* the anesthetic is injected into the outside membranes of the spinal cord. A *caudal* is similar, but it is done lower in the back. The epidural and caudal numb sensations in the body below the point of injection. The woman, however, still is able to use her muscles in this region to assist delivery. A *spinal anesthetic* is injected into the space just under the membranes surrounding the spinal cord near the center of the back, whereas a *saddle block* is injected into the same space lower in the back. The spinal and saddle blocks numb both sensory and motor nerves below the point of injection, so the woman is unable to use the muscles of this region to help with delivery. In a *paracervical,* the anesthetic is injected into both sides of the cervix. Finally, in a *pudendal block* the anesthetic is given into the pudendal nerve on each side of the vagina. The paracervical and pudendal blocks are used to ease the process of the baby passing through the birth canal.

Analgesics are sometimes used during labor to ease pain. They can be given intramuscularly (injection into the muscle) or intravenously (injection into a vein). Analgesics promote relaxation between uterine contractions and relieve discomfort.

When any medication is used during labor and delivery, the choice of which to use and how it is given should be the result of communication between the physician and the patient. Often, a woman can consult with her physician before labor begins, and together they can decide on the choice of medication if needed. Questions to be asked about a medication are: "Is it safe for the mother?" Is it safe for the fetus?" and "Will it lengthen the duration of labor?"

Although we do not have much information on the subject, it is possible that certain medications can affect the newborn adversely. For example, lack of oxygen *(anoxia)* can harm the newborn,[27] and some anesthetics depress the newborn's respiration. Moreover, medications used during labor can cross the placenta into the fetus. When this occurs, the fetus is not able to inactivate the drugs efficiently because its kidneys and liver, two organs that are involved in breakdown and excretion of drugs, are not as functional as are those of adults. Also, the fetal brain is not fully developed and may be susceptible to drug damage. There is disturbing evidence of adverse effects from drugs used during labor on the development and behavior of infants and even children up to seven years after they are born. These adverse effects are seen on heart rate, suckling and feeding behavior, language development, alertness, mother-infant interactions, and cognitive development. Controversy, however, surrounds these findings, [28] and we need to know much more about the potential effects of these medications on infant development. Certainly, medication is necessary in difficult birth situations. But because use of medications during labor can prevent the mother from totally experiencing the birth of her baby, and because of possible adverse effects of these drugs on the mother and fetus, new methods of prepared childbirth have become popular in recent years.

NEW BIRTHING METHODS

In the middle 1900s, most women requested medication during their labor. The father, instead of participating in the birth process, was usually banished to a waiting room where he nervously and helplessly paced the floor. In 1933, Dr. Grantly Dick-Read first presented a new concept that he called "natural childbrith." [29] He proposed that childbirth without drugs was better for the mother and infant. In the past, most women were convinced that childbirth was filled with pain, suffering, and danger. Even the Bible says, "In sorrow thou shalt bring forth children" (Genesis 3:16). But labor means "hard work," not suffering, and it has been found that much of the childbirth pain is intensified by the fear of pain. If the mother is well-versed in the biological events during the birth process and is physically and psychologically prepared for the process, the pain is greatly lessened and more manageable; often no medications are needed during "natural" labor.

Recently, many birthing methods have appeared that can be called *prepared* (or "controlled," "natural," "cooperative") *childbirth*. Of these, the *Lamaze method* has been the most popular. This method was introduced to the Western world in 1951 by Fernand Lamaze (1890–1957), a French physician. It previously was used in Russia. In this method, the mother is taught not only the biology of the birth process but also the techniques of controlled breathing and relaxation to manage discomfort and pain. Her partner plays an important role in these exercises and in the delivery itself. Medications are used during Lamaze births if the situation requires them, but the parents have an intelligent choice in the matter. Besides the advantages of avoiding use of medications, the mother is wide awake during the whole process, and both parents can more directly experience the birth of their child.

Some women are now choosing to combine some prepared childbirth method with delivery in the comfortable surroundings of their own homes. The most common problems are lack of oxygen for the fetus or newborn (usually due to compression of the umbilical cord) and hemorrhage of the mother. *Home births* may not be prepared to cope with these and other unforeseen circumstances, which is why many physicians are opposed to home births. It is recommended that these home births should be attended by a physician or *nurse midwife* (a registered nurse with further education in midwifery). If home birth is planned, arrangements should be made for medical help in case of emergencies. Because of the potential dangers of home births, which now make up about 1 percent of births in the United States, many hospitals have constructed *birthing rooms* modeled after the home. Both labor and delivery can occur in these rooms.

CHAPTER SUMMARY

The average length of pregnancy is 280 days, but parturition can occur before or after this due date. There are slight seasonal and daily cycles in the number of human births. In sheep, the fetal adrenal glands secrete cortisol and an androgen, both of which travel to the placenta and increase placental secretion of an estrogen and decrease placental secretion of progesterone. These high blood levels of estrogen in the female sheep interact with oxytocin and prostaglandins to initiate labor. Evidence that a similar system operates in humans is circumstantial. Physicians can induce labor artificially with synthetic oxytocin or prostaglandins.

Labor progresses in three stages: (1) cervical effacement and dilatation, (2) expulsion of the fetus, and (3) expulsion of the placenta. Labor lasts longer in primiparous than in multiparous women. Causes of premature births can relate to hormonal factors in the fetus or mother, as well as to fetal or maternal diseases or abnormalities. Multiple births produce small-sized newborns, who often are premature. Forceps delivery can be performed if the fetus is in the breech presentation, if fetal distress exists, or if some difficulties are encountered in delivery because of maternal disease or abnormality. Cesarean delivery can be performed when the fetus is in the breech or transverse position, when fetal distress is present, or if labor is not proceeding normally.

Medications used for reduction of discomfort during labor include anesthetics and analgesics. The choice of drug or method used should result from communication between physician and patient. New methods of prepared childbirth incorporate education about labor and birth with controlled relaxation and breathing techniques and assistance of a partner. Home births can be a joyful experience, but participants should be aware of potential dangers. Medical assistance by a physician or midwife is advisable in such births.

NOTES

1. J. D. Palmer, "Human Rhythms," *Bioscience,* 27, no. 2 (1977), 93–99.
2. R. T. W. L. Conroy and J. N. Mills, pp. 112–14, *Human Circadian Rhythms* (London: Churchill, 1970).

3. W. Binns et al., "Toxicosis of *Veratrum californicum* in Ewes and Its Relationship to a Congenital Deformity in Lambs," *Annals of the New York Academy of Science,* 111 (1964), 571–76.

4. G. C. Liggins et al., "The Mechanism of Initiation of Parturition in the Ewe," *Recent Progress in Hormone Research,* 29 (1973), 111–50.

5. J. R. G. Challis, "Physiology and Pharmacology of PGs in Parturition," *Population Reports,* ser. G, no. 5 (1974), pp. 45–56.

6. Ibid.

7. A. C. Turnball et al., "Significant Fall in Progesterone and Rise in Oestradiol Levels in Human Peripheral Plasma before Onset of Labor," *Lancet,* i (1974), 101–4.

8. R. M. Pinto et al., "Action of Estradiol-17β at Term and at the Onset of Labor," *American Journal of Obstetrics and Gynecology,* 98 (1967), 540–46.

9. A. B. M. Anderson et al., "Fetal Adrenal Weight and the Cause of Premature Delivery in Human Pregnancy," *Journal of Obstetrics and Gynecology, British Commonwealth,* 78 (1971), 481–88.

10. V. A. Frandsen and G. Stakemann, "The Site of Production of Oestrogenic Hormones in Human Pregnancy. II. Experimental Investigations on the Role of the Foetal Adrenal," *Acta Endocrinologica, Copenhagen,* 43 (1963), 184–94.

11. U. Nwoso et al., "Possible Role of the Fetal Adrenal Glands in the Etiology of Post-Maturity," *American Journal of Obstetrics and Gynecology,* 121 (1975), 366–70.

12. A. I. Csapo and C. Wood, "The Endocrine Control of the Initiation of Labour in the Human," in *Recent Advances in Endocrinology,* 8th ed., ed. V. H. T. James, pp. 207–39 (London: Churchill, 1968).

13. S. M. M. Karim, "Appearance of Prostaglandin $F_{2\alpha}$ in Human Blood during Labour," *British Medical Journal,* 4 (1968), 618–21.

14. S. C. Sharma et al., "Prostaglandin $F_{2\alpha}$ Concentrations in Peripheral Blood during First Stage of Normal Labor," *British Medical Journal,* 1 (1975), 709–11; and J. R. G. Challis et al., "Maternal and Fetal Prostaglandin Levels at Vaginal Delivery and Cesarean Section," *Prostaglandins,* 6 (1974), 281–88.

15. M. J. M. C. Keirse and A. C. Turnbull, "E Prostaglandins in Amniotic Fluid during Late Pregnancy and Labor," *Journal of Obstetrics and Gynecology, British Commonwealth,* 80 (1973), 970–73.

16. R. B. Lewis and J. D. Schulman, "Influence of Acetyl Salycyclic Acid, an Inhibitor of Prostaglandin Synthesis, on the Duration of Human Gestation and Labour," *Lancet,* ii (1973), 1159–61.

17. E. Alsat and L. Cedard, "Stimulatory Action of Prostaglandins on Production of Oestrogens by Human Placenta Perfused in Vitro," *Prostaglandins,* 3 (1973), 145–53.

18. J. A. Coch et al., "Oxytocin Equivalent Activity in the Plasma of Women in Labor and during the Puerperium," *American Journal of Obstetrics and Gynecology,* 91 (1965), 10–17.

19. J. S. Roberts and L. Share, "Effects of Progesterone and Estrogen on Blood Levels of Oxytocin during Vaginal Distention," *Endocrinology,* 84 (1969), 1076–81.

20. Coch et al. (1965), op. cit.

21. A. Fuchs et al., "Oxytocin Receptors and Human Parturition: A Dual Role for Oxytocin in the Initiation of Labor," *Science,* 215 (1982), 1396–98.

22. F. J. Zlatnik and F. Fuchs, "A Controlled Study of Ethanol in Threatened Premature Labor," *American Journal of Obstetrics and Gynecology,* 112 (1972), 610–12.

23. G. C. Liggins et al., "Control of Parturition in Man," Biology of Reproduction, 16 (1977), 39–56; and M. Serón-Ferré and R. B. Jaffe, "The Fetal Adrenal Gland," *Annual Review of Physiology,* 43 (1981), 141–62.

24. Challis (1974), op. cit.

25. R. L. T. Raja et al., "Endocrine Changes in Premature Labour," *British Medical Journal,* 4 (1974), 67–71.

26. G. B. Kolata, "NIH Panel Urges Fewer Cesarean Births," *Science,* 210 (1980), 176–77.

27. Anonymous, "Anoxia May Underlie Much of Neonatal Deafness," *Journal of American Medical Association,* 241 (1979), 2360; and W. F. Windle, "Brain Damage by Asphyxia at Birth," *Scientific American,* 221 (1969), 76–84.

28. S. H. Broman, "Obstetrical Medication Study," *Science,* 205 (1979), 446; and Y. Brockbill, "Answer to Broman," *Science,* 205 (1979), 447–48.

29. G. Dick-Read, *Natural Childbirth* (London: Heinemann, 1933).

FURTHER READING

CEKAN, Z., "Steroid Biosynthesis in the Human Foeto-Placental Unit," *Research in Reproduction,* vol. 4, no. 3 (1972), map.

CHALLIS, J. R. G., "Physiology and Pharmacology of PGs in Parturition," *Population Reports,* ser. G, no. 5 (1974), pp. 45–56.

EWY, D., AND R. EWY, *Preparation for Childbirth. A Lamaze Guide.* Boulder, Colo.: Pruett Publishing Co., 1972.

FINDLAY, A. L. R., "The Control of Parturition," *Research in Reproduction,* vol. 4, no. 5 (1975), map.

GUTTMACHER, A. F., *Pregnancy, Birth and Family Planning.* New York: Signet, 1973.

KARMEL, M., *Thank you, Dr. Lamaze.* Garden City, N.Y.: Dolphin Books, 1975.

LIGGINS, G. C., "The Fetus and Birth," in *Reproduction in Mammals, Book 2, Embryonic and Fetal Development,* eds. C. P. Austin and R. V. Short. Cambridge, England: Cambridge University Press, 1972. Pp. 72–109.

LIGGINS, G. C., et al., "Control of Parturition in Man," *Biology of Reproduction,* 16 (1977), 39–56.

MILINAIRE, C., *Birth.* New York: Harmony Books, 1974.

WINDLE, W. F., "Brain Damage by Asphyxia at Birth," *Scientific American,* 221 (1969), 76–84.

11 *The Neonate and the New Parents*

15. Describe the effects of suckling on endocrine glands of the mother.
16. Discuss the effectiveness of nursing as a contraceptive measure.
17. List possible reasons why a woman may choose not to breast-feed her baby.
18. List several advantages to the infant of breast milk over artificial infant formulas or cow's milk.
19. List some disadvantages to the infant of breast-feeding.
20. List some possible advantages and disadvantages to the mother of breast-feeding and bottle-feeding.

INTRODUCTION

The newborn child (*neonate*) is expelled into an unfamiliar and in some ways harsh environment. A few hours before birth, it was resting comfortably in a warm, quiet bath of amniotic fluid and was receiving its nutrition and oxygen from its mother. Suddenly, it is forced out of this peaceful existence and must adapt to life outside. That is, it now must breathe with its lungs, urinate and defecate, digest external nutrient sources, and combat the microorganisms and fluctuating temperatures of its new environment. Also, it must develop behaviors that elicit care from others. Similarly, the mother has undergone a joyful yet stressful experience. She now must adjust to the marked physiological and anatomical changes of the *postpartum* (after birth) period, and must face the responsibilities, along with her mate, of caring for a helpless infant. In this chapter, we discuss the condition and care of the newborn and the adjustments of the parents to this new arrival.

TREATMENT OF THE NEWBORN

In the previous chapter, we reviewed how the umbilical cord is cut, mucus is removed from the airways, and a substance such as silver nitrate is put in the newborn's eyes to prevent infection. The baby now is checked for possible bone breakage or dislocations due to its stressful passage through the birth canal, and its length, head circumference, heart rate, and rectal temperature are recorded. Average newborn measurements are presented in Table 11–1.

Apgar Score

The *Apgar score* is named after Virginia Apgar who invented the procedure. It is a rating of the general level of well-being of the newborn. Numerical values, from 0 to 2, are given to five responses of the newborn (Table 11–2). These five are (1) heart rate, (2) respiratory effort, (3) muscle tone, (4) reflex irritability, and (5) color of the skin. Thus, a maximal score of 10 can be obtained. A baby with a score of 7 to 9 is normal or only slightly depressed, one with a score of 4 to 6 is moderately depressed, and a score of 0 to 3 reflects severe health problems. About 80 percent of newborns in

TABLE 11-1. THE AVERAGE NEWBORN

Measurements	Average
Weight	3400 g (7.5 lb)
Height/length	50 cm (20 in.)
Head circumference	33 cm (13 in.)
Chest circumference	30 cm (11.8 in.)
Heart rate	120–160 beats/min
Body temperature	35–36°C (97–99°F)

Source: Adapted from *Maternal Nutrition and the Course of Pregnancy. Summary Report,* U.S. Department of Health, Education and Welfare, Publication (HSM) 72–5600, p. 8 (Washington, D.C.: U.S. Govt. Printing Office, 1971). Used with permission.

the United States receive a score of 7 or above, and usually no alarm is raised for scores of 6 or above. Newborns with low scores require intensive care immediately after delivery.

Leboyer Method

The *Leboyer method* of newborn care is becoming increasingly popular in the United States today.[1] This method assumes that most hospital delivery rooms are stressful and harmful to the infant because of the presence of cold temperatures, loud noises, and bright lights. With the Leboyer method, the delivery room is kept at a warmer temperature than usual, noise is kept to a minimum, and lights are dimmed. The newborn also is immersed in a warm bath. Close physical contact between the mother and her baby is encouraged. There is no strong scientific evidence that this method improves the emotional or physical health of the newborn. In fact, early stress may be

TABLE 11-2. THE APGAR NEWBORN SCORING SYSTEM

Sign	Score		
	0	1	2
Heart rate	Not detectable	Below 100	Above 100
Respiratory effort	Absent	Slow (irregular)	Good (crying)
Muscle tone	Flaccid	Some flexion of extremities	Active motion
Reflex irritability	No response	Grimace	Vigorous cry
Color[a]	Pale	Blue	Pink

[a] If the natural skin color of the child is not white, alternative tests for color are applied, such as color of mucous membranes of mouth and conjunctiva, lips, palms, hands, and soles of feet.

beneficial to the infant (Chapter 6). Early mother-infant contact, however, has proven to be important to the development of the mother-infant bond, as we see later in this chapter.

Circumcision

Circumcision has been, and still is, a cultural and religious custom in many regions of the world. This procedure involves surgical removal of the prepuce of the penis of a male infant; some traditions practice this procedure later in life. Females are also circumcised in some cultures; this involves removal of the prepuce of the clitoris, or even the entire clitoris, labia minora, and part of the labia majora. In Egypt, for example, about 95 percent of women have been circumcised.[2] In the Hebrew religion, male infants are circumcised on the eighth day of life, as a symbol of the convenant between God and the Jewish people (Genesis 17:9-27). Choosing this day is medically sound since it is the first day that the infant produces its own vitamin K, which helps blood to clot.

A majority of male infants in the United States are circumcised. The reasons parents choose this procedure are complex. Circumcision allows better hygiene of the glans penis; that is, smegma and bacteria can be removed more easily (young boys, however, easily can be taught how to clean their penis). There is also some evidence that circumcised men develop less cancer of the penis and their wives less cancer of the cervix.[3] Others feel that the absence of the prepuce will allow more sexual pleasure during coitus (but there is no difference between circumcised and uncircumcised men in this regard).[4] Additionally, some parents have their son circumcised because they feel he (or they) will be embarrassed if he is not and his peers are.

Male circumcision is done anytime in the first week of life, usually on the second day. No anesthetic is used because it may harm the infant. After the baby is physically immobilized, the prepuce is removed with a knife or scissors, and the region is bandaged. The procedure is painful to the baby, and some feel it is an unnecessary trauma. In general, there is increasing concern about the value of male circumcision in the United States.[5]

ADAPTATIONS OF THE NEWBORN

As mentioned, the newborn is thrust from the uterus into a new environment, and now we discuss some ways that it adapts to this abrupt change.

The Respiratory System

The newborn takes its first breath immediately after delivery. The air sacs in its lungs are coated with *surfactant,* which is a lipoprotein that affects surface tension in the lungs. Presence of this surfactant is necessary so that the lung air sacs do not collapse during expiration. If little or no surfactant is present, the newborn's lungs are unable to fill with air, leading to *respiratory distress syndrome*. In the previous chapter, we saw how the fetal adrenal glands secrete cortisol soon before birth. This hormone, besides playing a role in labor, causes secretion of the surfactant in the fetal lungs. Premature babies often have not secreted enough surfactant, but this condition can be alleviated by administration of corticotropin (ACTH), which causes the adrenal glands to secrete cortisol.

The newborn is quite capable of sneezing and hiccoughing, and will do so often, sometimes to the distress of parents. These phenomena, however, are normal reflexes that clear the respiratory tract of congestion. Newborns also are loud criers! In fact, this crying serves to elicit care-giving behavior by adults and increases the development of lung efficiency. It is noteworthy that analyses of the sounds of crying newborns can detect certain abnormalities. That is, sick babies cry differently from healthy ones![6]

The Circulatory System

In Chapter 9, we saw how the cardiovascular system of the fetus has adaptations that permit use of the placenta as a respiratory, excretory, and nutritional organ. These adaptations include the ductus arteriosus, ductus venosus, and foramen ovale. Only later do these structures close anatomically, but they are closed functionally soon after birth. That is, as soon as the umbilical cord is tied, changes in blood pressure allow blood to flow to the newborn's lungs and through its liver, which causes functional closure of the fetal shunts. If the structures fail to close, the lungs will not receive enough blood, and thus the tissues of the infant will not have enough oxygen. For example, *patent ductus arteriosus* is when this fetal shunt fails to close. This can be due to a birth defect, prematurity, or the mother having ingested too much aspirin when she was pregnant (Chapter 9). Patent ductus arteriosus leads to a bluish color of the newborn skin, a condition called cyanosis ("blue baby"). Blue babies can be given anti-prostaglandins to close the ductus arteriosus.[7] If this fails, immediate heart surgery is necessary.

The Digestive Tract

The primary nutrient of the fetus is glucose supplied by the mother. However, after the baby is born, the placenta can no longer act as a source of this nutrient. Instead, the newborn's heart and liver release glucose into the blood for energy, but these sugar stores are depleted within a few hours. Then, fat is broken down into fatty acids and glycerol, which in turn are converted to more glucose. This energy should last until the baby receives milk from the mother or a bottle, but often sugar water is fed to the infant to supplement energy (and water) requirements. Meanwhile, the newborn's digestive tract must begin to secrete digestive enzymes to process external food sources. A dark green, sticky feces (meconium) is defecated by the newborn up to the fourth day after birth. The feces change to a yellowish or brownish color when the infant receives milk.

Phenylketonuria (PKU) is an extremely rare inherited condition that can lead to mental retardation. It is a disorder of protein metabolism that results in high levels of the amino acid *phenylalanine* in the newborn's blood and urine (Chapter 12). Every baby is tested for PKU by pricking its heel and dabbing a drop of blood onto a special test paper. If the test is positive, the infant usually is put on a phenylalanine-free diet to control the disorder.

Thermoregulation

The temperature of the delivery room often is several degrees below the temperature of the womb, and the infant's capacity to maintain a normal body temperature is not yet fully developed. Therefore, one way that the infant maintains its body temperature is by heat production from stores of *brown fat* around its vital organs and blood vessels. Brown fat differs from white fat in that its cells contain very high numbers of mitochondria, the energy-producing organelles. When brown fat is broken down immediately after birth, the mitochondria release large amounts of heat that warm the newborn's body. It is interesting that hibernating mammals use a similar mechanism to warm their bodies after their winter sleep.[8]

The Nervous System

The newborn can hear, see light and dark as well as movement, smell, and respond reflexively. Several reflexes can be tested to detect if the nervous system is functioning normally. One example is the *grasp reflex*. If you press an object on the infant's palm, near where the fingers join, the infant will flex its fingers and grasp the object. Similarly, its toes will flex if you stroke the sole of the foot; this is the *plantar reflex*. If the infant is subjected to a jolt or loud noise, it will respond with the *Moro* (or *startle*) *reflex*. That is, it will stiffen its body, draw up its legs, and fling its arm up and out. Then, it will bring its arms forward. These reflexes appear to be primitive adaptations to hold onto the mother. If you rub a finger on the hungry infant's mouth or cheek, it will turn its head to keep in contact with the finger. This *rooting reflex* is used to stimulate the infant to find the breast. If you touch the infant's lips, it will display the *sucking reflex*. Walking reflexes also exist in the newborn, and exercise of these reflexes leads to an earlier onset of walking.[9]

WHAT A NEWBORN LOOKS LIKE

Some parents, upon seeing their new baby, wonder if it should have stayed in longer! Its head often is egg-shaped because the unfused skull bones were squeezed together as its head passed through the birth canal. However, the head rounds out in a few days, although the skull bones remain unfused for a year or more. The openings between the bones are "soft spots," or *fontanels,* on the head (Fig. 11–1). These are normal and slowly change to bone in the first year of life. Sometimes a thick, soft swelling occurs on the part of the scalp that once rested against the cervix. This *caput*

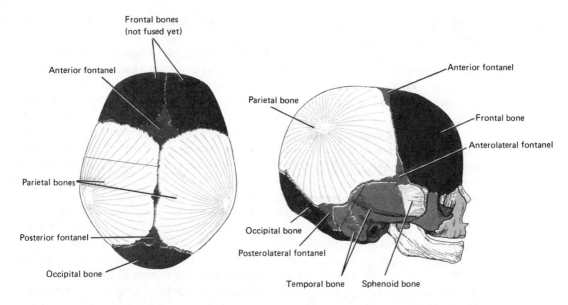

Frontal bones
(not fused yet)

Anterior fontanel

Parietal bones

Posterior fontanel

Occipital bone

Anterior fontanel

Parietal bone

Occipital bone

Posterolateral fontanel

Temporal bone   Sphenoid bone

Anterior fontanel

Frontal bone

Anterolateral fontanel

Figure 11-1. Fontanels of the newborn skull. These "soft spots" on the newborn's head are places where the skull bones have not yet fused. Adapted from *Principles of Anatomy and Physiology,* 3rd ed., by Gerard J. Tortora and Nicholas P. Anagnostakos (New York: Harper & Row, Publishers, Inc., 1981), Fig. 7–3, p. 149. © 1981 by Gerard J. Tortora and Nicholas P. Anagnostakos. Reprinted by permission of Harper & Row, Publishers, Inc., and Dr. Gerard J. Tortora.

(Latin for "head") normally goes away in a few days. Some newborns may also have a blood clot between the skull bone and its covering membrane, due to pressure from the birth canal. This clot, called a *cephalohematoma,* often appears a day or two after birth and usually goes away in several weeks. It does not damage the brain. The newborn's soft nose frequently has been squashed during delivery, and its shape is not what it will look like in a few days. In addition, use of forceps can temporarily alter head shape and appearance. Or the forceps can press on the facial nerves, which can lead to temporary paralysis of the eye and mouth muscles. This is quite disturbing to parents until they are reassured that normal function will develop soon.

The eyes of newborn Caucasians are usually gray-blue, those of blacks are brown or black, and those of Orientals are green-blue. Only later will pigment change in the iris, resulting in the various eye colors. Some infants' eyes will be bloodshot, but this will disappear in a few days. Also, a puslike secretion may exude from the eyes, especially in those babies who received silver nitrate. This secretion will go away in about a day. Many parents are alarmed to see that their babies are cross-eyed. This is because the newborn is unable to focus its eyes properly, an ability that is gained over time. Also the tear glands are not functional until up to three months of age, so the cries of a newborn have no tears! A disease called *retrolental fibroplasia* at one time afflicted 10 to 15 percent of premature babies, and this disorder caused partial or complete blindness. Now we know that it was caused by administration of too much oxygen to premature babies, and precautions against this practice have virtually eliminated this problem.

The skin of newborn Caucasians usually is pale pink to dark pink or red; that of blacks, a pale pink to pinkish brown; of Orientals, a rosy tint; and of Latins, olive or yellow. Vernix caseosa, the cheesylike protective secretion on the fetal skin, may still be present, especially in premature babies. Fetal downy hair (lanugo) is present on the shoulders, back, forehead, and temples of some newborns. This hair is usually shed by the end of the first week. The skin of newborns is thin and sensitive, and may exhibit temporary blemishes. Tiny white spots on the nose, chin, and cheeks are clogged pores, which go away soon. *Strawberry marks* are regions of dilated blood vessels in the skin, and they are common on the forehead, nose, upper eyelids, and nape of the neck. These usually disappear in a few months.

About one-third of babies develop *physiological jaundice,* which occurs on about the second to the fifth day of postpartum. This condition, which appears as a transitory yellowish tint to the skin and eyes, is caused by excessive amounts of bilirubin. It is more common and severe in premature babies. Newborns with this condition can be treated with ultraviolet light in an incubator. Jaundice can also indicate erythroblastosis from Rh disease (Chapter 9), but this jaundice develops on the first, not the second, day of life.

The reproductive tracts of male and female newborns are fully formed but inactive (Chapter 5). The breasts of some males and females enlarge on the third or fourth day after birth. In fact, some secretion of fluid, called *witch's milk,* can occur. During the medieval period, this fluid was believed to have wonderful healing powers. This breast enlargement, which goes away in a few days, is due to exposure of the fetus to placental hormones during late pregnancy. Sometimes a female newborn will have a little bloody vaginal discharge for the same reason.

DISORDERS OF THE NEWBORN

Premature babies are those born alive and weighing between 0.99 kg (2 lb, 3 oz) and 2.49 kg (5.5 lb). The chances of a 2-lb, 3-oz baby surviving, even with special intensive care, is about 50 percent. This percentage, however, increases to 90 percent in 4-lb babies. Elaborate hospital care, including incubation units that are almost like "artificial wombs," is increasing this survival rate.

In previous chapters, we have seen how developmental abnormalities can occur in the embryo or fetus. These can be due to chromosomal anomalies as a result of fertilization errors, inherited disorders (Chapter 12), the effects of teratogens or mutagens, or maternal disorders such as diabetes mellitus or toxemia (Chapter 9). A newborn also can suffer damage during difficult births. For example, it could have been deprived of oxygen because of premature separation of the placenta or a prolapsed umbilical cord. Many fetuses suffering from developmental abnormalities are lost during pregnancy (spontaneous abortion), but some are born. Such newborns are said to have *congenital defects* (*birth defects*).

Nevertheless, most babies born in the United States are healthy, with no defect (87 percent), or only a mild congenital disorder (11 percent). Of the remaining 2 percent, about half have a major defect and half die at birth or soon after. Some of the latter are stillborn; stillbirths occur in about 1 of every 200 deliveries. Other babies with severe defects suffer *neonatal death,* which means they die in the first month of life. Most of these die in the first 24 hours after birth. Prematurity accounts for about one-half of the neonatal deaths in the United States. Finally, other infants with severe birth defects can grow up to lead normal and happy, albeit handicapped lives, and some of these handicaps can be treated with medical and psychological procedures.

Sudden-Infant-Death Syndrome (SIDS) is the sudden, unexpected death of an apparently healthy infant for which no known cause of death can be found. Many of these deaths occur while infants are sleeping in their cribs, which is why the syndrome often is called *crib death.* In the United States, this happens to about 1 in 500 infants, or about 10,000 infants a year. It is most frequent between the ages of one month and one year, with most occurring at about three months of age. Naturally, this frightens most parents, and it is unfortunate that we only have some vague inklings about what causes this disorder. There is nothing, however, that a parent can do to prevent SIDS, except to avoid use of certain drugs during pregnancy.

Most of the present evidence indicates that SIDS infants suffer from an abnormality of the sensory and brain structures that control respiration.[10] Before they suddenly die, many of these babies display periods of *apnea* (temporary absence of breathing) at night, and it is thought that they finally are unable to recover from one of these episodes. But what leads to these respiratory abnormalities? There is no evidence that they are related to oxygen deprivation, trauma, or use of anesthetics at birth. There are, however, several factors that are correlated with occurrence of SIDS. These are: presence of blood type B in the infant, bacterial infection of the amniotic fluid when the mother was pregnant, smoking cigarettes during pregnancy, taking of barbiturates during pregnancy, maternal anemia, and crowded housing. Also, bottle-fed babies are more likely to develop SIDS than are breast-fed babies. There may also be an inherited component to this disorder, since parents of SIDS children tend to exhibit abnormalities in respiratory control mechanisms. Finally, recent evidence has revealed that SIDS victims have higher levels of thyroid hormone in their blood at the time of death than do healthy infants or those dying from other causes.[11] It is possible, but not proven, that these high thyroid hormone levels cause the respiratory abnormalities leading to SIDS.

CONDITION OF THE NEW MOTHER

As mentioned, the new mother has just undergone the stress of delivery and now must face the responsibilities as well as the joys of motherhood. Also, her body is rapidly changing back to the nonpregnant condition following birth, the postpartum period, or *puerperium.*

Physical Changes

After delivery, a woman's reproductive tract begins to return to the nonpregnant condition. Her uterus, which weighs about 2 lb after birth, regresses to about 3 oz in about 6 weeks. Most multiparous and a few primiparous mothers experience *after-pains* for about 3 days after delivery. These painful uterine contractions often are increased by nursing, which causes secretion of oxytocin (see below). During the postpartum period, most women have a vaginal discharge called *lochia*. This discharge is bright red for the first 3 or 4 postpartum days, turning pale pink or yellow-white a week or so after delivery. This discharge lasts about 10 days in nursing mothers and up to 30 days in nonnursing mothers. A woman, therefore, should wear a sanitary napkin (not a tampon) during this time. Frequency of urination is high in the first week after delivery because a lot of fluid is retained in a woman's body during late pregnancy. Normally, a woman can resume coitus about six weeks after delivery.

Psychological Changes

The human mother-infant bond is very important to the development of the child. Besides providing warmth, protection, and nourishment, the mother's behavior provides visual and tactile stimuli that have important effects on the child's developing emotional and social behavior.[12] Newborns, for example, form a specific attachment and preference for their mother's voice.[13]

The period of pregnancy is a time during which the mother-to-be must come to accept the fact that she is pregnant. She must do this not only on an intellectual level, but also on all levels of mental functioning, especially the subconscious. The acceptance of pregnancy leads to acceptance of the neonate. If a woman has accepted pregnancy and is prepared for the appearance of the infant, she also is well prepared for the commencement of maternal behavior. However, problems can arise when the woman, for various reasons, has failed to accept her pregnancy. This may mean that she will be improperly prepared to accept the newborn. Many women have some negative feelings such as anger toward their newborn. Although most women eventually adjust to the demands of motherhood, some have a more difficult time doing so. These individuals experience emotional trauma, usually manifest some degree of denial or rejection of the infant, and may reject the man, who is held responsible. Hence, the entire relationship between the woman and her family may suffer.

A woman typically remains in the hospital for two or three days after delivery. During the first two days, she often is elated but exhausted. Some women, however, experience depression and periods of crying on about the third day postpartum. This depression, which can last a day or two or up to several weeks, is called *postpartum depression*.[14] Minor postpartum depression occurs in anywhere from 25 percent to 67 percent of women, depending on which study you read. Severe mental disturbances, called *postpartum psychiatric syndrome,* occurs in about 1 out of 400 mothers. Most

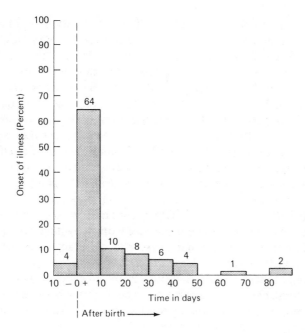

Figure 11–2. This graph indicates when postpartum psychiatric syndromes appeared in 100 women. Note that most appear during the first ten days after delivery, as is also the case for less severe postpartum depression. Adapted from F. T. Melges, "Postpartum Psychiatric Syndromes," *Psychosomatic Medicine,* 30 (1968), 95–108, by permission of the publisher. © 1968 by the American Psychosomatic Society, Inc.

women who develop these syndromes do so during the first ten days after birth (Fig. 11–2).

There may be many causes for postpartum depression. The new mother may be depressed because she is in an unfamiliar hospital environment, which is why some people favor home deliveries. She has just undergone the stress of labor, and she is undergoing marked physiological changes. Furthermore, she must now face, with her mate, the responsibilities of child care. A marriage also faces many adjustments to the new arrival.

Many women who suffer from postpartum depression have a history of severe premenstrual tension during the menstrual cycle. It is interesting that both the premenstrual period and the postpartum period are characterized by a sudden drop in blood levels of progesterone, and administration of progestogens can alleviate symptoms of both these syndromes.[15] An interesting but perhaps far-fetched idea is that human females tend to be depressed after bearing a child because they do not eat their placenta! Such an act, called *placentophagia,* is common in most mammalian mothers, and even in some human cultures.[16] The placenta contains high levels of progesterone, and eating it would perhaps prevent the abrupt lowering of progesterone levels during the postpartum period. When mother rats eat the placenta, something in this tissue increases levels of progesterone and prolactin in their blood.[17] It is noteworthy that many scientists feel that the receptive and motherly moods of women in the luteal phase of the menstrual cycle are related to the high levels of progesterone at this time.

Why should a change in maternal hormone levels in the brief period after birth affect the maternal mood of a new mother? It may be that a certain physiological state in this brief period sensitizes the mother to respond to her newborn, and her experiences during this state may have lasting effects on her later maternal behavior. Is there such a sensitive period immediately after birth when a woman must experience her newborn in order to feel motherly later on? The few studies relating to this question answer yes! For example, one study divided 28 pregnant women into two groups. In one group, the women only briefly glimpsed their children right after they were born. Six to 12 hours later, they experienced another brief glimpse for identification purposes. In the next three days, they fed their children five times a day, each session lasting 20 to 30 minutes. In the other group, the women, besides briefly glimpsing the children immediately and 3 hours after birth, experienced 5 continuous hours of contact (visual, tactile, etc.) each day for three days. One month and one year later, the research group determined the maternal behavior of these women by interviews and films. The second group of mothers, the ones that had extended contact with their infants in the few days after birth, were more affectionate in that they gave their children more eye contact and touched and kissed them more. In general, they were more effective, caring mothers than the other group.[18] The investigators concluded that a maternal sensitive period exists in humans soon after birth. Another study, however, failed to demonstrate such a sensitive period.[19]

The usual practice of hospitals has been to separate the infant from the mother immediately after birth, and subsequent mother-infant contact while in the hospital is often limited to brief daily feeding periods. Is this separation preventing the mothers from experiencing their child during a maternal sensitive period? Could this be one reason why there is a high incidence of at least mild forms of depression a few days after a woman has a child? Does this separation have something to do with the fact that, according to one study, women first seeing their children felt that their babies were subhuman, and only 59 percent felt positively toward their infants? Is there a clue here that may help explain why the number of women breast-feeding their infants is low in the United States, as we see later in this chapter? More and more hospitals now have policies that allow increased early mother-infant contact. These *rooming in* policies provide that the infant actually sleeps in the same room as the recovering mother.[20] I hasten to say that learning can offset lack of infant contact during the sensitive period. For example, women that adopt children do not have early postpartum contact with their child, but many are excellent mothers.

This discussion would not be complete if we did not mention *paternal behavior,* which means care-giving behavior by the father. In our species, certainly, the father participates in caring for the child. However, some fathers do not participate to any great extent, especially in the early stages of the infant's life. The usual reason these fathers give for not getting involved is related to the masculine mystique; i.e., child caring is a feminine role. During World War II, fathers who were absent when their first child was born, or were absent during the child's first year, later were more critical of its personality and were worried more about its eating, eliminating, and sleeping habits.[21] Perhaps there is a sensitive period for paternal attachment in fathers, and maybe we should reevaluate the need for extensive early contact for the infant with both parents.

BREAST-FEEDING

Postpartum Endocrine Changes and Lactation

As noted in Chapter 9, hormones of pregnancy cause growth of the mammary gland tissue and ducts. These hormones include estrogens and progesterone. During late pregnancy, the high levels of estrogens and progesterone in the mother's blood suppress secretion of prolactin from her anterior pituitary gland, so that the mammary glands secrete colostrum but do not undergo *lactation* (milk secretion). When the placenta is expelled, the levels of estrogens and progesterone drop rapidly to very low levels, which allows prolactin secretion to increase. This hormone then causes the primed mammary tissue to secrete milk. Colostrum is the first secretion of the mammary glands after birth, with milk appearing in two or three days.

When the infant *breast-feeds* (suckles the breast), sensory receptors in the nipples are stimulated, and the nerve impulses travel to the woman's brain. These nerve impulses cause release of the hormone oxytocin from the posterior pituitary gland, which is carried by the blood to the mammary glands. Here, oxytocin causes contraction of the myoepithelial cells, squeezing the mammary alveoli and causing milk ejection from the nipples and into the baby's mouth. This entire process is called the *milk-ejection reflex* (Fig. 11–3). Stressors, such as distraction, embarrassment, fear, or anxiety, can cause the nervous system to constrict blood vessels in the mammary gland. Thus, delivery of oxytocin is lowered, and milk ejection slows or does not occur. The milk-ejection reflex also can be conditioned to the sight, smell, or sound of the baby, so that after a while, these stimuli alone can cause milk ejection. Nursing mothers often have been amazed at milk ejection from their nipples when they hear their baby cry!

Suckling the nipples by the baby not only causes milk ejection, but also nerve impulses reach the hypophysiotropic area of the hypothalamus and inhibit GnRH and PRIH release (Chapter 2). Thus, suckling inhibits secretion of FSH and LH but stimulates secretion of prolactin from the anterior pituitary gland. Blood levels of PRL increase 2- to 20-fold within 30 minutes after suckling begins. Not only does nursing inhibit FSH and LH secretion, but PRL also decreases the responsiveness of the ovaries to these gonadotropins.[22] Therefore, nursing can inhibit ovulation.

Can nursing be an effective contraceptive measure? The answer is no because of its relatively high failure rate (Chapter 13). Women who nurse on a regular schedule usually do not ovulate for six to nine months postpartum, as compared to one to four months in women who do not nurse.[23] After six to nine months of nursing, however, menstrual cycles resume even if nursing is continued. Furthermore, the first postpartum ovulation occurs before the first menses. Therefore, it usually is recommended that nursing women use contraceptive measures beginning three to six months post-

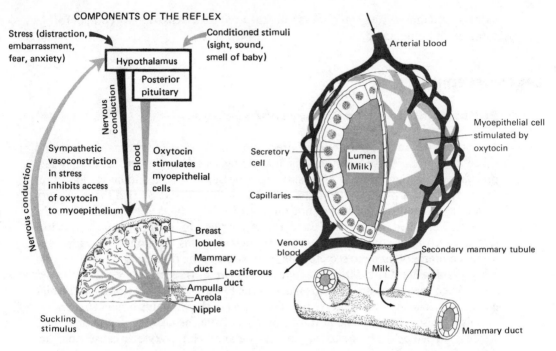

COMPONENTS OF THE REFLEX

Stress (distraction, embarrassment, fear, anxiety)

Conditioned stimuli (sight, sound, smell of baby)

Hypothalamus

Posterior pituitary

Nervous conduction

Blood

Sympathetic vasoconstriction in stress inhibits access of oxytocin to myoepithelium

Oxytocin stimulates myoepithelial cells

Nervous conduction

Breast lobules

Mammary duct

Lactiferous duct

Ampulla

Areola

Nipple

Suckling stimulus

Arterial blood

Myoepithelial cell stimulated by oxytocin

Secretory cell

Lumen (Milk)

Capillaries

Venous blood

Milk

Secondary mammary tubule

Mammary duct

Figure 11–3. The milk-ejection reflex. The baby suckles the nipple, and nervous impulses reach the hypothalamus. Then, oxytocin is released into the blood from the posterior pituitary gland and travels to the mammary glands. Here, oxytocin causes contraction of myoepithelial cells surrounding each alveolus, and milk is ejected from the nipple. The entire reflex takes about 30 seconds. Stress can activate neurons of the sympathetic nervous system, which causes constriction of mammary gland blood vessels, less delivery of oxytocin and inhibition of milk ejection. Conditioned stimuli can replace suckling as an initiator of milk ejection. Adapted from A. L. R. Findlay, "Lactation," *Research in Reproduction,* vol. 6, no. 6 (1974), map. © 1974; used with permission of the International Planned Parenthood Federation and Dr. A. L. R. Findlay.

partum. It is unwise, however, to use oral contraceptives at this time since the estrogens in these pills can enter the breast milk[24] and may have adverse effects on the infant.[25] Therefore, only injectable progestogens and progestogen-only pills as well as condoms, IUDs, spermicides, and diaphragms (Chapter 13) should be used by nursing women.

The frequency of nursing may have a lot to do with its effectiveness as a contraceptive measure. For example, the !Kung hunter-gatherers of Botswana and Namibia have an extremely long interval between births, averaging 44.1 months. This is so despite the fact that nursing women have coitus frequently and do not use any other contraceptive measure. These women nurse briefly and frequently, about every 13 minutes during the day. It is this frequent nursing schedule, plus the fact that they do not wean their children until they are 3½ years old, that very effectively keeps prolactin secretion high and FSH and LH secretion low, and prevents ovulation.[26]

Advantages and Disadvantages of Breast- and Bottle-Feeding

About two-thirds of the mothers in the world breast-feed their babies. In underdeveloped countries, women often will nurse for more than two years because if they quit, their infant may become malnourished.[27] In developed countries such as the United States, however, breast-feeding is less frequent. About one third of the mothers in the United States begin to breast-feed their infants, even though 95 percent of them are physically capable of doing so. And only about 20 percent of them are still nursing after two months. The relative infrequency of breast-feeding in the United States is due to such things as the commercial production of synthetic infant formulas as well as the increase in employment of women. Some women also may choose to bottle-feed their babies for psychological or social reasons. They may feel embarrassed about nursing or may think that it is not the right thing to do. In fact, women with ambivalent or negative feelings about nursing who nurse anyway will produce less milk and will quit earlier (Table 11–3). Certainly, a woman should not breast-feed her baby if she suffers from an infection, kidney impairment, heart trouble, or anemia. This is because nursing would put a dangerous stress on her diseased system. A woman, however, can breast-feed after a cesarean delivery. Today, breast-feeding is on the rise in the United States. This is because more women are beginning to realize its benefits. For some, it has become almost a status symbol to nurse their infants.

There are several advantages of breast-feeding for the infant. Besides always being sterile and at the right temperature, human breast milk is a nutritionally com-

TABLE 11-3. RELATION OF ATTITUDES TOWARD BREAST-FEEDING TO BREAST-FEEDING PERFORMANCE

Verbalized Attitude toward Breast-Feeding	Mean Amount (grams) of Milk Obtained by Baby at 4th-Day Feedings	Successful Nursing by Mothers[a] (%)
Positive	59	74
Ambivalent	42	35
Negative	35	26

Source: From N. Newton and M. Newton, "Psychologic Aspects of Lactation," *New England Journal of Medicine,* 297 (1967), p. 1180. Reprinted by permission of the *New England Journal of Medicine;* © 1967.

[a] As judged by no need for bottle supplement after 4th hospital day.

plete infant diet. Table 11–4 compares the contents of human colostrum, human milk, and cows' milk. In general, human milk contains all of the nutrients necessary for the infant in the first six months of life and most of those nutrients required after six months. Lactose sugar, fats (including fat-soluble vitamins), and proteins (casein, lactalbumins, lactoferritin) are present. Also, electrolytes such as sodium, chloride, calcium, and bicarbonate are abundant. Some nutrients in breast milk, such as certain amino acids, fatty acids, and cholesterol, are absent in cows' milk and infant formulas.

There is good evidence that breast-feeding has health benefits to the infant. Colostrum contains antibodies that protect the baby against bacterial infection of the

TABLE 11-4. APPROXIMATE CONCENTRATION OF THE MORE IMPORTANT COMPONENTS PER 100 ml WHOLE MILK UNLESS OTHERWISE STATED

	Human Colostrum	Human Milk	Cows' Milk
Water, g	—	88	88
Lactose, g	5.3	6.8	5.0
Protein, g	2.7	1.2	3.3
Casein: lactalbumin ratio	—	1:2	3:1
Fat, g	2.9	3.8	3.7
Linoleic acid	—	8.3% of fat	1.6% of fat
Sodium, mg	92	15	58
Potassium, mg	55	55	138
Chloride, mg	117	43	103
Calcium, mg	31	33	125
Magnesium, mg	4	4	12
Phosphorus, mg	14	15	100
Iron, mg	0.09[a]	0.15[a]	0.10[a]
Vitamin A, μg	89	53	34
Vitamin D, μg	—	0.03[a]	0.06[a]
Thiamine, μg	15	16	42
Riboflavine, μg	30	43	157
Nicotinic acid, μg	75	172	85
Ascorbic acid, mg	4.4[b]	4.3[b]	1.6[a]

Source: From A. L. R. Findlay, "Lactation," *Research in Reproduction,* vol. 6, no. 6 (1974), map. Reprinted with permission of the International Planned Parenthood Federation and Dr. A. L. R. Findlay; © 1974.

[a] Poor source.

[b] Just adequate.

digestive tract. In addition, the lactoferritin present in human milk binds to iron and prevents bacteria from multiplying in the newborn's gut. Also, breast-fed babies develop less atherosclerosis, hypertension, allergies, childhood diabetes, and middle-ear infections than do bottle-fed babies.[28] Artificial milk formulas contain more sodium and chloride than does breast milk (Table 11–4), and some feel that this is one reason that the incidence of SIDs is about 20 times higher in bottle-fed babies.[29] Some infant formulas also contain more sugar than does breast milk, and the fact that bottle-fed babies tend to be more obese than breast-fed babies may relate to sugar addiction.[30] Another cause of the greater obesity in children who were bottle-fed may be that, unlike breast-fed babies who stop sucking when they are full, bottle-fed babies often are forced to eat more than they want. Finally, there may be important things in breast milk that we are not aware of. For example, only recently has it been shown that *epidermal-growth factor* is present in human milk and colostrum, and this substance may help the growth of infant tissues.[31] Also, GnRH is present in human breast milk, but its role in maturation of the infant reproductive system is not yet known.[32]

One disadvantage of breast-feeding to the infant is that harmful substances ingested by the mother can pass into the breast milk. We already have seen that contraceptive estrogens are present in breast milk. Also such things as nicotine, polychlorinated biphenyls (PCBs) and other pollutants, barbiturates, and anesthetics can pass into the milk and affect the infant. Included in this list is alcohol, which causes excess hormone secretion from the infant's adrenal gland.[33]

There are some advantages of breast-feeding to the mother. The oxytocin released during the milk ejection reflex causes the uterus to contract and facilitates its postpartum recovery. Some women even experience sexual pleasure during nursing, which is normal and part of the joy of the mother-infant contact. Resumption of sexual activities also occurs sooner in nursing mothers.[34] Finally, breast-feeding is more convenient than bottle-feeding because the mother does not have to purchase and prepare sterile formulas. Bottle-feeding is also more expensive than breast-feeding.

There are, however, some disadvantages of breast-feeding to the mother. A nursing woman who is secreting 400 ml of milk a day is supplying 5 g of fat, 4 g of protein, and 28 g of lactose to her infant. Thus, breast-feeding is a nutritional drain on the mother, and she must watch her diet. Also, some women do not like the discomfort that nursing sometimes causes, since the breasts become engorged with milk and the nipples need care to avoid their becoming sensitive and painful during nursing. Finally, a husband can feed a bottle baby but not a breast-fed one, so nursing mothers are often the ones getting up during the middle of the night. A father who can help bottle-feed his child can have quality time with the infant that one with a breast-fed infant does not have.

From a biological point of view, breast-feeding is better for the infant and mother than bottle-feeding because breast milk is already "mixed" and is a complete nutritional package for the infant. There are, however, many happy and healthy bottle-fed babies. In the final analysis, the choice to breast-feed or bottle-feed an infant is an individual one, and should be based on the psychological and physical well-being of both the mother and the infant.

CHAPTER SUMMARY

The newborn is expelled into a new environment, to which it must adapt. The Leboyer method of childbirth attempts to reduce the stress of this environmental shock. The newborn's condition is checked by giving it an Apgar score, and body measurements are taken. Many young males are circumcised in the first week of life.

The newborn has special adaptations. These include the secretion of lung surfactant, crying, and closure of the ductus arteriosus and venosus as well as the foramen ovale. The infant's cells release glucose for energy, and heat is released from stores of brown fat to maintain its body temperature. It has instinctive reflexes that are adaptations for maternal contact, nursing, and walking.

The newborn's appearance is alarming to some who first view it. Its skull may be egg-shaped, with soft fontanels. Swellings, blood clots, and strawberry marks can be present on its head. Its eyes, which can see light and dark, are not the final color and may be crossed and often exude a puslike secretion. Its skin color varies in different races, but a bluish tinge (cyanosis) or yellowish tint (jaundice) may indicate need for medical treatment. Its reproductive tract is fully formed but inactive, except that slight breast enlargement and secretion of fluid (witch's milk) may occur.

Unfortunately, some newborns exhibit abnormalities. Premature newborns need special intensive care, but many survive and are healthy. Congenital defects can cause stillbirth or neonatal death. Other congenital defects can be minor. Sudden-infant-death syndrome is a major cause of infant death during the first year of life.

The new mother has just undergone a sometimes joyous yet stressful experience, and now must adapt to marked biologic changes and the responsibility of raising (along with her husband) the new arrival. Some women experience postpartum depression, which may be related to endocrine and psychological factors. There is good evidence that early and frequent mother-infant contact is important to the development of maternal behavior.

Breast-feeding stimulates endocrine responses in a woman that cause lactation, milk ejection, postpartum infertility, and uterine changes. Breast-feeding, however, is at best an ineffective contraceptive measure. Some women may choose not to breast-feed their babies for medical, sociological, or psychological reasons. Breast milk, however, is better for infants than most if not all artificial substitutes. Breast milk protects against digestive tract infections, atherosclerosis, hypertension, allergies, diabetes, middle-ear infections, and even sudden-infant-death syndrome. It is always sterile and at the right temperature. The nursing mother also benefits in that her physical recovery from childbirth occurs faster, and breast-feeding is more convenient and cheaper than bottle-feeding. Breast milk, however, can contain harmful substances ingested by the mother. Also a husband is unable to help the mother as much when she is breast-feeding her baby.

NOTES

1. F. Leboyer, *Birth without Violence* (New York: Random House, Inc., 1975).
2. M. Mahran, "Medical Dangers of Female Circumcision," *IPPF Medical Bulletin,* 15, no. 2 (1981), 1–3.
3. J. R. Hand, "Surgery of the Penis and Urethra," in *Urology,* vol. 3, eds. M. F. Campbell and J. H. Harrison (Philadelphia: W. B. Saunders Company, 1970).
4. W. H. Masters and V. E. Johnson, *Human Sexual Response* (Boston: Little, Brown & Company, 1966).
5. E. Wallerstein, *Circumcision: An American Health Fallacy* (New York: Springer Publishing Co., Inc., 1980).
6. P. F. Ostwald and P. Peltzman, "The Cry of the Human Infant," *Scientific American,* 230, no. 3 (1974), 84–90.
7. L. Mahony et al., "Prophylactic Indomethacin Therapy for Patent Ductus Arteriosus in Very Low Birth-Weight Infants," *New England Journal of Medicine,* 302 (1982), 506–10.
8. M. J. Dawkins and D. Hull, "The Production of Heat by Fat," *Scientific American,* 213, no. 2 (1965), 62–67.
9. P. R. Zelago et al., " 'Walking' in the Newborn," *Science,* 176 (1972), 314–15.
10. R. L. Naeye, "Sudden Infant Death," *Scientific American,* 242 (1980), 56–62.
11. Anonymous, "Hormone Levels High in SIDS," *Science News,* 120, no. 20 (1981), 310.
12. M. H. Klaus et al., "Human Maternal Behavior at the First Contact with Her Young," *Pediatrics,* 46 (1970), 187–92; and R. M. Restak, "Newborn Knowledge," *Science 82,* 3, no. 1 (1982), 58–65.
13. A. J. DeCasper and W. P. Fifer, "Of Human Bonding: Newborns Prefer Their Mothers' Voices," *Science,* 208 (1980), 1174–76.
14. F. T. Melges, "Postpartum Psychiatric Syndromes," *Psychosomatic Medicine,* 30 (1968), 95–108.
15. D. A. Hamburg et al., "Studies of Distress in the Menstrual Cycle and the Postpartum Period," in *Endocrinology and Human Behavior,* ed. R. F. Michael, pp. 94–116 (London: Oxford University Press, 1968).
16. K. Janszen, "Meat of Life," *Science Digest* (November/December 1980), pp. 71–81, 122.
17. M. S. Blank and H. G. Friesen, "Effects of Placentophagy on Serum Prolactin and Progesterone Concentrations in Rats after Parturition and Superovulation," *Journal of Reproduction and Fertility,* 60 (1980), 273–78.
18. M. H. Klaus et al., "Maternal Attachment: Importance of the First Post-Partum Days," *New England Journal of Medicine,* 286 (1972), 460–63; and J. H. Kennell et al., "Maternal Behavior One Year after Early and Extended Post-Partum Contact," *Developmental and Medical Child Psychology,* 16 (1974), 172–79.

19. M. J. Svejda et al., "Mother-Infant 'Bonding': Failure to Generalize," *Child Development,* 51 (1980), 775–99.

20. M. Greenberg et al., "First Mothers Rooming-in with Their Newborns; Its Impact upon the Mother," *American Journal of Orthopsychiatry,* 43 (1973), 783–88.

21. F. Rebelsky and C. Hawks, "Fathers' Verbal Interaction with Infants in the First Three Months of Life," *Child Development,* 42 (1971), 63–68; and E. Spelke et al., "Father Interaction and Separation Protest," *Developmental Psychology,* 9 (1973), 83–90.

22. J. E. Tyson et al., "Human Lactational and Ovarian Response to Endogenous Prolactin Release," *Science,* 177 (1972), 897–900.

23. R. Buchanan, "Breast-Feeding—Aid to Infant Health and Fertility Control," *Population Reports,* ser. J, no. 4 (1975), pp. 49–67.

24. S. Nilsson and K. Nygren, "Transfer of Contraceptive Steroids to Human Milk," *Research in Reproduction,* 11, no. 1 (1979), 1–2; and F. A. Kincl, "Debate on the Use of Hormonal Contraceptives during Lactation," *Research in Reproduction,* 12, no. 2 (1980), 1.

25. E. Curtis, "Oral Contraceptive Feminization of a Normal Male Infant," *Obstetrics and Gynecology,* 23 (1964), 295–96.

26. M. Konner and C. Worthman, "Nursing Frequency, Gonadal Function, and Birth Spacing among !Kung Hunter-Gatherers," *Science,* 207 (1980), 788–91.

27. Buchanan (1975), op. cit.

28. Ibid.

29. J. L. Emergy et al., "Hypernatraemia and Uraemia in Unexpected Death in Infancy," *Archives of Childhood Disease,* 49 (1974), 686–92.

30. E. E. Eid, "Follow-Up Study of Physical Growth of Children Who Had Excessive Weight Gain in the First Six Months of Life," *British Medical Journal,* 2 (1970), 74–76.

31. G. Carpenter, "Epidermal Growth Factor Is a Major Growth-Promoting Agent in Human Milk," *Science,* 210 (1980), 198–99.

32. T. Baram and Y. Koch, "Gonadotropin-Releasing Hormone in Milk," *Science,* 198 (1977), 300–301.

33. A. Binkiewicz et al., "Pseudo-Cushing Syndrome Caused by Alcohol in Breast Milk," *Journal of Pediatrics,* 93 (1978), 965–67.

34. Masters and Johnson (1966), op. cit.

FURTHER READING

ANONYMOUS, "More U.S. Women Breastfeeding Babies for a Longer Duration," *Family Planning Perspectives,* 248 (1982), 840–46.

BUCHANAN, R., "Breast Feeding—Aid to Infant Health and Fertility Control," *Population Reports,* ser. J, no. 4 (1975), pp. 49–67.

CARPENTER, G., "The Importance of Mother's Milk," *Natural History Magazine,* 9, no. 8 (1981), 6–14.

FINDLAY, A. L. R., "Lactation," *Research in Reproduction,* vol. 6, no. 6 (1974), map.

JANSZEN, K., "Meat of Life," *Science Digest* (November/December 1980), pp. 78–81, 122.

KLAUS, M. H., and J. H. KENNELL, *Maternal-Infant Bonding*. St. Louis: The C. V. Mosby Company, 1976.

KLOPFER, P. H. et al., *Maternal Care in Mammals,* Addison-Wesley Module in Biology, no. 4. Reading, Mass.: Addison-Wesley Publishing Co., Inc., 1973.

LEBOYER, F., *Birth without Violence.* New York: Random House, Inc., 1975.

NAEYE, R. L., "Sudden Infant Death," *Scientific American* (April 1980), pp. 56–62.

OSTWALD, P. F., and P. PELTZMAN, "The Cry of the Human Infant," *Scientific American,* 230, no. 3 (1974), 84–90.

RESTAK, R. M., "Newborn Knowledge," *Science 82,* 3, no. 1 (1982), 118–24.

WINDLE, W. F., "Brain Damage by Asphyxia at Birth," *Scientific American,* 221, no. 4 (1969), 76–84.

12 *Human Genetics*

LEARNING GOALS

Having read this chapter, you should be able to:

1. Describe the molecular and three-dimensional structure of the DNA molecule.
2. Discuss how DNA replicates during cell division.
3. Describe how DNA codes for synthesis of a polypeptide or protein.
4. Define a mutagen, and describe how a mutation occurs.
5. Describe variations of a simple Mendelian cross involving a trait controlled by a dominant and a recessive allele.
6. Give three examples of inherited human disorders involving an autosomal recessive or dominant genotype.
7. Describe, using an example, the phenomenon of incomplete dominance.
8. Discuss the inheritance of ABO blood type.
9. Give an example of a sex-linked trait, and discuss examples of sex-linked disorders.
10. List two kinds of chromosomal disorders, and describe how they arise during meiosis.
11. Discuss possible ramifications of human eugenics and genetic engineering.

INTRODUCTION

Gregor Mendel was born to a Moravian peasant family in 1822. After being educated in Vienna, he joined a monastery in Brünn (now Brno), Czechoslovakia, and began work on the inheritance of traits in pea plants. Mendel's work, first published in 1869, showed that inherited factors are carried by discrete units, which we now know as *genes* (Greek *genos* = "birth" or "race"). His research formed the basis of the science of genetics and led to our understanding of inherited human variation and disorders.

Genes are instructions that code for our physiological and physical make-up and interact with environmental factors to control our behavior. These instructions are like a musical score for the biological melody that is the individual. Furthermore, when each cell divides, it must pass on genetic instructions to its daughter cells. The gametes (sperm and eggs) also carry this code; and when they combine, the new individual has genes from both parents. Many inherited disorders are the products of faulty instructions; that is, there are missing or errant notes in the musical score. In this chapter, we discuss how genes work, how they are passed on from generation to generation, and how inherited disorders occur.

GENES AND CELLULAR PROCESSES

Before we discuss the inheritance of human genetic traits, you must understand how genetic information controls cellular biologic processes. In 1953, J. D. Watson and F. H. C. Crick described how genes on the chromosomes dictate cellular function. Chromosomes contain *deoxyribonucleic acid* (DNA). Each human chromosome has a single coiled DNA molecule, which is about 3 to 4 cm long when uncoiled. Therefore, the cells of your body contain about 25 billion km of DNA! Also present in the chromosomes are other proteins, such as *histones*. Each molecule of DNA contains hundreds of *nucleotides*, each of which consists of a *nitrogenous base* (there are four nitrogenous bases: *thymine, cytosine, adenine, guanine*) connected to a phosphate and sugar (deoxyribose) "backbone" (Fig. 12–1).

There are two nucleotide chains in each DNA molecule, and these two are twisted in the form of a spiral staircase, or *double helix* (Fig. 12–1). The two chains of nucleotides are connected by weak bonds between the nitrogenous bases. Thymine always binds to adenine, and cytosine to guanine (Fig. 12–1). When the DNA helix replicates (copies) itself before mitotic cell division, the double chain comes apart, and each chain serves as a template for formation of a new partner chain that is a complement to itself and identical to its old partner. This process of *DNA replication* results in daughter cells with DNA that is identical to the parental cell (Fig. 12–2).

How does DNA, with its four-letter alphabet (the four nitrogenous bases), control the multitude of different cellular processes? One chain of the DNA molecule serves as a template for construction of a complementary chain of another nucleic

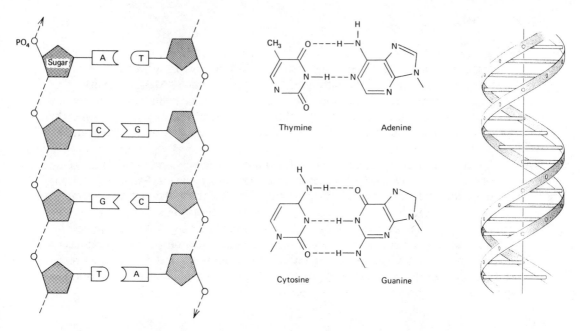

Figure 12–1. The DNA molecule is a double helix. Each strand includes a "backbone" of sugar and phosphate units in a vertical sequence, with lateral pairs of the nitrogenous bases thymine-adenine or cytosine-guanine. Any base pair can be stacked vertically adjacent to any other base pair throughout the length of the double helix. This contributes to DNA variety and gene diversity. Reproduced from C. J. Avers, *Biology of Sex* (New York: John Wiley & Sons, Inc., 1974). © 1974; used with permission.

acid, *ribonucleic acid* (RNA), a process called *transcription* (Fig. 12–3). RNA is like DNA with some important exceptions. It also has four nitrogenous bases, but instead of thymine it has the base *uracil*. Also, the sugar in RNA is ribose, not deoxyribose. The RNA made from a chain of DNA in the nucleus is called *messenger RNA* (mRNA) because this molecule leaves the nucleus and enters the cytoplasm, bringing a chemical message from the DNA to the cytoplasm. Messenger RNA travels to the *ribosomes,* which are present on cell organelles, the *endoplasmic reticulum.*

There are other RNA molecules called *transfer RNA* (tRNA) floating in the cytoplasm of the cell. This RNA is made in the nucleus by a chain of DNA, and then travels to the cytoplasm. Each tRNA molecule consists of about 80 nucleotides, to which are attached specific amino acids. A sequence of three bases in each tRNA molecule recognizes three complementary bases in mRNA. These tRNA-amino acid complexes then bind to the mRNA on the ribosomes, a process called *translation.* The code in the mRNA causes the tRNA molecules with their amino acids to line up in a certain sequence. Then, the amino acids bind together to form polypeptides or proteins, which function as enzymes, hormones, or structural components of the cell.

How is a specific message from DNA for the synthesis of a specific protein dic-

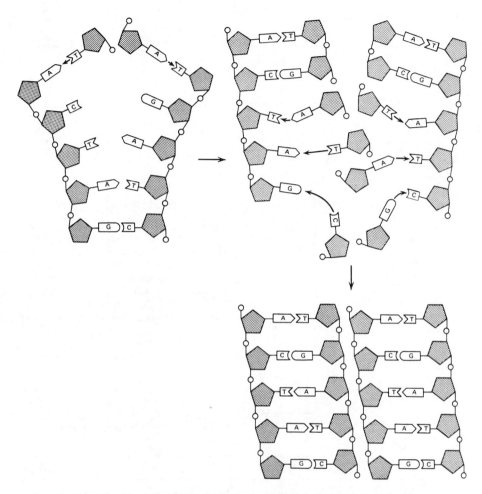

Figure 12–2. The DNA molecule replicates (copies) itself in a fashion that leads each original strand to direct the making of a new strand that is complementary to itself. This replication can occur because of complementary pairing of base pairs. DNA replication helps to explain genetic continuity from generation to generation. Reproduced from C. J. Avers, *Biology of Sex* (New York: John Wiley & Sons, Inc., 1974). © 1974; used with permission.

tated? Each chain of DNA, as we have seen, contains hundreds of nitrogenous bases. A sequence of three nitrogenous bases (triplet) in a chain codes for a specific amino acid. This triplet, which is called a *codon,* then is transcribed to a complementary triplet on mRNA. For example, a codon on a DNA chain of adenine-adenine-adenine would result in a triplet of uracil-uracil-uracil on the mRNA. Similarly, a DNA codon of thymine-adenine-cytosine would produce an mRNA triplet of adenine-uracil-guanine (see Fig. 12–4). Thus, when mRNA arrives at the ribosomes, it has a

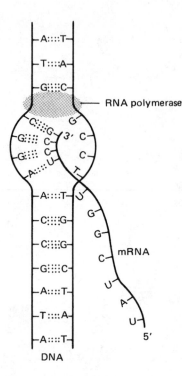

Figure 12–3. Transcription of the DNA code to mRNA. Nucleotide sequences in the mRNA are assembled using one strand of DNA as a template. RNA polymerase is an enzyme that splits the DNA. A similar process occurs in formation of tRNA. Adapted from Helena Curtis, *Biology,* 3rd ed., p. 276 (New York: Worth Publishers, 1979). © 1979; used with permission.

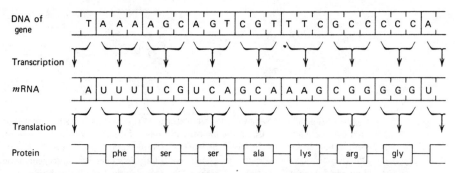

Figure 12–4. The information contained in the sequences of coded DNA is transcribed to molecules of mRNA, which travel from the nucleus to the cytoplasm. The translation of code words into amino acids of the proteins occurs in the ribosomes, according to the original instructions in the DNA blueprints. Reproduced from C. J. Avers, *Biology of Sex* (New York: John Wiley & Sons, Inc., 1974). © 1974; used with permission.

complementary message for the DNA chain. Each tRNA molecule contains a triplet of nitrogenous bases, called an *anticodon,* that recognizes a specific codon in mRNA. There are 20 amino acids, and there are at least 50 different tRNAs. Each specific tRNA with its amino acid then lines up in a sequence dictated by complementary triplets on the mRNA. Thus, the amino acids are lined up in a sequence dictated by the DNA chain that started the whole process. In this way, DNA controls the synthesis of specific polypeptides or proteins.

In the way of summary, these are the steps in the genetic control of cellular processes (see Fig. 12–5):

1. DNA in the nucleus contains a series of nitrogenous bases that, in sets of three, serve as triplet codes (codons), a triplet being a sequence of three bases.
2. One strand of DNA is a template for formation of a single chain of mRNA that contains complementary triplet bases.

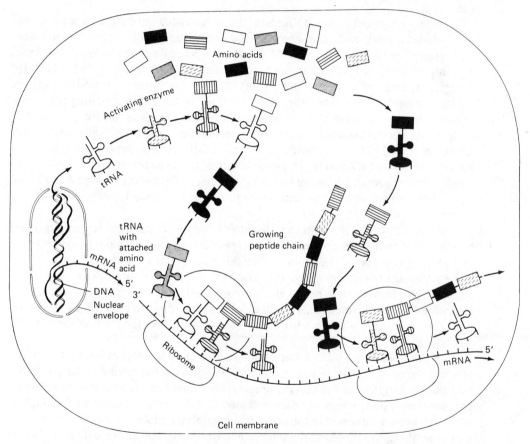

Figure 12–5. Diagram showing how a protein is made. More than 20 different kinds of tRNA molecules are formed, each attached to a specific amino acid. Each tRNA contains an anticodon that fits an mRNA condon for that specific amino acid. After the tRNA molecules attach to the mRNA, and tRNA detaches from the complex, leaving a chain of amino acids that bind together to form a specific protein. Adapted from Helena Curtis, *Biology,* 3rd ed., p. 280 (New York: Worth Publishers, 1979). © 1979; used with permission.

3. mRNA leaves the nucleus and travels to the cytoplasm, where it becomes associated with the ribosomes.
4. tRNA molecules, consisting of a chain of about 80 nucleotides attached to specific amino acids, match their anticodons to complementary triplets on the mRNA.
5. The sequence of amino acids then bind together, forming a specific polypeptide or protein.

The four-letter alphabet of DNA can form 64 different three-letter words, and this language directs the biological sentences (proteins) that allow us to grow and function.

A series of codons on a DNA chain that instruct for synthesis of a specific protein is called a gene, and genes are the fundamental units of inheritance. How does one explain the fact that there are 64 possible triplet codons on the DNA molecule, but only 20 amino acids? Research has shown that a single amino acid can be coded for by one, or more than one (up to six), triplets. Sixty-one of the 64 triplets are codons for amino acids. The other three serve as codons for punctuating the amino acid sentences, such as termination of a gene message. In fact, a rare form of anemia, *thalassemia,* appears related to defects in punctuation of the mRNA message.[1] There are about 50,000 genes that control synthesis of different polypeptides or proteins in human cells. We now know the location of at least one gene on each of the 23 kinds of human chromosomes, including the *loci* (locations) of genes controlling ABO blood type, red-green colorblindness, sensitivity to herpes and polio viruses, and synthesis of many enzymes.[2]

Since all of the cells in each individual's body have identical triplet messages, how is it that some cells do one thing whereas others do other things? One reason is that about 95 percent of the genes in each cell are repressed at any one time. That is, some genes are *regulatory genes* that control the rate of transcription of a DNA molecule by synthesizing a *repressor* molecule. Furthermore, repression of gene action varies among different cells so that, for example, one cell type will be synthesizing a certain protein while another cell type will not. This is the phenomenon of *differential gene action.*

A *point mutation* occurs when a single or a small number of base pairs on the DNA are altered. It is as if a note or series of notes were deleted or changed in the musical score so that the biological melody is incomplete or in error. An example of a past point mutation, which we talk about later in this chapter, is *sickle-cell anemia,* in which only one amino acid in the normal hemoglobin molecule has been altered by a past mutation. Environmental factors that cause mutations are called mutagens (Chapter 9). It has been estimated that 10 percent of spontaneous mutations are caused by natural ultraviolet radiation and radioactivity. Certain chemicals also are mutagens. An example is nitrous acid, which is made from sodium nitrite by your stomach. Sodium nitrite is used as a food preservative in hot dogs, bacon, and luncheon meats! Mustard gas (nitrogen mustard) also is a chemical mutagen, as are the organophosphate pesticides that are derivatives of this substance. It is suspected by some that some cancers result from mutations in somatic cells caused by chemical mutagens. Some mutagens can cause *chromosomal aberrations,* which we also discuss later in this chapter.

AUTOSOMAL GENETIC INHERITANCE

As we saw in Chapter 5, there are 22 pairs of autosomes and 1 pair of sex chromosomes in our cells. The autosomes carry most of our genes, which are arranged in linear fashion in the DNA molecules contained in homologous pairs of chromosomes. Each of a homologous pair of chromosomes contains a gene for a certain inherited trait. The genes for the same trait present on each member of a homologous pair of chromosomes are called *alleles* (Greek *allelon* = "of one another"). If the alleles are identical on each member of the pair, the individual is *homozygous* for the alleles. If, on the other hand, the alleles are different on each member of the pair, a *heterozygous* condition is present. The alleles determine the genetic characteristic (*genotype*) of the individual for a given trait. The overt physical expression of the genotype is called the *phenotype.*

One allele can be *dominant* over the other (*recessive*) allele. That is, when the dominant allele is present, the recessive phenotype is not expressed. The concept of dominant and recessive inherited traits was first proposed by Mendel. Say that we have a pair of alleles that control whether the ear lobes are attached or free, and that the allele for free ear lobes (*E*) is dominant over that for attached ear lobes (*e*). (Note that a capital letter is used for the dominant allele and a lower-case letter for the recessive allele.) If an individual's cells contain the dominant allele (*E*) on one member of a homologous chromosome pair and another dominant allele (*E*) on the other member, that person will have a *homozygous dominant* genotype and will have free ear lobes. Another person may have the alleles *E* and *e*. This genotype is heterozygous because the cells contain both a dominant allele and a recessive allele. But this person's phenotype will be "free ear lobes" because the *E* allele prevents the *e* allele from being expressed. This is called a *heterozygous dominant* genotype (*Ee*), and these people are said to be *carriers* of the recessive allele. Only if a person has a *homozygous recessive genotype* (*ee*) will the recessive phenotype (attached ear lobes) be expressed.

Simple Autosomal Inheritance

What happens when fertilization occurs? During meiosis, the gametes (sperm or eggs) end up with only one set of each pair of homologous chromosomes (the haploid condition). The fact that homologous chromosomes separate during cell division is

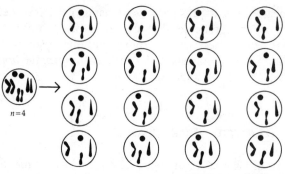

POSSIBLE GAMETES

Figure 12-6. Possible distributions of chromosomes at meiosis, illustrating the principles of segregation and independent assortment. As you can see, chromosomes of maternal and paternal origin are assorted into gametes independently of one another. If the original number of chromosome pairs is 4 (2N = 8), as in the above illustration, 16 different combinations are possible. Because maternal and paternal chromosomes differ in some alleles, these 16 kinds of gametes differ genetically. A human, with 46 chromosomes, is capable of producing 2^{23} kinds of gametes, or 8,388,608 different chromosome combinations. This is similar to the number of people in New York City! Crossing-over, which is not included here, increases the number of different kinds of gametes even more. Reproduced from Helena Curtis, *Biology,* 3rd ed., p. 218 (New York: Worth Publishers, 1979). © 1979; used with permission.

called *segregation*. The choice of which member of a homologous pair of chromosomes ends up in a given gamete is random and occurs independently of the other chromosomes, a principle of *independent assortment* first proposed by Mendel (Fig. 12-6). Since only one member of each homologous pair of chromosomes is present in each gamete, only one allele of each genetic trait is present. When the sperm and egg nuclei fuse, however, the resulting diploid nucleus contains a member of a homologous chromosome pair from both the mother and father, and thus an allele from both the mother and father. This combining of maternal and paternal alleles at fertilization is called *recombination*.

A couple of points should be made about the segregation of alleles in the gametes. First, if alleles for two different traits are on the same chromosome, the alleles show *linkage*. But sometimes two linked alleles will suddenly become unlinked. This is because of a phenomenon called *crossing-over*. That is, two segments of homologous chromosomes will exchange positions.

Given the above information, what are the genotypes and phenotypes of offspring produced by mating when we are looking at a dominant and recessive allele for a given trait? Let us say that an *EE* male mates with an *ee* female. All of the sperm will be *E,* and all of the eggs will be *e.* Thus, the only possible genotype of the baby for this trait is *Ee,* so all children of this couple will be heterozygous dominant and

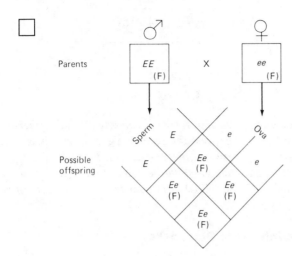

Figure 12-7. A basic mendelian cross involving the alleles controlling ear lobes. The father is homozygous dominant (*EE*) and has free ear lobes (F), and the mother is homozygous recessive (*ee*) and has attached ear lobes (A). All offspring will be heterozygous dominant (*Ee*) and have free ear lobes.

therefore have free ear lobes. An example of how to determine outcomes of such matings is shown in Fig. 12–7.

Now, suppose the male was *Ee* and the female *Ee*. The offspring of this mating would have a 25 percent chance of being *EE* (free ear lobes), a 25 percent chance of being *ee* (attached ear lobes), and a 50 percent chance of being an *Ee* (free ear lobes) carrier. Thus, the offspring would have a 75 percent chance of having the dominant free ear lobe, and a 25 percent chance of having the recessive attached ear lobe (Fig. 12–8). As a challenge, you figure out the particular probabilities of the different genotypes and phenotypes of offspring from the following matings:

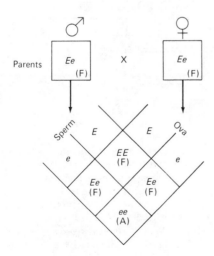

Figure 12-8. A basic mendelian cross involving ear lobes. The father and mother both are heterozygous dominant (*Ee*) and have free lobes (F). The genotypic ratio of their offspring will be 25 percent *EE*, 50 percent *Ee*, and 25 percent *ee*. Three-quarters of their offspring will have free lobes (F), and one-quarter will have attached lobes (A).

Ee (free) $\quad \times\ ee$ (attached)

EE (free) $\quad \times\ EE$ (free)

ee (attached) $\times\ ee$ (attached)

EE (free) $\quad \times\ Ee$ (free)

You can see from the above discussion that heterozygous individuals are carriers of a recessive allele but do not express the recessive genotype. The chance of a recessive phenotype appearing in a series of matings can be determined using a chart of the phenotypes of these matings called a *pedigree*. Close marriages (those between genetically related individuals) are more likely to produce recessive phenotypes (Fig. 12–9).

Autosomal Recessive Inherited Disorders

Most recessive phenotypes are not harmful (Table 12–1). Some, however, can result in severe illness or death. Many inherited recessive disorders result in spontaneous abortion, but less deleterious disorders can produce children born with minor or major genetic defects. These inherited errors are like errors in the notes in the musical

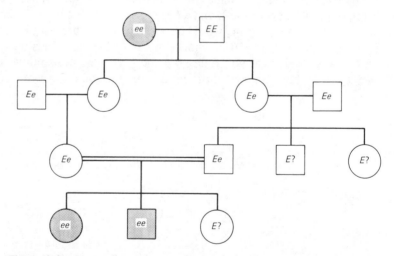

Figure 12–9. In a pedigree chart, males are depicted as squares and females as circles. Shaded circles and squares are "afflicted" individuals; in this case, attached ear lobes. A horizontal line between a circle and square represents a mating, and a double line is a mating between cousins. A vertical line leads to children who are placed along a horizontal line if more than one. Since cousins are more likely to carry the same recessive alleles, their offspring are more likely to show a recessive phenotype. Adapted from Sylvia S. Mader, *Human Reproductive Biology* (Dubuque, Iowa: William C. Brown Company, Publishers, 1980). © 1980; reprinted with permission.

TABLE 12-1. SOME HUMAN DOMINANT AND RECESSIVE
AUTOSOMAL TRAITS

Trait	Dominant	Recessive
Type of ear lobes	Free	Attached
Hair color	Dark	Light
Tongue mannerism	Roller	Nonroller
Freckles	Present	Absent
Widow's peak	Present	Absent
Hitchhiker's thumb	Present	Absent
Interlocking fingers	Right over left	Left over right
Nose shape	Convex	Straight
Dimples	Present	Absent
Shape of little finger	Bent	Straight
Length of second finger	Shorter than fourth	Longer than fourth
Mid-digital hair	Present	Absent
PTC[a]	Taster	Nontaster

Source: Adapted from Sylvia S. Mader, *Human Reproductive Biology* (Dubuque, Iowa: William C. Brown Company, Publishers, 1980). © 1980; reprinted by permission.

[a] The chemical phenylthiocarbamide is an antithyroid drug that prevents the thyroid from incorporating iodine into the thyroid hormone. The ability to taste PTC is associated with pathology of the thyroid gland.

score, whereas developmental errors caused by teratogens (Chapter 9) are like errors the musicians make with a perfect score in front of them. Some examples of inherited disorders brought about by a homozygous recessive condition are discussed below.

Phenylketonuria (PKU) is an inherited homozygous recessive condition. It represents an *inborn error of metabolism,* which is a genetic disorder causing absence of a critical enzyme. Inborn errors of metabolism were first described by an English physician, Sir Archibald Garrod, in 1908. Cells of individuals with PKU are not able to convert the amino acid phenylalanine to the amino acid tyrosine. This results in high levels of phenylalanine in the blood and products of phenylalanine breakdown (phenyl acetic and phenyl pyruvic acid) in the urine. It is thought that the high levels of phenylalanine in the blood cause the blindness and mental retardation that develop in these children beginning about one year after birth. These children also tend to be light-skinned because the brown pigment melanin is synthesized from tyrosine at low levels in the blood of PKU individuals.

Phenylketonuria, if left untreated, causes death, usually before the age of 30. This disorder occurs in about 1/15,000 births, but is more common in the Pennsylvania Amish and in fair-skinned Northern Europeans, exemplifying that mating between genetically related individuals results in a greater appearance of a recessive trait. PKU can be detected in a newborn by a simple test (Chapter 11) or in the fetus by amniocentesis (Chapter 9). Phenylalanine is one of the eight amino acids that our bodies obtain from the diet, so an affected newborn can be put on a restricted phenylalanine diet until he or she is about five years old, when the brain has completely formed.

Tay-Sachs disease is another inherited, homozygous, recessive disorder. This disorder occurs in about 1/300,000 births, but is mostly found (1/3600) in Ashkenazie Jews of Central European ancestry. About 90 percent of American Jews fall into this category, and about 1 in 25 carries the recessive gene. This disorder is an inborn error of metabolism characterized by deterioration of the nervous system beginning at about eight months of age, followed by death by the fifth year. These individuals lack an enzyme that breaks down a certain lipid-sugar complex, which results in accumulation of this complex around nerve cells in the brain. There is no known cure, but affected individuals carrying the recessive gene can be identified with a blood test. Fetuses with this trait can be detected by amniocentesis.

Albinism is another inherited recessive trait found in 1/30,000 whites and 1/22,000 blacks. This inborn error of metabolism involves the absence of an enzyme that converts tyrosine to melanin. Thus, albinos have white hair, pink skin, pink or only lightly colored eyes, and are very sensitive to sunlight.

Inherited recessive traits also can produce physical defects. An example is the occurrence of six-fingered dwarfs in at least 33 families of the Old Order Pennsylvania Amish, a religious order living in Lancaster County, Pennsylvania. Apparently, a heterozygous carrier for this recessive trait passed the recessive allele to his or her children. Because of the high frequency of inbreeding within this sect, several homozygous recessive individuals are born.

Cystic fibrosis is an inherited homozygous recessive inborn error of metabolism that is rather common, especially in Caucasians. About 1200 children are born each year with cystic fibrosis in the United States. This disorder is characterized by presence of an abnormal enzyme (NADH dehydrogenase), which results in overproduction of mucus that clogs ducts of the pancreas, respiratory system, and sweat glands.[3] It can be fatal, the average life expectancy being 19 years. The cells of parents can be examined to test for presence of the abnormal form of the enzyme, and fetuses can be checked for presence of the abnormal enzyme in the amniotic fluid.

Autosomal Dominant Inherited Disorders

Not all rare inherited traits are the product of recessive alleles. About 450 known human traits are the product of dominant alleles. Most people with these traits are heterozygous dominant because homozygous dominant individuals usually are lost through spontaneous abortion. Examples of dominant traits that are not harmful to health are the presence of wooley hair in Norwegian families, a cleft chin, polydactyly (extra digits), and brown teeth. Sometimes, presence of a dominant trait does not mean it is expressed fully in the phenotype. That is, the *gene expressivity* of a dominant genotype depends on the presence of other genes. An example is *Marfan's syndrome*. When fully expressed, this disorder involves an abnormal position of the lens in the eye, longer than normal digits and limbs, and defects in blood vessels. Abraham Lincoln, with his long legs, may have been a case of Marfan's syndrome at its least expressivity. Some dominant disorders do not appear until years after birth. An example is *Huntington's disease* (previously called *Huntington's chorea),* in which there is a progressive deterioration of the nervous system leading to writhing, thrashing, and insanity. This disorder usually begins when a person is 35 to 45 years old, and there is no known cure.

Incomplete Dominance

In some cases, one allele is not completely dominant over the other allele, a phenomenon called *incomplete dominance.* For example, a person with straight hair has the genotype *SS,* and one with curly hair *CC.* An individual that is heterozygous for these alleles (*SC*) has wavy hair (Fig. 12–10).

Some inherited disorders also exhibit incomplete dominance. For example, *sickle-cell anemia* is an inherited disorder of hemoglobin in the red blood cells of black people, first described in 1910. Hemoglobin is the protein-iron complex found in red blood cells that binds to oxygen and carries it to tissues. Individuals with the genotype *AA* have normal hemoglobin A. Individuals with the *SS* genotype have abnormal hemoglobin S, which causes sickle-cell anemia. Hemoglobin S, which differs from hemoglobin A in one out of 287 amino acids, is insoluble in water. Therefore, the molecules bind together, which results in sickle-shaped red blood cells. This disorder leads to poor circulation, anemia, jaundice, abdominal pain, and poor resistance to infection. Persons with sickle-cell anemia rarely live beyond 40 years of age. Individuals with the heterozygous genotype (*AS*) have *sickle-cell trait.* These people have both hemoglobin A and S, and their hemoglobin is normal unless they

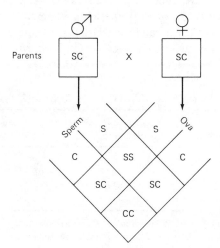

Figure 12–10. Example of a mendelian cross relating to alleles that show incomplete dominance. *SS* = straight hair; *SC* = wavy hair; *CC* = curly hair. Neither curly nor straight hair is dominant. When both alleles are present, the individual has wavy hair.

are exposed to low oxygen. When this occurs, they develop the symptoms of sickle-cell anemia. It is estimated that about 14 percent of American blacks have sickle-cell trait, and about 1 percent have sickle-cell anemia. Sickle-cell trait is most common in a broad area in West Africa, and this trait actually increases resistance to malaria. Presence of sickle-cell anemia can be checked using amniocentesis or fetoscopy.

Codominance

One type of blood grouping, the *ABO blood type,* is influenced by a gene with three alleles—*A, B,* and *O.* Alleles *A* and *B* exhibit *codominance;* that is, they both are dominant over *O* but not over each other. People with type A blood are of the genotype *AA* or *AO,* those with type B blood are *BB* or *BO,* those having type AB blood are *AB,* and those with type O blood have the genotype *OO.* The frequency of these blood types in human populations is given in Table 12–2. Alleles *A* and *B* cause the presence of a specific polysaccharide on the surface of red blood cells. Allele *O* produces no polysaccharide. The A or B polysaccharide acts as an antigen. If a person receives blood with a foreign polysaccharide, his body will form antibodies that cause the red blood cells to agglutinate (clump together).

Knowledge of ABO blood type is important if a blood transfusion is required (Fig. 12–11). A type A person can receive type A or type O blood. A type B person can receive type B or type O blood. Type AB people can receive any blood, but can give blood only to other type AB people. Thus, type AB people are called "universal recipients." Type O people can donate blood to anyone (that is they are "universal donors") but can receive blood only from other type O individuals. There is some evidence that incompatibility of ABO blood type in the mother and her fetus can cause spontaneous abortion.[4]

Knowledge of ABO blood type can also be useful in determining paternity. Suppose a woman with type A blood (*AA* or *AO*) seeks child support from a man with type O (*OO*) blood. If the child is type AB, there is no way that the man is its father. If, however, the child is type B (*BB* or *BO*), there is a 50 percent chance that the man is its father. (Why?) ABO blood type can be used to disprove paternity, but not prove it. Table 12–3 summarizes the genetic bases of ABO blood types.

TABLE 12-2. FREQUENCY OF BLOOD TYPES IN THE UNITED STATES

Phenotype	Genotype	Frequency in Caucasian Population (percent)	Frequency in Negro Population (percent)
Type AB	*AB*	3	3.7
Type B	*BO,BB*	10	21
Type A	*AO, AA*	42	26
Type O	*OO*	45	49.3

Source: From Sylvia S. Mader, *Human Reproductive Biology* (Dubuque, Iowa: William C. Brown Company, Publishers, 1980). © 1980; reprinted by permission.

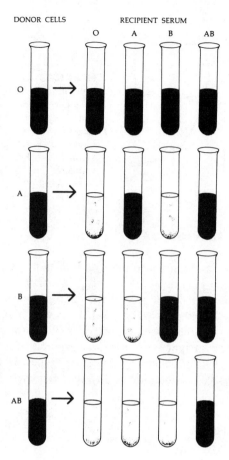

DONOR CELLS   RECIPIENT SERUM
                O   A   B   AB

O

A

B

AB

Figure 12–11. Severe and sometimes fatal reactions can occur following transfusions of blood of a different ABO type from the recipient. These reactions result in agglutination of the blood cells caused by antibodies present in the recipient's serum. Blood-group reactions can be shown in test tubes as demonstrated here. When agglutination occurs, the tubes become more clear. Reproduced from Helena Curtis, *Biology,* 3rd ed., p. 304 (New York: Worth Publishers, 1979). © 1979; used with permission.

TABLE 12-3. GENETIC BASES FOR ABO BLOOD TYPES

Phenotypes of Parents	Children Possible	Children Not Possible
A × A	A, O	AB, B
A × B	A, B, AB, O	—
A × AB	A, B, AB	O
A × O	A, O	AB, B
B × B	B, O	A, AB
B × AB	A, B, AB	O
B × O	B, O	A, AB
AB × AB	A, B, AB	O
AB × O	A, B	O, AB
O × O	O	A, B, AB

Source: From Helena Curtis, *Biology,* 3rd ed., p. 304 (New York: Worth Publishers, 1979). © 1979; used with permission.

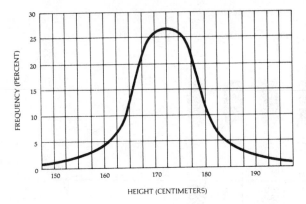

Figure 12–12. Height distribution in males in the United States. Height and weight are examples of polygenic inheritance. A graph of the distribution of such traits takes the form of a bell-shaped curve, with the average falling at the midpoint. Reproduced from Helena Curtis, *Biology,* 3rd ed., p. 236 (New York: Worth Publishers, 1979). © 1979; used with permission.

Polygenic Autosomal Inheritance

Most of our inherited traits are controlled by several alleles of a gene or by several genes, a phenomenon called *polygenic inheritance.* For example, body height and weight are affected by several genes; a plot of the distribution of such traits in the population exhibits a bell-shaped curve (Fig. 12–12). As with many inherited traits, environmental factors can influence the expressivity of genes. Nutrition, for instance, can influence the body height and weight. An example of simple polygenic inheritance is skin color, which is influenced by at least two genes, each with two alleles. The following is the phenotypic skin color of different genotypes:

AABB = black
AABb or *AaBB* = dark
AaBb, AAbb, or *aaBB* = mulatto
Aabb or *aaBb* = light
aabb = white

Several inherited disorders are the product of the polygenic action of several genes, and it often is difficult to predict their chances of occurrence. These disorders include cleft palate, club foot, congenital dislocation of the hip, certain spinal abnormalities, and congenital high blood pressure.

SEX-LINKED INHERITANCE

The human X chromosome contains at least 100 genes, whereas the Y chromosome has no definitely proven genes except those controlling differentiation of the testes and production of H–Y antigen.[5] Genes on the X chromosome are said to be sex-linked, and the resultant phenotypic effects of these genes are *sex-linked traits.* An example is red-green colorblindness, in which the allele X^c produces normal vision, and the allele X^c causes colorblindness. In fact, the gene for colorblindness on the X chromosome was the first gene to be located on any human chromosome, in 1911. The table below shows the phenotypes of men and women having different genotypes for this sex-linked trait.

Vision	Female	Male
Normal	$X^C X^C$	$X^C Y$
Normal (carrier)	$X^C X^c$	—
Colorblind	$X^c X^c$	$X^c Y$

You can see that more males than females are colorblind, but females can be carriers. Figure 12–13 is an exhibit of the results of several matings in relation to red-green colorblindness. Because of the presence of genes on only the X chromosome, sex-linked characters can skip a generation, appearing in a grandfather and his grandson but not in his son.

Recall from Chapter 5 that one X chromosome in female cells is genetically inactive, and that the particular X chromosome in tissues that is inactivated is a random

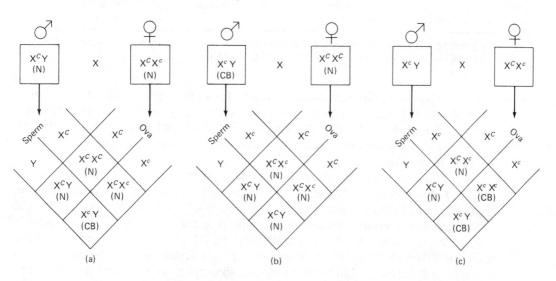

Figure 12–13. Results of three different matings on the sex-linked trait of red-green colorblindness. N = normal vision; CB = colorblind. (a) When the male parent is normal and the female is a carrier, daughters have a 50 percent chance of being a carrier and sons have a 50 percent chance of being colorblind. (b) When the male parent is colorblind and the female parent is normal, all daughters are carriers and all sons are normal. (c) When the male parent is colorblind and the female parent is a carrier, both sons and daughters have a 50 percent chance of being colorblind. Adapted from Sylvia S. Mader, *Human Reproductive Biology* (Dubuque, Iowa: William C. Brown Company, Publishers, 1980). © 1980; reprinted by permission.

TABLE 12-4. SEX–LINKED RECESSIVE GENETIC DISEASES

Name	Symptoms	Defect	Comments
Red-green color-blindness	Unable to see all colors	Receptors in eyes	Can lead normal life.
Classical hemophilia	Blood does not clot	Factor VIII deficiency	Lifetime treatment; carriers can be detected; afflicted fetuses can be detected by blood test following fetoscopy.
Duchenne's muscular dystrophy	Muscle weakness	Muscle chemistry is faulty	Carriers can sometimes be detected; afflicted fetuses can be detected by blood test following fetoscopy.
Lesch-Nyhan syndrome	Self-mutilation; mental retardation	High blood level of uric acid	Carriers can be detected; afflicted fetuses can be detected following amniocentesis.
Agammaglobulinemia	Constant and severe infections	Unable to make antibodies	Some immunity genetic diseases are not sex-linked.

Source: Adapted from Sylvia S. Mader, *Human Reproductive Biology* (Dubuque, Iowa: William C. Brown Company, Publishers, 1980). © 1980; used by permission.

event. Therefore, different tissues of a female can have one or the other sex-linked allele active. This can produce a *genetic mosaic* phenotype. An example is the calico cat, which always is female. There are sex-linked alleles for coat color in cats, and random inactivation of the X chromosome produces an orange-black mosaic of fur color in these animals.

There are other more severe inherited traits that are sex-linked (Table 12–4). For example, there are 12 factors that are necessary for normal blood clotting, and people with *classical hemophilia* ("bleeder's disease") lack factor VIII. This form of hemophilia is recessive and sex-linked, appearing mostly in males. It was present in the royal families of Europe, especially England (Fig. 12–14). Fortunately, synthetic factor VIII now is available to treat sufferers of this disease. Another example of a sex-linked disorder is *Duchenne's muscular dystrophy*. This is a recessive sex-linked disorder leading to degeneration of skeletal muscle and death as a teenager. Another

Figure 12–14. As this pedigree shows, Queen Victoria was the original carrier of the recessive allele for classical hemophilia that has afflicted males of the royal families of Europe since the nineteenth century. The present British royal family escaped the disease because King Edward VII, and consequently all of his progeny, did not inherit the recessive allele. Reproduced from Helena Curtis, *Biology,* 3rd ed., p. 306 (New York: Worth Publishers, 1979). © 1979; used with permission.

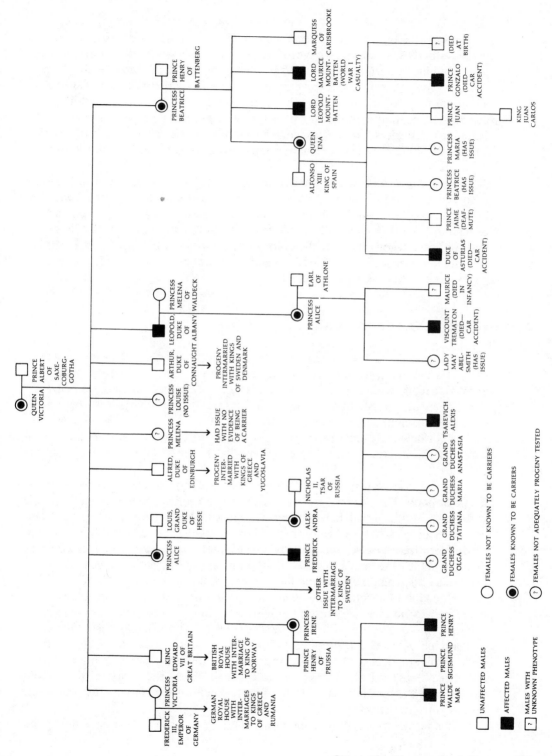

QUEEN VICTORIA PRINCE ALBERT OF SAXE-COBURG-GOTHA

PRINCESS VICTORIA FREDERICK III, EMPEROR OF GERMANY — GERMAN ROYAL HOUSE WITH INTERMARRIAGES TO KINGS OF GREECE AND RUMANIA

KING EDWARD VII OF GREAT BRITAIN PRINCESS ALEXANDRA — BRITISH ROYAL HOUSE WITH INTERMARRIAGE TO KING OF NORWAY

PRINCESS ALICE LOUIS, GRAND DUKE OF HESSE

PRINCE HENRY OF PRUSSIA PRINCESS IRENE — OTHER ISSUE WITH INTERMARRIAGE TO KING OF SWEDEN

PRINCE WALDE-MAR PRINCE SIGISMUND PRINCE HENRY

PRINCE FREDERICK

ALEXANDRA NICHOLAS II, TSAR OF RUSSIA

GRAND DUCHESS OLGA GRAND DUCHESS TATIANA GRAND DUCHESS MARIA GRAND DUCHESS ANASTASIA GRAND TSAREVICH ALEXIS

ALFRED, DUKE OF EDINBURGH — PROGENY INTERMARRIED WITH KINGS OF GREECE AND YUGOSLAVIA

PRINCESS MELENA — HAD ISSUE WITH NO EVIDENCE OF BEING A CARRIER

PRINCESS LOUISE (NO ISSUE)

ARTHUR, DUKE OF CONNAUGHT — PROGENY INTERMARRIED WITH KINGS OF SWEDEN AND DENMARK

LEOPOLD, DUKE OF ALBANY PRINCESS MELENA OF WALDECK

PRINCESS ALICE EARL OF ATHLONE

LADY MAY ABEL-SMITH (HAS ISSUE) VISCOUNT TREMATON (DIED—CAR ACCIDENT) MAURICE (DIED IN INFANCY)

PRINCESS BEATRICE PRINCE HENRY OF BATTENBERG

QUEEN ENA ALFONSO XIII KING OF SPAIN

LORD LEOPOLD MOUNTBATTEN LORD MAURICE MOUNTBATTEN (WORLD WAR 1 CASUALTY) MARQUESS OF CARISBROOKE

DUKE OF ASTURIAS (DIED—CAR ACCIDENT) PRINCE JAIME (DEAF-MUTE) PRINCESS BEATRICE (HAS ISSUE) PRINCESS MARIA (HAS ISSUE) PRINCE JUAN KING JUAN CARLOS PRINCE GONZALO (DIED—CAR ACCIDENT) (DIED AT BIRTH)

○ FEMALES NOT KNOWN TO BE CARRIERS

◉ FEMALES KNOWN TO BE CARRIERS

⒴ FEMALES NOT ADEQUATELY PROGENY TESTED

□ UNAFFECTED MALES

■ AFFECTED MALES

⸦?⸧ MALES WITH UNKNOWN PHENOTYPE

Chap. 12 Human Genetics 283

sex-linked recessive disorder is *agammaglobulinemia,* in which a person is unable to make antibodies. These males have to live their entire lives protected from germs, in a plastic bubble or sterile suit. *Lesch-Nyhan syndrome* is a sex-linked recessive disorder of the nervous system. Lack of an enzyme involved in synthesis of nucleotides causes abnormal brain function, leading to mental retardation, involuntary writhing, and compulsive self-mutilation (biting of one's own lips and finger tips). Many of these sex-linked disorders can be detected by amniocentesis. In the case of classical hemophilia, blood of the fetus can be sampled (following fetoscopy) for the presence of factor VIII.

    Sex-limited traits occur mostly in one sex. In these conditions, a certain genotype must be present, but the action or expressivity of these genes is controlled by hormones. An example is baldness. Males require only one allele for baldness to be bald because they have high circulating levels of testosterone. Females, however, must by homozygous for bald alleles for baldness to occur because there is little circulating testosterone in their blood. Therefore, the gene for baldness is dominant in males but acts as a recessive in females.

CHROMOSOMAL ABERRATIONS

Errors of meiosis or fertilization can produce embryos with chromosomal aberrations. Fortunately, greater than 90 percent of these embryos are spontaneously aborted, usually within the first trimester. In fact, 42 percent of embryos or fetuses that are spontaneously aborted have chromosomal abnormalities.[6] A few fetuses with chromosomal defects, however, are born; about 1 out of every 100 newborns has such a defect. It must be emphasized that some of these disorders are not inherited in the strictest sense because the genes of the parents do not govern their occurrence.

    In rare cases, one sperm will fertilize the ovum, and a second sperm will fertilize the polar body. The two fertilized cells then form an embryo that is a genetic mosaic in that half of its cells will have a different genetic make-up from the other half. This condition also can occur when the haploid ovum divides into two cells, and each cell then is fertilized by a separate sperm. For example, half a person's hair could be blond, and the other half brown. Or if an X and a Y sperm were involved, half of an embryo's cells would be male and half female, resulting in an intersex (Chapter 5).

    One kind of chromosomal aberration occurs when the sperm head fails to activate the second meiotic division in the ovum. Thus, no pronuclear fusion occurs, and the embryo develops with only one set of chromosomes (haploid cells) and genes of the male only. This process of embryonic formation is called *androgenesis.* A similar situation occurs when the ovum pronucleus develops normally, but the sperm pronucleus does not form. In this case, called *gynogenesis,* the embryo also is haploid but has only the female's genes. Both of these conditions are lethal after only a few cell divisions in the embryo.

    In contrast to the above conditions, some embryos may develop with triploid cells (3N) that have 69 chromosomes (three complete sets). *Triploidy* can occur in at least three ways. First, the sperm penetrating the ovum may be the product of a failure of reduction of division during meiosis in the testis, and thus it has 46 instead of the normal 23 chromosomes. When this sperm fertilizes a haploid ovum, a triploid embryo develops. Second, even though mechanisms to prevent polyspermy are present (Chapter 8), these mechanisms are not fail-safe. Thus two haploid sperm can penetrate a single ovum (polyspermy), and both of their pronuclei then fuse with the haploid ovum pronucleus. Finally, reduction division (meiosis) may not have occurred in the oocyte, and the resultant diploid female pronucleus then fuses with a haploid sperm pronucleus to produce a triploid zygote.

The excess dosage of genes in triploid embryos tends to be less destructive than when there are too few genes as in androgenesis or gynogenesis. Most triploid embryos develop to about the third month of pregnancy before spontaneously aborting. The very few triploid fetuses that survive to term are malformed and are stillborn or die soon after birth. Only 1 percent of all human embryos are triploid.

Another error in fertilization results in embryos with either one too many (47) or one two few (45) chromosomes in their cells; these conditions collectively are called *aneuploidy*. This happens when there is aberrant chromosome movement during the first or second meiotic division in the testis or ovary, or in the first cleavage division of the zygote. That is, a pair of chromosomes fails to separate during division, both members going to one daughter cell (an error called *nondisjunction*). The resultant cell has 47 chromosomes, and the cell coming up short has only 45. Thus, the aneuploid condition can be either *monosomic* (45 chromosomes) or *trisomic* (47 chromosomes); see Fig. 12–15.

Most monosomic embryos spontaneously abort early in their development. An exception, however, is when monosomy for a sex chromosome occurs (Fig. 12–15). That is, each cell has only a single sex chromosome, either an X or a Y. About 98 percent of these embryos abort, but a few with one X (XO condition; Turner's syndrome) are born as sterile females with short stature and physical defects (Chapter 5). Only 1 in 3500 living females has this syndrome.

Most trisomic embryos die in the second or third month of pregnancy and spontaneously abort; 20 percent of miscarried fetuses are trisomic. Some, however, are born with severe physical and mental defects. The most common trisomic condition in infants is *Down's syndrome,* also called *Mongolism,* a condition in which cells of the individual are trisomic for chromosome number 21. Children with Down's syndrome exhibit abnormal body development and severe mental retardation.

For some as yet unknown reasons, the gametes of older men and women are more likely to produce trisomic embryos. The chances are 1/1000 for having a trisomic embryo for women under 35, but are 1/200 for 35-year-old women, and 1/15 for 45-year-old women. Women over 35 have only 15 percent of all babies but 50 percent of all Mongoloid children. Therefore, it is recommended that women in their midthirties should consider having the cells of their fetus examined, by amniocentesis, for evidence of chromosomal abnormalities. If certain chromosomal aberrations are found, induced abortion might be considered (Chapter 14). It used to

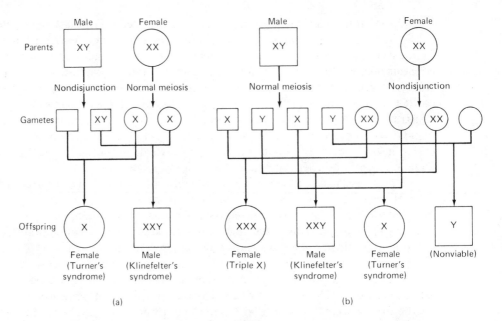

Figure 12–15. (a) Nondisjunction, the failure of a pair of chromosomes to separate during meiosis, can cause chromosomal abnormalities. This figure shows the effects of nondisjunction of sex chromosomes in the father's sperm. The fertilized egg will have two X chromosomes and one Y chromosome (Klinefelter's syndrome) or only an X chromosome (Turner's syndrome). (b) Nondisjunction in the mother produces ova with two X chromosomes or none. Depending on which sperm (X or Y) fertilize, such ova can give rise to four possible chromosomal anomalies. Three can be viable, and one never is viable.

be thought that errors in meiosis in oocytes of older women were the main cause of trisomy. Recently, however, we have become aware that about one-fifth of trisomic infants are caused by chromosomal abnormalities in the sperm of older men.[7]

Nondisjunction of sex chromosomes can produce males with trisomic cells of an XXY or XYY make-up. In the former condition, called Klinefelter's syndrome (Fig. 12–15), males are sterile and have femalelike breasts. About 1 out of 600 males is born with this condition. In the latter "supermale" condition (XYY), males are very tall and often have acne. These males tend to exhibit mental and social adjustment problems at a higher percentage than do normal XY males. One in 2000 males has XYY cells. There is some statistical evidence that the percentage of XYY males (1.8 percent to 12.0 percent) in penal institutions is greater than their percentage (0.14 percent to 0.38 percent) in the general population.[8] Some controversy, however, surrounds these studies, and it is not clear if the greater maladaptive behavior of XYY males is a direct result of their chromosomal abnormality or is due to social problems they had when growing up because of their unusual physical appearance. Apparently the elevated crime rate of XYY men is not related to aggression but may be related to low intelligence.[9] Women with nondisjunction of the X chromosome have cells that are XXX (Fig. 12–15). These women are female but sterile. Cases in which males have several X chromosomes (XXXY) are due to penetration of the ovum by more than one sperm.

Sometimes, a gamete contains a chromosome with an extra piece from another chromosome attached to it; this is the result of *chromosomal translocation*. The chromosome from which the piece was "stolen" thus suffers from *chromosomal deletion*. An example of a disorder resulting from chromosomal deletion is the *Cri-du-chat* (French for "cry of the cat") syndrome, in which a piece of chromosome 5 is missing. These children are born with a small head, widely separated eyes, low-set ears, and mental retardation. When they cry, it sounds like a hungry kitten. Human kidney cancer has also been linked to an inherited chromosomal translocation in which a piece of chromosome 3 is hooked onto chromosome 8.[10]

Recently, it has been found that an inherited disorder of the X chromosome is the second leading cause of mental retardation. In these people, the X chromosome (in either sex) has an abnormally long, fragile arm. In this disorder, mental retardation is less severe in females than in males.[11]

DETECTION OF INHERITED DISORDERS

In Chapter 9, we discussed how the procedure of amniocentesis is used to detect inherited disorders of the fetus. More than 60 of the 1600 inherited human disorders can now be detected using this method.[12] These include the Lesch-Nyhan syndrome, Tay-Sachs disease, the adrenogenital syndrome (Chapter 5), sickle-cell anemia, several chromosomal disorders, and many other enzyme deficiencies. The procedure of fetoscopy can be used to sample fetal blood (Chapter 9), and disorders such as classical hemophilia and Duchenne's muscular dystrophy can be detected in this way. Also, some inherited defects such as PKU can be detected in the neonate (Chapter 11). Several of the congenital inherited disorders are treatable. For example, *neural tube defects,* which occur in 1/1000 newborns, can even be treated while the fetus is still within the uterus.[13] Examples of these defects are *hydrocephalus* (excessive fluid accumulation in the brain) and *spina bifida* (protrusion of the spinal cord because of unfused vertebrae). Individuals with other disorders may be doomed to serious illness and often death. Parents of a fetus that has been shown to have a severe inherited disorder have the always difficult choice of having the pregnancy aborted (Chapter 14) or raising a handicapped child.

MANIPULATION OF HUMAN GENES

Our growing knowledge about human inheritance, combined with the ability to detect fetal genetic disorders and the possibility of manipulating human genes, has increased our ability to influence human genetics.

Genetic Counseling

Because of our increasing understanding of human genetics, it is possible to predict with some degree of reliability the probability of certain inherited disorders in the offspring of parents with a well-known family history (pedigree). Couples can seek *genetic counseling,* during which the pedigree of the man and woman is examined, and the probability of their fetus having an inherited disorder is estimated. Once this is done, the couple then has a choice about bearing a child. However, with the increasing availability of genetic counseling, some legal problems have arisen. Parents could sue the conselor, for example, if they were told that they were not carriers of a specific genetic defect and they then had an infant with that defect.[14] In addition, a physician failing to give genetic counseling to a couple could be liable for lifelong care of a child born with an inherited disorder.[15]

Eugenics is the attempt to maintain or delete certain genes from a population by selective breeding.[16] It is "positive" eugenics if a desirable trait is selected for, and "negative" eugenics if an undesirable trait is selected against. The problem occurs when one defines "desirable" and "undesirable"! Genetic counseling is a form of negative eugenics only if a couple chooses not to bear a child as a result of this counseling.

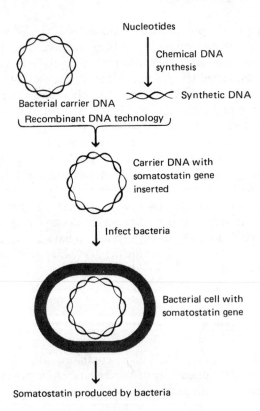

Figure 12–16. It now is possible to splice synthetic genes onto the DNA of bacteria. In this case, the bacteria receive synthetic DNA that codes for a human hormone, somatostatin, that causes secretion of growth hormone. The bacteria then serve as "factories" for making this useful molecule. Adapted from Helena Curtis, *Biology,* 3rd ed., p. 325 (New York: Worth Publishers, 1979). © 1979; used with permission.

Genetic Engineering

Remarkable scientific advances in the past few years have brought genetic engineering into the limelight. One kind of genetic engineering is the technique of *recombinant DNA*. Basically, this means the artificial introduction of foreign genes (DNA) into the DNA of an organism. For example, the human hormone *somatostatin,* which among other things inhibits secretion of growth hormone from the anterior pituitary gland (Chapter 2), is now being made by bacteria! That is, synthetic DNA that codes for somatostatin synthesis has been spliced to the DNA of a bacteria, and these bacteria then synthesize large amounts of somatostatin (Fig. 12-16). In a similar manner, human insulin is now being made in large quantities by bacteria, and now is available to diabetics.

The potential health benefits of genetic engineering are mind-boggling. Recently, an attempt was made to introduce synthetic DNA into the cells of people suffering from an inherited defect of hemoglobin, the new DNA coding for the normal form of hemoglobin.[17] The recombinant DNA procedure, however, must be tested thoroughly, and it holds some frightening possibilities. For example, a genetic accident could produce a new bacterial strain that causes a plague! Also, many people worry about the possible misuse of genetic engineering and its potential effects on human population genetics. For these reasons, future research on recombinant DNA is done under very strict federal guidelines.

CHAPTER SUMMARY

Deoxyribonucleic acid (DNA) in our chromosomes replicates itself during cell division. Genes are composed of a series of nitrogenous-base triplets in the DNA molecules. These genes code for transcription of messenger RNA (mRNA), which carries the message to the ribosomes in the cell cytoplasm. Transfer RNA (tRNA) molecules bound to specific amino acids recognize the code in mRNA, and thus a specific chain of amino acids, a polypeptide or protein, is formed. Even though all of an individual's cells have the same genes, differential gene action may occur because of the action of regulatory genes. Mutations can occur in the DNA, spontaneously or because of exposure to mutagens.

The autosomes carry most of our genes. Two (or more) genes influencing the same trait, each being on a different member of a pair of homologous chromosomes, are called alleles. Often, one allele will be dominant over the other (recessive) allele. During meiosis, homologous chromosomes segregate, and the distribution of chromosomes within gametes is random and independent of one another. Alleles on the same chromosomes are linked unless crossing-over occurs. Recombination of maternal and paternal chromosomes at fertilization produces predictable genotypic and phenotypic ratios in the offspring.

Autosomal recessive traits only appear when both alleles are recessive. Many of these traits are harmless, but some cause inborn errors of metabolism or physical defects. Such disorders include phenylketonuria, Tay-Sachs disease, albinism, and six-fingered dwarfism. Autosomal dominant disorders include Huntington's disease, polydactyly, and Marfan's syndrome. In some cases, incomplete dominance of alleles leads to an intermediate phenotype in heterozygous individuals. This is the case with sickle-cell anemia and inherited disorders of hemoglobin. Some traits are influenced by several genes or alleles. Examples are body height and weight. The inheritance of ABO blood type is an example of codominance.

The human X chromosome carries about 100 genes, whereas the Y chromosome contains only the gene or genes for sex determination. Genes on the X chromosome are sex-linked, and the phenotypic expressions of these genes are called sex-linked traits. A well-known example is red-green colorblindness. Sex-linked disorders include classical hemophilia, Duchenne's muscular dystrophy, agammaglobulinemia, and Lesch-Nyhan syndrome. Sex-limited traits are influenced by the levels of sex hormones in the blood, an example being baldness.

Inherited chromosomal abnormalities can result from nondisjunction, translocation, or deletion. The Cri-du-chat syndrome results from deletion of chromosome 5, and kidney cancer is influenced by translocation of a piece of chromosome 3 onto chromosome 8. Nondisjunction produces either a monosomic (45 chromosomes) or trisomic (47 chromosomes) condition. An example of monosomy is Turner's syndrome. Examples of trisomy are Down's syndrome, Klinefelter's syndrome, and XXY or XYY males. Increasing age of parents increases the likelihood of having offspring with chromosomal abnormalities.

Many inherited disorders can be detected in fetuses by amniocentesis or fetoscopy. Parents who are concerned about the genotype of their potential offspring can utilize genetic counseling. Eugenics is the selective breeding of humans with the purpose of increasing the prevalence of desirable genes or decreasing the frequency of undesirable genes. In the future, some inherited disorders may be corrected by using recombinant DNA, but there are serious moral and medical concerns about this procedure.

NOTES

1. G. B. Kolata, "Thalassemias: Models of Genetic Diseases," *Science,* 210 (1980), 300–302.
2. V. A. McKusick and F. H. Ruddle, "The Status of the Gene Map of the Human Chromosomes," *Science,* 196 (1977), 390–405.
3. Anonymous, "Cystic Fibrosis: Genetic Error Found," *Science News,* 116 (August 1979), 118.
4. A. E. Szulman, "ABO Incompatibility and Foetal Wastage," *Research in Reproduction,* 5, no. 1 (1973), 3–4.
5. McKusick and Ruddle (1977), op. cit.

6. D. H. Carr, "Chromosomal Abnormalities in Human Foetuses," *Research in Reproduction,* 4, no. 2 (1972), 3–4; and J. G. Lawutsen, "The Cytogenetics of Spontaneous Abortion," *Research in Reproduction,* 14, no. 3 (1982), 3–4.

7. J. Stene et al., "Paternal Age Effect in Down's Syndrome," *Annals of Human Genetics,* 40 (1977), 299–306.

8. L. I. Gardner and R. L. Neu, "Evidence Linking an Extra Y Chromosome to Sociopathic Behavior," *Archives of General Psychiatry,* 26 (1972), 220–22.

9. H. A. Witkin et al., "Criminality in XYY and XXY Men," *Science,* 193 (1976), 547–55.

10. A. J. Cohen et al., "Hereditary Renal-Cell Carcinoma Associated with a Chromosomal Translocation," *New England Journal of Medicine,* 301 (1979), 592–94.

11. Anonymous, "A Fragile X and Female Retardation," *Science News,* 18, no. 13 (1980), 199.

12. T. Friedmann, "Prenatal Diagnosis of Genetic Disease," *Scientific American,* 225, no. 5 (1971), 34–42.

13. Anonymous, "Preventing Hydrocephalus in Infants," *Science News,* 120, no. 12 (1981), 181; and A. Rosenfield, "The Patient in the Womb," *Science 82,* 3, no. 1 (1982), 18–24.

14. Anonymous, "Wrongful Life," *Science 80,* 1, no. 8 (1980), 7.

15. B. J. Culliton, "Physicians Sued for Failing to Give Genetic Counseling," *Science,* 203 (1979), 251.

16. F. Osborn, "The Emergence of a Valid Eugenics," *Scientific American,* 61 (1973), 425–29.

17. Anonymous, "Gene Therapy," *Science 80,* 1, no. 8 (1980), 6.

FURTHER READING

ALLISON, A. C., "Sickle Cell and Evolution," *Scientific American,* 195, no. 2 (1956), 87–94.

CERAMI, A., AND C. M. PETERSON, "Cyanate and Sickle-Cell Disease," *Scientific American,* 232, no. 4 (1975), 44–50.

COHEN, S. N., "The Manipulation of Genes," *Scientific American,* 233, no. 1 (1975), 24–33.

FRIEDMANN, T., "Prenatal Diagnosis of Genetic Disease," *Scientific American,* 225, no. 5 (1971), 34–42.

GERMAN, J. L., "Studying Human Chromosomes Today," *American Scientist,* 58 (1970), 182.

KARP, L. E., "Genetic Crossroads," *Natural History Magazine,* 87 (1978), 8–22.

MCKUSICK, V. A., "The Mapping of Human Chromosomes," *Scientific American,* 224, no. 4 (1971), 104–13.

MITTWOCH, U., "Sex Differences in Human Cells," *Scientific American,* 209 (July 1963), 54.

RUDDLE, F. H., AND R. S. KUCHERLAPATI, "Hybrid Cells and Human Genes," *Scientific American,* 231, no. 1 (1974), 36–44.

13 *Contraception*

LEARNING GOALS

Having read this chapter, you should be able to:

1. Describe the ingredients of the combination pill and how they work as a contraceptive agent.
2. List some adverse side effects of the combination pill.
3. List advantages and disadvantages of the combination pill as a contraceptive measure.
4. Describe the make-up and mechanisms of action of the minipill, the progestogen implant, and injectable progestogen.
5. Discuss the use and drawbacks of the postcoital pill.
6. Describe the mechanisms of action of intrauterine devices, and give some disadvantages of IUD usage.
7. Describe the chemical ingredients of spermicides, and tell how they work and what forms are available.
8. Discuss how the diaphragm and cervical cap work, and list some advantages and disadvantages of diaphragm and cap use.
9. List two precautions one should be aware of when using a condom.
10. Compare coitus interruptus, coitus reservatus, and coitus obstructus.
11. Describe how a woman can calculate the "fertile" and "safe" periods in her menstrual cycle using the calendar method.

12. Describe how basal body temperature and cervical mucus can be used to indicate the fertile and safe periods.

13. Answer the question, "Is breast-feeding an efficient method of contraception?"

14. Differentiate among laparotomy, minilaparotomy, laparoscopy, culdoscopy, and colpotomy as operations used for tubal sterilization.

15. Describe how a vasectomy is done, and list at least five possible complications of a vasectomy.

16. Describe three possible future methods of contraception.

17. Give five psychological reasons why a couple would experience an unwanted pregnancy because of the misuse or avoidance of contraception.

INTRODUCTION

Advances in reproductive science and medicine have made it possible to manipulate human fertility. For couples that desire to avoid or terminate an unwanted pregnancy, contraceptive measures and induced abortion (Chapter 14) are available. In this chapter, we review current methods of contraception, including their usage, advantages and disadvantages, mechanisms of action, effectiveness, and minor and major adverse side effects. In addition, we discuss possible future methods of contraception.

CONTRACEPTION

Contraception literally means "against conception," and many contraceptive devices prevent fertilization of an ovum by a sperm. For our discussion, however, we also include contraception methods that prevent transport of a developing embryo through the oviduct or implantation of an embryo in the uterus. Contraceptive use has increased throughout the world as people have begun to realize the problems of overpopulation (Chapter 20) and as couples come to understand the effects of an unwanted pregnancy on their lives (Table 13–1).

The Combination Pill

Scientific research on the possible use of hormones for human contraception began in 1897 when John Beard, an anatomist at the University of Edinburgh, surmised that the corpus luteum inhibited ovulation during early pregnancy. In 1898, an endocrine function of the corpus luteum was proposed by the Frenchman August Prenant. Then, in 1931, the Austrian physiologist Ludwig Haberlandt, while doing research on the role of the corpus luteum in reproduction of mice and rabbits at the University of Innsbruck, suggested that hormonal sterilization could be an effective contraceptive method in humans. Soon after, research on laboratory mammals clarified that a

TABLE 13-1. ESTIMATED NUMBER OF COUPLES USING BIRTH CONTROL, WORLDWIDE, BY METHOD, 1970 AND 1977

	1970 (millions)	1977 (millions)
Voluntary sterilization	20	80
Oral contraceptives	30	50
Condom	25	35
IUD	12	15
Other methods[a]	60	65
Total	147	250

Source: Adapted from C. P. Green, "Voluntary Sterilization: World's Leading Contraceptive Method," *Population Reports,* ser. M, no. 2 (1978), pp. 37–70; with permission of the Population Information Program, The Johns Hopkins University, Baltimore, Md.

[a] Diaphragm, spermicides, rhythm, withdrawal, etc.

certain ratio of estrogen and progestogen in the blood during the luteal phase of the estrous cycle blocks ovulation (Chapter 7). This was an exciting revelation that led to discussions about the possible use of these steroid hormones for human contraception. The three natural estrogens (estradiol-17β, estriol, and estrone) were identified in 1929 and 1930, and the natural progestogen (progesterone) in 1934. In 1937, Sir Charles Dodd synthesized an artificial estrogen (stilbestrol), and this was followed by the introduction of more synthetic estrogens such as ethinyl estradiol and synthetic progestogens such as ethisterone and norethynodrel. An important development in the 1940s was the finding by Russel Marker in the United States that a cheap source of material to make synthetic steroid hormones was the sweet potato! Thus, the idea of human steroidal contraception, as well as a cheap source of synthetic steroid hormones, was available by 1950. In the early 1950s, Margaret Sanger, a pioneer in the establishment of family planning in the United States (Chapter 20), along with the scientists Gregory Pincus and M. C. Chang, began serious discussions about the possibility of steroidal contraception.

By the mid-1950s, clinical investigations began on the effectiveness of the progestogen norethynodrel as a contraceptive agent by Gregory Pincus as well as by John Rock and Celso Ramon-Garcia. Then, clinical trials of norethynodrel in pill form occurred in Puerto Rico in 1956, organized by John Rock, Gregory Pincus, and Edris Rice-Wray, a feminist and birth control advocate. The original preparation of norethynodrel contained small amounts of mestranol, an estrogen, and it was discovered that the presence of this estrogen made the pill more effective. Thus, the Federal Drug Administration (FDA) approved the use and marketing of the *combination pill* containing an estrogen *plus* a progestogen in 1960. This first combination pill contained 10 mg of norethynodrel and 150 μg of mestranol. By 1963, the progestogen content was reduced to 2 mg or less, and the estrogen to 100 μg or less. The combination pill became the most popular contraceptive measure in the United States as well as in many other parts of the world. In 1975, 64 million prescriptions for combination pills were obtained in the United States. Although this number had dropped by 23 percent (to 49 million) by 1982 because of possible side effects (to be discussed later), the combination pill is still the most popular reversible contraceptive method in the United States and in several other countries.

Ingredients. Today, there are about 37 different brands of combination pills. All of these contain a synthetic estrogen and progestogen. The brands differ in the relative and absolute amounts of these compounds in each pill. The estrogen is either mestranol or ethinyl estradiol, and the progestogen can be one of five types (ethynodiol diacetate, norethindrone, norethindrone acetate, northynodrel, or norgestrel). Some synthetic estrogens and progestogens are more potent than others, and the choice of what brand to use depends on each individual woman's physical make-up.

How the Combination Pill Works. The combination pill prevents conception by inhibiting ovulation. The combination of estrogen and progestogen prevents the surge of FSH and LH secretion from the pituitary that normally causes ovulation during the middle of the menstrual cycle (Chapter 7). In essence, this pill mimics the negative feedback effects of estrogen and progestogen present during the luteal phase of the menstrual cycle and pregnancy. Even if the pill fails to prevent ovulation, conception may not occur because the pill renders the cervical mucus hostile to sperm transport. And even if some sperm reach the egg and fertilization occurs, the pill renders the uterine endometrium unreceptive to implantation of the embryo.

Use of the Combination Pill. The pills come in cycle packages; a woman takes one hormone-containing pill a day beginning on the Sunday after her last menses began. She then continues for 21 days. During this time, ovulation is suppressed. She then either stops taking any pills for seven days, or switches to an inert (sugar) pill for the remaining seven days of her cycle. After stopping the hormonal pill, menstruation begins. A woman is unlikely to get pregnant during the seven-day period when she is not taking the hormonal pills, because she is menstruating and an embryo would not implant. If she misses one day of taking a hormonal pill, she can take two the next day with no loss of protection. If, however, she misses two or more days, she should stop taking the hormonal pills for seven days and then begin a new pill cycle; other contraceptive protection (foam, condom, diaphragm) should be used during this seven-day waiting period, as pregnancy certainly is possible.

It should be mentioned that another form of pill was available in the past. In this *sequential pill,* a woman took pills containing only an estrogen for 15 to 16 days after menstruation. She then took pills containing both an estrogen and a progestogen for 5 to 6 days, after which she stopped to allow menstruation. The sequential pill was withdrawn from the market in 1978 because of its low use and effectiveness as well as some evidence that it increased the incidence of breast cancer in dogs.

Failure Rate.   Because of the combination pill's prevention of ovulation, conception, and implantation, it is a very effective contraceptive method. In this discussion, we refer to failure rates of contraceptive measures in 100 woman-years, or 100 WY (*Pearl's formula*). A *woman-year* is the use of a contraceptive device by a woman for one year (12 potential menstrual cycles). Thus, 100 woman-years of use could be, for example, the use of a contraceptive by 1 woman for 100 years, or 10 women for 10 years, or 100 women for 1 year. A failure rate of 1/100 WY means that 1 woman out of 100 would become pregnant using this device for 1 year. Failure rates of a contraceptive measure are given in a range, with the lower figure representing its theoretical effectiveness if used properly and the larger figure representing the actual failure rate. The latter figure is higher because people sometimes do not follow instructions or are not conscientious in their use of the device. The failure rate of the combination pill is 0.7 to 3.0/100 WY, with the higher figure present because some women forget to take, or choose to not take, a pill every day. This failure rate is very low in relation to rates of other contraceptive devices, which are shown in Fig. 13–1.

Adverse Side Effects.   Many women experience adverse side effects when taking the combination pill. Estrogen-related adverse side effects can include nausea or vomiting, bloating, fluid retention and weight gain, irritability, nervousness, headaches, engorgement of veins with blood, breast tenderness, *chloasma* (brown spots on the face), dysmenorrhea, changes in vision, an increase in blood sugar and fat levels, minor blood clotting, an increase in blood pressure, and suppression of lactation. Some women also develop a persistent yeast infection in the vagina due to the increase in blood sugar levels. And women who take estrogen in the pill for more than $2\frac{1}{2}$ years can develop gallstones. Progestogen-related adverse side effects can be erratic menstrual bleeding, amenorrhea, breast shrinkage, an increase in appetite, breakthrough bleeding, depression, and (rarely) loss of hair. Not all women develop these symptoms, but a woman experiencing one or more of them may want to consult with her doctor about switching to a brand with a different dosage or a different kind of estrogen or progestogen. It should be mentioned that some of these "common" side effects of the combination pill have not been verified scientifically. For example, one careful experimental study revealed that the combination pill did not affect nervousness, body weight, or the incidence of depression.[1]

About 70 percent of women who terminate the combination pill are fertile by 3 months or less,[2] although some require 18 months or more to become fertile again. There is no evidence that use of the combination pill produces permanent infertility, but a few women may develop prolactin-secreting pituitary tumors after using the combination pill, which can lead to infertility.[3]

Much has been written in the popular press about possible serious side effects of the combination pill, and this probably explains the above-mentioned 23 percent decrease in its use in the United States in recent years. Does the combination pill cause cancer? The present scientific evidence suggests a tentative no.[4] Although there is a higher incidence of noncancerous (benign) liver tumors in pill users, there is no clear evidence that the incidence of uterine, cervical, or breast cancer is higher in women

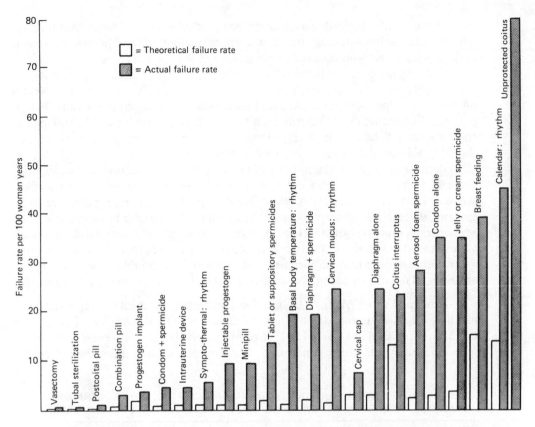

Figure 13–1. Failure rates of various contraceptive measures. The theoretical rates are based on biological effectiveness of the measures, whereas the actual failure rate is that during actual use by the population. The greater the difference between the theoretical and actual rates, the greater the effect of human error on the method. These failure rates are approximate and are based on a composite of the literature (see text).

taking the pill. Cervical precancerous growths (dysplasia), however, do tend to spread faster in combination pill users.[5]

There is good evidence that the death rate due to *cardiovascular system disease* in combination pill users (25.8/100,000 WY) is about five times higher than in nonpill users (5.5/100,000 WY).[6] In women who have used the combination pill for five years or more, the death rate from cardiovascular system disease is ten times greater than for nonpill users.[7] Cardiovascular disease produces deaths from heart attacks, blood clots in the lungs (pulmonary embolism), or bursting of a blood vessel in the brain (cerebral hemorrhage). The combination pill raises levels of triglycerides and cholesterol in the blood, which therefore increases the chance of developing cardiovascular system disease.[8] Exercise may reduce the chance of clots. These effects of combination pill use on cardiovascular disease persist (to a lesser degree) even if a woman stops taking the pill. Smoking cigarettes, especially in women over 35, greatly increases these adverse cardiovascular effects[9] (see Table 13-2). Despite warnings on pill containers about smoking and pill use, approximately 30 to 40 percent of women in the United States who take the combination pill also smoke.

The steroid hormones in the pill also enter breast milk,[10] but there is no evidence that consumption of these steroids by the infant harms its development. Nevertheless, a lactating woman should avoid taking the combination pill because it suppresses milk production. Furthermore, although there is no evidence that taking the combination pill before or during pregnancy harms the fetus, pregnant women should avoid use of the combination pill as a precaution.

Because of the potential minor and serious adverse side effects of the combination pill, women should consider some other contraceptive measure if they smoke, are over 35, or have a history of blood clotting, liver disease, high blood pressure, epilepsy, diabetes, or abnormal menstrual bleeding. The combination pill also is inadvisable if a woman is greater than 30 percent overweight, if a parent died of a heart attack at less than 50 years of age, or if the individual is a teenager with less than three menstrual periods. Many physicians advise women to stop taking the combination pill every 18 months or so and use another contraceptive measure for 2 or 3 months.

Beneficial Side Effects. Despite these potentially harmful effects of the combination pill, the maternal death rate from pill use is still less than that of pregnancy and birth[11] (see Table 13-2). In fact, the combination pill may have beneficial side effects other than preventing conception. For example, women using the pill have a

TABLE 13-2. DEATHS PER 100,000 WOMEN ASSOCIATED WITH CONTRACEPTIVE MEASURES AND AGE IN A DEVELOPED COUNTRY SUCH AS THE UNITED STATES

	15–19	20–24	25–29	30–34	35–39	40–44
Associated with pregnancy and birth	5.3	6.1	7.4	13.8	21.0	21.6
Combination pill (without predisposing conditions)[a]	—	1.2	1.2	1.9	4.0	—
Combination pill with predisposing conditions	—	1.4	1.4	11.4	28.8	—
Combination pill— nonsmokers	0.6	1.1	1.6	3.0	9.1	17.7
Combination pill— smokers	2.1	4.2	6.1	11.8	31.3	60.9
Intrauterine devices	0.8	0.8	1.0	1.0	1.4	1.4

Note: The death rates shown in this table are approximate, and are derived from notes 4 and 6 at the end of this chapter.

[a] Predisposing conditions are those discussed in the text that indicate a woman should not take the combination pill.

decreased incidence of ovarian cancer, pelvic inflammatory disease, and rheumatoid arthritis. Also, the monthly menstrual flow averages less in volume in pill users as opposed to nonpill users.

Use of the combination pill usually has no effect on a woman's sex drive, although some women using the pill report a lowered interest in sex.[12] At present, it is very difficult to determine if changes in sex drive with use of the combination pill are due to direct effects of the hormones in the pill or to psychological effects related to pill use.

Costs and Benefits. One must consider the costs and benefits of every contraceptive device. The combination pill has the advantage of being convenient in that its use does not interfere with ongoing sexual activity. Another benefit is its very low failure rate. The yearly cost of combination pill use is relatively low ($60) but more expensive than some other means. On the negative side of the ledger, the combination pill has potential minor and serious adverse side effects. It also must be used daily. A couple must weigh such costs and benefits when choosing any contraceptive method, including the combination pill.

The Minipill

The *minipill* first was marketed in 1973 as another reversible oral contraceptive measure.[13] This pill contains only a small amount of progestogen and is taken daily, even during menstruation. It often is recommended that a woman use a spermicide during her first one or two months of minipill use because the contraceptive action of this pill may take a while to be effective. The minipill may or may not block ovulation; its major effects are to render the cervical mucus hostile to sperm transport and to disrupt transport and implantation of the early embryo.[14] One advantage of this pill over the combination pill is that there are fewer adverse side effects, although there is an increase in breakthrough bleeding, some variation in cycle length, and amenorrhea in some women. Also, the failure rate of the minipill (1 to 10/100 WY) is higher than that of the combination pill. For this reason, use of the minipill is not widespread. In situations where use of estrogens are ill-advised, however, a prescription for this pill may be indicated. This requires special permission in the United States, however, because the minipill has not yet been approved for general use.

Intradermal Progestogen Implants

Intradermal (''within the skin'') *progestogen implants* are being used as a contraceptive measure in various parts of the world and are now undergoing clinical trials in the United States.[15] These implants contain a synthetic progestogen in a silastic (silicone polymer) tube, and the hormone is released in small amounts for up to six years. The mechanism of action of the progestogen implant is similar to that of the minipill. Some minor side effects include irregular or prolonged menstrual bleeding, amenorrhea, weight gain, acne, and headaches in some women. Also, there is evidence that regular menstrual cycles do not return for several months after the implant is removed.[16] These implants, however, have some distinct advantages. They are convenient, and some evidence suggests that they lower blood cholesterol and triglycerides. Also, the implants do not interfere with, and even increase, milk production in lactating women. The failure rate of these implants is about 2 to 4/100 WY.

Injectable Progestogens

Injectable progestogens (Depo-provera, Levonorgestrel, or Norigest) have been used in 64 countries since the late 1950s but are not yet approved for use in the United States.[17] The progestogen is injected every 90 days and works by blocking the LH surge.[18] The failure rate of these injections is about 1 to 10/100 WY. The advantages, disadvantages, and side effects are similar to those of the intradermal progestogen implant.

The Postcoital Pill or Injection

The *postcoital pill* (*morning-after pill*) or injection contains a large (25 to 50 mg) dosage of an estrogen, either diethylstilbestrol (DES) or ethinyl estradiol.[19] The treatment (either injected or taken by mouth) must begin within 72 hours of coitus and be repeated each day for five days. This treatment is a very effective contraceptive agent, with a failure rate of 0 to 1/100 WY); it works by speeding up embryo transport in the oviduct, thus inhibiting implantation because the embryo arrives before the uterine endometrium is ready to receive it.[20] However, about two-thirds of women receiving the treatment experience severe nausea, vomiting, headache, breast tenderness, dizziness, and/or diarrhea. Because of these unpleasant side effects and because of potential adverse effects of DES on the embryo if the treatment fails to prevent pregnancy (Chapter 9), this method usually is recommended only in cases of rape or incest.

Intrauterine Devices

An *intrauterine device* (IUD) consists of a coil of flexible plastic placed through the cervical canal into the uterine cavity. Some newer IUDs have a wrapping of copper wire around the plastic or contain a progestogen.[21] Probably the first IUDs were pebbles placed in the uterus of camels by Arabs and Turks. One of the first IUDs made for human use was invented by the German physician Richard Richter in 1909. It was ring-shaped and made of silkworm gut. In the 1920s, Grafenberg (another German) introduced the first widely used IUD, a ring of gut and silver wire. Now, IUDs are the most widely used reversible contraceptive device in the United States after the combination pill. Intrauterine devices come in a variety of shapes and sizes (Fig. 13–2). The shape of the IUD does not appear to influence its effectiveness, but the T-shaped and 7-shaped IUDs appear to be retained better in the uterus. The most popular IUD in the United States is the Lippes loop (Fig. 13–2).

An IUD must be obtained by prescription and costs $25 to $100. After a pelvic

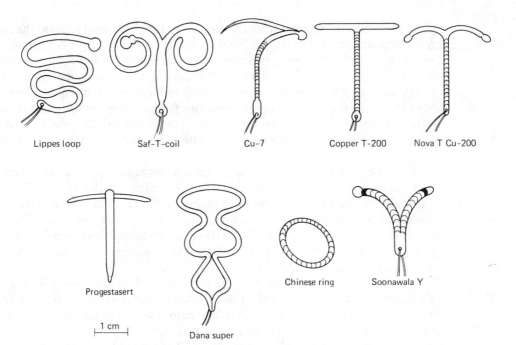

Figure 13–2. Kinds of IUDs. Lippes loop was one of the first modern IUDs and is used worldwide. The Saf-T-Coil also is used worldwide. The Cu-7 and Copper T-200 contain copper wire and are used worldwide. TCu-200 is another copper IUD used in Spain and Austria. Progestasert is an IUD that releases a progestogen and is used in the United States as well as in West Germany and Italy. The Chinese ring is made of stainless steel wire and is the principal IUD used in China. The Dana Super is used in Eastern Europe and Vietnam. The Soonawala y contains copper wire and is used in Bombay and India. Redrawn from P. T. Piotrow et al., "IUDs—Update on Safety, Effectiveness, and Research," *Population Reports,* ser. B, no. 3 (1979) pp. 49–98.

exam, the physician prescribes an IUD and places it in the uterus through the cervix. Some IUDs have an attached string that protrudes through the external cervical os. The woman should feel for this string after each menses to check if the IUD is in place. The string usually is not felt by the male during coitus. An IUD can be inserted at any time, although many physicians prefer to insert it immediately after cessation of menstruation.

We are still not sure how an IUD works. There is some evidence that the presence of an IUD, being a foreign body, causes inflammation and thus increases the number of white blood cells in the uterus; these cells then may block implantation.[22] Recent research indicates that in some women with an IUD, implantation occurs but the IUD causes early abortion of the embryo.[23] In rats, IUDs cause release of prostaglandins from the uterus.[24] Some newer *copper IUDs* (Cu-7; Tcu-200) were once proposed to be more effective than plastic-only IUDs. However, the copper diffusion from the IUD is gone from the blood within several weeks,[25] and the failure rates and expulsion rates of copper IUDs are similar to those that do not contain copper.[26] Copper IUDs, however, appear to cause less uterine bleeding. If the copper is effective, it may be because it interferes with the role of zinc in implantation.[27] Copper IUDs are also being employed postcoitally to block implantation.[28] A new IUD called *Progestasert* contains progesterone. This IUD not only has the effects of other IUDs but the progesterone prevents implantation and inhibits sperm transport through the cervical mucus. The expulsion rate of Progestasert, however, is high (20 percent).

The use of IUDs has several side effects. A woman receiving an IUD may suffere from abnormally high menstrual flow and spotting, especially in the first weeks after insertion, and the average monthly menstrual flow in IUD users is more than in non-IUD users. Abdominal cramps, especially in the first few weeks after insertion, are common. Also, the presence of an IUD increases the incidence of pelvic infection three to five times. Women with repeated or severe venereal disease infections in the past should not use an IUD.

Severe pelvic infection can take the form of *pelvic inflammatory disease,* with its dangers of scarring of the uterus and oviducts, and infertility (Chapters 15 and 16). There are 500,000 cases of pelvic inflammatory disease in the United States each year, and 22 percent of these are related to use of an IUD.[29]

In about 20 percent of women, the IUD is expelled from the uterus, often because of a poor fit (Fig. 13–3). About one out of four women with a Lippes loop is not using it after 12 months because it was expelled, or removed because of pain or bleeding, or because of accidental pregnancy. Speaking of accidental pregnancy, the incidence of ectopic pregnancies is greater with an IUD.[30] If a uterine pregnancy occurs with an IUD in place, there is a 50 percent chance that a miscarriage will occur because of uterine infection (*endometritis*). Thus an IUD should be removed once pregnancy is verified, even though IUD removal itself carries with it a minor risk of miscarriage.

In the early 1970s, an IUD called the Dalkon Shield was used widely. It was withdrawn, however, by the FDA in 1974 because of a higher than usual incidence of pelvic infections during pregnancies. In June 1980, a woman in Denver, Colorado was awarded $6.8 million in a law suit against the company that manufactured Dalkon Shield. The woman used this IUD but later became pregnant and suffered a serious, almost fatal spontaneous abortion caused by uterine infection. The failure rate of IUDs (1 to 6/100 WY) is relatively low; the higher figure is due to expulsion of the device without further protection.

Intrauterine devices may be especially useful to teenagers who may not have the motivation or experience to use the condom or diaphragm. Also, use of the combination pill by prepubertal females may disrupt the normal pubertal process. But recent studies show that as many as one-half of teenagers fitted with an IUD fail to retain it after nine months, and they suffer from much pain and bleeding.[31] About one out of four of these teens develops pelvic infections, which could cause infertility.

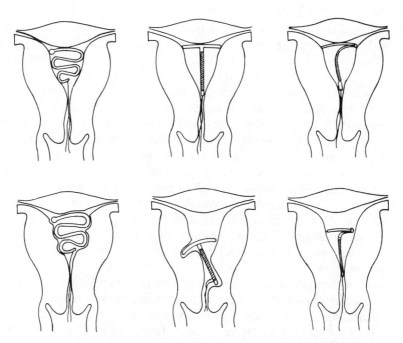

Figure 13–3. In the top row, left to right, a Lippes loop, TCu-200, and Cu-7 IUD are shown in proper position high in the uterine fundus. In the bottom row are a Lippes loop that is too large for the uterus and is pressing into and irritating the endometrium, a TCu-200 that is perforating the cervix, and a Cu-7 that is too small and has slipped out of position because of uterine contractions. The IUDs in the bottom row would be likely to produce higher rates of pregnancy, expulsion, and removal because of bleeding and pain. Adapted from P. T. Piotrow et al., "IUDs— Update on Safety, Effectiveness, and Research," *Population Reports,* ser. B, no. 3 (1979), pp. 49–98. With permission of the Population Information Program, Johns Hopkins University, Baltimore, Md.

Spermicides

The word *spermicide* literally means "killing of sperm." Actually, spermicides provide both a mechanical barrier to sperm transport as well as having adverse side effects on sperm. In the past, substances such as gum, animal dung, and various acidic compounds were used as spermicides. In late nineteenth-century England, a combination of quinine sulfate, cocoa butter, and lactic acid was used widely. Present-day spermicides consist of an inert base and an active ingredient. The base provides a mechanical barrier to sperm movement and suspends the active ingredient. The active ingredient is usually nonoxynol-9 or octoxynol in present-day spermicides; these are surfactants that disrupt the structure of the sperm membrane.[32]

    Spermicides are placed high in the vagina next to the external cervical os. They can come in the form of creams and jellies ($5 per tube), aerosol foams ($3–$4 for 20 applications), and foaming tablets or suppositories ($4–$5 for 12). Creams, jellies, and foams should be applied one hour or less before coitus. Tablets and suppositories should be inserted between 15 minutes and one hour before coitus. Douching should only be done six to eight hours after coitus to allow the spermicide to be effective. A new application of spermicide should precede subsequent sexual activity.

The failure rate of spermicides ranges from 2 to 29/100 WY (aerosols), 4 to 36/100 WY (jellies and creams), and 2 to 14/100 WY (tablets and suppositories). There are several reasons for the high failure rates of spermicides under actual use:[33] e.g., failure to remove the wrapping from suppositories and tablets, inserting a spermicide into the rectum, or not placing the substance high enough into the vagina. Also, spermicides should be applied before any penetration by the male because sperm can be present in the small drop of semen during stage 1 of ejaculation (Chapter 8). Other reasons for failure of spermicides are ejaculation before the spermicide has dispersed (sperm can reach the cervix in 90 seconds or less), coitus without reapplication of the spermicide, or lack of use during the "safe" period of the menstrual cycle.

A recent research finding sheds a dark cloud over the free and easy use of spermicides as a conceptive measure.[34] If a woman becomes pregnant in spite of the use of spermicides near or at the time of conception, her fetus has a 2.2 percent chance of having a severe congenital defect, as compared to a 1 percent chance without the use of spermicides. These defects include Down's syndrome, malignant brain tumors, and limb deformities. The incidence of late spontaneous abortions also is higher if spermicides were in use. It is possible that the spermicidal chemicals in these contraceptive formulas can enter the woman's blood stream and can harm the fetus, but more research is needed to verify this possibility. One immediate side effect can be local irritation of the genitals.

From a cost-benefit point of view, spermicides are safe and cheap and do not require a prescription; they can also prevent spread of some sexually transmitted diseases (Chapter 16). These benefits, however, are balanced by the relatively high failure rate and the fact that their use can interfere with sexual interaction. Spermicides are most effective when used with other contraceptive devices such as a condom or diaphragm.

Douching (bathing the vagina with an acidic solution) is not an effective contraceptive measure because it usually is done after sperm have already entered the cervical canal. In fact, persistent douching may lead to vaginal irritation or infection *(vaginitis),* and can even lead to pelvic inflammatory disease.[35]

The Diaphragm

A diaphragm is a shallow cup of thin rubber stretched over a flexible wire ring; this device is placed in the vagina so that it covers the external cervical os;[36] see Fig. 13–4. The diaphragm thus prevents sperm from entering the cervical canal. A spermicide should be used with the diaphragm for complete protection. In the distant past, diaphragms consisted of gums, leaves, fruits, seed pods, and sponges. Casanova

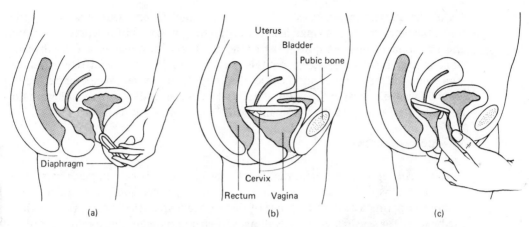

Figure 13-4. Insertion, proper position, and removal of a diaphragm. In (a), the diaphragm is inserted manually into the vagina (inserters also can be used with some kinds of diaphragms). In (b), the diaphragm is in proper position, fitting snugly between the posterior aspect of the pubic bone and the posterior vaginal fornix, completely covering the cervix. In (c), a diaphragm is removed by hooking a finger under the forward rim (the edge behind the pubic bone).

(1725–1798) recommended using a squeezed half of a lemon as a diaphragm; it may have worked in some cases because of the acidity of lemon juice! Use of the diaphragm decreased in the United States between 1965 and 1974, but now is increasing again because it appears to be safer than the combination pill or IUD. In 1975, 503,000 prescriptions for diaphragms were obtained in the United States, but this number was up 140 percent, to 1,205,000, in 1978.

A diaphragm is fitted and prescribed by a physician ($3–$10 plus fees for physician). A refitting should be done after delivery of an infant or if the woman's weight changes by 10 lb or more. Diaphragms can be one of three types. The *coil-spring* type can be compressed anywhere along its circumference while being inserted, either manually or with an inserter. The *flat-spring* type can only be compressed in one place before inserting manually or with an inserter. The third type, called an *arcing spring,* forms an arc when squeezed, and it can only be inserted manually. Most women find manual insertion easier with all types of diaphragms. Spermicidal jellies or creams can be applied to the inner side of the rubber dome and rim. The diaphragm should be inserted less than 1 hour before coitus and left in at least 6 hours after coitus to allow the spermicide to work. It must be removed, however, within 24 hours after coitus to avoid irritation of the vagina.

A diaphragm has several advantages: (1) It usually is not felt by the male during coitus; (2) it is relatively safe, although some users develop bladder infections; (3) it can be used during menstruation; (4) the spermicide on the diaphragm increases lubrication during coitus; and (5) a diaphragm can be used by a lactating woman.

The relatively high failure rate of a diaphragm with a spermicide (2 to 20/100 WY) or without a spermicide (3 to 25/100 WY) is due to several factors. For example, use of the diaphragm requires high motivation since it may disrupt sexual interactions. In some couples, the man inserts the diaphragm as part of foreplay, which may alleviate some of this disruption. Use of the diaphragm also may be avoided by some women because it is messy, the genitals must be handled, or it is difficult to insert. And some failures may occur because of not applying the spermicide or because of removing the diaphragm too soon. Finally, the penis may dislodge a diaphragm during coitus, especially in the "woman-above" position (Chapter 17).

The Cervical Cap

The *cervical cap* is a small cup that blocks the cervix. It is held in place by suction. Most such caps are made of clear, rigid plastic or rubber and can be used with or without a spermicide. An advantage of the cervical cap over the diaphragm is that it can be worn for seven days. Disadvantages when compared to the diaphragm are that it is often more difficult to insert, should not be worn during menstruation, and sometimes causes an unpleasant vaginal odor. Its failure rate is 8/100 WY. Cervical caps are not yet approved for general use in the United States although certain clinics can prescribe them, at a cost of $40 to $80.[37]

Condoms

Condoms (from Latin *condus* for "receptacle") are sheaths of latex rubber or (less commonly) animal cecum (a pouch on the intestine) that are placed over the penis to prohibit semen from entering the vagina.[38] Condoms of animal ceca or silk have been used for hundreds of years. After the discovery of vulcanization of rubber by Goodyear and Hancock in 1944, latex rubber condoms were introduced. Condoms come in a standard 7-inch length; they can come with a round or teated top, and can be dry or prelubricated with silicone or a spermicide. One can even purchase condoms of different colors or those with rubber bristles or flaps ("French ticklers"). These novel condoms, however, often break and are not good contraceptive agents.

A condom is placed on the erect penis, either by the man or woman, before coitus. About one-half inch is left at the end for collection of semen. After ejaculation, the penis should be withdrawn while it is still erect, holding the base of the condom with fingers to prevent loss of semen into the vagina. Petroleum jelly should not be used as a lubricant because it could damage the condom and irritate the vagina. Instead, water-soluble lubricants such as surgical jelly or saliva can be used with dry condoms, if desired.

Condoms have advantages. They are cheap and easily available without prescription. Condoms are manufactured with rigorous governmental inspection for damage, but they can develop cracks or holes if very old or if kept at a high temperature (like in the glove compartment of a car!). A definite advantage of condoms is that they are a prophylactic against the transfer of sexually transmitted diseases; use of condoms, however, does not totally eliminate the possibility of this transfer (Chapter 16). The high failure rate of condoms (3 to 36/100 WY) is due to the fact that they might not be used in the passion of the moment, or that they are improperly put on or taken off. Use of a spermicide with a condom reduces the failure rate to 1 to 5/100 WY, roughly the same as the IUD.[39] The rubber condoms may decrease sensation by the penis; this is less likely with the use of cecum condoms or the new thin plastic types.

Coitus Interruptus

Coitus interruptus, or withdrawal of the penis before ejaculation, is a time-honored attempt at contraception and still is a common method in some European countries. Withdrawal is condemned by some religions, which may stem from the book of Genesis 38:8–9, where Onan was killed for "spilling semen on the ground." Disadvantages of this method include the fact that it requires high motivation and is highly frustrating to some couples. There is no sound scientific evidence, however, that coitus interruptus harms the couple either physically or emotionally. Another disadvantage is that any sperm deposited before withdrawal, or left on the vulva wall during withdrawal, could reach the cervix. These factors account for the high failure rate of coitus interruptus; it is about 15/100 WY for experienced couples and 20 to 25/100 WY for inexperienced couples.

Coitus Reservatus and Coitus Obstructus

More rare is the practice of *coitus reservatus,* which allows loss of an erection while the penis still is inserted, with no ejaculation. This is an inefficient method, however, because it takes high motivation, and pregnancy may occur from sperm present in the initial drop of the ejaculate. Even less common is *coitus obstructus,* in which pressure is placed by squeezing on the base of the spongy urethra, causing retrograde ejaculation into the bladder (Chapter 17). This is relatively ineffective as a contraceptive measure and may cause bladder infection.

Rhythm Methods

Obviously, one way to avoid conception is to abstain from coitus, a practice called *celibacy* (Chapter 18). Periodic abstinence also is used in *rhythm methods* of contraception,[40] and is the only contraceptive method condoned by the Roman Catholic Church without special permission. This method is based on restriction of coitus to a "safe period" of the menstrual cycle. The "fertile" period of the cycle, when coitus is avoided, is determined by the number of days sperm remain capable of fertilization (a maximum of four days) and the time that the ovum remains capable of being fertilized (up to three days). Therefore, one must be able to estimate the time of ovulation to use this method, and there are several ways that couples can do this. Before reviewing these methods, be aware that use of rhythm methods of contraception requires instruction, daily monitoring, and a strong commitment from both sexual partners.

Calendar Method.   Use of the *calendar method* of predicting the safe period began with studies by Knaus in Austria and Ogino in Japan in the 1930s. This method is based on the predicted time of ovulation (14 days before day 1 of the next cycle), taking into account individual variation in menstrual cycle length.[41] To use the calendar method, a woman should keep track of her cycle lengths for at least a year. Then, 18 days are subtracted from the shortest cycle recorded and 11 days from the longest cycle recorded. Thus, the fertile period is defined. For example, suppose a woman, after recording her cycle lengths for a year, had a longest cycle of 35 days and a shortest cycle of 25 days. Her possible fertile period then would be from days 7 to 24 of her cycle (Fig. 13–5). This obviously does not leave much time for coitus! The failure rate using the calendar rhythm method is 14 to 47/100 WY, a high rate due to the inaccuracy of the method, nonadherence by excited couples, and the possibility of coitus-induced ovulation (Chapter 7).

Basal Body Temperature Method.   There is a 0.3 to 0.5°C (0.5–1.0°F) rise in *basal body temperature* immediately after ovulation during the menstrual cycle (Chapter 7). A special thermometer (with a 96°–100°F range) or a standard oral thermometer can be used. Recently, a device called a "sexometer" has been developed that measures a woman's body temperature by placing a small electronic temperature sensor in the mouth. A green light shines on a recorder if she is "safe." Temperature should be measured soon after awakening in the morning, before becoming active or eating. The most effective way to use basal body temperature measurement as a contraceptive method is to limit coitus to a time beginning 3 days after the temperature rise and extending to day 1 of the next cycle. The failure rate is 1 to 7/100 WY if coitus is restricted to this time period, but is 1 to 20/100 WY if coitus occurs before as well as after the rise in body temperature. Failures occur because it is difficult to detect such a small rise in body temperature and because there can be variation in the pattern of this rise in one individual for different cycles.[42]

Cervical Mucous Method.   The amount and consistency of the cervical mucus change throughout the menstrual cycle (Chapter 7). During the "wet days," immediately before and after ovulation, the mucus becomes more abundant, becomes clear (like egg white), and has a high degree of threadability as detected by placing it between two fingers. Previous to this stage, it is cloudy yellow or white and sticky. After the wet days, the mucus volume decreases, and it is cloudy. Coitus should be avoided from the time that the wet days begin until the fourth day after the wet days end. The *cervical mucous method* is relatively inefficient, the failure rate being 1 to 25/100 WY.

Sympto-Thermal Method.   The *sympto-thermal method* (sometimes called "natural family planning") uses several indicators to detect the fertile period. These include basal body temperature, cervical mucus, breast tenderness, vaginal spotting, Mittelschmertz (pain of ovulation), and the degree of opening of the external cervical os. Because this method combines several indicators of impending or recent ovulation, its failure rate (about 1 to 7/100 WY) is less than for the methods using basal body temperature or cervical mucus alone.

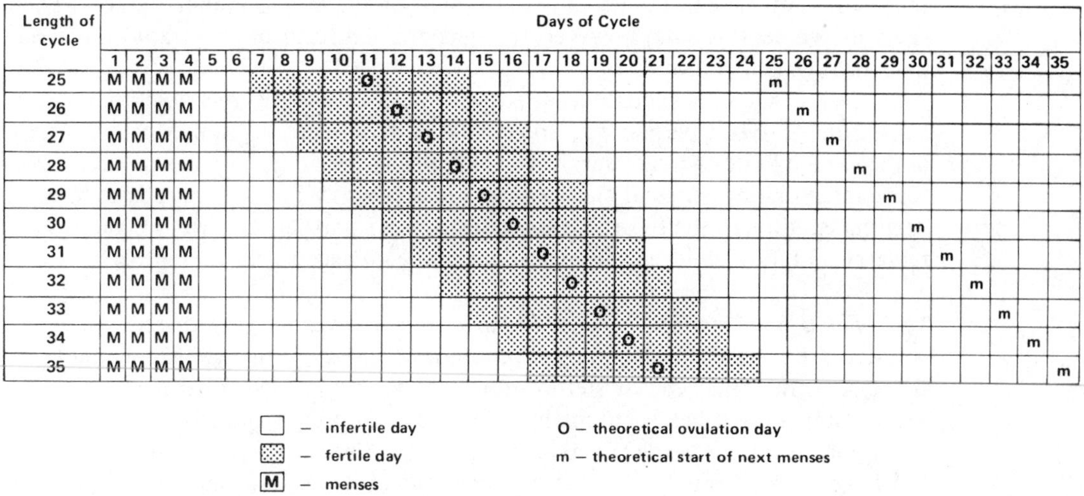

Length of cycle	Days of Cycle																																		
	1	2	3	4	5	6	7	8	9	10	11	12	13	14	15	16	17	18	19	20	21	22	23	24	25	26	27	28	29	30	31	32	33	34	35
25	M	M	M	M							O														m										
26	M	M	M	M								O														m									
27	M	M	M	M									O														m								
28	M	M	M	M										O														m							
29	M	M	M	M											O														m						
30	M	M	M	M												O														m					
31	M	M	M	M													O														m				
32	M	M	M	M														O														m			
33	M	M	M	M															O														m		
34	M	M	M	M																O														m	
35	M	M	M	M																	O														m

□ — infertile day   O — theoretical ovulation day

▦ — fertile day   m — theoretical start of next menses

M — menses

Figure 13–5. An ovulation calendar for a woman whose menstrual cycles over 12 months varied from 25 to 35 days. Theoretically, ovulation occurs 14 days before the next menses, but actually it may occur at any time from the 12th to the 16th day before menses. The fertile period for this woman would be day 7 (25 minus 18) to day 24 (35 minus 11) of each future cycle. Adapted from C. Ross and P. T. Piotrow, "Birth Control without Contraception," *Population Reports,* ser. I, no. 1 (1974), pp. 1–19. With permission of the Population Information Program, Johns Hopkins University, Baltimore, Md.

Is Breast-Feeding a Contraceptive Measure?

Breast-feeding can inhibit ovulation and produce postpartum amenorrhea if done frequently and consistently (Chapter 11). The inhibition of ovulation by even a rigorous breast-feeding schedule, however, becomes ineffective after about six to nine months. After this time, ovulation and pregnancy can occur even if breast-feeding continues. In addition, ovulation occurs in 3 to 10 percent of breast-feeding women before their first postpartum menstruation. Thus, a breast-feeding woman should use another form of contraception. Combination pills, however, are not recommended for lactating women because they suppress milk production, and the steroids can enter the breast milk (Chapter 11). Other devices such as an IUD, diaphragm, or condom (with spermicide) are better for lactating women. The failure rate for breast-feeding as a contraceptive measure is about 15 to 40/100 WY.

Surgical Sterilization

At present, surgical sterilization is the leading method of contraception in the United States and elsewhere.[43] In the United States, the number of those choosing to be sterilized has increased from 3 million in 1970 to over 13 million in 1980. Today, more than 40 million have been surgically sterilzed in China, and 25 million in India. There is no government to date, however, that has ordered surgical sterilization for a segment of their population.

Even though some methods of surgical sterilization are reversible, they should be considered to be a permanent, irreversible form of contraception. Most physicians in the United States will not surgically sterilize anyone under 21 years of age. A person contemplating surgical sterilization should ask: What are the possibilities of wanting children in the future? If divorced or your partner dies, would you want to remarry and have children? Would adoption be an acceptable alternative?

Tubal Sterilization. *Tubal sterilization* involves surgery during which the oviducts are excised, buried, plugged, cauterized, or tied so that gametes cannot pass through them. This type of sterilization is commonly called tubal ligation, which means "tying the tubes." Originally, the tubes were tied with thread, the first such operation being performed in England in 1823. Today, several methods of tubal blockage are performed, as well as several methods of surgically exposing the tubes.[44] In the United States, tubal sterilization is the leading method of contraception in women married 10 years or more and in women over 30. Over 600,000 such operations are performed annually in the United States (at a cost of $800–$1000 each).

Several methods are used to expose and view the oviducts prior to the actual sterilization procedure.[45] One method uses a 10-cm abdominal incision, an operation called a *laparotomy*. This operation is done under a spinal or general anesthetic. After sterilization, the woman has to be hospitalized for two to five days, and recovery takes several weeks. In 1961, a variation of this operation, called a *minilaparotomy* was introduced. In this method (also called the band-aid method), a small (2.5 cm) incision is made in the abdominal wall just above the pubic bone using local anesthesia. The complete operation (exposure of the oviducts as well as sterilization) is done in about 10 minutes, and the patient, after a brief rest, goes home. Hospitalization usually is not required except with complications. In both laparotomy and minilaparotomy, gas (carbon dioxide or nitrous oxide) is pumped into the abdominal cavity using a syringe to render the tubes more visible to the surgeon, so that the tubal sterilization can be performed. The gas later is removed.

Laparoscopy is a newer technique used for exposure of the oviducts (Fig. 13–6). This method, introduced in 1967, involves insertion of an optical tube (laparoscope) into the abdominal cavity to view the oviducts after gas is injected. Attachments to manipulate, anesthetize, and block the tubes can be contained in the optical instrument (Fig. 13–7). Laparoscopy and subsequent sterilization takes about 20 minutes and requires a brief (one or two nights) hospital stay.

Two other new methods for surgical approach to the tubes also use an optical instrument: this is inserted through a 3- to 5-cm incision through the posterior vaginal fornix and into a blind peritoneal pouch between the anterior wall of the rectum and the posterior uterine wall. The oviducts then are pulled from their natural location into the region of incision and the sterilization is performed. This procedure is called *culdoscopy*.[46] A local anesthetic is used, and no exterior scar remains. A *colpotomy* is

Figure 13–6. Laparoscopy offers a safe and simple way to perform tubal sterilization. Light from an external source is transmitted through glass fibers (enclosed in the tube in the upper part of the illustration). The surgeon has a clear view of the abdominal organs. The abdomen is distended with an inert gas to give the surgeon room to maneuver his instruments. In the surgeon's right hand is a specially designed forceps to manipulate the tissue. Another type of laparoscope (Fig. 13-7) contains an optical viewer, forceps, anesthetizer, and a clip or band applicator, all in one instrument. Reproduced from R. G. Edwards, "Control of Human Development," in *Reproduction in Mammals, Book 5, Artificial Control of Reproduction,* eds. C. R. Austin and R. V. Short, p. 94 (Cambridge, England: Cambridge University Press, 1972). © 1972; used with permission.

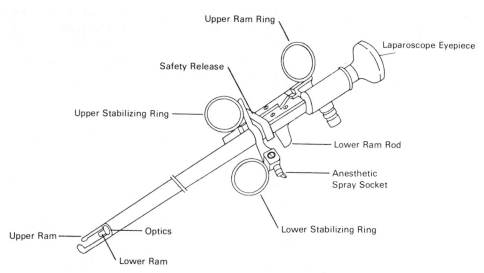

Figure 13–7. A clip applicator for use in tubal sterilization includes a laparoscope and two metal rams that open and close the clip. An anesthetic channel is also included, through which a local anesthetic is sprayed at the point of clip application on the oviduct. Reproduced from J. Wortman et al., "Laparoscopic Sterilization with Clips," *Population Reports,* ser. C, no. 4 (1974), pp. 45–52. © 1974; used with permission of the Population Information Program, Johns Hopkins University, Baltimore, Md.

like a culdoscopy except that the tubes are not brought through the incision with forceps.

Once the tubes are exposed, they can be blocked in several ways and in several places.[47] The infundibulim of the oviduct can be excised, buried, plugged, or capped. The ampulla and isthmus can be tied, cut, excised, clipped, banded, or buried. Near the uterotubal junction, the oviducts can be clogged or blocked with chemicals or plugged with an inert material like silicone. The most common method is that of Pomeroy, first developed in 1930. In this method, each tube is cut, and the free ends are folded back on themselves and tied (Fig. 13–8). Tubes can also be burned (electrocautery), or compressed by the use of clips or bands such as a silastic band called the "Falope ring." Occlusion of the tubes with a chemical, quinacrine, is now undergoing clinical trials.[48]

Although a tubal sterilization should be considered permanent, in reality it is reversible in some cases. Success of reversibility is close to 50 percent, and is most successful when the original operation involves use of clips or rubber plugs.[49] A woman may desire reversal, for example, because of divorce or remarriage, death of children, improvement of economic situation, or an adverse psychological reaction to the fact of being sterile.

Advantages of tubal sterilization as a contraceptive measure are that its failure rate is low (0.4/100 WY), and it can be permanent, but disadvantages include that it is expensive and that there are some dangers to health. The latter include an occurrence of heart irregularities in 7.4/1000 women due to the gas injected into the abdomen. Very few of these cases, however, result in death. Another problem is a 6 percent risk of ectopic pregnancy in women with their tubes blocked. This is because some sperm may jump the gap in the oviduct and fertilize an ovum, the embryo then implanting in the oviduct. Tubal ectopic pregnancies can be dangerous (Chapter 9). In addition to

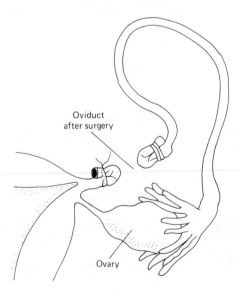

Oviduct
after surgery

Ovary

Figure 13–8. Tubal sterilization using the Pomery method. Each oviduct is cut and tied off.

these ectopic pregnancies, there are some rare failures of tubal sterilization due to incomplete blocking of the tubes or regrowth and attachment of the free ends of the tubes.

Hysterectomy.   Hysterectomy (surgical removal of the uterus) is done for various medical reasons, such as uterine cancer. In many instances in the past, however, hysterectomy was done as a method of sterilization alone. This is considered inadvisable now because it is an operation carrying all of the risks of major surgery. Tubal sterilization is safer and less expensive. One advantage of hysterectomy is that its failure rate is zero, although some women continue to use contraceptives to avoid ectopic pregnancies.

Vasectomy.   A *vasectomy* is sterilization of the male by excising and tying the vasa deferentia. Such operations were first done in England and Sweden in 1894, and now are becoming a popular form of contraception throughout the world.[50] About 1 million vasectomies are performed each year in the United States, and greater than 6 million per year in India.

Vasectomies are done in a doctor's office or clinic, using a local anesthetic (at a cost of $100 to $300). A small incision is made in the middle or each side of the scrotum to expose the vasa deferentia. Once exposed, a small section of each vas is removed, and the loose ends are burned and tied back on themselves (Fig. 13–9). Trials of newer methods of occluding the vasa, such as the use of clips, plugs, or valves, have met with only limited success. After the vasa are blocked, the incision is sutured and the patient can leave. It may take several months, however, for the sperm present in the vasa to leave the body. Thus, a man should abstain from unprotected coitus until his sperm count (checked after about 15 ejaculations) is very low or zero.

About 2 percent of men receiving a vasectomy experience some minor side effects of the operation such as skin discoloration, bruising, swelling, a blood clot in the area of incision, or a slight infection of the scrotum or epididymides (Fig. 13–10). These discomforts usually are gone in two weeks.

Most evidence suggests that a vasectomy either has no effect on sexual motivation or actually improves it; there is no effect of vasectomy on levels of testosterone in the blood.[51] Also, semen volume and the experience of orgasm virtually are unchanged in vasectomized men. The failure rate of vasectomy is very low (about 0.4/100 man-years); failure is due to not cutting the vas or to anastomosis (growing together) of the cut ends of the vasa (Fig. 13–9).

After a vasectomy, the testes continue to manufacture new sperm, which then enter the vasa deferentia but are blocked from passing from the body during ejaculation. The blocked sperm die and are resorbed. In some men, however, sperm may leave the vas deferens and enter the abdominal or scrotal cavity. In these cases, a man can form antibodies to his own sperm. Antisperm antibodies are found in one-third to two-thirds of vasectomized men; they are also present in about 2 percent of non-vasectomized men.[52] If sperm leave the male tract at the site of the incision, a man may develop *autoallergic orchiditis,* an allergic reaction at the point of incision.[53] The lump formed *(sperm granuloma)* may have to be removed surgically. We do not

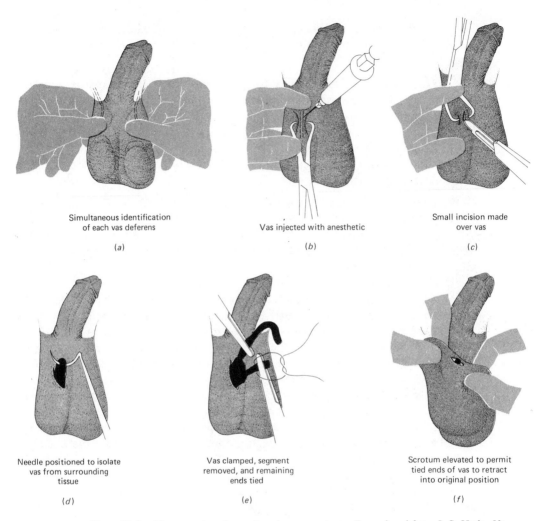

Simultaneous identification of each vas deferens	Vas injected with anesthetic	Small incision made over vas
(a)	(b)	(c)

Needle positioned to isolate vas from surrounding tissue	Vas clamped, segment removed, and remaining ends tied	Scrotum elevated to permit tied ends of vas to retract into original position
(d)	(e)	(f)

Figure 13–9. The procedure for performing a vasectomy. Reproduced from J. S. Hyde, *Understanding Human Sexuality* (New York: McGraw-Hill Book Company, 1979). © 1979; used with permission.

know if there are general adverse effects of the presence of antisperm antibodies on a man's future health. In Rhesus monkeys, however, these antibodies hasten the development of *atherosclerosis* (hardening of the arteries).[54]

For reasons similar to those discussed for women desiring reversal of sterilization, a man may decide to have this vasectomy reversed. Even though a vasectomy should be considered permanent, about half are reversible.[55] The more extensive the pieces of vasa removed, the less likely that reversal can occur.[56] Future development of valves to block the vasa may render this operation more reversible. An interesting

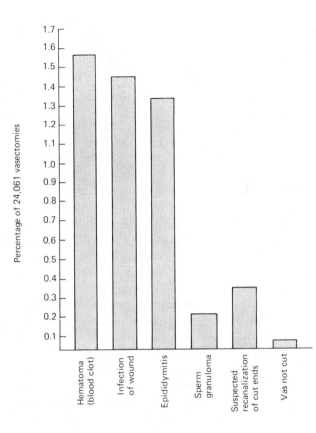

Figure 13-10. Percentage of vasectomy complications in 24,961 procedures. Adapted from J. Wortman, "Vasectomy—What Are the Problems?" *Population Reports,* ser. D, no. 2 (1975), pp. 25-39. © 1975; used with permission of the Population Information Program, Johns Hopkins University, Baltimore, Md.

fact is that, of men who have had successful reversal of their vasectomies, only about one-half are able to conceive.[57] Whether this is because of their possession of antibodies to their own sperm is not clear.

Future Contraceptive Measures

Ongoing scientific research will no doubt contribute to new human contraceptive methods in the future. Much of this research is on laboratory mammals, and you should realize that findings must be verified scientifically and must undergo rigorous clinical trials for effectiveness and safety before governmental approval is obtained for use by humans. Of the many possible future methods, some of the more promising are discussed below.

Contraceptive Pills for Males. More and more research is being done on a possible contraceptive pill for males. One difficulty in developing such a pill is that spermatogenesis usually requires several weeks to cease. Even after spermatogenesis stops, there are still living sperm in the epididymides and vasa deferentia for several weeks.

A pill for men, containing an androgen and an estrogen, has been effective in clinical trials.[58] The estrogen in this pill inhibits spermatogenesis by having a negative feedback effect on FSH secretion. The androgen inhibits LH secretion but also maintains male secondary sex structures (Chapter 4). After stopping use of this pill, men become fertile again in about 100 days. Some men taking this pill experience mild nausea, and some liver damage may occur due to the androgen used. Thus, more studies are needed on this pill's effectiveness and side effects before it could be made available to the public.

An antiandrogenic chemical, *cyproterone acetate,* blocks the effects of testosterone on target tissues and reduces fertility in some laboratory mammals. Its main effect is to inhibit motility of sperm present in the lower portions of the epididymis. The difficulty in using this drug as a male contraceptive pill is that it also blocks the action of androgens on male secondary sexual structures and on sex drive.

Gossypol is a yellowish compound extracted from the seeds, stems, and roots of cotton plants (its name is derived from *Gossypium,* the scientific name for the cotton plant). In the late 1970s, Chinese physicians reported this compound to be an effective, reversible contraceptive pill for 98.8 percent of 10,000 men.[59] If a man takes 20 mg of gossypol daily, he will be infertile in about two months. After this, a maintenance amount of 150 to 220 mg per month maintains infertility. Fertility is restored within three months after discontinuing this treatment. The Chinese report that gossypol does not influence LH or testosterone secretion. About 13 percent of Chinese males, however, suffered from low blood pressure and increased loss of potassium in the urine after gossypol treatment. American scientists now are doing research on gossypol and have found, for example, that this compound inhibits a key enzyme (lactate dehydrogenase) necessary for spermatogenesis. We probably can look for a major contraceptive role for gossypol in the future.

Other Inhibitors of Gondotropin Secretion. Inhibitory analogs of GnRH (LHRH) have been synthesized in the laboratory (Chapter 2). Some of these inhibitory GnRH analogs actually cause gonadotropin (FSH and LH) secretion if given in very low dosages several times a day, and therefore can be used to treat infertility (Chapter 15). If high dosages are given, however, they inhibit FSH and LH secretion from the pituitary gland and therefore have a contraceptive effect. In one study, 27 women, ages 21 through 37, sprayed one of these inhibitory GnRH analogs into their nose for three to six months. Ovulation was inhibited in all but 2 of the 89 treatment months. No serious side effects were detected.[60] This and more recent studies show that inhibitory GnRH analogs offer much hope as a contraceptive measure for both men and women.[61]

One problem with using GnRH inhibitory analogs as a contraceptive in males is that this treatment not only reduces sperm production in the testes but also lowers testosterone levels in the blood; the latter effect could reduce a male's sex drive and cause impotence. Therefore, a combination treatment of an inhibitory GnRH analog with testosterone has been proposed for men.[62]

Inhibin reduces FSH secretion in male and female laboratory mammals

(Chapters 3 and 4). This substance is found in follicular fluid and testicular fluid, and has potential as a contraceptive agent in males and females. However, it has yet to be characterized chemically.[63]

The pineal gland secretes melatonin and other antigonadotropic substances, which could in the future be of use to limit human fertility. Much needs to be discovered, however, about these substances before they could become effective contraceptive measures in humans.

Contraceptive Antibodies. Human chorionic gonadotropin (hCG), secreted by the blastocyst and placenta, plays a role in implantation and corpus luteum function during early pregnancy (Chapter 9). If one could inactivate hCG, it might be possible to prevent implantation or to cause early abortion. Indeed, antibodies to hCG have been shown to prevent human implantation in clinical trials.[64] A single injection of hCG antibody prevents pregnancy for one year, after which a booster vaccination lasts up to five years. Further studies on the effectiveness and side effects of this method, however, are needed.

In laboratory mammals, tissue antibodies made against the ovary, testes, epididymides, sperm, or zona pellucida may have use in the future as contraceptive agents.[65] Likewise, antibodies to GnRH may be effective in inhibiting gamete production in males and females by lowering secretion of LH and FSH (Chapter 2).

Inhibitor of Fertilization. Acrosin, an enzyme released from the sperm head, plays an important role in sperm penetration and fertilization (Chapter 8). Recently, a synthetic acrosin inhibitor has been shown to be effective against human acrosin.[66] Such an inhibitor could be a contraceptive agent if it were made a component of vaginal creams.

New Contraceptive Methods Involving Progesterone. A plastic *intracervical ring* containing progesterone placed within the cervix can alter the cervical mucus and render it hostile to sperm transport. Such a device is now in the product development stage in the United States. A doughnut-shaped, progestogen-containing *vaginal ring* is also being developed.

A very recent study in France has indicated that an *antiprogesterone pill* (RU–486) can induce menstruation or can prevent implantation in women. There appear to be no side effects, and this method has potential as a contraceptive measure.

New Barrier Contraceptive. A new form of *sponge barrier contraceptive* may soon be available without prescription.[67] One version of this device is a collagen sponge, about $2\frac{1}{2}$ by 1 in. in size. This is moistened and placed in the vagina next to the cervix. The sponge holds up to 3 oz of semen (average amount of ejaculate is 0.2 oz), and it catches the semen and prevents penetration of sperm into the uterus. The sponge also is acidic and has a spermicidal effect. It is inserted before coitus and must be removed by one hour after coitus. These sponges can be washed and reused up to six times. The developers of this sponge say that there is no danger of toxic shock syndrome (Chapter 7) with this device.

THE PSYCHOLOGY
OF CONTRACEPTIVE AVOIDANCE

A few couples can experience an unwanted pregnancy even though they consistently use a contraceptive measure. This may be due to the inherent imperfection in the measure itself. Others, however, experience an unwanted pregnancy because of misuse or avoidance of contraception. This can happen because a couple is ignorant of contraceptive use or availability. In other instances, however, psychological patterns are responsible for contraceptive misuse or avoidance.[68] For example, the man or woman may deny the possibility of pregnancy, uttering statements like, "I don't understand how it happened." Or they may deny responsibility for a pregnancy. Love may even be an excuse for an unwanted pregnancy in that a person may feel that giving one's partner a child is a "gift of love."

Some couples may not use contraceptive measures because of guilt. They may feel that contraceptive use labels them as promiscuous, and they may even accept pregnancy as a punishment for their supposed excessive sexual activities. Others, perhaps with certain religious beliefs, will feel that having a child is the only natural and condoned result of coitus, and that to use contraception is a sin. Some couples may avoid use of contraceptives out of shame. For example, they may avoid birth control counseling because it would expose their "shameful" sex life or their ignorance about sexual matters. Other people may relate their fertility to their self-esteem; a woman may feel that being fertile is being feminine and attractive, and a man may feel that his fertility is a reflection of his virility and masculinity.

Unwanted pregnancies also can result from hostility in that pregnancy can be used as revenge against a partner or parents or to recapture a lost partner. People even avoid or misuse contraception for the thrill of risk-taking, because the passion of the moment is the only important experience, or to prove that they are immoral, unfit persons. Finally, couples may not use contraception because they fear the possible adverse side effects on their physical health or sex life.

Anyway, abortion is available, isn't it? The answer is yes, but induced abortion is psychologically a more upsetting event than contraceptive use; it puts more responsibility on the woman; and medically it can be more involved than the use of contraceptives and is a lot more expensive. Induced abortion should never be considered a form of contraception.

CHOOSING A CONTRACEPTIVE

The ultimate contraceptive would be one that: (1) could be used by either sex, (2) is free, (3) never fails, (4) does not interfere with sexual activities, (5) does not require conscientious adherence to the method, and (6) has no adverse side effects. As we have seen, no such contraceptive exists at the present time, each measure having its unique portrait of advantages and disadvantages. When a person chooses to use a certain contraceptive, he/she must consider failure rate and possible adverse side effects. When a measure interferes with comfortable sexual activity, the problem often can be overcome by caring communication between sexual partners. Although finances may interfere with the use of certain measures, the potential financial costs of a pregnancy are even more burdensome. Finally, a couple must realize that adherence to conscientious use of the method they choose is important if avoidance of pregnancy is a primary concern.

CHAPTER SUMMARY

Contraceptive measures either prevent fertilization or disrupt embryo transport or implantation. These measures are used to prevent unwanted pregnancies. The combination pill is the most popular reversible contraceptive measure. This pill contains a synthetic estrogen and progestogen, and it inhibits ovulation, sperm transport, and implantation. It has a very low failure rate (0.7–3.0/100 WY). Minor side effects of this pill are common, but the only proven serious side effect relates to higher incidence of cardiovascular system disease and deaths in pill users, especially those who smoke or are over 35 years of age. The minipill, intradermal progestogen implant, and injectable progestogen methods have fewer side effects than the combination pill, but they also have higher failure rates. The postcoital pill contains a high amount of estrogen, and because of severe side effects, it is usually used only in cases of rape or incest.

Intrauterine devices are flexible plastic coils placed within the uterus; some contain copper or a progestogen. These devices work by causing an inflammatory response in the uterus, thus preventing implantation or causing embryonic abortion. Side effects such as abdominal cramping, increased menstrual bleeding, and pelvic infection are seen in some women who use IUDs. Also, IUDs are spontaneously expelled in a few women. The failure rate of IUDs (1 to 6/100 WY) is relatively low.

Spermicides are placed in the vagina and act both as barriers to sperm transport and as sperm-killing agents. The failure rates of spermicides are relatively high, and they are best used with other contraceptive measures such as the diaphragm or condom.

The diaphragm is a shallow cup of thin rubber stretched over a flexible wire ring, and is placed in the upper vagina so as to block the cervix. The effectiveness of this barrier to sperm transport is enhanced by simultaneous use of a spermicide. The failure rate of diaphragms with a spermicide is 2 to 20/100 WY. Failure usually is the result of improper use. A major disadvantage of the diaphragm is that its insertion can disrupt sexual interaction. Another barrier method is use of a condom, a sheath of latex rubber or animal cecum placed over the erect penis before penetration. The failure rate of condoms is high (3 to 36/100 WY) mainly because people do not use it properly. Use of a spermicide with the condom greatly decreases its failure rate, down to 1 to 5/100 WY.

"Natural" contraception can involve inhibition of ejaculation by coitus interruptus (ejaculation outside the female), coitus reservatus (absence of ejaculation while still inside the vagina), and coitus obstructus (retrograde ejaculation into the male bladder). Rhythm methods of contraception utilize prediction of a safe period for coitus and avoidance of coitus during the fertile period of the menstrual cycle. Means for predicting the fertile and safe periods include the calendar method, basal body temperature method, cervical mucous method, and the sympto-thermal method. Breast-feeding can be another form of natural contraception. All of these natural contraceptive measures, however, have very high failure rates.

Surgical sterilization is an effective, popular contraceptive measure. In females, the oviducts can be exposed by surgery using laparotomy, minilaparotomy, laparascopy, culdoscopy, or colpotomy. Tubes can then be blocked with cutting, electrocautery, tying, or use of clips or bands. Hysterectomy should not be used as a contraceptive measure. In males, the vasa deferentia can be cut, burned, or tied, a procedure called vasectomy. Surgical sterilization may be reversible in both sexes, but still should be considered a permanent contraceptive measure. Side effects of these procedures are minor, and the failure rates (0–4/100 WY or "man-years") are very low.

New contraceptive measures may be developed in the future. These include use of antibodies to hCG, GnRH inhibitory analogue, an acrosin inhibitor, a male pill containing androgen and estrogen, gossypol, inhibin, intracervical and vaginal progestogen rings, collagen sponges, an antiprogesterone pill, and pineal gland substances.

Experiencing an unwanted pregnancy can result from inherent failure of a contraceptive measure. More often, however, it is the result of misuse or lack of use of contraception for many, often subtle, psychological reasons. A couple must weigh the advantages and disadvantages of the various contraceptive measures, with special attention to failure rates and adverse side effects, before choosing one for them.

NOTES

1. J. W. Goldzieher, "Nervousness and Depression Attributed to Oral Contraceptives: A Double-Blind, Placebo-Controlled Study," *American Journal of Obstetrics and Gynecology,* 111 (1971), 1013–20.

2. Anonymous, "Resumption of Fertility after Discontinuation of Contraception," *Research in Reproduction,* 10, no. 2 (1978), 1.

3. B. M. Sherman et al., "Pathogenesis of Prolactin-Secreting Pituitary Adenomas," *Lancet,* ii (1978), 1019–21.

4. W. Rinehart and J. C. Felt, "Debate on Oral Contraceptives and Neoplasia Continues: Answers Remain Elusive," *Population Reports,* ser. A, no. 4 (1977), pp. 69–100.

5. E. Stern et al., "Steroid Contraceptive Use and Cervical Dysplasia: Increased Risk of Progression," *Science,* 196 (1977), 1460–62.

6. W. Rinehart and P. T. Piotrow, "OCs—Update on Usage, Safety, and Side Effects," *Population Reports,* ser. A, no. 4 (1979), pp. 134–86.

7. Royal College of General Practitioners, "Oral Contraceptive Study. Mortality among Oral Contraceptive Users," *Lancet,* ii (1977), 727–31.

8. D. D. Bradey, "Serum High-Density-Lipoprotein Cholesterol in Women Using Oral Contraceptives, Estrogens, and Progestins," *New England Journal of Medicine,* 299 (1978), 7–20.

9. Rinehart and Piotrow (1979), op. cit.

10. S. Nilsson and K. Nygren, "Transfer of Contraceptive Steroids to Human Milk," *Research in Reproduction,* 10, no. 1 (1978), 2; and F. A. Kincl, "Debate on the Use of Hormonal Contraceptives during Lactation," *Research in Reproduction,* 12, no. 2 (1980), 1.

11. R. A. Hatcher et al., *Contraceptive Technology, 1978–1979,* 9th rev. (New York: Irvington Press, 1980).

12. F. J. Kane, "Evaluation of Emotional Reactions to Oral Contraceptive Use," *American Journal of Obstetrics and Gynecology,* 126 (1976), 968–72.

13. W. Rinehart, "Minipill—A Limited Alternative for Certain Women," *Population Reports,* ser. A, no. 3 (1975), pp. 53–67.

14. H. W. Rudel et al., "Role of Progestogens in Hormonal Control of Fertility," *Fertility and Sterility,* 16 (1965), 158.

15. Anonymous, "Contraception by Intradermal Implants," *Research in Reproduction,* 11, no. 3 (1979), 1–2.

16. E. Weiner and E. D. B. Johansson, "Plasma Levels of D-Norgestrel, Estradiol and Progesterone during Treatment with Silastic Implants Containing D-Norgestrel," *Contraception,* 14 (1976), 81–92.

17. W. Rinehart and J. Winter, "Injectable Progestogens—Officials Debate but Use Increases," *Population Reports,* ser. K, no. 1 (1975), pp. 1–15.

18. D. R. Mishell et al., "Contraception with an Injectable Progestin," *American Journal of Obstetrics and Gynecology,* 101 (1968), 1046.

19. W. Rinehart, "Postcoital Contraception—An Appraisal," *Population Reports,* ser. J, no. 9 (1976), pp. 141–54.

20. M. J. Morris, "Postcoital Contraception," *Annual Reviews of Internal Medicine,* 73 (1970), 650.

21. P. T. Piotrow et al., "IUDs—Update on Safety, Effectiveness, and Research," *Population Reports,* ser. B, no. 3 (1979), pp. 49–98; and L. Liskin and G. Fox, "IUDs—An Appropriate Contraceptive for Many Women," *Population Reports,* ser. B, no. 4 (1982), pp. 101–35.

22. D. M. Smith et al., "Effects of Polymorphonuclear Leukocytes on the Development of Mouse Embryos Cultured from the Two Cell Stage to Blastocysts," *Biology of Reproduction,* 4 (1971), 74.

23. B. B. Saxena and R. Landesman, "Does Implantation Occur in Presence of an IUD?" *Research in Reproduction,* 10, no. 3 (1979), 1–2.

24. Anonymous, "Prostaglandins and the Mode of Action of IUDs," *Research in Reproduction,* 7, no. 3 (1975), 4.

25. L. Randic et al., "Copper Levels in Cervical Mucus of Women with Copper-Bearing and Noncopper-Bearing Intrauterine Devices," *Biology of Reproduction,* 8 (1973), 499–503.

26. F. B. Orlans, "Copper IUDs—Performance to Date," *Population Reports,* ser. B, no. 1 (1973), pp. 1–20.

27. K. Hagenfeldt, "Intrauterine Contraception with the Copper T Device. 1. Effect on Trace Elements in the Endometrium, Cervical Mucus and Plasma," *Contraception,* 6 (1972), 37.

28. Rinehart (1976), op. cit.

29. D. A. Eisenbach et al., "Pathogenesis of Acute Inflammatory Disease: Role of Contraception and Other Risk Factors," *American Journal of Obstetrics and Gynecology,* 128 (1977), 838–49.

30. M. P. Vessey et al., "Outcome of Pregnancy in Women Using an Intrauterine Device," *Lancet* i (1974), 495–8.

31. Anonymous, "The Use of IUDs in Nulligravid Schoolgirls," *Research in Reproduction,* 10, no. 4 (1978), 3.

32. "An Analysis of the Value of Spermicides in Contraception," *Research in Reproduction,* 11, no. 6 (1979), 1–2.

33. S. Coleman and P. T. Piotrow, "Spermicides—Simplicity and Safety Are Major Assets," *Population Reports,* ser. H, no. 5 (1979), pp. 77–118.

34. H. Jick et al., "Vaginal Spermicides and Congenital Disorders," *Journal of the American Medical Association,* 245 (1981), 1329–32.

35. H. H. Neuman and A. DeCherney, "Douching and Pelvic Inflammatory Disease," *New England Journal of Medicine,* 295 (1976), 783.

36. J. Wortman, "The Diaphragm and Other Intravaginal Barriers—A Review," *Population Reports,* ser. H. no. 4 (1976), pp. 57–75.

37. Anonymous, "Cervical Cap Study Finds Eight Pregnancies per 100 Users per Year, Continuation Rate of 67 Percent," *Family Planning Perspectives,* 14 (1982), 215–16.

38. J. J. Dumm et al., "The Modern Condom—A Quality Product for Effective Contraception," *Population Reports,* ser. H., no. 2 (1974), pp. 21–36; and J. D. Sherris et al., "Update on Condoms—Products, Protection, Promotion," *Population Reports,* ser. H, no. 6 (1982), pp. 121–55.

39. Anonymous, "Condoms," *Consumer Reports,* 44, no. 10 (1979), 583–88.

40. C. Ross and P. T. Piotrow, "Birth Control without Contraception," *Population Reports,* ser. I, no. 1 (1974), pp. 1–19; and L. S. Liskin, "Periodic Abstinence: How Well Do New Approaches Work?" *Population Report,* ser. I, no. 3 (1981), pp. 33–71.

41. Ibid.

42. Ibid.

43. C. P. Green, "Voluntary Sterilization: World's Leading Contraceptive Method," *Population Reports,* ser. C, no. 2 (1978), pp. 37–70; N. Lauersen, "A Closer Look at Sterilization," *Medical World News* (May 28, 1979), p. 87; and J. Stepan et al., "Legal Trends and Issues in Voluntary Sterilization," *Population Reports,* ser. E, no. 6 (1981), pp. 73–103.

44. Ibid.

45. J. Wortman, "Tubal Sterilization—Review of Methods," *Population Reports,* ser. C, no. 7 (1976), pp. 73–95.

46. J. Wortman, "Female Sterilization Using the Culdoscope," *Population Reports,* ser. C, no. 6 (1975), pp. 61–71.

47. Wortman (1976), op. cit.

48. J. Zipper et al., "Human Fertility Control by Transvaginal Application of Quinacrine on the Fallopian Tube," *Fertility and Sterility,* 21 (1970), 581.

49. J. Wortman et al., "Laparoscopic Sterilization with Clips," *Population Reports,* ser. C, no. 4 (1974), pp. 45–52; and J. F. Hulka, "Current Status on the Reversibility of Sterilization," *Research in Reproduction,* 10, no. 5 (1978), 1–2.

50. J. Wortman, "Vasectomy—What Are the Problems?" *Population Reports,* ser. D, no. 2 (1975), pp. 25–39.

51. R. E. Hackett and K. Waterhouse, "Vasectomy—Reviewed," *American Journal of Obstetrics and Gynecology,* 116 (1973), 438.

52. L. E. Bradshaw, "Vasectomy Reversibility—A Status Report," *Population Reports,* ser. D, no. 3 (1976), pp. 41–60.

53. Anonymous, "Some Consequences of Vasectomy and Testicular Biopsy," *Research in Reproduction,* 6, no. 5 (1974), 3.

54. N. J. Alexander and D. J. Anderson, "Vasectomy: Consequences of Autoimmunity to Sperm Antigens," *Fertility and Sterility,* 32 (1979), 253–60; and G. B. Kolata, "Vasectomies May Increase the Risk of Atherosclerosis," *Science,* 211 (1981), 912–13.

55. Bradshaw (1976), op. cit.

56. Ibid.

57. Ibid.

58. Anonymous, "Oral Steroid Contraceptive for Men," *Research in Reproduction,* 7, no. 1 (1975), 1.

59. Anonymous, "Gossypol, A New Contraceptive for Men," *Research in Reproduction,* 11, no. 3 (1979), 3.

60. Anonymous, "A Sniff a Day Keeps Pregnancy Away," *Science News,* 116, no. 8 (1979), 133.

61. K. L. Sheehan et al., "Luteal Phase Defects Induced by an Agonist of Luteinizing Hormone—Releasing Factor: A Model for Fertility Control," *Science,* 215 (1982), 170–72.

62. D. Heber and R. S. Swerdloff, "Male Contraception: Synergism of Gonadotropin-Releasing Hormone Analog and Testosterone in Suppressing Gonadotropin," *Science,* 209 (1980), 936–38.

63. B. P. Setchell, "Inhibin," *Research in Reproduction,* 12, no. 2 (1980), 3.

64. V. C. Stevens, "Potential Control of Fertility in Women by Immunization with hCG," *Research in Reproduction,* 7, no. 3 (1975), 1–2; and R. G. Edwards, "Immunization against hCG-β for Suppressing Human Fertility," *Research in Reproduction,* 8, no. 3 (1976), 1–3.

65. Chang, M. C., "Effects of Antisera on Mammalian Fertilization," *Research in Reproduction,* 10, no. 1 (1978), 2.

66. L. J. D. Zaneveld et al., "Acrosin Inhibitors as Vaginal Contraceptives in the Primate and Their Acute Toxicity," *Biology of Reproduction,* 20 (1979), 1045–54.

67. C. James, "The Birds and the Beeswax," *Science 81,* 2, no. 3 (1981), 83–84.

68. E. C. Sanberg and R. I. Jacobs, "Psychology of the Misuse and Rejection of Contraception," *American Journal of Obstetrics and Gynecology,* 110 (1971), 227–40.

FURTHER READING

ATKINSON, L., et al., "Prospects for Improved Contraception," *Family Planning Perspectives,* 12 (1980), 173–80.

DRYFOOS, J. L., "Contraceptive Use, Pregnancy Intentions and Pregnancy Outcomes Among U.S. Women," *Family Planning Perspectives,* 14 (1982), 81–94.

HATCHER, R. A., et al., *Contraceptive Technology (1978–1979),* 9th rev. New York: Irvington Press, 1980.

JACKSON, H., "Chemical Methods of Male Contraception," in *Reproduction in Mammals, Book 5, Artificial Control of Reproduction,* eds. C. R. Austin and R. V. Short. Cambridge, England: Cambridge University Press, 1972. Pp. 67–86.

MARSHALL, J., "Cervical Mucus and Basal Body Temperature Method of Regulating Births," *Lancet,* ii (1976), 282.

PORTER, J. F., "The Regulation of Fertility," *Research in Reproduction,* vol. 5, no. 4 (1973), map.

POTTS, D. M., "Limiting Human Reproductive Potential," in *Reproduction in Mammals, Book 5, Artificial Control of Reproduction,* eds. C. R. Austin and R. V. Short. Cambridge, England: Cambridge University Press, 1972. Pp. 32–66.

ZELNIK, M., AND J. F. KANTNER, "Contraceptive Patterns and Premarital Pregnancy among Women Aged 15–19 in 1976," *Family Planning Perspectives,* 10 (1978), 135–42.

ZELNIK, M., AND J. F. KANTNER, "Sexual Activity, Contraceptive Use and Pregnancy Among Metropolitan Area Teenagers: 1971–1979," *Family Planning Perspectives,* 12 (1980), 230–37.

14 *Induced Abortion*

LEARNING GOALS

Having read this chapter, you should be able to:

1. Describe past and current trends in governmental policies on induced abortion around the world.
2. List important recent legal rulings concerning abortion in the United States, and discuss their meaning in relation to the incidence of abortion in America.
3. List reasons why a woman or couple who want a child would choose to abort a pregnancy.
4. List reasons why a woman or couple would not want a child and thus would terminate the pregnancy by induced abortion.
5. Discuss how the concept of "ensoulment" has related to the controversy about the morality of induced abortions.
6. Describe the procedures used to abort an embryo, and list the advantages and disadvantages of these early first trimester abortions.
7. Describe the procedures used to abort an early fetus, and list the advantages and disadvantages of these late first trimester abortions.
8. Discuss the advantages and disadvantages of cervical dilatation using metal dilators or laminaria tents.
9. List reasons why a woman or couple would wait until the second or even third trimester of pregnancy before requesting an induced abortion.

10. Describe the procedures used to abort a fetus during the second trimester, and discuss the advantages and disadvantages of these procedures.
11. Describe a hysterotomy, and discuss the disadvantages of third trimester abortion using this procedure.
12. Discuss the effects of an induced abortion on future fertility.

INTRODUCTION

Induced abortion is the termination of pregnancy by artificial measures. In this chapter, we discuss the history of induced abortion in the world and, more specifically, in the United States. In addition, we review reasons why a woman or couple would want an abortion and some of the controversies surrounding this procedure. Finally, we summarize the medical procedures used to induce abortion and some benefits and problems of these procedures.

HISTORY OF INDUCED ABORTION

Induced abortions have been used to terminate unwanted pregnancies for centuries. A work by the Chinese Emperor Shen Nung, 4600 years ago, contained a recipe for inducing abortion using mercury. In ancient Greece, Hippocrates recommended violent exercise to terminate a pregnancy. Many often dangerous "folk methods" have been used to induce abortion, and some are still used in certain parts of the world. These include physical trauma to the abdomen and introduction of chemicals or sharp objects into the uterus. Throughout history, induced abortion, along with *infanticide* (killing the newborn), was practiced in various parts of the world to adjust sex ratios, control population pressure, eliminate deformed children, or terminate pregnancies resulting from incest or adultery.[1] In more recent times, laws of different governments about abortion have ranged from permissive to highly restrictive. Since 1966, however, abortion has changed from a largely disputed practice to a medically and socially acceptable procedure in many countries.[2]

In today's world, abortion laws vary widely among different countries.[3] About one-third of the world's population live in countries with nonrestrictive abortion laws, allowing abortion on request up to the tenth to twenty-fourth week of pregnancy, with the usual time being the twelfth week. The United States, as we see later in this chapter, is such a country at present. "Abortion on request" means that one can obtain an abortion simply by wanting one, regardless of the reasons. Another third of the world's population live in countries with moderately restrictive abortion legislation. Abortion on request is not allowed in these countries, but it is permitted for a wide range of medical, psychological, and socioeconomic reasons. Finally, about one-third of the people on earth live under restrictive abortion laws. That is, abortion is illegal unless there is a threat to the woman's health or life. Illegal abortions are frequent in these countries and present a major health hazard.

INDUCED ABORTION IN THE UNITED STATES

History of Abortion Legislation in the United States

The Comstock Law of 1873 made induced abortion illegal in the United States, and this was the situation until the late 1960s. Then, because of the increasing concern about overpopulation and the high frequency of dangerous illegal abortions, some states liberalized their abortion laws. From 1967 to 1970, about a dozen states enacted legislation allowing abortion in cases where pregnancy posed a substantial risk to a woman's mental or physical health, when the fetus had grave physical or mental defects, or if the pregnancy resulted from rape or incest. In 1970, four states enacted even more liberal laws, allowing early abortion simply on request.

In 1973, the U. S. Supreme Court legalized abortion for all states under certain conditions. According to their decision, pregnancies in the first trimester (first 12 weeks) can be done simply by request, after communication between the woman and her physician. Second trimester abortions (thirteenth to the twenty-fourth week), however, were restricted to cases of danger to the woman's physical or mental health, and states could regulate the qualifications of the people doing the procedures and the facilities and places where these abortions occur. States, however, must allow second trimester abortions related to fetal defects. In the case of third trimester abortions (after the twenty-fourth week), individual states could regulate if and how these are done, except that these late abortions must be permitted in all states if there is danger to a woman's physical or mental health.

In 1977, the Hyde Amendment to this law (sponsored by Representative Henry Hyde) was supported by the U. S. Supreme Court. This amendment prohibits use of federal funds (Medicaid) for abortions except to preserve a woman's life. In early 1980, however, U. S. District Judge J. F. Dooling declared this amendment unconstitutional, primarily because it made abortion less available to poor people. About one-third of the more than one million legal abortions performed each year in the United States are done on women receiving welfare. Then, on June 30, 1980, the Supreme Court ruled against Dooling's ruling and supported the Hyde Amendment by a 5 to 4 vote. The Court also allowed Medicaid abortions only in cases of rape or incest or when at least two physicians say childbirth would endanger a woman's life or cause severe and long-lasting danger to her physical health. Thus, abortion is now legal (under the conditions noted above) in the United States but is more available to people who can pay for it.

Needless to say, the legal aspects about induced abortion could change in the future. At the time of this writing, for example, the U.S. Congress is debating a bill that states, "The Congress finds that present-day scientific evidence indicates significant likelihood that actual human life exists from conception." If this bill passes, the intrauterine devices and postcoital pill, as well as possibly some other current contraceptive measures, would be prohibited because they interfere with biologic processes that occur after conception.

Present-Day Abortion Statistics in the United States

In America today, induced abortion is an increasingly more common choice to end an unwanted pregnancy among married and unmarried people. In 1978, for example, a survey by the Alan Guttmacher Institute reported that 29 percent of all established pregnancies were terminated in this manner. This means that about 1.37 million pregnancies were terminated by induced abortion in 1978. In the early 1980s, this number increased to nearly 1.5 million. American women of all ages are having abortions, but induced abortion is especially common in teenagers. About 1 in 10 teenaged women in the United States becomes pregnant each year, and about 30,000 of these are under 15 years of age. In 1977 and 1978, about one-third of all abortions were performed on teenagers, and about 3 of every 4 abortions were performed on unmarried women.

WHY WOMEN HAVE ABORTIONS

There are several reasons why a woman who desires a child would want her pregnancy terminated by induced abortion. First of all, her physical health may be in danger because of being pregnant, so the abortion is needed because of the threat of severe illness or death. Maternal disorders that threaten a pregnant woman include diabetes, kidney disease, and disorders of the circulatory system. *Therapeutic abortion* is an induced abortion to prevent harm to the woman.

Second, amniocentesis, fetoscopy, ultrasound, or fetal blood tests may have indicated grave fetal abnormalities, and a woman or couple may choose to terminate the pregnancy to avoid birth of a severely handicapped infant. Unfortunately, these tests are performed most effectively around the sixteenth week of pregnancy (Chapter 9), and it can take several weeks for the results of some of these tests to be known. Therefore, induced abortion in the late second trimester or early third trimester must be done at some discomfort or even risk to the woman, as we discuss later.

Although some feel that even a deformed or handicapped child has a right to life and happiness, others feel that the impact of the birth of a grossly deformed child on the parents' lives and finances, on siblings, and on the social institutions that care for such children justifies pregnancy termination. Inherent in the difficulty of choosing abortion in these cases is determining the potential severity of the fetal problem. Some defects are minor and some severe, with many somewhere in between.

In another category is termination of unwanted pregnancies. Why would a pregnancy be unwanted? One reason could be because it occurred "by accident" as a result of coitus between a couple with no long-term relationship, or in extreme cases, it could have been the result of incest or rape. Or a couple that originally wanted a child could change their minds; perhaps their financial situation worsened or a divorce is imminent. An unwanted pregnancy often occurs because a contraceptive measure failed or was not used properly. Thus, termination of a potentially healthy fetus carried in a physically healthy woman can occur because of threats to the emotional health and happiness of the potential parents.

In considering an abortion, a woman or couple must weigh the various reasons for or against this choice. These reasons can be religious, psychological, moral, legal, or medical. In respect to medical considerations, the maternal death rate due to childbirth and that due to abortion both increase with the age of the woman. In general, abortion before the seventeenth week of pregnancy carries with it a lower maternal death rate than does childbirth. The maternal death rate for abortions after the seventeenth week, however, is greater than for childbirth. Abortions are expensive, usually costing somewhere between $200 and $500, but this is inexpensive considering the cost of childbirth at a hospital or birthing center. On the positive side of the ledger, abortion allows a woman to choose to be a parent or not after pregnancy has occurred, and it gives her a choice about raising a child she may not want. You can read the book *Taking Chances* by Kristin Luker that is listed in the Further Reading section of this chapter if you are interested in the abortion decision-making process.

THE ABORTION CONTROVERSY

As you know, people have different beliefs about the morality of induced abortion. Much of this controversy relates to the concept of *ensoulment,* or when in human development an embryo or fetus acquires a soul and becomes a person. The "event" of ensoulment (or in nonreligious terms, the acquisition of "humanhood") is morally important because killing a fetus before this time would simply be preventing a potential human life, whereas pregnancy termination after this time would be, in some people's opinion, murder. Different religious and other groups have defined ensoulment at various points in the cycle of pregnancy: (1) at the moment of conception, (2) at implantation (about day 7), (3) during the ninth week of pregnancy (when the fetus begins to look like a person), (4) at the sixteenth to the twentieth week of pregnancy (when "quickening" occurs), (5) at the twenty-fourth week (when about 10 percent of fetuses can survive with special care if born), or (6) at the twenty-eighth week (when most fetuses can live with special care if born). Some people even believe that a human does not acquire a soul until it interacts with another human after birth, and some do not believe there is a soul.

The Roman Catholic Church and other conservative sections of Christianity as well as some parts of Judaism are restrictive about induced abortion. In contrast, religions such as Buddhism, Islam, and Hinduism are more tolerant of induced abortion. In the mid-1800s, the Roman Catholic Church allowed abortion anytime up to day 40 of pregnancy, a point where it was believed ensoulment occurred. In 1869, however, Pope Pius IX declared that ensoulment occurred at conception, and thus disallowed abortion at any stage of pregnancy. Pope Paul VI reaffirmed this opinion in his encyclical, *Humanae vitae,* in 1968. In this statement, he proclaimed that

> the direct interruption of the generative process already begun and, above all, abortions, even for therapeutic reasons, are to be absolutely excluded as lawful means of controlling the birth of children.

In reality, it is very difficult if not impossible to describe a point in human development when an embryo or fetus becomes a person, although a leading reproductive scientist recently argued that a new individual occurs at the end of two weeks of development.[4] Indeed, it is more accurate to speak of the potential for human life, which is present in the egg and sperm as well as at all stages of embryonic and fetal development. Thus, many abortion decisions weigh the potential life of the fetus against the potential length and quality of the woman's life. These always are difficult decisions, regardless of one's beliefs.

FIRST TRIMESTER INDUCED ABORTIONS

As mentioned, the risk of maternal death due to induced abortion increases as pregnancy advances, but until the seventeenth week of pregnancy, the risk is still less than that of childbirth. Induced abortion during the first trimester of pregnancy (up to and including the twelfth week) is relatively safe and simple from a medical viewpoint. In the United States, about three-quarters of all induced abortions are performed during the first trimester. We consider here termination of pregnancy during the embryonic period (up to the eighth week), and the early fetal period (eighth to the twelfth week) of the first trimester.

Before any induced abortion procedure, a complete medical history of the woman should be taken. Certain previous or present conditions such as diabetes, circulatory disorders, epilepsy, venereal disease, uterine infection, excessive uterine bleeding, or previous cesarean section may require hospitalization for even the most simple abortion procedures. In addition, tests for pregnancy and the rhesus factor should be performed. If the woman is Rh– and the partner is Rh +, the fetus may be Rh +. In these cases, fetal blood could mix with the mother's blood during the abortion procedure, and the antibodies formed by the mother could damage the fetus of future pregnancies. Therefore, such a woman is given Rhogam to prevent antibody formation (Chapter 9).

Menstrual regulation (or *menstrual extraction*) is a relatively simple method to induce abortion of the embryo.[5] This method usually is done within two weeks after a woman's missed menstrual period (Fig. 14–1). No positive pregnancy test is needed for menstrual regulation, and it is the least painful of all abortion procedures. Menstrual regulation involves placing a flexible plastic tube (cannula) through the vagina and cervical canal and into the uterus. The other end of the tube is attached to

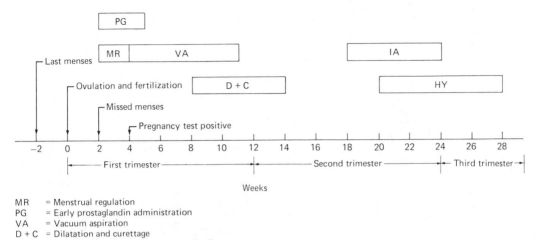

Figure 14–1. The usual periods of pregnancy during which various methods of induced abortion are utilized.

MR = Menstrual regulation
PG = Early prostaglandin administration
VA = Vacuum aspiration
D + C = Dilatation and curettage
IA = Intraamniotic saline, urea, or prostaglandin
HY = Hysterotomy

a syringe, and the embryo and endometrial tissue are sucked into the cannula using this syringe. Many women do not require an anesthetic during this procedure, although some request one. After a 10- to 30-minute rest in the clinic or physician's office, the woman can leave for home.

Some women experience minor adverse reactions to menstrual regulation, including postoperative uterine infection, uterine cramping, or excessive uterine bleeding. In a few cases, some of the embryonic or placental products are not removed, or the embryo is missed altogether and the pregnancy continues.[6] Complication rates of abortion procedures are presented in Fig. 14–2; note that this rate is relatively low for menstrual regulation. The maternal death rate due to menstrual regulation is 0.4/100,000, a very low figure.[7]

A new method of early first trimester abortion involves the use of prostaglandins.[8] These drugs can be administered orally or as a vaginal suppository up to three weeks after a missed menstrual period (Fig. 14–1). Prostaglandins cause contraction of the uterine muscle and abortion of the fetus. Some advantages of this method are that it is 100 percent effective if done less than three weeks after the last menstrual period, and it has little risk to the woman. No pregnancy test is required for this procedure. One disadvantage of this method is that prostaglandins can cause contraction of smooth muscle in the digestive tract, leading to nausea and diarrhea, but pretreatment with a digestive-tract relaxant can alleviate this problem.[9] Another disadvantage in some women is that uterine bleeding begins about 3 to 6 hours after the procedure and can last 10 to 14 days.[10] Also, the abortion can take several hours. However, newer and more potent prostaglandin analogs are now being tested.[11] In the future, vaginal prostaglandin suppositories could be used by a woman in her home to induce early abortion.

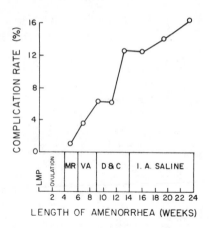

LENGTH OF AMENORRHEA (WEEKS)

Figure 14–2. Percentage of women who develop complications after abortions at different stages of pregnancy and after various procedures. Complications of each procedure are discussed in the text, and include excessive bleeding, uterine perforation, retained tissue, and uterine infection. LMP = last menstrual period; MR = menstrual regulation; VA = vacuum aspiration; D & C = dilatation and curettage; I.A. saline = intraamniotic saline; Amenorrhea = absence of menstruation as a measure of stage of pregnancy. Note that the complication rate increases as pregnancy advances, suggesting that early first trimester abortions are the safest. Reproduced from S. L. Chaudry et al., "Pregnancy Termination in Midtrimester—A Review of Major Methods," *Population Reports,* ser. F, no. 5 (1976), pp. 65–83. Used with permission of the Population Information Program, Johns Hopkins University, Baltimore, Md.

Dilatation and curettage (D & C) usually is done 6 to 12 weeks after a woman's missed menstrual period (Fig. 14–1). In this procedure, the cervix is dilated using methods discussed below, and then a metal scraper (*curette*) is inserted through the cervical canal and into the uterus. The fetus and endometrium then are scraped and removed (*curettage*).[12] The entire procedure lasts about 10 to 15 minutes. Some women require hospitalization for this method, whereas others have a D & C in a clinic or physician's office. A paracervical or general anesthetic often is used during this procedure because it can be painful to some women.

In the past, metal instruments (*dilators*) were used to expand the cervix so that a curette could be inserted into the uterus. In about 5 percent of cases, however, this procedure causes laceration of the cervix or perforation of the uterine wall. Also, this type of cervical dilatation can lead to an incompetent cervix, wherein future pregnancies are premature because the cervix is unable to support the growing fetus. There is about a 13 percent chance of a future birth being premature if a woman has had one previous D & C; this percentage is up to 17 percent with two previous D & Cs, and 21 percent with three previous D & Cs. Cervical dilatation with metal instruments also can lead to a higher incidence of stillbirths and miscarriages in future pregnancies.[13]

For the above reasons, cervical dilatation is more commonly done in other ways. In the past, items such as sponges, slippery elm bark, and inflatable rubber tubes were used to dilate the cervix. Prostaglandins also can be used to dilate the cervix.[14] Today, however, the safest and most common method of cervical dilatation is the use of *laminaria tents*.[15] These tents are cylinders of dried and sterilized seaweed (*Laminaria japonica* or *Laminaria digitata*) that swell to three to five times their original size when placed in the cervix. They are inserted in the cervical canal (Fig. 14-3) for a minimum of 3 to 5 hours before curettage. The woman is usually free to go home during this waiting period. The tent should be removed within 24 hours after surgery to avoid uterine infection. The tent must be inserted properly, or it could be expelled or enter the uterus. Some woman experience uterine cramping while the tent is in place. However, laminaria tents are safe, with no danger of cervical laceration, uterine perforation, or incompetent cervix. They also are less painful and cause less bleeding than metal dilators. Laminaria tents are used not only for cervical dilatation before curettage abortion but also for other abortion procedures and for medical therapy related to dysmenorrhea or infertility because of a blocked cervical os. It should be mentioned that a new method of dilating the cervix could replace use of laminaria tents in the future, especially for second trimester abortions. This device involves the use of a rubber catheter connected to an inflatable, pear-shaped balloon.[16]

Vacuum aspiration (or *vacuum curettage*) is replacing dilatation and curettage as a safer and easier method of first trimester abortion.[17] It is commonly done from the third to the ninth week after a menstrual period is missed (Fig. 14-1). This method is quicker than dilatation and curettage; it causes less uterine bleeding and is less painful; there is a lesser risk of uterine infection; and it does not cause an incompetent cervix. Before vacuum aspiration, the cervix is dilated with a laminaria tent. Then, an anesthetic usually is injected into the cervical wall, and a tube (*vacurette*) is placed into the uterus. This tube is connected to a suction device attached to the collection bottles (Fig. 14-4). After the tube is inserted so as to touch the amniotic sac, the suction is turned on for 1 to 1½ minutes, during which time the collection bottles are examined for evidence of embryonic and placental tissue. If this material is not com-

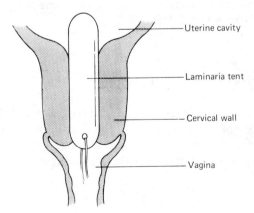

Uterine cavity

Laminaria tent

Cervical wall

Vagina

Figure 14-3. Laminaria tent properly inserted in the cervix.

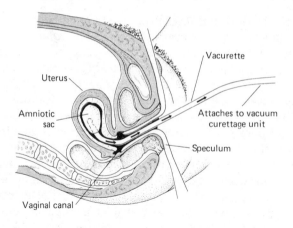

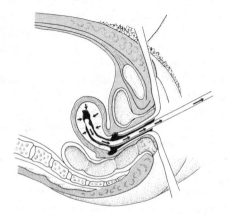

Figure 14–4. A vacuum aspiration abortion. A tube (vacurette) is inserted through the cervix and into the uterus, after the vagina is opened using a speculum (top). The embryo and uterine lining then are sucked into a bottle (not shown) by using a suction pump (bottom). Reproduced from J. S. Hyde, *Understanding Human Sexuality* (New York: McGraw-Hill Book Company, 1979). © 1979; used with permission.

pletely removed by suction, the endometrium can then be scraped with a curette. Some women (about 3/1000) experience minor complications such as uterine cramping, bleeding, or infection after vacuum aspiration, but these complications occur with less frequency than after a D & C (Fig. 14–2).

SECOND TRIMESTER INDUCED ABORTIONS

Induced abortions in the second trimester of pregnancy (thirteenth to the twenty-fourth week) are more complicated and risky to the woman. After the seventeenth week of pregnancy, the maternal death rate due to abortion is greater than that for childbirth. Still, over 100,000 second trimester abortions are performed in the United States each year.

Why A Second Trimester Abortion?

Why would a woman wait until the second trimester to have an abortion? One answer may be that she did not have the money for a first trimester abortion, or perhaps she lacked information on the availability of abortion. Also she may not have realized that she was pregnant, or may have denied this fact, or thought the problem would "go away." More important, she may have developed some maternal disorder related to her pregnancy that did not appear until this time. Or it may have been diagnosed during midpregnancy that the fetus had a severe abnormality. Finally, she, alone or with her partner, may have decided against childbirth for various emotional or psychological reasons.

From the twelfth through the fifteenth week of pregnancy, the uterine wall is thin and susceptible to perforation. Also, the placenta is now highly vascular, and a D & C or vacuum aspiration could cause excessive uterine hemorrhage. Thus, a physician often (but not always) advises a woman to wait until the sixteenth week for an abortion if her pregnancy is past the twelfth week, although some physicians will do a D & C abortion up to the fifteenth week (Fig. 14–1).

Intraamniotic Saline

One common method used to induce abortion in the second trimester is injection of *intraamniotic saline,* sometimes called "salting out."[18] In this method, usually done in a hospital using a spinal or general anesthetic, the cervix first is dilated with a laminaria tent. Then, a long needle is inserted through the abdominal and uterine wall and into the amniotic sac. After about 250 ml of amniotic fluid is removed, a syringe is used to inject saline solution into the amniotic sac (Fig. 14–5). The saline solu-

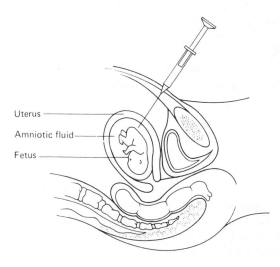

Uterus

Amniotic fluid

Fetus

Figure 14–5. Diagram of administration of intraamniotic saline, urea, glucose, or prostaglandin to induce second trimester abortion.

tion kills the fetus and induces delivery within 24 to 48 hours.[19] In some cases, a solution of urea or glucose is injected instead of saline. Saline, urea, and glucose cause uterine contractions by inducing uterine secretion of prostaglandins.[20] Oxytocin also can be administered to facilitate a saline-, urea-, or glucose-induced abortion.

There are several dangers to this method of induced abortion. In 10 to 16 percent of saline-induced abortions, the placenta is not delivered with the fetus and must be scraped out using a curette. In other cases, the fetus is born alive but soon dies of respiratory failure. Also about 34/1000 women develop complications due to misinjection of saline into a blood vessel, uterine infection, or uterine hemorrhage. Delivery of the fetus after saline, urea, or glucose administration can be similar to normal labor with its risk of maternal complications. The maternal death rate after saline abortions is about 12 to 18/100,000.

Prostaglandin Injections

Prostaglandin injections were introduced in 1970 as a method for second trimester abortions and are now the most widely used procedure in the United States.[21] These drugs work best from the fifteenth to the twentieth week of pregnancy. The procedure is similar to that used during saline-induced abortions, except that no amniotic fluid is withdrawn and only a small volume of fluid is injected. The prostaglandin can be injected into the amniotic fluid or between the fetal membrane and the uterine wall. The latter "extraamniotic" route, using a needle passed through the cervical canal, requires a lower amount of prostaglandin and has fewer side effects than the "intraamniotic" route.[22] Vaginal suppositories containing prostaglandins are also being used for second trimester abortions and are considered 100 percent effective.[23] Urea, saline, and oxytocin can be used with prostaglandins.

Use of prostaglandins for second trimester abortions has certain advantages over the use of saline. Primarily, it is faster, with delivery occurring after 6 to 8 hours if the cervix is dilated with a laminaria tent.[24] Also, there are less complications. However, some women develop diarrhea and nausea because the prostaglandins cause contraction of smooth muscle in the digestive tract. Another disadvantage of prostaglandin-induced second trimester abortion is that the fetus often is born alive.

Other Methods

In Japan, a chemical called *Rivanol* (ethacridine lactate) is injected into the amniotic sac to induce second trimester abortions. This drug, however, is not used in the United States because it takes a long time for delivery of the fetus. The Japanese also use intrauterine insertion of a slender cylindrical instrument (*bougie*) or a rubber bag (*metreurynter*) to induce abortion, but neither is used in the United States because they require prolonged hospitalization, delivery is slow, there is a high risk of uterine infection, and many fetuses are born alive.

THIRD TRIMESTER INDUCED ABORTIONS

Third trimester abortions (after the twenty-fourth week of pregnancy) are rare in the United States, mainly because after this time the fetus, with special care, could survive on its own. *Hysterotomy* is the usual procedure used to induce abortion in the third trimester, although saline or prostaglandins also are used. Hysterotomies can be performed from the middle of pregnancy to the early part of the third trimester. After the fetus can usually survive on its own (twenty-eighth week on), induced abortion by any means is rare.[25] This operation is major surgery done under anesthesia of the pelvic region or general anesthesia. It often is called a "mini-cesarean" since the procedure is similar to a cesarean delivery. Hospitalization is required for 4 to 6 days. Minor or major complications of this surgery occur in about 23 to 51 percent of women, and the mortality rate is about 45 to 271/100,000. One negative aspect of these abortions is that the fetus is delivered alive and must be left to die (often after several hours) of respiratory failure, a disturbing task for all persons involved.

FERTILITY AFTER AN INDUCED ABORTION

An induced abortion usually has no effect on a woman's future fertility. As mentioned, however, certain abortion procedures that dilate the cervix with metal instruments can lead to an incompetent cervix and can increase the risk of spontaneous abortions or premature births in future pregnancies.[26] Ovulation usually occurs in two or three weeks after a first trimester abortion.[27] This delay is longer after second or third trimester abortions, and is similar to the delay after a normal delivery in a nonnursing woman (about three months).

CHAPTER SUMMARY

Induced abortion is the termination of a pregnancy by artificial measures. Governments can be permissive or restrictive in their legislation about abortion. Induced abortion is legal in the United States today but is less available to poor people.

A woman or couple with a wanted pregnancy may choose an abortion because of danger to the woman's physical or mental health or because the fetus is gravely deformed. Also some women simply do not wish to be pregnant and terminate pregnancy as a means of fertility control or to avoid birth of an unwanted child. There is a moral controversy about induced abortion that relates to the concept of "ensoulment." Choosing or not choosing to have an abortion usually is a difficult decision and involves various religious, moral, psychological, legal, and medical considerations.

Induced abortion procedures are safer and simpler the earlier the stage of pregnancy. During the embryonic period (early first trimester), abortion can be induced by menstrual regulation or administration of prostaglandins. During the early fetal period (late first trimester), pregnancy can be terminated by dilatation and curettage or, more commonly, by cervical dilatation using a laminaria tent followed by vacuum aspiration. Common methods of inducing abortion during the second trimester are administration of intraamniotic saline, urea, glucose, or prostaglandins, alone or in combination. Extraamniotic prostaglandin also is used. Induced abortion of a third trimester fetus is rare; when done, it usually involves hysterotomy.

Most abortion procedures do not influence future fertility. Menstrual cycles resume in about two or three weeks after a first trimester abortion, and in about three months after a second or third trimester abortion.

NOTES

1. C. Tietze and S. Lewit, "Abortion," *Scientific American,* 220, no. 1 (1969), 21–27.
2. C. Tietze and S. Lewit, "Legal Abortion," *Scientific American,* 236, no. 1 (1977), 21–27.
3. Ibid.
4. C. Grobstein, "When Does Human Life Begin?," *Science 82,* 3, no. 2 (1982), 14.
5. T. Vander Vlugt and P. T. Piotrow, "Menstrual Regulation Update," *Population Reports,* ser. F, no. 4 (1974), pp. 49–64.
6. Ibid.
7. S. L. Chaudry et al., "Pregnancy Termination in Midtrimester—A Review of Major Methods," *Population Reports,* ser. F, no. 5 (1976), pp. 65–83.
8. R. Shalita, "Prostaglandins Promise More Effective Fertility Control," *Population Reports,* ser. G, no. 6 (1975), pp. 57–62.
9. Ibid.
10. M. Bygdeman and S. Bergstrom, "Clinical Use of Prostaglandins for Pregnancy Termination," *Population Reports,* ser. G, no. 7 (1976), pp., 65–75.
11. L. J. Gail, "The Use of PGs in Human Reproduction," *Population Reports,* ser. G, no. 8 (1980), pp. 77–118.
12. Chaudry et al. (1976), op. cit.
13. E. R. Ott, "Cervical Dilatation—A Review," *Population Reports,* ser. F, no. 6 (1977), pp. 85–103.
14. Gail (1980), op. cit.
15. Ott (1977), op. cit.
16. E. L. Southern, "A New Device for Preoperative Cervical Dilatation," *International Planned Parenthood Federation Medical Bulletin,* 14, no. 2 (1980), 3–4.
17. Chaudry, Hunt, and Wortman (1976), op. cit.
18. Ibid.
19. Gail (1980), op. cit.
20. Ibid.

21. Shalita (1975), op. cit.

22. Bygdeman and Bergstrom (1976), op. cit.

23. Gail (1980), op. cit.

24. Chaudry et al. (1976), op. cit.

25. C. R. Austin, "The Ethics of Manipulating Human Reproduction," in *Reproduction in Mammals, Book 5, Artificial Control of Reproduction,* eds. C. R. Austin and R. V. Short, pp. 141–52 (Cambridge, England: Cambridge University Press, 1972).

26. Anonymous, "No Increased Risk of Spontaneous Abortion Found Among Women with a Previous Induced Abortion," *Family Planning Perspectives,* 13 (1981), 238–39.

27. Anonymous, "The Resumption of Menstrual Cycles after Abortion," *Research in Reproduction,* 10, no. 2 (1978), 1.

FURTHER READING

AUSTIN, C. R., "The Ethics of Manipulating Human Reproduction," in *Reproduction in Mammals, Book 5, Artificial Control of Reproduction,* eds. C. R. Austin and R. V. Short. Cambridge, England: Cambridge University Press, 1972. Pp. 141–52.

CATES, W., JR., et al., "The Effect of Delay and Method Choice on the Risk of Abortion Morbidity," *Family Planning Perspectives,* 9 (1977), 266–73.

EDWARDS, R. G., "Control of Human Development," in *Reproduction in Mammals, Book 5, Artificial Control of Reproduction,* eds. C. R. Austin and R. V. Short. Cambridge, England: Cambridge University Press, 1972. Pp. 87–113.

GROBSTEIN, C., "When Does Human Life Begin?" *Science 82,* 3, no. 2 (1982), 14.

HENSHAW, S. K., et al., "Abortion Services in the United States, 1979 and 1980," *Family Planning Perspectives,* 14 (1982), 5–15.

LUKER, K., *Taking Chances—Abortion and the Decision Not to Contracept.* Berkeley, Calif.: University of California Press, 1975.

MILUNSKY, A., and L. H. GLANTZ, "Abortion Legislation: Implications for Medicine," *Journal of the American Medical Association,* 248 (1982), 833–34.

TIETZE, C., and S. LEWIT, "Legal Abortion," *Scientific American,* 236, no. 1 (1977), 21–27.

15 *Infertility*

LEARNING GOALS

Having read this chapter, you should be able to:

1. Define when a couple is infertile.
2. List procedures used to test if a woman or man is infertile.
3. Discuss possible causes of failure to ovulate, and list possible treatments of this form of female infertility.
4. Describe some causes of tubal blockage and possible treatments of this condition.
5. Describe what can be done for a woman who is infertile because of faulty implantation.
6. List some causes of oligospermia or azoospermia in men and possible treatments for these conditions.
7. List kinds of infertility in men that involve malfunction of the sex accessory ducts and glands.
8. Define "AID" and "AIH" artificial insemination.
9. List some reasons why a couple would resort to artificial insemination.
10. List five questions that still remain about artificial insemination.
11. Describe the procedure utilized to produce a "test-tube baby," and discuss the legal and moral controversy about this procedure.
12. Discuss how surrogate pregnancies are being used by infertile couples.
13. Describe how someone goes about adopting a child.

INTRODUCTION

Couples that utilize contraception wish to avoid unwanted pregnancies. In contrast, many couples wish to bear children but are unable to conceive. Current estimates place couples suffering from *infertility* in the United States at about 3.5 million, or 1 out of 5 married couples of childbearing age. [1] A couple is considered infertile if they have participated in unprotected coitus for a year without becoming pregnant. In about 40 percent of these couples, the problem is with the female, in about 40 percent with the male, and the remainder with both partners.

Problems of infertility are more common with increasing age, although younger couples also can suffer this fate. A recent French study indicated a rise in infertility of women between ages 31 and 35 (27 percent are infertile), and a sharper rise over 35 (47 percent are infertile). [2] This has many women concerned, especially those who have waited until their mid-to-late thirties to have children. The French study, however, was based on rates of pregnancy after artificial insemination. More accurate infertility rates would be 4.1 percent at ages 20 to 24, 5.5 percent at ages 25 to 29, 9.4 percent at ages 30 to 34, and 19.7 percent at ages 35 to 39. [3]

Hope is ever present for infertile couples, because over 80 percent of cases of infertility can be diagnosed, and 70 percent can be successfully treated. In those situations where treatment fails, a couple may want to adopt a child or seek help in other ways. We now look at some causes and treatments of infertility.

SEEKING MEDICAL HELP FOR INFERTILITY

When infertile couples seek medical help, procedures can be used to diagnose the cause of their problem. In the case of the female, several tests can be administered. [4] Her menstrual cycle, if occurring, can be followed using the sympto-thermal method and/or hormone assay. If it is found that she is ovulating, administration of dyes, gas, or laparoscopy can be used to check if her uterus or oviducts are blocked. An endometrial biopsy can be done to check the condition of her uterine lining, and her reproductive tract can be tested for infection. Finally, the husband's semen can be cultured with her tissues to see if she is producing antibodies to his sperm.

In the case of the male, semen can be collected and a sperm count done. If his count is fertile (greater than 20 million sperm/ml of semen), the sperm will be checked for normalcy of motility and structure. His semen will also be checked to see if it is ejaculated in normal amounts (1 to 6 ml) and if it contains enough fructose ($>1200 \mu g$/ml). If his sperm count is low, levels of hormones (LH, androgens) will be assayed in his blood or urine, and the blood supply to his testes may be checked. Also, a sample of testicular tissue may be taken to see if spermatogenesis is normal; this is called a *testicular biopsy*. Finally, the condition of the sex accessory glands and ducts can be checked, and immunological tests done to see if he is producing antibodies to his own sperm.

Infertile couples can suffer from psychological and emotional problems related to their condition. These problems can be caused by feelings of inadequacy and frustration over their situation. Alternatively, their infertility may be psychosomatic in origin.[5] Therefore, it is often wise for infertile couples to seek counseling about their problem.

FEMALE INFERTILITY

Failure to Ovulate

Failure to ovulate is the leading cause of infertility in females. A woman may not ovulate because her hypothalamus or pituitary gland is not fully functional. For example, her hypothalamus may not be secreting enough GnRH to stimulate an LH and FSH surge and ovulaton. In this case, administration of GnRH analogs in small dosages has been effective in inducing ovulation.[6] But suppose her pituitary gland is not capable of secreting enough LH and FSH in response to adequate levels of GnRH. In this case, one common treatment is administration of the drug *clomiphene* (or Clomid). Clomiphene is an antiestrogen, and its administration inhibits the negative feedback action of estrogens on FSH and LH secretion. Thus, FSH and LH levels increase when clomiphene is given, and 30 to 50 percent of infertile women so treated will ovulate and become pregnant.

One effect of clomiphene is that the chance of twinning is 5 percent, as compared to 1 percent in untreated women. This is because the increased gonadotropin levels may hyperstimulate the ovary, causing ovulation of two eggs. Also, there is evidence in rats that clomiphene could cause reproductive tract abnormalities in both the mother and potential offspring.[7]

Another treatment for a woman who is secreting insufficient FSH and LH is the administration of human menopausal gonadotropin (Perganol), which contains FSH and LH, followed by administration of hCG (Fig. 15-1). This treatment causes ovulation in 50 to 70 percent of cases, so it is highly effective. The incidence of twins, however, is 15 percent, and for triplets or more is 5 percent. For these reasons, the gonadotropin treatment is used only as a last resort unless, of course, twins or triplets are desired. For some unknown reason, vitamin B_6 also can help some infertile women to ovulate.[8]

Another cause of infertility in relation to malfunction of the pituitary gland is too much pituitary secretion of prolactin. Some infertile women have high prolactin levels in their blood; that this condition is causing their problem is shown by the fact that a drug called *bromocriptine*, which inhibits prolactin secretion, restores fertility in many of these women.[9]

In other women, failure to ovulate may not be caused by malfunction of the hypothalamus or pituitary, but instead the ovary may be incapable of responding to an LH and FSH surge and ovulating. This may be due to the presence of ovarian cysts, tumors, or scars caused by ovarian infection. In these cases, ovarian surgery may be

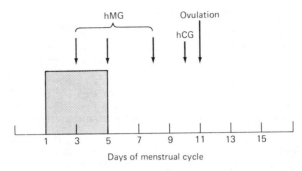

Figure 15-1. Treatment of an infertile woman with human menopausal gonad-otropin (hMG) followed by human chorionic gonadotropin (hCG) to induce ovula-tion. In the figure, the shaded area represents menses. Adapted from R. G. Edwards, "Control of Human Development," in *Reproduction in Mammals, Book 5, Artificial Control of Reproduction,* eds. C. R. Austin and R. V. Short (Cambridge, England: Cambridge University Press, 1972). © 1972; used with permission.

needed to restore fertility. Some women, however, may have permanently malfunctioning ovaries. The ovaries of women with Turner's syndrome (XO), for example, lack follicles (Chapter 5).

Tubal Blockage

In some infertile women, ovulation occurs but the sperm fail to reach the upper reaches of the oviduct to fertilize the ovum because of a blockage in one or both oviducts. Tubal blockage is the second leading cause of infertility in females, occurring in 30 to 35 percent of infertile women. Tubal blockage can be caused by a kink in the tube or by scarring due to past venereal disease infection (Chapter 16). Also, a piece of the uterine endometrium may have become displaced from the uterus and lodged in the oviduct. This endometriosis (Chapter 3) causes sterility because the uterine tissue grows and blocks the tubes.

Tubal blockage often can be repaired by introducing a gas (CO_2 or air) into the tubes. If this fails to open the tubes, they usually can be repaired surgically. A new method (perfected on monkeys) to circumvent tubal blockage may be available to humans in the near future. In this surgery called *low tubal ovum transfer,* the ovulated egg is removed just before it enters the oviduct and is then inserted back into the oviduct below the point of blockage.[10]

Absence of Implantation

In some women, the embryo may reach the uterus, but implantation does not occur. Priming of the uterus by estrogen and progesterone is needed for implantation to occur (Chapter 9), and this priming may be inadequate in some women. Some of these cases can be treated by administration of steroid hormones (estrogens or progestogens) to render the uterine endometrium more receptive to the embryo.

Other causes of infertility may be due to damage to the endometrium. Perhaps fibroids or scars from pelvic infection are present (Chapter 4). Or a previous abortion using dilatation and curretage (Chapter 14) may have damaged the uterine lining.

Reduced Sperm Transport or Antibodies to Sperm

Finally, the female tract may not allow transport or survival of the male's sperm. The woman's vagina, for example, may be highly acidic, which can be treated by using alkaline douches. Alternatively, her cervical mucus may be hostile to sperm movement, a condition perhaps alleviated by estrogen administration. If the cervix has been damaged, as sometimes occurs after artificial dilatation during an abortion or as a result of infection, this damage may be corrected surgically. In rare cases, a woman is producing antibodies to her husband's sperm, and in this case the couple may choose to become pregnant using donor artificial insemination (see below).

MALE INFERTILITY

An infertile man may have a low sperm count *(oligospermia)* or the absence of sperm *(azoospermia)* in the semen because his hypothalamus or pituitary gland is functioning below normal levels. A low sperm count is the leading cause of infertility in males. Treatment with GnRH analogs or gonadotropins may restore fertility in these men. Some infertile men have high prolactin levels in their blood, and treatment with bromocriptine can make them fertile.[11]

Sometimes the testes themselves may be incapable of responding to gonadotropins. For example, there may be structural abnormalities or permanent damage to the testes. The latter could occur because of exposure to mumps or radiation, or because of old age (Chapter 4). A varicose vein in the scrotum (a condition called *varicocele)* can raise testis temperature and cause infertility. When testes are damaged, the treatment may be artificial insemination of the women with donor sperm. Some structural problems, however, can be corrected by testicular surgery.

In about 8 to 13 percent of infertile men, their problem is caused by the production of antibodies to their own sperm. This occurs because some sperm inadvertently enter a man's body outside the reproductive tract. Treatment with adrenal hormones has been shown to alleviate this problem in some men.[12]

In some infertile men, the secondary accessory ducts or glands are not functioning properly. The vasa deferentia, for example, may be occluded by an enlarged testicular vein pressing on it (varicocele). The vasa also can be blocked by scar tissue caused by venereal disease infection (Chapter 16). Many of these cases can be corrected with surgery. Finally, a sex accessory gland may be malfunctioning or inactive.

If the glands are simply underdeveloped, this condition can be treated by the administration of an androgen. Infection of the prostate can lead to sterility in some men.

Some men may be temporarily infertile because of environmental factors. Smoking, for example, can decrease sperm motility and increase the number of structurally abnormal sperm in the ejaculate. Testosterone levels in the blood also are lower in men who smoke.[13] Accumulation of some environmental pollutants, such as PCBs, can reduce sperm count (Chapter 4).

GAMETE STORAGE AND ARTIFICIAL INSEMINATION

Artificial insemination is when sperm are introduced into a woman's reproductive tract by means other than coitus. The sperm can come from a "donor" man who is not the woman's husband ("artificial insemination donor," or AID). Or the sperm can come from the husband ("artificial insemination husband," or AIH). In both cases, the man contributes semen into a vial by masturbating. Several ejaculations are required to pool enough sperm to be effective. Donors are classified, anonymously, by their physical characteristics. More than 6000 babies are born each year in the United States using artificial insemination, about two-thirds being AIH. Usually, the sperm are chilled before use, but they also can be frozen at $-196°C$ ($-321°F$) for up to ten years. Chilled or frozen sperm are stored in "sperm banks" run by hospitals or private businesses.

Artificial insemination can be useful for infertile couples because of some problem with the husband such as a low sperm count or other kinds of infertility. In such cases, semen can be used from a donor male. Sperm storage also can be useful for a man who is going to have a vasectomy, just in case he may later want to father a child.

Anyone who is contemplating artificial insemination should be aware, however, that there are biological questions to be answered. For example, how does freezing affect sperm? After freezing, many sperm lack vigor and motility, and up to 85 percent are abnormally shaped.[14] This may explain why the rate of spontaneous abortions after artificial insemination is five times the normal rate.[15] And why are more male babies than usual born by this method?[16]

There also are moral, ethical, and legal considerations about artificial insemination. For example, who is the legal father of a child born using semen from a male other than the husband? If a couple having a child by this method is divorced, is the ex-husband responsible for child support? If a child conceived in this manner has a birth defect, can the physician be charged with malpractice? What are the inheritance rights of a child conceived by sperm from a man other than its legal father? Should semen be used for selective breeding for certain inherited traits, and what are the moral implications of selective breeding?[17] An example of such an attempt at selective breeding is a sperm bank in California using sperm donated by Nobel prize winners! Artificial insemination is codified only in a few states, and the legal aspects of this practice lag far behind our scientific abilities in this area.

In 1977, the first scientific report appeared showing that, in the laboratory mouse, ovulated ova (oocytes) could be collected from the oviduct and stored at −1960°C for six years! They then were thawed, fertilized in a dish, implanted into the uterus of a female mouse, and developed into normal mouse pups.[18] This may be possible, with human ova in a few years, and there well could be "ova banks" along with sperm banks. In addition, the future may see freeze-drying of human embryos for later thawing and implantation in a woman's uterus. Such a procedure has been successful in cattle and mice.

TEST-TUBE BABIES

On July 25, 1978, a 5 lb, 12 oz baby girl named Louise Joy Brown was born to Mr. and Mrs. Gilbert J. Brown of Bristol, England. Louise, who is a perfectly normal little girl, is a special person in that she was the world's first *test-tube baby*. That is, her father's sperm fertilized her mother's ovum in a dish, and the early embryo resulting from this *in vitro* (Latin for "in glass") fertilization was then put back into Mrs. Brown's uterus to develop and be born. Thus, Louise was not "born" in a test tube, but she was conceived in a dish. By the end of 1982, about 150 test-tube babies had been born in England, Australia, and the United States, including twins! This procedure, however, costs $4,000–$8,000!

This remarkable achievement was the culmination of a long history of research in mammalian reproduction. The first human egg to be fertilized by in vitro fertilization (*external fertilization*) was achieved by J. Rock and M. Menkin in 1944.[19] Drs. P. C. Steptoe and R. G. Edwards of England then refined this technique and eventually determined how to implant the embryo back into the mother's uterus.[20] Women who are infertile because of a condition like a tubal blockage are chosen for this procedure.

A day or two before the woman's predicted time of ovulation (see Chapter 5), she is monitored every 3 hours for levels of luteinizing hormone (LH) in her urine. About 20 to 21 hours after the beginning of the LH surge in the menstrual cycle, LH levels reach their peak (Chapter 7). (Some laboratories are injecting clomiphene and hCG to cause ovulation instead of waiting for a woman's natural LH surge.) Then an oocyte is recovered from the large follicle in the woman's ovary. The operation is done by laparoscopy, a technique by which a small incision is made near the navel (under general anesthetic) and a tiny telescope and illuminator are inserted through the abdominal wall. Once the ovary is located and examined, the oocyte is removed by suction from the large follicle. Before laparoscopy is done, the husband produces (by masturbation) fresh sperm, which are placed in a dish under wax. The dish contains a fluid that nourishes and capacitates the sperm. The oocyte (which is now arrested in the second meiotic division) then is added to the dish with the sperm, and fertilization usually occurs within 12 to 14 hours.

After fertilization, the embryo is transferred to a new culture dish. It takes three days for the embryo to reach the 8-cell stage and four days to reach the 16-cell stage. The 8- or 16-cell embryo then is placed into a tiny tube, which is inserted into the woman's uterus through her vagina. Finally, the embryo is squirted into the uterus, and implantation occurs. A new technique of using the oviduct of a laboratory animal, instead of a dish, to fertilize human ova with human sperm may appear in the future.[21]

Of 79 women first accepted for this procedure in England in the late 1970s, 68 underwent laparoscopy (the other 11 women had unsatisfactory menstrual cycles). Of the 68 women, 44 produced ova in the correct state, and 32 of the ova were fertilized and developed into 8- or 16-cell embryos, which were then transplanted back into the mothers. But only 4 of the 32 women became pregnant, meaning that only four had successful embryonic implantation. And only two of these pregnancies produced normal infants at term; the other two fetuses were aborted early in pregnancy and were found to exhibit chromosomal abnormalities.[22] With recent refinements of the technique, pregnancy occurs in about 20 percent of attempts. Also, recent "test-tube" births suggest that there is no increase in congenital defects over those found in normal births.

Moral and ethical questions surround the test-tube baby procedure.[23] Will it lead to selective breeding? Would rich people pay poor people to carry their babies, and if this happened, who would be the parents? Is there danger of scientists growing human embryos in vitro later than the 16-cell stage, and if so, would death of these embryos in the laboratory be abortion or murder?

It is these kinds of questions that prompted the U.S. Government to ban research on external fertilization in 1975 while an Ethics Advisory Board of the Department of Health, Education and Welfare looked into the matter. In March 1979, this board sanctioned such research in America with the ultimate goal of overcoming the problem of human infertility. In June 1979, the board approved of research on external fertilization with certain stipulations, some of which were: (1) no embryos fertilized in the laboratory can be maintained for over 14 days (the maximum time it takes for a normal embryo to implant fully in the uterus); (2) people donating eggs or sperm must be notified about, and must approve of, the purposes and uses of their cells; (3) if the embryo is to be implanted back into the uterus, the gametes must be obtained from married couples; (4) the public must be informed about incidences of abnormal offspring resulting from these procedures.

The achievement of Drs. Steptoe and Edwards was indeed a landmark of reproductive medicine, and their method has great promise in helping couples to have children who otherwise are infertile because of a block in transport of the egg or sperm through the oviduct. However, the above ethical considerations must be discussed, and new research must make this method successful and safe for the fetus. Even if in vitro fertilization becomes a pathway to childbearing for many infertile couples, it is very doubtful that true test-tube babies, developed and born in "human hatcheries" as described by Aldous Huxley in his book *Brave New World,* will be a reality in the near future, if ever.

SURROGATE PREGNANCIES

A woman who is infertile or who has had a hysterectomy now is able to hire another woman to carry her child. In these *surrogate pregnancies,* the hired woman is artificially inseminated with the sperm of the infertile woman's husband.[24] She then becomes pregnant, delivers the child, and gives it to the original couple. Usually, a contract is written between the couple and the surrogate woman. Stipulations can include that the surrogate be married with children of her own, have a physical and psychological examination, be willing to surrender the child, and not use tobacco, alcohol, and other drugs while she is pregnant. Her fee for this favor can be $13,000 or more.

ADOPTION

Infertile couples, of course, have the option of adopting a child. Adopted children usually are the product of unwanted pregnancies or conditions in which the biological parents want their child but are unable to care for it. The people that give up a child for adoption are not told who adopted their child. A person (single man or woman) or a couple who desire to adopt should first contact an adoption agency. Then, the agency determines if the couple or person fits their standards. What are common standards? For many agencies, a couple must be infertile and not be more than 40 years older than the child they adopt. Additionally, a home study is done to determine if the family will provide adequate physical and emotional care for the adopted child. If the couple is approved, they may have to wait up to five years for an American child; this wait is less for a foreign child. Adoption is more expensive than pregnancy and delivery, costing from $2000 to $5000.

In general, adopted children are as happy and healthy as are children with their biological parents. Some adopted children, especially when over two years of age when adopted, have adjustment problems that last about a year. Adopted parents also may have adjustments to make. In the long run, however, families with adopted children are no different from those with biological parents and offspring.

CHAPTER SUMMARY

A woman can be infertile for several reasons. If her pituitary gland is not secreting enough FSH and LH to cause ovulation, she can be treated with GnRH, clomiphene, or hMG followed by hCG. If her ovaries are not responding to FSH and LH because of the presence of cysts, tumors, or scars, ovarian surgery can help. If her oviducts are blocked, they can be opened by forcing gas into them or by surgery. If her uterus does not support implantation, treatment with an estrogen and a progestogen can induce implantation. If her cervix is hostile to her husband's sperm, artificial insemination may be necessary.

An infertile male with a low sperm count can be helped by administration of gonadotropins or GnRH analogs. If his testes are damaged, or if he is producing antibodies to his own sperm, artificial insemination may be the answer. Nonfunctional male sex accessory ducts or glands can be made functional with androgen treatment or surgery.

Artificial insemination utilizes chilled or frozen sperm from the husband (AIH) or a donor (AID). One problem with this procedure is the introduction of nonmotile or abnormally shaped sperm. "Test-tube" babies are produced by removing an oocyte from a woman, fertilizing it with the husband's sperm in a dish, and placing the embryo back into the woman's uterus. In the United States, definite regulations govern this procedure.

Some infertile couples now can hire another woman to bear a child for them, utilizing artificial insemination with the husband's sperm. Infertile couples also can adopt a child.

NOTES

1. M. D. Mazor, "Barren Couples," *Psychology Today* (May 1979), pp. 101–12.
2. F. Cecos et al., "Female Fecundity as a Function of Age," *New England Journal of Medicine,* 306 (1982), 404–6.
3. J. Bongaarts, "Involuntary Childlessness with Increasing Age," *Research in Reproduction,* 14, no. 4 (1982), 1–2.
4. R. Homburg and V. Insler, "Evaluation of Female Infertility," *Karger Gazette* (December 1978), pp. 6–7.
5. Mazor (1979), op. cit.
6. Anonymous, "Pulsatile Treatment with LH-RH in Amenorrhoeic Women," *Research in Reproduction,* 14, no. 4 (1982), 3.
7. S. McCormack and J. H. Clark, "Clomid Administration to Pregnant Rats Causes Abnormalities of the Reproductive Tract in Offspring and Mothers," *Science,* 204 (1979), 629–31.
8. Anonymous, "Infertile Women Conceive after Vitamin B_6 Therapy," *Medical World News* (March 19, 1979), p. 43.
9. H. G. Friesen, "Prolactin and Human Reproduction," *Research in Reproduction,* 8, no. 3 (1976), 3–4.
10. Anonymous, "Highlights," *Science 80,* 1, no. 8 (1980), 9.
11. Friesen (1976), op. cit.
12. Anonymous, "Steroids for Male Infertility," *Science News,* 116, no. 14 (1979), 233.
13. Anonymous, "Hazards of Tobacco Smoking in Human Reproduction," *Research in Reproduction,* 11, no. 3 (1979), 2.
14. Anonymous, "Sperm in Poor Shape Have No Future," *New Scientist,* 78 (1978), 591.
15. T. Mann, "Semen: Metabolism, Antigenicity, Storage, and Artificial Insemination," in *Frontiers in Reproduction and Fertility Control,* eds. R. O. Greep and M. A. Koblinsky (Cambridge, Mass.: M.I.T. Press, 1977), pp. 427–33.

16. S. J. Kleegman, "Female Sex Problems," *Sexology,* (November 1964), pp. 226–29.

17. F. Osborn, "The Emergence of a Valid Eugenics," *Scientific American,* 61 (1973), 425–29.

18. D. C. Whittingham, "Fertilization in Vitro and Development to Term of Unfertilized Mouse Oocytes Stored at − 196°C," *Journal of Reproduction and Fertility,* 49 (1977), 89–94.

19. J. Rock and M. F. Menkin, "In Vitro Fertilization and Cleavage of Human Ovarian Eggs," *Science,* 100 (1944), 105–7.

20. C. Grobstein, "External Human Fertilization," *Scientific American,* 240 (1979), 57–67; and Anonymous, "Dr. Steptoe's Full Report—At Last," *Medical World News* (February 1979), pp. 10–19.

21. F. J. DeMayo et al., "Fertilization of Squirrel Monkey and Hamster Ova in the Rabbit Oviduct (Xenogenous Fertilization)," *Science,* 208 (1980), 1468–69.

22. Anonymous (1979), op. cit.

23. R. G. Edwards, "Fertilization of Human Eggs in Vitro: Morals, Ethics and the Law," *Quarterly Review of Biology,* 49 (1974), 3–26.

24. J. Seligmann and R. Curry, "Pregnancy by Proxy," *Newsweek* (July 7, 1980), p. 72.

FURTHER READING

GROBSTEIN, C., "External Human Fertilization," *Scientific American,* 240 (1979), 57–67.

JEFFRIES, M., "Birth on Ice," *Omni* (July 1981), pp. 77–80, 94.

LENARD, L., "Hi-Tech Babies," *Science Digest* (August 1981), pp. 86–89, 116.

MAZOR, M. D., "Barren Couples," *Psychology Today* (May 1979), pp. 101–12.

MOSER, W. D., "Infertility Trends Among U.S. Couples: 1965–1976," *Family Planning Perspectives,* 14 (1982), 22–27.

16 *Sexually Transmitted Diseases*

LEARNING GOALS

Having read this chapter, you should be able to:

1. List the causative agents, transmission, common symptoms, diagnosis, and treatment of the following sexually transmitted diseases:
   (a) Gonorrhea
   (b) Syphilis
   (c) Herpes genitalis
   (d) Nonspecific urethritis
   (e) Chancroid
   (f) Lymphogranuloma venereum
   (g) Granuloma inguinale
   (h) Monilia vaginitis
   (i) Hemophilus vaginitis
   (j) Trichomonas vaginitis
   (k) Genital warts
   (l) Molluscum contagiosum
   (m) Viral hepatitis, type B
   (n) Acquired immune deficiency syndrome
   (o) Pediculosis pubis
   (p) Scabies

2. Discuss the possible effects of having a sexually transmitted disease on one's emotional and sexual life.

3. List precautions one can take against contracting a sexually transmitted disease.

INTRODUCTION

Sexually transmitted diseases are those that are transmitted from one person to another during coitus or other genital contact. This is why they sometimes are called *venereal diseases,* after Venus, the Roman goddess of love. Incidence of these diseases is disturbingly high. At least 8 to 10 million Americans contract some form of sexually transmitted disease each year.

Organisms causing sexually transmitted diseases usually do not live and reproduce on dry skin surfaces. Instead, they require the moist environments of membranes in the so-called "transitional zones" of the body—those that occur at openings between the external and internal body surfaces. These transitional zones include the vulva, vagina, and urethra of the female; the penis and urethra of the male; and the mouth, oral cavity, eyes, and anus of both sexes. These zones usually are where the sexually transmitted diseases first gain a foothold; from there they can invade other body tissues. Although the body forms antibodies to many of the sexually transmitted disease organisms, immunities are slow to develop or may never occur.

The purpose of this chapter is to discuss the incidence, causes, transmission, symptoms, and diagnosis and treatment of the more common sexually transmitted diseases as well as some of the more uncommon ones. Although each disease is discussed separately, it often is the case that a person has more than one sexually transmitted disease at the same time.

GONORRHEA

Gonorrhea ("Clap," "drip," "strain") is a sexually transmitted disease that has been afflicting humans for centuries. For example, references are made to this infirmity in ancient Chinese and Hebrew writings. Since the early 1950s, the frequency of gonorrhea in the United States has shown a steady increase (Fig. 16–1) and now can be considered an epidemic,[1] being the third most common communicable disease after the common cold and herpes genitalis. It is estimated that 2 to 3 million new cases will appear this year in the United States, and 100 million in the world. Most new cases are in the 15- to 29-year-old age group, although this disease can occur in people of any age. The incidence of gonorrhea is especially high in teenagers and male homosexuals.

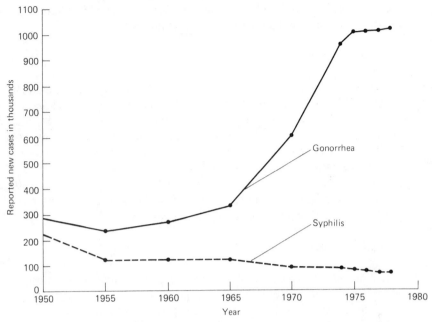

Figure 16–1. Incidence of reported cases of gonorrhea and syphilis in the United States. Note that the actual incidence is much higher since many cases are not reported. Therefore, this graph shows the trends through time but not the actual number of cases. Redrawn from *Statistical Abstract of the United States, 1980,* 101st ed. (Washington, D.C.: U.S. Bureau of the Census, 1981).

Cause of Epidemic

What has caused this gonorrhea epidemic in the United States? One factor is the increase in sexual activity and in the number of sexual partners of young people (Chapter 19). Furthermore, many sexually active teenagers are ignorant of precautions against, and symptoms of, gonorrhea. The condom, diaphragm, and spermicidal foams, creams, and jellies offer some protection against transmission of this disease, but the increasing preference for the combination pill and intrauterine devices over these methods in recent years (Chapter 13) has decreased this prevention. The combination pill may actually increase a female's susceptibility to gonorrhea infection because it increases moisture and pH of the vagina, conditions favorable to the gonorrhea organism. There is no good scientific evidence, however, that the incidence of gonorrhea infection is increased by use of the pill alone.[2] Another factor causing the increase in incidence of gonorrhea in the United States has been the Vietnam War. After every war, there is an increase in sexually transmitted diseases because people carry new strains of the organisms to other geographic regions. Finally, we see later in this chapter that new strains of gonorrhea are appearing that are resistant to traditional treatment with antibiotics.

Cause of Gonorrhea

Gonorrhea is caused by the bacterium *Neisseria gonorrhoeae,* named after the scientist Albert Neisser, who identified it in 1879. The term *gonorrhoeae* is derived from a Greek work that means "flow of seed." Neisseria gonorrhoeae is a gram-negative, diplococcus bacterium. In 1883, Christian Gram of Denmark invented a stain that differentiated gram-positive from gram-negative bacteria. Gram-positive bacteria stain more darkly because of differences in their cell wall structure. "Diplococcus" means that these bacteria occur in pairs, with their adjacent sides flattened. A common name for the bacterium *Neisseria gonorrhoeae* is *gonococcus,* and you often will see gonorrhea referred to as a "gonococcal infection."

Transmission

There are several strains of this *Neisseria gonorrhoeae,* some more damaging than others. All strains die quickly when exposed to dry air and sun, so it is virtually impossible to catch this disease by touching toilet seats, and only very rarely can it be transmitted by moist towels, clothes, or hands. The main way this bacterium is transmitted is during heterosexual or homosexual coitus. The bacteria also can be transmitted to the mouth or anus during oral or anal coitus. The gonorrhea bacteria thrive in the moist membranes of the urogenital tract, as well as in the mouth and oral cavity, anus, and eyes. Once the bacteria are introduced into one of these regions, the *incubation period* (time it takes before symptoms appear) is usually two to five days, but it can be as soon as one day or as long as eight days.

Symptoms in Females

About 75 percent of females that acquire gonorrhea are asymptomatic (show no symptoms).[3] There are, at present, about 800,000 females in the United States with undiagnosed gonorrhea. This presents a problem because they are carriers of the disease without knowing it, and the disease can reach an advanced stage before a female knows that she has the affliction. In the remaining 25 percent of women who exhibit symptoms, the first sign usually is the appearance of a clear or whitish fluid discharge from the vagina (infection or inflammation of the vagina, called *vaginitis*). This discharge soon changes to a yellowish or greenish color; that is, it becomes a *purulent* (pus-filled) discharge. The vaginal wall can become quite irritated at this time. Eventually, the infection can reach the cervix; infection and inflammation of this organ (cervicitis) contribute to the purulent discharge. A discharge also can come from the urethra (urethritis). Urination can become difficult and painful when urethritis is present. Also, the bacteria can reach the urinary bladder, causing infection (*cystitis*).

Complications in Females

If left untreated, the bacteria can infect the uterus (endometritis) and can reach the oviducts in two to ten weeks after the initial infection. Inflammation and infection of the oviducts (salpingitis) can lead to infertility (Chapter 15). If left untreated, the bacteria can spread to other pelvic and abdominal organs, resulting in a dangerous condition called pelvic inflammatory disease, which we discuss later in this chapter. The bacteria also can cause inflammation of the heart, brain, spinal cord membranes, eyes, skin, and joints in both sexes.[4] As mentioned, oral coitus with a person carrying the bacteria in his or her genital region can lead to infection of the oral cavity. Anal coitus can transmit the disease into the anus and cause inflammation of the rectum (*proctitis*).

Pregnancy and Gonorrhea. Pregnant women who have gonorrhea can pass the disease to their fetus. The bacteria can enter the fetal blood across the placenta. If this happens in the first trimester, there is an increased risk of spontaneous abortion. In addition, *Neisseria gonorrhoeae* in the birth canal can infect the eyes of the newborn, a condition called *gonococcal ophthalmia neonatorum*. Untreated, the newborn's eyes develop a purulent discharge within 21 days of birth, and the eyes can eventually be destroyed. Treatment of the newborn's eyes with silver nitrate or an antibiotic (Chapter 10) can prevent this from occurring.

Symptoms in Males

Most men (70 to 90 percent) develop recognizable symptoms of gonorrhea.[5] The first signs usually are a purulent discharge from the urethra and redness of the glans of the penis. Urination can become painful and difficult, and scar tissue can form in the urethra. Men also can have painful erections, pain in the groin region, and a low fever. If not treated, the infection can spread in about three weeks to the urinary bladder and prostate gland and can infect these organs. The epididymides also can become infected, and in some cases the testes themselves become infected and inflamed (orchiditis), sometimes leading to infertility.

Diagnosis

The symptoms of gonorrhea are not especially useful in diagnosing the disease. This is because they are similar to symptoms of some other kinds of sexually transmitted disease and because many females and some males are asymptomatic. Unfortunately, there is as yet no reliable blood test for gonorrhea, and direct microscopic observation only reveals the organisms in about half of the infected individuals. Therefore, swabs of the urethra, cervix, and/or rectum are made and are cultured in a special (Thayer-Martin) medium. The colonies of bacteria growing in this medium then are examined with a microscope for the presence of *Neisseria gonorrhoeae,* and several biochemical tests are used to confirm the presence of this organism. This culture test takes about 24 to 48 hours. Unfortunately, in about 15 to 20 percent of women who have the disease, the gonorrhea bacteria are not visible in a cervical culture.[6] Therefore, it is a good idea to have repeat cultures done.

Treatment

Treatment with an injection of penicillin has been the standard treatment for gonorrhea since the 1940s. Lately, however, some strains of *Neisseria gonorrhoeae* have appeared that produce an enzyme (*penicillinase*) that destroys penicillin. These penicillin-resistant strains, called "supergonorrhea," are becoming more common in the United States, and unfortunately produce fewer and milder early symptoms. The treatment of choice is still probenecid (oral) plus procaine penicillin G (injection), or probenecid plus ampicillin (both given orally).[7] The drug probenecid is necessary because it reduces excretion of the antibiotics by the kidneys. If a person is allergic to penicillin or has contracted a penicillin-resistant strain of gonococcus, the antibiotics spectinomycin (injected) or tetracycline (oral) can be given. A follow-up culture should be done one week after antibiotic treatment. Preliminary trials with a gonorrhea vaccine[8] and antibody treatment[9] have been encouraging.

SYPHILIS

Syphilis is a serious sexually transmitted disease caused by a bacterium, *Treponema pallidum*. About 50 million people in the world now have syphilis, and its incidence has been slowly increasing. In the United States, the incidence of reported cases of syphilis, after peaking in 1975, is decreasing slightly (Fig. 16-1). Currently, there are about 500,000 people with untreated syphilis in the United States. The actual incidence, however, is higher because many cases are not reported.

Theoretical Origins

Columbian Theory. There are at least two theories about the origin of syphilis.[10] The "Columbian theory" proposes that Christopher Columbus and his crew contracted the disease from natives on their first voyage to the West Indies, in 1493. They then introduced the disease to Europe. The first documented epidemic of syphilis occurred in Western Europe at the end of the fifteenth century. People in one country usually blamed foreigners for introduction of the disease. Thus, it was called the "Neopolitan disease" in France, the "French pox" by Italians, and the "French or Spanish disease" by the English. In 1520, after the epidemic was over, an Italian physician and philosopher, Hieronymous Fracastorius, wrote a poem in which the people of the earth were given a horrible disease by the sun god, Apollo, because a shepherd encouraged his people to worship the king instead of Apollo. The shepherd's name was Syphilis.

Evolutionary Theory. The "evolutionary theory" of the origin of syphilis proposes that this disease is related to other nonvenereal diseases such as yaws and nonvenereal syphilis. *Yaws,* a nonvenereal tropical disease of the skin, is the most primitive of all the diseases. It is caused by a bacterium closely related to *Treponema pallidum,* called *Treponema pertenue.* When people migrated to cooler, drier northern climates, their skin became drier, and *Treponema pertenue,* favoring moist regions, migrated to the axilla, mouth, nostrils, crotch, and anus and caused *nonvenereal syphilis.* This disease (also called *endemic syphilis*) is a childhood infection in arid regions of the world. It is transmitted by direct (nonsexual) body contact, in drinking water, or by eating utensils. The bacteria that cause this disease are not distinguishable from *Treponema pallidum* and probably evolved from *Treponema pertenue.* Later, *Treponema pallidum* began to favor the even more moist areas of the genitals and became the sexually transmitted affliction we now know as syphilis.

Syphilis as Distinct from Gonorrhea

For years, it was thought that syphilis and gonorrhea were caused by the same organism. An English physician, John Hunter (1728–1793), once attempted to prove that these diseases were caused by different organisms by injecting pus from a gonorrhea patient into himself. Unfortunately, and unbeknownst to him, the patient also had syphilis. Hunter contracted both diseases, perpetuating the misconception that they were caused by the same organism, and he died of his self-inflicted syphilis! In 1838, The Frenchman Phillipe Ricord determined that gonorrhea and syphilis were caused by different bacteria.

Transmission

Treponema pallidum is a corkscrew-shaped (spirochete) bacterium. It thrives in moist regions of the body and will survive and reproduce only where there is little oxygen present. It is killed by heat, drying, and sun. Therefore, one cannot catch syphilis from contacting toilet seats, bath towels, or bedding. It can, however, live in collected blood for up to 24 hours at 4°C, and thus, in rare cases, is transmitted during blood transfusion. Nine out of ten cases of syphilis transmission occur during sexual intercourse, although it also can be introduced into an open wound in the skin. Fortunately, only about one in ten people exposed to the bacteria develops syphilis.

Stages of the Disease

Primary Stage. The symptoms of untreated syphilis occur in four stages.[11] The *primary stage of syphilis* usually appears as a single sore called a *chancre* (pronounced shang'ker) at the place where the bacteria first entered the body. This is a round ulcerlike sore with a hard raised edge and a soft center. It looks like a crater, about 1/2 to 1 in. in diameter. This chancre, which for all its awful appearance is painless, appears in 10 to 90 days after entry of the bacteria. Because the chancre is painless and may be in a location not readily noticed, a person may not realize that he or she is infected. In males, the chancre usually occurs on the glans or corona of the penis, but it can occur anywhere on the penis or on the scrotum. In females, it usually appears on the vulva, but sometimes can appear on the cervix or vaginal wall. After oral coitus with an infected person, it can appear on the lips, tongue, or tonsils, and it can appear in the anus after anal coitus with an infected person. Inguinal lymph nodes enlarge a few days after the sore appears. The chancre heals in one to five weeks, and the primary stage is then over. Meanwhile, the bacteria are traveling in the blood or lymphatic system to other parts of the body, and eventually will cause the secondary stage of syphilis if the person is not treated.

Secondary Stage. The *secondary stage of syphilis* occurs in two weeks to six months after the primary stage. This stage is characterized by a rash that appears on the upper body, arms, and hands, which then spreads to other skin regions. In white-skinned people, the rash appears as cherry-colored blemishes or bumps that change to a coppery-brown color. In dark-skinned people, the blemishes are grayish blue. Larger bumps can develop and burst, especially in the inguinal region. The rash does not itch and is painless, but the syphilis bacteria are present in great numbers in these sores, and contact with the sores is very infectious to other people. Other symptoms of the secondary stage include hair loss, sore throat, headache, loss of appetite, nausea, constipation, pain in the joints and abdominal muscles, a low fever, and swollen lymph glands. The symptoms are minor and cause little inconvenience in about 60 percent of untreated individuals in the secondary stage and thus can be completely overlooked. The secondary stage goes away in two to six weeks, and the untreated individual then enters the latent stage of syphilis.

Latent Stage. During the *latent stage of syphilis,* which can last for years, a person exhibits few or no symptoms. After about a year of the latent stage, the individual can no longer transmit the bacteria to another person (except to a fetus, as we discuss below). About half of the people that enter the latent stage never leave it, even if not treated. The other half eventually enter the tertiary stage of syphilis if not previously treated with penicillin.

Tertiary Stage. Entrance into the *tertiary stage of syphilis* occurs because the bacteria have invaded tissues throughout the body. The tertiary stage is characterized by large tumorlike sores (*gummas*) that form on tissues of skin, muscle, the digestive tract, liver, lungs, eyes, nervous system, heart, or endocrine glands. Infection of the heart (cardiovascular syphilis) can cause severe damage to the heart and its valves. Invasion of the bacteria into the central nervous system causes "neurosyphilis," and the brain and spinal cord can be severely damaged. People with neurosyphilis can develop partial or total paralysis, blindness, or psychotic and unpredictable behavior. People in the tertiary stage are not infectious, but about 4000 die annually of tertiary syphilis in the United States.

Congenital Syphilis

As noted above, a person is not infectious in most of the latent stage nor in the tertiary stage of syphilis. This is true except in the case of an infected pregnant woman, who can pass on the bacteria to her fetus at any stage of syphilis. The placenta protects the fetus against invasion of the syphilis bacteria up to the sixth month of pregnancy, after which time the *Treponema pallidum* organism passes through the placental membranes into the fetal bloodstream. Then, the fetus can contract the disease from the mother. If this happens, about 30 percent of the fetuses miscarry, and 70 percent are born with *congenital syphilis.* The latter children are contagious in their first and second year, and go through all the stages of syphilis if left untreated. About 23 in 100 such cases develop tertiary syphilis in 10 to 20 years. Symptoms of tertiary congenital syphilis include damage to the eyes, deafness, flattening of the bridge of the nose ("saddle nose"), and central incisor teeth that are spread apart and notched ("Hutchinson's teeth"). Many of these individuals die from this affliction.

Diagnosis

Several of the symptoms of syphilis can be confused with those of other sexually transmitted diseases. Also, it has proven difficult to grow cultures of *T. pallidum* in the laboratory. Therefore, other tests are necessary to see if a person has contracted the disease. Diagnosis of syphilis can be accomplished in several ways. A blood test for the presence of antibodies to the disease can be done a week or so after the primary chancre appears. Several such tests are available, including the Venereal Disease Research Laboratory (VDRL) test, the Rapid Plasma Reagin (RPR) test, and the Syphla-Chek Test. All are equally sensitive, but Syphla Chek seems to be the best for primary syphilis.[12] The older "Wasserman Test" for syphilis has been displaced by these newer methods. False positive results occur in 1/3000 of these blood tests, and more importantly, false negatives occur about 25 percent of the time. Because of this error factor, an individual's tissues also should be checked for presence of live *Treponema pallidum,* based on their characteristic shape and movement, using a dark-field microscope to confirm the diagnosis.

Treatment

Once it has been determined that a person has syphilis, treatment with one of several antibiotics is effective.[13] Most commonly, benzathine penicillin G is given as a single injection each day for eight days if the person is in the primary stage, and for three to four weeks at higher dosage if the person is in a more advanced stage of the disease. Tetracycline or erythromycin can be used if a person is hypersensitive to penicillin. It must be emphasized that syphilis, like gonorrhea, is a curable disease. However, individuals with tertiary syphilis, though treated, still may have suffered permanent tissue damage.

HERPES GENITALIS

Herpes genitalis is a sexually transmitted disease that is reaching epidemic proportions in the United States. This year, about 5 million people in the world and more than 500,000 in the United States will catch this disease. Because herpes is, at present, incurable, it is estimated that there are 10–25 million sufferers in the United States today. Herpes genitalis has overtaken gonorrhea and now is the most common venereal disease.[14] It is most prevalent in teenagers and young adults, especially in poorer regions, but it can infect anyone. There may be an inherited resistance to herpes viruses since a gene for herpes virus sensitivity is present on chromosome 3 in humans.[15]

Cause

This disease is caused by *herpes simplex type 2 virus*. There are 25 herpes viruses, which cause diseases such as fever blisters and cold sores (*Herpes simplex type 1 virus*), chicken pox in children or shingles in adults (*varicella-zoster virus*), infectious mononucleosis (*Epstein-Barr virus*), and cytomegalic inclusion disease, which affects the fetus and newborn and results in enlargement of the liver and spleen (*cytomegalovirus*). The herpes simplex type 2 virus usually affects the body below the waist (e.g., the genitals, thighs, and buttocks), whereas type 1 usually invades areas above the waist. About 10 percent of herpes infections of the genital region, however, are caused by herpes simplex type 1, usually the consequence of oral coitus with an infected person. Similarly, type 2 occasionally is isolated from mouth sores. It should be emphasized that herpes genitalis can be transmitted by nonsexual contact with an infected person.

Symptoms

Once a person contracts the herpes genitalis virus, usually through genital contact, clusters of tiny blisters develop that change to painful round sores in four to seven days. Two or three days later, these take the form of multiple, small, round, itchy ulcers. Severe ulcers, however, only occur in about 10 percent of infected people; in the remaining 90 percent, the sores are minor and often go unnoticed.[16] In males, the sores occur mainly on the penis (shaft, foreskin, glans, urethral meatus), especially in the uncircumcised.[17] The primary symptoms usually are more painful in females, sores appearing on the labia, clitoral hood, cervix, vaginal introitus, urethral meatus, or perineum. Urination and coitus can be painful. More severe but less common symptoms in both sexes include fever and enlargement of the inguinal lymph nodes. In general, the symptoms are more severe in people who have never been exposed to any herpes virus. If a person touches an open sore and then his or her eyes, he or she can develop a virus infection that can lead to blindness. There is a disturbing positive correlation of females developing cervical cancer after previously having herpes genitalis, the incidence being two to four times higher than in those who have not had herpes genitalis.[18] The sores, if they develop, heal in one to six weeks. Individuals can infect other people when sores are present, but usually are not infectious after the sores and scabs disappear.

After the herpes sores have healed, the virus migrates up sensory nerves to clumps of nerve cells near the spinal cord. They lay dormant there for several days, weeks, or months. Then, they become active again, migrate back to the skin, and cause recurrence of the symptoms. Such recurrence can on different parts of the penis or scrotum in males, or the vulva, vagina, or cervix in females. The recurrent attack often is accompanied by enlargement of the lymph nodes in the groin, as well as fever and headaches. Then, the symptoms go away in one to four weeks. Recurrence of symptoms can occur frequently (e.g., once a month), or there may be several months between attacks. Some individuals can have up to more than ten attacks, and these often are associated with times of stress, or, in females, with menstruation. Eventually, antibodies are formed that alleviate or stop recurrences, and a few people may never have a second attack. Nevertheless, herpes genitalis can stay with many people for their entire lives.

Herpes simplex type 2 virus in the blood of a pregnant female can cross the placenta and damage the fetus. This, however, is rare. More often, the fetus is exposed to the virus during birth,[19] especially if the woman has open sores during delivery. About 25 percent of newborns exposed to herpes virus type 2 develop blindness or brain damage, and another 25 percent of exposed newborns develop skin lesions. For this reason, the fetuses of women with herpes infection often are delivered using cesarean section (Chapter 10).

Diagnosis

Besides noting the above-described symptoms, herpes genitalis can be diagnosed by taking a smear from the cervix, culturing it with tissue cells, and examining the cells for presence of the virus. Also, a blood test for antibodies is available, but it does not differentiate between antibodies to herpes simplex type 1 and type 2 viruses.

Treatment

There is no reliable cure for herpes genitalis, which explains why the actual number of people having the virus is higher than those having gonorrhea. There are, however, some procedures that have potential to reduce the severity and frequency of recurrence of the herpes symptoms. Proper hygiene has such an effect. Also, vitamin A seems to be helpful because it stimulates the immune system. Zinc sulphate administered in a special tampon prevented recurrence and spread of the disease in a preliminary trial.[20] Topical application of a new drug, 5-iodo-deoxyuridine, seems to relieve the symptoms.[21] Also a drug called 2-deoxy-D-glucose can combat herpes genitalis.[22] Another new drug, acyclivor (Zovirax) interferes with reproduction of the herpes virus. Acyclivor recently was approved by the FDA, and seems to improve herpes symptoms.[23] There is evidence, however, that new strains of the herpes virus resistant to acyclivor can appear.[24] Also, exposure of the herpes sores to a certain dye and fluorescent light can increase healing rate and decrease recurrence,[25] but there is a possibility that such exposure increases the risk of cervical cancer.[26] Wet compresses or hot sitz baths should not be used since they can spread the sores.[27] Herpes sufferers can get consultation and information on new treatments by contacting the Herpes Resource Center, Box 100, Palo Alto, California 94302; their hotline is 415-328-7710.

NONSPECIFIC URETHRITIS

Nonspecific urethritis is any sexually transmitted infection or inflammation of the urethra and genital organs that is not caused by *Neisseria gonorrhoeae*. This is why it sometimes is called *nongonococcal urethritis*. We now know that a leading cause of this affliction is the gram-negative, bacteriumlike microorganism *Chlamydia trachomatis*. This organism is a member of a group of very small bacteria that were once thought to be viruses. *Ureaplasma urealyticum* also has been implicated in nonspecific urethritis. This is a small bacterium without cell walls, and is in a group called T-strain mycoplasms. *Hemophilus vaginalis,* a bacterium that also is a cause of vaginitis, produces some cases of nonspecific urethritis. About half of the cases of nonspecific urethritis in males and females are caused by *Chlamydia*. An increase in sexual activity increases the transmission of this organism.[28] The frequency of occurrence of nonspecific urethritis is increasing and may be exceeding that of gonorrhea and herpes genitalis.

The symptoms of nonspecific urethritis are similar to those of gonorrhea but generally are milder. In males, a thin, clear, or slightly white discharge occurs from the urethra. In females, the cervix, vagina, urethra, and vulva become reddened and irritated. *Chlamydia* can travel to other reproductive organs in the female and can cause salpingitis and consequent infertility.[29] Also, *Chlamydia* can be passed to the eyes by touching infected regions, and this leads to eye infection. The T-strain mycoplasms and *Chlamydia* are treated with tetracycline, ampicillin, or another broad-spectrum antibiotic because they do not respond to penicillin.

CHANCROID

Chancroid (or "soft chancre") is a sexually transmitted disease caused by a very small, gram-negative bacterium, *Hemophilus ducreyi*.[30] It is rare in the United States, being most common in the tropical Far East. However, it has been appearing more frequently in the United States. Once the bacteria enter a cut in the skin or invade the mucous membranes of the genital region, a small *papule* (small elevation on the skin) appears in 12 to 24 hours. This papule, commonly found on the penis or vulva, then bursts, and an ulcer forms in one or two days. This ulcer looks like the chancre of primary syphilis except that it has soft edges. It also is painful, whereas the hard chancre of primary syphilis is painless. Multiple soft chancres can develop as the organism spreads, even to the thighs. Another common symptom is swollen lymph glands in the groin region. Many females do not show symptoms but are carriers of the bacteria. Anal coitus can lead to soft chancre in the anus. Chancroid can be diagnosed by examining a smear or culture under a microscope and looking for *Hemophilus ducreyi*. Also, presence of *H. ducreyi* can be detected by a skin test (Ducrey's Skin Test), which becomes positive one to two weeks after a person is infected. Standard treatment is with a sulfa drug such as sulfonamide.

LYMPHOGRANULOMA VENEREUM

Lymphogranuloma venereum is a tropical sexually transmitted disease that is rare in the United States, but its incidence is on the rise, partly because of being carried to this country by Vietnam veterans.[31] It is caused by *Chlamydia trachomatis,* the same organism (but a different strain of it) that is one cause of nonspecific urethritis. Although most people catch this disease after close sexual contact, it also can be transmitted on clothing. A newborn's eyes also can be infected with this organism if it is present in the vulva of a delivering mother. After an incubation period of 7 to 21 days, a small, painless blister forms. In males, this blister occurs on the penis or scrotum, and in females on the vagina, vulva, cervix, or urethra. The anus, tongue, or lips can also be infected. Other symptoms can include swelling of inguinal lymph nodes, fever, chills, backache, abdominal and joint pain, and loss of appetite. In a few cases, the spreading organisms can cause a serious complication called *Reiter's syndrome,* which is characterized by rheumatism, arthritis, conjunctivitis, and heart valve defects.[32] Diagnosis of this disease is by a skin test (Frei Test) or by a test for antibodies. Both tests, however, only become positive three to four weeks after the onset of the disease. Treatment is with tetracycline or the sulfonamides.

GRANULOMA INGUINALE

Granuloma inguinale is caused by a gram-negative bacterium by the name of *Calymmatobacterium granulomatis*[33] (formerly called *Donovania granulomatis).* This is an extremely rare sexually transmitted disease, about 100 cases occurring annually in the United States. It is more common in South India and New Guinea. From one to several weeks after the bacteria enter the body, a tiny blister occurs on the penis or vulva, which then develops into an open, bleeding sore. The skin around the sore becomes swollen and red. One can diagnose this disease by looking for the bacteria in a small piece of skin taken from the edge of the sore, using a microscope. Treatment is with tetracycline.

VAGINITIS

Vaginitis can be a symptom of any of the aforementioned sexually transmitted diseases. Vaginitis, however, also can be caused by other organisms, such as *Candida albicans, Trichomonas vaginalis,* and *Hemophilus vaginalis,* all of which can occur normally in the vagina (Chapter 3). Common physical symptoms of vaginitis include vaginal itching, burning, tenderness, and painful or difficult coitus. A vaginal discharge usually accompanies vaginitis. You should realize, however, that a non-purulent vaginal discharge occurs normally. The normal cervical mucus mixes with vaginal cells and bacteria to produce a whitish discharge, which changes to pale yellow when exposed to air. The discharges associated with vaginitis have a color and odor different from this normal discharge.

Monilia Vaginitis (Yeast Infection)

Monilia vaginitis is a yeast infection of the vagina, also known as "candida," "moniliasis," or "vaginal thrush." It accounts for about one-half of the cases of vaginitis. It is caused by the gram-positive fungus *Candida albicans,* which normally is present in the vagina of 40 percent of women (Chapter 3). When *Candida albicans* multiplies rapidly, the signs of monilia vaginitis appear. *Candida albicans* also can be present in males and can be transmitted during coitus.

This fungus (a yeast) thrives in a moist environment with a high pH, so that anything that makes the vagina more basic could lead to spontaneous appearance of monilia vaginitis. Bacteria called *Döderlein's bacilli (Lactobacillus acidophilus)* normally are found in the vagina (Chapter 3). These bacteria metabolize glycogen (a long-chain sugar) to lactic acid, and thus help maintain an acidic vaginal condition. If antibiotics are given to a female, these bacteria may die, the vaginal pH increases, and a spontaneous monilia infection may occur. Such infections also can occur in the premenstrual period, during pregnancy, and in diabetics. About one-third of females with venereal disease of another type also have monilia.[34] Tension and anxiety also can lead to yeast infection, and frequent coitus can aggravate the symptoms. There appears to be no increase in incidence of yeast infection with the use of combination pills.[35]

In a female with monilia vaginitis, the vagina is itchy and raw, and the vagina, vulva, and anal regions can become reddened and swollen. White patches ("thrush") appear on the vulva, vaginal wall, and cervix, and a thick, white, cottage-cheeselike vaginal discharge with a yeasty odor can occur. Coitus can be painful under these circumstances. Monilia can be transmitted by oral-genital contact and sometimes can occur in the mouth, accompanied by white patches, swelling, and pain. A fetus can contract *Candida* as it passes through the birth canal. This leads to presence of the organisms in the newborn's digestive tract, causing thrush and digestive disorders. Males contracting *Candida albicans* are usually asymptomatic.

Presence of *Candida albicans* can be detected by taking a swab of the vaginal discharge and examining it under a microscope. Also, this yeast can be grown on a culture medium, and the colonies examined. This affliction can be treated with fungicides such as Nystatin (Mycostatin) in vaginal tablets, candicidin (Vanolud) in vaginal tablets or ointment, or with micronazole nitrate (Monistat cream). Periodic acidic douches can help, as can painting the area with a 1 percent gentian violet solution or using a tampon containing this dye. The resultant bright blue color wears off in a few days. Application of a yogurt douche containing the bacterium *Lactobacillus acidophilus* also may be helpful. [36]

Hemophilus Vaginitis

Hemophilus vaginalis is a gram-negative bacterium that can occur spontaneously in the vagina (Chapter 3) or can be transmitted sexually; it also can occur in males. This bacterium causes *Hemophilus vaginitis,* the third leading kind of infectious vaginitis. In females, presence of large numbers of *Hemophilus vaginalis* produces a scanty vaginal discharge accompanied by mild burning and itching. Then, a profuse discharge occurs, which is dirty white or gray and foul-smelling. Urination and coitus are painful at this stage. About 80 percent of males are asymptomatic. *Hemophilus vaginitis* can be treated with sulfa creams or the antibiotics ampicillin or tetracycline.

Trichomonas Vaginitis

A flagellate protozoan, *Trichomonas vaginalis,* is the cause of *Trichomonas vaginitis,* or *trichomoniasis.* This is the second leading kind of vaginitis, after monilia. The single-celled organism is pear-shaped and has fine flagella that it uses to swim with (Fig. 16–2). These organisms normally are present in the vagina (Chapter 3), and about one in five women who have this protozoan in their vagina do not have trichomoniasis. It is only when the organisms multiply in large numbers that vaginitis occurs. There is a profuse, watery, frothy, odorous (stale or musty) discharge, rang-

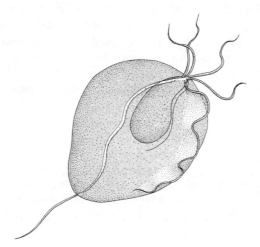

Figure 16–2. *Trichomonas vaginalis,* a flagellated protozoan that causes vaginitis. Reproduced from J. S. Hyde, *Understanding Human Sexuality* (New York: McGraw-Hill Book Company, 1979). © 1979; used with permission.

ing from grayish white to yellowish brown, that develops in 4 to 28 days. The vulva can swell, become red, and itch, and the swelling and redness can spread to the cervix, thighs, urethra, and urinary bladder. Most men who have trichomoniasis have no symptoms, but it is responsible for 5 to 15 percent of nongonococcal urethritis and cystitis in men. In this case, a fluid discharge will be exuded from the urethra. This organism can be sexually transmitted. It is also one of the few sexually transmitted afflictions that also can be contracted from bath water, towels, swimming pools, or hot tubs. Trichomoniasis, however, does not survive in the mouth or rectum. Therefore, it is not transmitted by homosexual men.

The presence of *Trichomonas vaginalis* can be detected by examining some of the discharge under a microscope. It can be treated with oral administration of metronidazole (Flagyl); recent studies have not substantiated the previous claim that this drug causes cancer.[37] Side effects of flagyl, however, can include an upset stomach, loss of appetite, diarrhea, hives, dizziness, and a bad taste in one's mouth.

Nonspecific Vaginitis

Nonspecific vaginitis means just what it implies; i.e., vaginitis for which a specific cause is not identified. Most often, however, the culprit is *Escherichia coli,* a bacterium commonly present in the digestive tract. Vaginitis caused by *Escherichia coli* is characterized by a gray, foul-smelling vaginal discharge. Poor hygiene often is the cause. Vaginitis also can be caused by estrogen withdrawal (as after menopause) or allergic responses to chemicals, and can even be psychological in origin.

Precautions against Vaginitis

A woman can take several precautions against vaginitis. These include proper hygiene (washing and drying the vulva thoroughly). After defecation, a female should never wipe the anus from back to front since this could facilitate passage of *Escherichia coli* to the vulva. Also a couple should never switch from anal to vaginal coitus without first washing the genitals, and a condom should be used for anal coitus. Use of feminine hygiene deodorant sprays should be avoided since they can irritate the vagina.[38] Wearing tight nylon pants, such as panty hose, can decrease air circulation and may favor development of vaginitis. A low carbohydrate diet reduces sugar levels, which decreases the chance of vaginitis. Finally, a woman can douche periodically with a mildly acidic solution.

GENITAL WARTS

Genital warts (Condyloma acuminata) can occur in the genital region because of the presence of a papilloma virus. This virus often is transmitted sexually, which is why the condition may be called "venereal warts." These warts also can appear spontaneously. (The kind of warts that occur on the skin in other body regions is caused by a different virus.) After the genital wart virus is contracted, the warts appear in three weeks to eight months. They are moist, soft, cauliflowerlike bumps occurring singly or in groups. They can be pink, red, or dark gray. Females often get them on the labia, vulva, or perineum, whereas in males they appear on the prepuce, glans, or coronal ridge of the penis. Genital warts can be treated with the medications podophyllin or trichloroacetic acid, dry ice, or liquid nitrogen; they usually dry up and fall off in a few days after being treated. If this does not work, the warts can be removed by surgery or heat cauterization.

MOLLUSCUM CONTAGIOSUM

Molluscum contagiosum is a disease caused by a virus related to the chicken pox virus. This virus can be sexually transmitted, but it also can be transmitted by skin contact. The condition is most commonly seen in children. The symptoms are small, pink, wartlike growths on the face, arms, back, umbilical region, or buttocks. This disease is relatively harmless, and the growths can be removed by freezing or burning.

VIRAL HEPATITIS B

Infection of the liver with hepatitis virus B *(viral hepatitis, type B)* often is transmitted by the use of an infected hypodermic needle. It can also be transmitted during sexual contact, or during other close contact with infected people.[39] The virus is present in saliva and semen, and can be transmitted during kissing or anal or oral intercourse. About 50 to 60 percent of these cases not attributed to injections using infected needles occur in homosexual men, although hepatitis type B also can be transmitted during heterosexual coitus.[40] There are about 80,000 to 100,000 new cases of hepatitis B in the United States each year, and 1 to 2 percent of these are fatal. Symptoms include an inflammed liver (*hepatitis*), fever, weakness, headache, and muscle pain. Preliminary trials with a vaccine have met with 92 percent success.[41]

ACQUIRED IMMUNE DEFICIENCY SYNDROME

Acquired immune deficiency syndrome (AIDS) is a new, sexually transmitted disease, which first appeared in 1981. The infectious organism (an unidentified virus) is transmitted by intimate bodily contact or by blood transfusion. Infected individuals lose resistance to other diseases because their immune system is not working. A particular kind of skin cancer also tends to appear. Thus far, the disease has appeared in male homosexuals, Haitians and their children, a small number of black Africans, drug addicts and their children, and female sex partners of bisexual men.

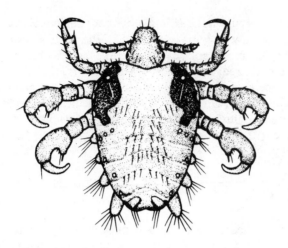

Figure 16–3. *Phthirus pubis,* a crab louse that causes pediculosus pubis (crabs). Reproduced from J. S. Hyde, *Understanding Human Sexuality* (New York: McGraw-Hill Book Company, 1979). © 1979; used with permission.

The incidence of this disease is increasing and it is dangerous; as of mid-March, 1983, 434 people have died of AIDS.[42]

PEDICULOSIS PUBIS

Pediculosis pubis, or "crabs," is caused by a tiny, parasitic, blood-sucking crab louse by the name of *Phthirus pubis* (Fig. 16–3).[43] This organism can be transmitted by direct body contact and also by contact with hair, clothing, or bedding. The organism can be seen at the base of hairs or as black spots visible on underwear. It lives in pubic, axillary, eyebrow, eyelash, and facial hair, but never in scalp hair. It needs the environment of human hair since it dies in 24 hours if removed. Female parasites lay tiny white eggs at the base of hairs, and the eggs hatch into larvae in seven to nine days. Adults or larvae cause itching, and scratching can lead to secondary infection of the skin or hair follicles. In World War II, parasitized people were shaved, and a 10 percent DDT powder was used for 24 hours. Now, an insecticide called gamma benzene hexachloride (Kwell) is applied as a cream, lotion, or shampoo. If one member of a household has crabs, all members should be treated. All underclothing and linens should be washed in hot water with bleach.

SCABIES

Scabies is caused by a mite named *Sarcoptes scabiei* and is transmitted by close contact, sexual or otherwise.[44] The female mites burrow under the skin and then lay their eggs. This leads to itchy rashes on the abdomen, flanks, thighs, genitals, buttocks, or forearms—anywhere except the face. All members of a household must be treated if one member has it. Treatment for crabs also kills scabies mites.

SOME GENERAL ASPECTS
OF VENEREAL DISEASE INFECTIONS

Inflammation and/or infection of the reproductive and urinary tracts is a common component of many sexually transmitted diseases. We have already discussed vaginitis; now we take a closer look at some others.

Vulvitis

Infection or inflammation of the vulva *(vulvitis)* can be caused by several sexually transmitted diseases. Infection with *Trichomonas* or *Candida* is a common cause of this malady, as can be the presence of parasites such as crabs or scabies. Infectious vulvitis can be accompanied by vaginitis or urethritis. Vulvitis also can be noninfectious, the vulval membranes becoming itchy and inflamed. A common cause of noninfectious vulvitis is exposure to chemicals in spermicides, douches, soap, or toilet paper.

Leukorrhea

Leukorrhea is a white vaginal or uterine discharge containing many white blood cells. It is a symptom of many diseases, such as inflammation or tumors of the reproductive tract, excessive douching, estrogen stimulation, use of tampons, or a diaphragm worn too long. Also, leukorrhea is common during infections with *Candida* or *Trichomonas*.

Urethritis and Cystitis

Urethritis is a common component of sexually transmitted diseases. The disease organisms can pass up the urethra to the urinary bladder, causing its infection or inflammation (*cystitis*). Bladder infections also can be caused by *Escherichia coli* or other bacteria normally resident in feces. This can be the result of poor hygiene since *Escherichia coli* commonly are present in fecal material. Because of a "milking" action of the penis on the urethra during coitus, frequent sexual activity can facilitate passage of *Escherichia coli* into the urethra and eventually into the urinary bladder. Manual or oral sexual stimulation, as well as use of objects such as vibrators or dildos, can also increase the risk of urethritis and cystitis in females.

Symptoms of cystitis include a desire to urinate frequently, a burning pain when urinating, hazy urine often tinged with blood, and backache. Cystitis can be treated with an antibiotic or a sulfa drug. A medication for pain often is given with the antibiotic, and this medication contains a dye that turns the urine bright orange-red. Urination soon after coitus seems to lesson the chance of developing urethritis and cystitis.

Cervicitis

Many sexually transmitted diseases can infect or inflame the cervix, a condition called *cervicitis*. Chronic low-level cervicitis is common in women after childbirth, accompanied by a mild purulent cervical discharge, backache, and pelvic pain. Women taking combination pills with high estrogen also tend to develop this affliction. Cervicitis usually can be treated with antibiotics, but it often disappears spontaneously. Of concern is the possibility that cervicitis may predispose a female to developing cervical cancer.

Pelvic Inflammatory Disease

Several sexually transmitted diseases can invade the uterus and eventually the oviducts. Inflammation and infection of the oviducts (salpingitis) are accompanied by low abdominal pain, vomiting, fever, and disturbed menstruation. Salpingitis can lead to the formation of scar tissue, which blocks the tubes and can lead to ectopic pregnancy and infertility (Chapters 9 and 15).

The infection then can spread to many pelvic organs, a condition called pelvic inflammatory disease. This condition is very serious and can cause death. Pelvic inflammatory disease can be caused by venereal disease organisms such as *Neisseria gonorrhoeae* and *Chlymadia trachomatis,* but it also can be caused by nonvenereal infections due to bacteria such as *Streptococcus, Staphylococcus, Escherichia,* and *Mycobacterium.* Gonorrhea, however, is responsible for 65 to 75 percent of cases of pelvic inflammatory disease. The incidence of pelvic inflammatory disease increases after a nonsterile abortion, or after coitus within a few days after an induced abortion, or during usage of an IUD (Chapter 13). Pelvic inflammatory disease can be treated with either tetracycline or penicillin plus ampicillin. This can kill the causative bacteria but does not eliminate scar tissue already formed.

Prostatitis

The prostate gland can become infected and inflamed, a condition called prostatitis (Chapter 4). About 80 percent of the time, bacterial prostatitis is caused by *E. coli,* but it can also be due to *Neisseria gonorrhoeae, Trichomonas vaginalis,* and other microorganisms. The prostate glands of many young men are chronically infected, often with no obvious symptoms. Other men develop symptoms such as urethral discharge, fever, chills, a tender groin, an increase in the frequency and desire to urinate, lower back pain, and perianal aching. Bacterial prostatitis is the most common cause of recurring urinary tract infection in males.[45] Men with bacterial prostatitis can suffer spasmodic, painful contractions of the prostate during orgasm.[46] Antibiotic treatment for infectious prostatitis must be specific for the particular bacteria present.

PSYCHOLOGICAL ASPECTS
OF SEXUALLY TRANSMITTED DISEASE

As we have seen, the effects of a sexually transmitted disease on a person's physical health will vary with the type of disease organism involved, ranging from little or no discomfort to lifetime impairment or death. Similarly, the adverse effects of having a sexually transmitted disease in regard to a person's emotional health will vary to some extent with the disease type. A person having syphilis may feel shame or guilt about contracting such a "filthy" problem, but he or she may not feel that way about having, for example, nonspecific urethritis. A person with a sexually transmitted disease may feel that he or she is a "bad person," or that he or she is being punished for "improper" behavior. Often, this relates to the individual's basic feeling about sexual activity.

A person's life frequently will be disrupted after contracting a sexually transmitted disease. Usually, the individual must refrain from sexual activity for a while to avoid transmitting the disease to other people. Even if protective devices such as barrier contraceptives are used, the presence of the disease may mean a change in sexual habits and interactions. Also the infection or pain associated with the infection can cause distraction during coitus or even avoidance of sexual activity. The person may be faced with the difficult task of telling a potential sex partner or spouse about the problem before having intimate contact. Even more difficult, a physician or other health professional must question a person with a venereal disease and determine recent sexual contact(s) in order for the other person(s) to be treated.

In other words, contraction of a sexually transmitted disease often is a major disruption to a person's personal life. The pain of this process, however, can be alleviated by: (1) having a clear understanding about the disease, (2) admitting one's responsibility in the matter, (3) using an open approach to medical and psychological help, and (4) dealing with guilt associated with contracting the disease. Have you ever had a sexually transmitted disease? If so, how did you feel about it? How did it affect your personal life? What did you do about it? If you now have such a disease, how is it changing your life, and what are you doing about it? If you never have had such a disease, what are you going to do to continue your good fortune? How would you react, and how would your life change, if you did contract such an affliction, and what would you do about it? How would you react if a close one told you that he or she had it? People having questions or concerns about venereal disease can call the national VD Hot-Line any day of the week (8–8, Mon.–Fri.; 10–6, Sat.–Sun.) at 800-227-8922.

PREVENTING SEXUALLY TRANSMITTED DISEASE

One sure way to avoid contracting a sexually transmitted disease is to abstain from sexual contact, an approach not acceptable to many people. So, what else can a person do to lessen the odds that he or she will suffer from one of these diseases? One important way is to limit your sexual activity to people whom you know well and can trust. This is why married couples, often but not always, are free of this problem. If you find it difficult to limit your sexual activity in this way, it is helpful to discuss the topic and even to inspect a sex partner's genital region before coitus; this even can be part of foreplay if approached in a tactful, caring manner. Good hygiene, including washing the genital region before coitus, can help, as can urination before or after coitus. As mentioned, some contraceptive devices, such as the condom, diaphragm, and spermicides, can help prevent transmission of the disease organisms.[47] None of these measures will guarantee freedom from these diseases, but they will lessen the incidence of them. All of these preventive measures require a thoughtful approach to one's sex life, and the passion of the moment often wins the battle with one's rationality!

One preventive measure that is not recommended is the use of antibiotics as a *prophylactic measure* (a measure that may prevent disease) because: (1) specific antibiotics often are required for each kind of disease organism, (2) organisms can develop an insensitivity to an antibiotic if exposed to it often enough, (3) the antibiotic may mask an infection, (4) the antibiotic may cause an infection by eliminating bacteria and allowing minor normal nonbacterial microorganisms like *Candida* and *Trichomonas* to become infectious, and (5) prolonged antibiotic use can be toxic.

Sexually transmitted diseases can have serious physical and psychological effects. Everyone should be concerned with decreasing the incidence of these diseases in the general population as well as preventing any spread to himself/herself or to his/her sexual partner(s). If anyone has any questions or thinks that he or she has contracted such a disease, he or she should contact a physician or one of the many venereal disease clinics that exist in each state.

CHAPTER SUMMARY

Sexually transmitted diseases are those that are transmitted during coitus or other genital contact. The incidence of several of these diseases is on the rise in the United States, some of which have reached epidemic proportions and present a major public health hazard. Sexually transmitted diseases, with the organisms causing them in parentheses, are: (1) gonorrhea *(Neisseria gonorrhoeae)*, (2) syphilis, *(Treponema pallidum)*, (3) herpes genitalis *(herpes simplex* virus, type 2), (4) nonspecific (nongonococcal) urethritis *(Chlamydia trachomatis, Ureaplasma urealyticum, Hemophilus vaginalis)*, (5) chancroid *(Hemophilus ducreyi)*, (6) lymphogranuloma venereum *(Chlamydia trachomatis)*, (7) granuloma inguinale *(Calymmatobacterium granulomatis)*, (8) monilia vaginitis *(Candida albicans)*, (9) hemophilus vaginitis *(Hemophilus vaginalis)*, (10) Trichomoniasis vaginitis *(Trichomonas vaginalis)*, (11) genital warts (a papilloma virus), (12) molluscum contagiosum (a pox virus), (13) viral hepatitis, type B (hepatitis virus B), (14) acquired immune deficiency syndrome (a virus), (15) pediculosis pubis *(Phthirus pubis)*, and (16) scabies *(Sarcoptes scabiei)*.

Symptoms of some of these afflictions in females include vulvitis, vaginitis, urethritis, cystitis, cervicitis, endometritis, and salpingitis. Infection can be widespread, resulting in pelvic inflammatory disease. In males, common symptoms include urethritis, prostatitis, and orchiditis. Some diseases, such as syphilis and herpes genitalis, are characterized by appearance of sores in the genital region. Advanced stages of some afflictions such as syphilis are dangerous and can kill.

Many of these diseases that involve bacterial infections can be treated with antibiotics. Parasitic infestations are treated with insecticides. For other afflictions, such as herpes genitalis, no cure has been developed.

Contracting a sexually transmitted disease can disrupt one's emotional health and personal relationships. An understanding of the disease, as well as seeking treatment and dealing with guilt, is advisable. Precautions against sexually transmitted disease include proper hygiene, limiting one's sexual activity, and use of barrier contraceptives.

NOTES

1. W. H. Smartt and A. G. Lighter, "The Gonorrhea Epidemic and Its Control," *Medical Aspects of Human Sexuality,* 5 (1971), 96–115.

2. I. Roberts and P. Kane, "VD and the Pill," *Inform,* no. 1 (London: Family Planning Association, 1973).

3. N. J. Fiumara, "The Diagnosis and Treatment of Gonorrhea," *Medical Clinics of North America,* 56 (1972), 1105–13.

4. S. J. Kraus, "Complications of Gonococcal Infection," *Medical Clinics of North America,* 56 (1972), 1115–25.

5. H. H. Handsfield et al., "Asymptomatic Gonorrhea in Men—Diagnosis, Natural Course, Prevalence, and Significance," *New England Journal of Medicine,* 290 (1975), 117–23.

6. Z. A. McGee and M. A. Melly, "Gonorrhea in 1979: Update on Diagnosis and Treatment," *Consultant,* 19, no. 4 (1979), 36–43.

7. Anonymous, "Gonorrhea: Treatment Recommendations from the CDC," *Modern Medicine* (May/June 1979), pp. 76–78.

8. L. Greenberg, "Field Trials of a Gonococcal Vaccine," *Journal of Reproductive Medicine,* 14 (1975), 34.

9. T. M. Buchanan et al., "Antibody Response to Gonococcal Pili in Patients with Gonorrhoea," *Journal of Clinical Investigation,* 72 (1972), 17A.

10. C. J. Hackett, "On the Origin of the Human Treponematoses," *Bulletin of the World Health Organization,* 29, (1963), 7–41.

11. P. F. Sparling, "Diagnosis and Treatment of Syphilis," *New England Journal of Medicine,* 284 (1971), 642–53.

12. J. D. Dyckman et al., "Clinical Evaluation of a New Screening Test for Syphilis," *American Journal of Clinical Pathology,* 70 (1978), 918–21.

13. Sparling (1971), op. cit.

14. F. Rapp, "Herpes Viruses, Venereal Disease, and Cancer," *American Scientist,* 66 (1978), 670–74.

15. V. A. McKusick and F. H. Ruddle, "The Status of the Gene Map of the Human Chromosomes," *Science,* 196 (1977), 390–403.

16. S. S. C. Yen et al., "Herpes-Simplex Infection in the Female Genital Tract," *Obstetrics and Gynecology,* 25 (1965), 479.

17. J. D. Parker and J. E. Banatvala, "Herpes Genitalis: Clinical and Virological Studies," *British Journal of Venereal Disease,* 43 (1967), 212.

18. Rapp (1978), op. cit.

19. J. B. Hanshaw, *"Herpesvirus hominis* Infections in the Fetus and Newborn," *American Journal of Diseases of Children,* 126 (1973), 546–55.

20. R. G. Marks, "New Hope in Herpes Genitalis," *Current Prescribing,* 5, no. 3 (1979), 27–34.

21. Anonymous, "A Novel Approach to Herpes Treatment," *Science News,* 115, no. 23 (1979), 375.

22. H. A. Blough and R. L. Guintoli, "Successful Treatment of Human Genital Herpes Infections with 2-Deoxy-D-Glucose," *Journal of the American Medical Association,* 241 (1979), 2525–26.

23. J. A. Trichel, "Drug against Surface Herpes Approved," *Science News,* 121, no. 15 (1982), 247; and A. E. Nilsen et al., "Efficacy of Oral Acyclivor in the Treatment of Genital Herpes," *Lancet,* ii (1982), 571–3.

24. W. H. Burns et al., "Isolation and Characterization of Resistant Herpes Simplex Virus after Acyclivor Therapy," *Lancet,* i (1982), 421–23.

25. T. D. Felber, "Photoinactivation May Find Use against Herpes Virus," *Journal of the American Medical Association,* 217 (1971), 270; and E. G. Friedrich, "Relief for Herpes Vulvitis," *Obstetrics and Gynecology,* 41 (1973), 74.

26. R. S. Berger and C. M. Papa, "Photodye Herpes Therapy—Cassandra Confirmed?" *Journal of the American Medical Association,* 217 (1977), 133–34.

27. T. Chang, "Local Dissemination of Herpes Simplex Following Soaking or Sitz Bathing," *American Journal of Obstetrics and Gynecology,* 131 (1978), 342–43.

28. W. M. McCormack et al., "Sexual Activity and Vaginal Colonization with Genital Mycoplasmas," *Journal of the American Medical Association,* 221 (1972), 1375–77.

29. P-A Mardh et al., *"Chlamydia trachomatis* Infection in Patients with Acute Salpingitis," *New England Journal of Medicine,* 296 (1977), 1377–79.

30. A. W. Hoke, "Chancroid, LGV, and GI—Part of the VD 'Differential,'" *Consultant* (June 1979), pp. 128–43.

31. Ibid.

32. J. A. H. Hancock, "Surface Manifestations of Reiter's Disease in a Male," *British Journal of Venereal Disease,* 36 (1960), 36.

33. Hoke (1979), op. cit.

34. R. N. Thin et al., "How Often Is Genital Yeast Infection Sexually Transmitted?" *British Medical Journal,* 2 (1977), 93–94.

35. W. N. Spellacy et al., "Vaginal Yeast Growth and Contraceptive Practices," *Obstetrics and Gynecology,* 38 (1971), 343.

36. T. E. Will, *"Lactobacillus* Overgrowth for Treatment of Moniliary Vulvovaginitis,"* *Lancet,* ii (1979), 452.

37. C. Beard et al., "Lack of Evidence for Cancer Due to Use of Metronidazole," *New England Journal of Medicine,* 301 (1979), 519–22.

38. Anonymous, "Should Genital Deodorants Be Used?" *Consumer Reports,* 37, no. 1 (1972), 39–41.

39. J. Heathcote and S. Sherlock, "Spread of Acute Type B Hepatitis in London," *Lancet,* i (1973), 1468; and K. S. Lim et al., "Role of Sexual and Non-Sexual Practices in the Transmission of Hepatitis B," *British Journal of Venereal Diseases,* 53 (1977), 190–92.

40. Ibid.

41. T. H. Maugh, II, "Hepatitis B Vaccine Passes First Major Test," *Science,* 210 (1980), 760–62.

42. P. Taulbee, "AIDS Update: Search for 'Agent x'," *Science News,* 123, no. 16 (1983), 245.

43. B. Russel, "Parasitic Infestations of the Skin," *Practitioner,* 192 (1964), 621.

44. Ibid.

45. T. A. Stamey, "Chronic Bacterial Prostatitis," *Hospital Practice,* 6, no. 4 (1971), 49–55.

46. J. E. Davis and D. T. Mininberg, "Prostatitis and Sexual Function," *Medical Aspects of Human Sexuality,* 10 (1976), 32–40.

47. D. Barlow, "The Condom and Gonorrhea," *Lancet,* ii (1977), 811.

FURTHER READING

COREY, L., "The Diagnosis and Treatment of Genital Herpes," *Journal of the American Medical Association,* 248 (1982), 1041–49.

HART, G., "Sexually Transmitted Diseases," *Carolina Biology Reader.* Burlington, N. C.: Carolina Biological Supply Co., 1976.

JAWETZ, E., et al., *Review of Medical Microbiology,* 13th ed. Los Altos, Calif.: Lange Medical Publications, 1978.

MORTON, B. M., *VD: A Guide for Nurses and Counselors.* Boston: Little, Brown & Company, 1976.

PORTER, J. F., and P. KANE, "Sexually Transmitted Diseases," *Research in Reproduction,* vol. 7, no. 4 (1975), map.

ROSEBURY, T., *Microbes and Morals.* New York: The Viking Press, 1971.

17   *The Human Sexual Response*

LEARNING GOALS

Having read this chapter, you should be able to:

1. Discuss individual and cultural variation in response to erotic stimuli.
2. List the physiological events during the four stages of the male sexual response cycle.
3. Explain the physiology of the erection reflex.
4. List the physiological events during the four stages of the female sexual response cycle.
5. Compare and contrast the female and male sexual response cycles.
6. Describe the more common coital positions, and list their advantages and disadvantages.
7. List evidence that hormones influence sex drive in women and men.
8. List evidence that pheromones play a role in human reproduction.
9. List the kinds of sexual dysfunction that can occur in women and men, and their physical or psychological bases.
10. Give evidence that the following kinds of therapeutic and nontherapeutic drugs affect human sex drive or performance:
    (a) Antihypertensive drugs
    (b) Autonomic nervous system drugs
    (c) Minor tranquilizers
    (d) Cantharides
    (e) Tetrahydrocannabinol
    (f) Lysergic acid diethylamıde

    (g) Amyl nitrite
    (h) Amphetamines
    (i) Cocaine
    (j) Yohimbine
    (k) Alcohol
    (l) Narcotics
    (m) Barbiturates

INTRODUCTION

To most of us, sexual fantasies and behavioral interactions are pleasurable and important parts of our personal lives. Many thoughts and stimuli can be sexually arousing to us, and this is especially true during the act of love-making. When a female or male is sexually aroused, and adequate stimuli remain present, the human sexual response cycle occurs. In this chapter, we discuss this response cycle and see how malfunction of the cycle can occur. We also discuss the act of coitus (and variations in this act) as a common component of this cycle. Finally, we see how hormones, pheromones, and therapeutic and nontherapeutic drugs can influence our sex drive and the sexual response cycle.

SEXUAL AROUSAL

One is sexually aroused when environmental factors or thoughts initiate a sexual response. *Erotic stimuli* are those factors in the environment that are sexually arousing to an individual. A detailed survey of the kinds of erotic stimuli in our species would be impossible because what is erotic differs greatly among individuals and is highly subject to past experience. What excites one person may not be sexually arousing to another.

Cultural Influence

There is certainly a cultural influence on what is perceived as erotic. In the United States, the stereotypes of the sexually attractive woman and man are perpetuated by the media, advertisements, and movies. But, what may be true in our culture may not be true in other cultures. For example, many American men are sexually aroused by the sight of female breasts, but in cultures where the women do not commonly wear clothing on their upper bodies, this may not be so. Also in our culture, being thin is considered sexier than being fat, but this is not true in some other cultures.

Individual Variation

Within a given culture, there can be great individual variation in what is perceived as erotic. One woman may be aroused by a man's open shirt and hairy chest, whereas another may especially like men's forearms. One man may be excited by the sight of a woman's ankles, while another is stimulated by long hair. Even such diverse stimuli as religious ceremonies, athletic events, a man's shoes, or a woman's purse can be erotic to different people. It is, by the way, not true that women are less aroused by fantasies, erotic scenes, etc., than are men.[1]

Erogenous Zones

Erotic stimuli can be perceived by all of our senses: vision, hearing, smell, touch, and even taste. Touch (or tactile) stimuli are important for sexual arousal in both sexes. The body is particularly sensitive sexually in certain regions; these are known as the *erogenous zones*. In males, examples are the glans, corona, and lower side of the penis, and in females, the clitoris, mons, labia minora, and lower third of the vagina. The upper two-thirds of the vaginal wall are relatively insensitive to touch. Erogenous zones in both sexes include the nipples, lips, tongue, ear lobes, anus, buttocks, inner thighs, and even the back of the knees, soles of the feet, center of the back, eyebrows, and teeth.[2,3] There is, of course, individual variation in the sensitivity of these organs.

Erotic Stimuli

Sound can be erotic; soft music can set the scene for sexual interaction, as can the rhythmic beat of hard rock. The taste of certain food or drink can be associated with past sexual encounters and can be sexually arousing. Although humans are not considered to rely on smell as much as other mammals, smells can be associated with past sexual encounters and can be arousing, as evidenced by the commercial sales of scents, perfumes, and colognes. Also, as we see later, certain smells exuded by our bodies may play a role in our sexual biology.

Kissing the lips can, of course, merely be a sign of affection and not have sexual overtones. Kissing, however, is a common form of sexual arousal in the United States and other countries. This can be done simply by a couple pressing their lips together with gentle movement. In "French kissing," the tongue of one or both partners is inserted and moved within the other's oral cavity. In other cultures, kissing is not a usual form of sexual arousal. People in Japan, China, and Polynesia, for example, did not kiss until they were Westernized. And in some cultures, such as the Thonga of Africa, kissing the lips is looked upon as repulsive.

Mild pain also can be an erotic stimulus. Gentle nibbling, biting, pinching, and scratching can be sexually arousing for many couples. In some cultures, minor pain often is associated with arousing interactions. For example, the Apinaye females of South America bite off bits of the partner's eyebrow as part of the sexual encounter, and Trukese women of the South Pacific poke a finger in the ear of their partner. In our culture, forms of sexual interaction that cause intense pain are considered deviant (Chapter 18).

The partners in a sexual interplay are responsible for giving effective erotic stimuli to each other in the process of *foreplay* or "petting." To be effective, foreplay requires open expression of emotions, communication, and consideration of place, pace, and style. During foreplay, several means of arousing one's partner can be utilized, some more common than others. For many young people, especially those wishing to avoid coitus, foreplay is the "only play." For many other couples, however, foreplay is a prelude to coitus. In one study, foreplay of married couples for 1 to 10 minutes led to 40 percent of the wives having orgasm during coitus. This percentage rose to 50 percent with 12 to 20 minutes of foreplay, and to 60 percent with greater than 20 minutes of foreplay.[4] It is, of course, not only the time spent in foreplay that is important. It is the quality of the stimuli, including a good dose of love and affection, that is especially satisfying and arousing.

THE SEXUAL RESPONSE CYCLE

In the not too distant past, the human sexual response was little understood and was not the subject of much scientific observation. Alfred Kinsey had opened the door to the scientific study of human sexuality,[5] but it was not until the studies of William Masters and Virginia Johnson that we began to know more about the physiology of the human sexual response.[6] Although their methods have been criticized by some,[7] their research has had a profound effect on the understanding of ourselves, on research in human sexuality, and on the treatment of sexual dysfunction. Most of what is said here about the sexual response cycle will, therefore, be based on the studies of Masters and Johnson.

In both sexes, the *sexual response cycle* can be divided into four phases: *excitement, plateau, orgasmic,* and *resolution.* Figure 17-1 depicts these phases in the female and male. Although these are described as distinct phases, you should realize that they flow into one another as a continuous cycle if effective erotic stimuli are present. If these stimuli are not adequate, however, the initial phases are not followed by the final phases. Our discussion now will focus on the sexual response cycle during coitus; you must realize, however, that the full cycle in either sex can occur during masturbation as well (Chapter 18).

The Female Sexual Response Cycle

The female sexual response cycle is similar to that of the male in that it is divided into four phases. As we see, however, there are some important differences between the male and female cycles.

Excitement Phase. As with the male, the female excitement phase (Fig. 17-2) is initiated by the presence of effective erotic stimuli. The first change, usually occurring within 10 to 30 seconds, is *vaginal lubrication;* that is, the membrane lining the vagina becomes more moist. It used to be thought that this was caused by secretions from the Bartholin's glands, but the work of Masters and Johnson showed this not to

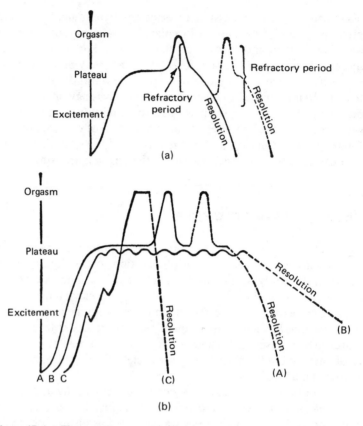

Figure 17-1. The sexual response cycle in men (a) and women (b). The curves represent degrees of sexual arousal. The three different female cycle patterns in (b) are discussed in the text. Reproduced from W. H. Masters and V. E. Johnson, *Human Sexual Response* (Boston, Mass.: Little, Brown & Company, 1966). © 1966 by W. H. Masters and V. E. Johnson; used with permission.

be the case.[8] Instead, the fluid leaks out of blood vessels present in the vaginal wall. Other responses occurring during the female excitement phase include:

1. The inner two-thirds of the vaginal barrel begins to increase in length and width. Thus, the vaginal cavity, which is closed at rest, begins to widen.

2. The body of the uterus ascends (the *tenting effect),* pulling the cervix away from the vagina and thus further increasing vaginal length. There can also be rapid, irregular uterine contractions (fibrillation). These uterine contractions are not painful. The size of the uterus also increases due to *vasocongestion* (pooling of blood in tissues).

3. The walls of the vagina become engorged with blood and become darker in color.

4. The shaft of the clitoris increases in diameter (but rarely in length), and there

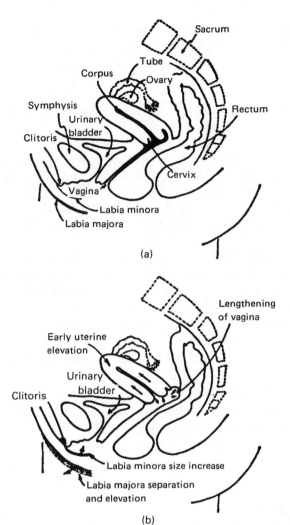

Figure 17-2. Female pelvic organs at rest (a) and during the excitement phase (b). Reproduced from W. H. Masters and V. E. Johnson, *Human Sexual Response* (Boston, Mass.: Little, Brown & Company, 1966). © 1966 by W. H. Masters and V. E. Johnson; used with permission.

may be a slight *tumescence* (swelling) of the clitoral glans due to vasocongestion.

5. The labia minora become engorged with blood, and their size increases considerably.

6. The labia majora, which at rest lie over the vestibule, flatten out and retract from the midline.

7. The nipples become erect, the areola becomes wider and darker, and the size of the breasts increases about 25 percent due to fluid accumulation.

8. A *sex flush* begins to appear in about 74 percent of women. That is, areas of skin become reddened due to dilatation of blood vessels. It looks like a rash,

and usually begins on the abdomen and throat and then spreads to the chest, face, and even the shoulders, arms, and thighs.

9. There is an overall increase in tension in voluntary and involuntary muscles *(myotonia)*.

Plateau Phase. During the female plateau phase (Fig. 17-3), the following changes occur if effective erotic stimuli are present:

1. The wall of the outer one-third of the vaginal barrel becomes greatly engorged with blood, so that the vaginal cavity is reduced from that in the excitement phase. Also, the labia minora become more engorged with blood and become larger. These changes in the outer third of the vaginal barrel and the labia minora are called the *orgasmic platform* because they indicate that orgasm is imminent.

2. The clitoris retracts to be completely covered by the clitoral hood, and its length decreases by about 50 percent. Thus, from here on, the clitoris can be directly stimulated only through the hood and indirectly stimulated by tension applied to the labia minora.

3. Uterine fibrillation continues and may increase in intensity. The uterus also elevates even further (the "tenting effect").

4. The nipples become even more erect and the areola darker; the breasts reach their maximal size.

5. The sex flush, if present, spreads and becomes more intense.

6. Heart rate, blood pressure, and the depth and rate of breathing increase.

7. There is a further increase in muscular tension.

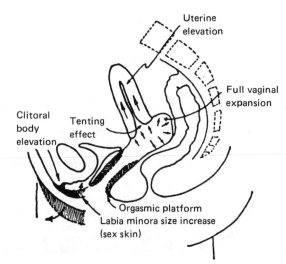

Figure 17-3. Female pelvic organs during the plateau phase. Reproduced from W. H. Masters and V. E. Johnson, *Human Sexual Response* (Boston, Mass.: Little, Brown & Company, 1966). © 1966 by W. H. Masters and V. E. Johnson; used with permission.

Orgasmic Phase. The female orgasmic phase occurs, if it occurs, usually 10 to 20 minutes after *intromission* (penetration of the penis into the vagina). The word *orgasm* ("climax") comes from the Greek word *orgasmos,* which means "to swell" or "be lustful." An orgasm in either sex is one of the most intense and pleasurable of human experiences. We talk more about the experience of orgasm later, but now let's look at some physiological changes that occur (Fig. 17-4):

1. There are strong muscular contractions in the outer one-third of the vaginal barrel. The first contraction lasts about 2 to 4 seconds, and is followed by

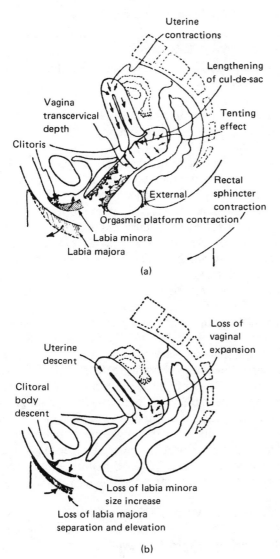

(a)

(b)

Figure 17-4. Female pelvic organs during the orgasmic phase (a) and the resolution phase (b). Reproduced from W. H. Masters and V. E. Johnson, *Human Sexual Response* (Boston, Mass.: Little, Brown & Company, 1966). © 1966 by W. H. Masters and V. E. Johnson; used with permission.

rhythmic contractions at intervals of 0.8 seconds, the same frequency as the contractions of the muscular contractions during male ejaculation. There can be 3 to 15 of these contractions, and the intensity of the initial ones is greater than that of later ones. The rectal sphincter also can exhibit rhythmic contractions at 0.8-second intervals.

2. The inner two-thirds of the vaginal barrel often expand, which facilitates movement of the penis within the barrel.

3. There are rhythmic contractions of the uterus, probably brought about by release of the hormone oxytocin (Chapter 8).

4. The sex flush, if present, peaks in intensity and distribution.

5. The heart rate, blood pressure, and depth and rate of breathing peak at rates similar to those during male orgasm (see below).

6. There may be strong involuntary muscle contractions and clutching or clawing motions of the hands and feet.

There also is great release of neuromuscular tension. The conditions of the labia minora, labia majora, clitoris, and breasts remain similar to those in the plateau phase.

One major difference between the female and male sexual response cycle is that the female does not have a refractory period right after orgasm, which the male has as we discuss later. Kinsey first reported that only about 14 percent of women in his study had multiple orgasms if effective stimuli were present, but probably many more women are physically capable of this. Such women fluctuate from orgasm, to plateau, to orgasm, to plateau, etc. They report that later orgasms in the sequence are more intense than the initial one. A few women can have *status orgasmus,* which is a sustained orgasm lasting up to a minute. [9]

The experience of orgasm can vary in one woman and among different women. [10] Hite reported that this experience often occurs in three stages. [11] First, women experience a sensation of "suspension," lasting only an instant, followed by a feeling of intense sensual awareness, oriented at the clitoris and radiating upward into the pelvis. In the second stage, there is a sensation of warmth, beginning in the pelvis and spreading to other parts of the body. Finally, there is pelvic throbbing, focusing in the vagina and lower pelvis. Other experiences, varying from one woman to another, include mild twitching of the extremities, body rigidity, facial grimacing, and uttering of groans, screams, laughter, or crying. Some women even report that they briefly lose consciousness. Most women say that orgasm is an intensely pleasurable event. Surveys of the experiences of orgasm in women and men suggest that the feelings of both sexes during orgasm are similar. [12] Orgasm in both sexes is stimulated by stretching of the pelvic muscles (due to vasocongestion) and by stimulation of the clitoris or penis. [13]

There are several kinds of orgasm in women. One kind, the *clitoral orgasm,* was described by Masters and Johnson. [14] This type results from stimulation of the clitoris during masturbation as well as coitus. *Vaginal orgasms,* on the other hand, are thought to be the result of direct stimulation of the vaginal wall. Recent evidence does suggest that there is a small region in the front wall of the vagina that, when stimulated, can produce sexual arousal and orgasm. This region is called the *Grafenberg spot*. The orgasm that results from stimulation of this spot involves intense contraction of the uterus and pubococcygeus muscle, and has been called an *A-frame orgasm* or *uterine orgasm*.[15] In reality, most orgasms probably involve a blend of the above kinds. A woman certainly should not suffer from "performance anxiety" if she does or does not have a particular kind of orgasm.

Does *female ejaculation* occur during orgasm? Recent studies indicate that about 10 percent of women expel a small amount of fluid into the vestibule during orgasm.[16] This fluid actually comes from the lesser vestibular (Skene's) glands near the urethral opening. These glands are homologous to the prostate gland of the male (Chapter 5).

It is not true that the size of a man's penis bears a relationship to sexual satisfaction in the female. This is because the vagina adapts to most penises. However, extremely small penises may not provide enough stimulation, whereas extremely large penises may cause some discomfort. It should also be mentioned that there is no benefit to simultaneous orgasm in a man and woman unless this is an achievable and pleasurable goal of a couple's sex life. In fact, if a woman enjoys multiple orgasms, it is necessary that the man delay his orgasm.

It is sometimes the case that the vagina becomes so relaxed that it leads to less sexual stimulation during coitus. This is a common complaint of women in their late thirties and forties who have had several children. In this case, the couple can try new coital positions, such as "woman-on-top" (see Fig. 17-10), and can exercise the pubococcygeus muscle (Chapter 3) to strengthen the vaginal wall.

Resolution Phase. After orgasm, and if there are no effective erotic stimuli present, the woman's system returns to normal during the resolution phase (Fig. 17-4). Some symptoms return to normal rapidly. In less than 10 seconds after orgasm, vaginal contractions cease, and the clitoris leaves its retracted position. The heart rate, blood pressure, and respiration quickly decline to resting levels. Also, the labia minora return to a pink color, usually within 2 minutes. The internal cervical os dilates immediately after orgasm, perhaps to allow sperm to move into the uterus (Chapter 8). Muscle tension decreases in about 5 minutes, and the breasts decrease in size in 5 to 10 minutes. Vasocongestion in the clitoris, vagina, and labia minora ebbs in 5 to 10 minutes, and the uterus usually returns to its normal size and position by this time. The labia major return to their resting condition in about an hour. About one-third of women sweat profusely after orgasm, and many have an intense desire to sleep.

Individual Variation. It is not true, as some believe, that all orgasms in one woman or among women are the same. Figure 17-1 shows three variations in the female sexual response cycle. In pattern A, a woman goes through a complete cycle, including multiple orgasm. In pattern B, a woman reaches plateau, approaches orgasm several times, then goes into resolution. This pattern often exists in inexperienced women or if inadequate stimuli are present. In pattern C, intense stimuli produce an early intense orgasm.

The Male Sexual Response Cycle

The sexual response cycle is very similar in all males, individual men differing in the duration more than the intensity of each phase. Also, the physiological changes in the different phases of the cycle are similar regardless of the nature of the stimuli present and regardless of whether the cycle is initiated by masturbation or by heterosexual or homosexual behavior.[17]

Excitement Phase. The excitement phase of the male sexual response cycle (Fig. 17-5) can be initiated by any effective erotic stimulus. The first thing that happens is that an *erection* begins. The penis stiffens, hardens, and increases in length and diameter. Thus, the penis is said to become tumescent. It should be noted that an erection also can occur without erotic stimuli being present. For example, it is very common for men to gain an erection about every 30 to 90 minutes at night, when rapid eye movement (REM) sleep occurs. Also, many times a man can wake up in the morning with an erection. The reason for this ''morning erection'' is not known, and probably is not caused by a full bladder. Spontaneous nonsexual erections also can occur if the urinary bladder or prostate gland is infected or inflamed, or they can occur in pubescent males (Chapter 6).

An erection involves a basic biological phenomenon that, as mentioned, also occurs in the female sexual response cycle. This phenomenon, vasocongestion, occurs when the flow of blood into a tissue in the arterial vessels is greater than the amount of blood that leaves the tissue in the venous drainage. This results in pooling and engorgement of blood in the tissue. Erotic stimuli initiate nerve impulses that travel directly to the spinal cord or to the brain and then to the spinal cord. This initiates an *erection reflex* by activating an *erection center* in the lower end of the spinal cord that contains neurons that control erection (Fig. 17-6). These neurons are part of the involuntary sympathetic and parasympathetic nervous systems. The sympathetic nerves release norepinephrine, and the parasympathetic nerves release acetylcholine. These neurons send their axons to the blood vessels (arterioles) that supply the erectile tissue in the penis. Erotic stimuli cause the parasympathetic nerves of the erection center to dominate, and these neurons release acetylcholine that causes the arterioles to dilate. This results in vasocongestion in the blood vessels contained in the corpora cavernosa and corpus spongiosum of the penis, and the engorgement of blood in these spongy tissues causes penile tumescence. Depending on the intensity and effectiveness of the stimuli, an erection may be gained partially and then lost a few times before a maximal response occurs.

Neurons from the brain send axons down the spinal cord that connect with the neurons of the erection center, and therefore the brain can inhibit the erection reflex (Fig. 17-6). This is the way, for example, that stress or fear can inhibit erection. The fact, however, that erection is basically a spinal cord reflex explains why men who

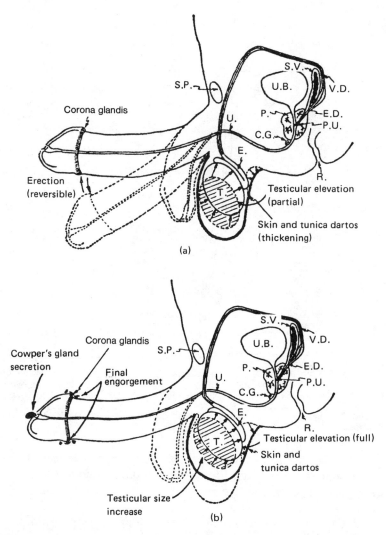

Figure 17–5. The male pelvic organs during the excitement phase (a) and plateau phase (b): S.V. = seminal vesicle, V.D. = vas deferens, U.B. = urinary bladder, E.D. = ejaculatory duct, P.U. = prostatic utricle, P = prostate, C.G. = Cowper's gland, S.P. = symphysis pubis, U = urethra, E = epididymis, T = testis, R = rectum. Reproduced from W. H. Masters and V. E. Johnson, *Human Sexual Response* (Boston, Mass.: Little, Brown & Company, 1966). © 1966 by W. H. Masters and V. E. Johnson; used with permission.

have had their spinal cord severed above the site of the erection center can still have an erection even though they are unable to feel it.

The ability to maintain an erection without ejaculating seems to vary with age. Kinsey found that males in their late teens or early twenties could hold an erection for

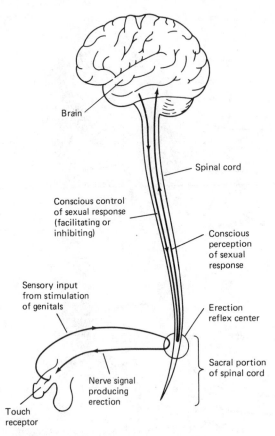

Labels on figure:
Brain
Spinal cord
Conscious control of sexual response (facilitating or inhibiting)
Conscious perception of sexual response
Sensory input from stimulation of genitals
Erection reflex center
Sacral portion of spinal cord
Nerve signal producing erection
Touch receptor

Figure 17–6. Nervous system control of erection. Note that the erection reflex occurs at the level of the spinal cord, but that the brain can facilitate or inhibit this reflex. Redrawn from J. S. Hyde, *Understanding Human Sexuality* (New York: McGraw-Hill Book Company, 1979). © 1979; used with permission.

up to an hour. This was reduced to 30 minutes in men from 45 to 50 years old.[18] It is interesting, however, that Masters and Johnson found the opposite; that is, older men take longer to gain an erection, but once this is achieved, they maintain an erection longer than younger men.[19]

Other physiological changes occur along with erection in the male excitement phase:

1. The urethral opening (urethral meatus) widens.
2. The scrotal skin becomes congested and thick, and thus the scrotal diameter is reduced.
3. The testes become elevated due to contraction of the cremaster muscle in the scrotum. Stroking of the inner thighs also can cause contraction of this scrotal muscle. This is the *cremasteric reflex.*
4. In about 60 percent of men, the nipples become more erect.
5. Areas of the skin become reddened due to dilatation of the blood vessels. This *sex flush* occurs in about 50 to 60 percent of men.

6. The heart rate, blood pressure, and depth and rate of breathing begin to increase.

7. There is an increase in tension of voluntary and involuntary muscles (myotonia).

Plateau Phase. The next phase in the male sexual response cycle, given the continued presence of erotic stimuli, is the plateau phase (Fig. 17-5). In this phase, an erection continues, and the following changes occur:

1. There is a slight increase in the size of the glans of the penis, and its color darkens. The coronal ridge (corona glandis) also tends to swell.

2. The *urethral bulb* (enlarged end of the urethra in males) enlarges to three times its normal size.

3. There may be preorgasmic emission of a few drops of semen from the penis. Although slight in volume, this first stage of ejaculation contains some sperm (Chapter 8).

4. The testes become even more elevated, rotate slightly, and come to lie closer to the groin. Also, the volume of the testes increases by about 50 percent due to accumulation of fluid.

5. The sex flush, if present, spreads and increases in intensity.

6. There is a further increase in heart rate, blood pressure, and the depth and rate of breathing.

7. There is even more tension of voluntary and involuntary muscles.

Orgasmic Phase. The male now enters the orgasmic phase (Fig. 17–7), many times within a few minutes of intromission.

1. There is a loss of voluntary control of muscles and a great release of neuromuscular tension. There may be clutching or clawing motions of the hands and feet.

2. *Ejaculation* is the expulsion of semen, and is controlled by an *ejaculation reflex.* There is an *ejaculatory center* in the spinal cord, located higher up than the erection center. When activated, this center sends sympathetic stimulation to the muscles at the base of the penis. Ejaculation occurs in two phases. There first is a specific sequence of contraction of smooth muscle in the walls of the testes, epididymides, vas deferens, ejaculatory duct, seminal vesicles, prostate gland, bulbourethral glands, and urethra. Simultaneously, a muscular sphincter that guards the opening of the urethra into the urinary bladder contracts, thus preventing urine from entering the urethra and semen from entering the bladder. This series of events constitutes the *emission stage of ejaculation,* and is experienced by a male as a sensation of imminent ejaculation or "coming." These contractions may be influenced by the hormone oxytocin and by the presence of prostaglandins in the seminal fluid.

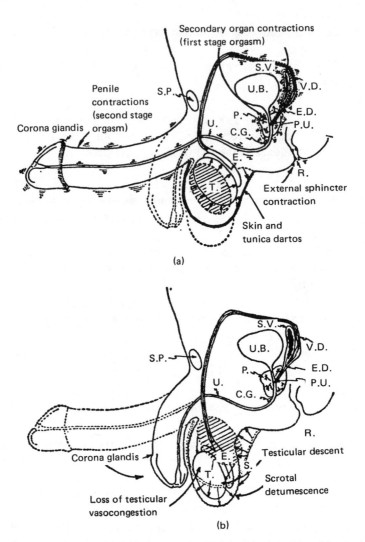

Figure 17-7. The male pelvic organs during the orgasmic phase (a) and the resolution phase (b). For abbreviations, see legend under Fig. 17-5. Reproduced from W. H. Masters and V. E. Johnson, *Human Sexual Response* (Boston, Mass.: Little, Brown & Company, 1966). © 1966 by W. H. Masters and V. E. Johnson; used with permission.

Next, the *expulsion stage of ejaculation* begins, with rhythmic contractions of the *bulbocavernosus muscle* which lies at the base of the penis. The first three or four of these contractions are intense and result in a forceful expulsion of the majority of the semen from the urethra. The contractions that follow are less intense and produce gentle spurts of semen. These expulsion contractions are 0.8 seconds apart.

3. The testes are at their maximal elevation.

4. The heart rate peaks as high as 180 beats per minute (from a resting rate of about 70). The blood pressure peaks at about 220 over 110, from a resting pressure of about 130 over 70. The respiratory rate peaks at about 41 breaths a minute, from a resting rate of about 12 per minute.

5. The sex flush, if present, peaks in intensity and distribution.

For most men, ejaculation is an essential component of the pleasure of orgasm. Orgasm, however, can occur without ejaculation. For example, during retrograde ejaculation (Chapter 13), there is the emission stage (but not the expulsion stage) resulting in a "dry orgasm" when the semen enters the urinary bladder. This can be due to physical damage to the urethra or to a relaxed urinary sphincter muscle. Also, coitus reservatus (Chapter 13) has been practiced as a birth control method by some people, for instance, in India. In this method, men learn to approach ejaculation repeatedly with no expulsion. It also is true that some men can repress the pleasure of orgasm, even if it occurs.[20]

Immediately after ejaculation, the male (unlike the female, as we see later) enters a *refractory period* (Fig. 17-1). During this period, potentially erotic stimuli are not effective in causing or maintaining an erection until sexual tension decreases to near resting levels. This refractory period may last only a few minutes in a young man but may take more than an hour in an older man. Thus, a younger man probably can have several orgasms, each separated by a few minutes. The amount of semen, however, is less in later ejaculations. The duration of the refractory period in young men is about 10 minutes, but can be influenced by fatigue and amount of sexual stimulation.[21] Kinsey found that 6 to 8 percent of the men he studied had more than one orgasm during one sexual encounter, and these men reported that the initial orgasm was the most pleasurable.[22]

Resolution Phase.   During and after the refractory period (and if no effective erotic stimuli are present), the male goes through the resolution phase, in which the arousal mechanisms return to a resting state (Fig. 17-7). In this phase, the erection is lost because the erection center is now dominated by the activity of sympathetic neurons. This causes the arterioles supplying the penile spongy tissue to constrict, thus reducing vasocongestion. About 50 percent of penis size is lost rapidly. Other responses that occur rapidly include disappearance of the muscle tension and sex flush and a lowering of heart rate, blood pressure, and respiratory rate (all usually in about 5 minutes). Other changes taking a longer time include final reduction in penis size, relaxation of the scrotum, descent of the testes, and loss of nipple erection. About one-third of men sweat over their body, and many experience an intense desire to sleep. The entire resolution phase can take up to 2 hours. Close physical contact with the partner, such as keeping the penis within the vagina, touching, and caressing, can delay male resolution. A desire or attempt to urinate, however, can speed up resolution.

COITUS

Coitus (Latin *coitio* = "a coming together") is, for most of us, a pleasurable experience that is a vehicle for the expression of emotion and intimacy. Strictly speaking, coitus (or *sexual intercourse)* is the penetration of the vagina by the penis, which can be called *vaginal coitus* (Fig. 17-8). However, the term coitus also is used for other forms of sexual contact, including *oral coitus* (oral-genital contact), *femoral coitus* (when the penis is inserted between the thighs), *mammary coitus* (when the penis is inserted between the breasts), and *anal coitus* (insertion of the penis into the rectum). There are many common slang phrases for coitus, such as "making love," "going to bed," and other more descriptive terms. Legally, *fornication* is the voluntary coitus between an adult man and woman who are unmarried. *Adultery* is voluntary coitus between two people, at least one of whom is married to someone else. *Sodomy* means different things in different states; it usually refers to anal or oral coitus, but also can mean "acts against nature" such as coitus with an animal (Chapter 18).

Vaginal Coitus

Given the flexibility of the human body, there are a wide variety of positions that are utilized during vaginal coitus. In Kinsey's day, about 70 percent of married couples used the "missionary position" (see Fig. 17-9). Couples in America today, however, are much more willing to experiment with, and derive pleasure from, a wider variety of coital positions.[23] Of the many possibilities, let us look at the most common positions of vaginal coitus and discuss some of their advantages and disadvantages.[24] One should realize that this act of intimacy between a man and woman is a wonderful

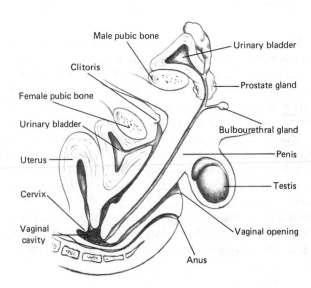

Figure 17–8. Representation of the erect penis inserted into the vagina during vaginal coitus.

thing, and to dissect it apart into "positions" is somewhat like listing the colors in a sunset; you can't quite capture the beauty of the total experience. Which positions a couple utilize is a matter of what gives the most pleasure and meets their needs for stimulation and enjoyment. Therefore a couple should communicate their likes and dislikes. Many couples enjoy experimenting with several positions to add variety.

Face-to-Face (Man Above). In the face-to-face (man above) position, the partners face each other with the man's body lying above the woman's. This "missionary position" is still the one most commonly used for vaginal coitus in the United States (Fig. 17-9). In this position, the man can be on top of the woman, supporting himself by his legs and arms. The woman's legs are spread, and his penis is inserted. In one variation, her legs can be placed together after intromission, thus allowing less vaginal penetration and greater friction on the penis. Or her legs can be drawn up, or be locked around his waist, or even hooked over the man's shoulders. Also, the woman can sit on the edge of a bed with her legs spread.

One advantage of the face-to-face (man above) position is that it allows more verbal communication and kissing between the partners. Also the woman's hands are free to stimulate and caress the man's body. One disadvantage of this position is that the man's weight can inhibit the woman's movement, especially if he is overweight. Also, he may get tired of supporting his own weight. In addition, it sometimes is difficult to get maximal clitoral stimulation in this position. This position is good if a couple wishes to conceive, especially if the penis is left in after ejaculation. It is not good, however, if the woman is in the later stages of pregnancy. This is because the penis can penetrate deeply within the vagina and can initiate uterine contractions or cause discomfort in the woman.

Face-to-Face (Woman Above). In the face-to-face (woman above) position, the woman is on top of the man. A recent survey indicates that about 75 percent of American married couples have used this position.[25] Most often, the woman faces the man's head (Fig. 17-10), but she also can face his feet. Her legs usually are spread, but they also can be stretched out and held together. Or the man can sit with the woman on his lap. These positions have the advantage that the woman has freedom of movement and can better control contact with the clitoris, depth of penetration, and tempo. Disadvantages include the possibility that having the woman on top may be psychologically bothersome to a man who feels that a woman should be passive during coitus or that this makes the female superior to him. Also, the man's movement is inhibited, and the penis has a tendency to slip out. These positions facilitate deep penile penetration, which can be pleasurable for the woman but also may be painful. The woman, however, has a greater control over depth of penile penetration when on top. Because of gravity, this position is not the best for conception.

Face-to-Face (Side-by-Side). In the face-to-face (side-by-side) positions, the man and woman are on their sides facing each other (Fig. 17-11). The legs can be in one of several positions. These positions are the least tiring to both the partners, and are good ones to use if either or both people are obese or if the woman is in the later

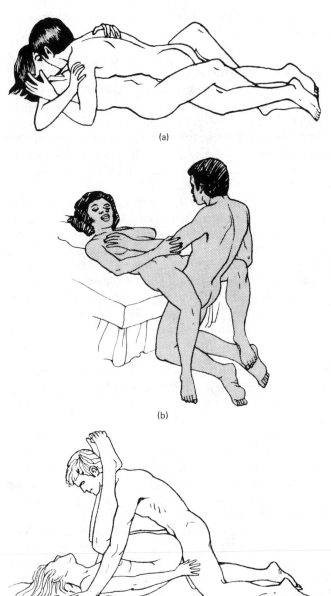

(a)

(b)

(c)

Figure 17–9. Three variations of the face-to-face (man above) position: (a) The usual face-to-face (man above) position, also called the "missionary position." (b) The face-to-face (man above) position, man kneeling. (c) The face-to-face (man above) position, woman's legs elevated.

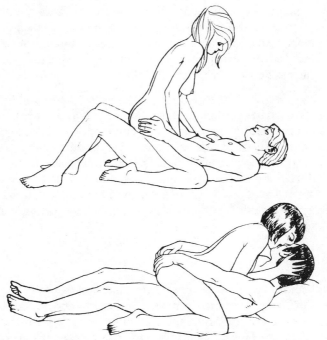

Figure 17–10. Two variations of face-to-face (woman above) coital position.

Figure 17–11. Two variations of the face-to-face (side-by-side) coital position. The lower position (lateral coital position) is recommended by Masters and Johnson to allow maximal stimulation of the female.

stages of pregnancy. The hands of both are free to caress. According to Masters and Johnson, the lateral coital position, a variation on this theme, is the most effective position for sexual satisfaction if there is mutual interest in coital enjoyment and a willingness for free sexual expression.[26] In their survey, three-quarters of the couples chose this position after trying it (Fig. 17-11).

Rear Entry. The rear entry position can take several forms (Fig. 17-12). Often, the woman lies on her stomach or supports herself on her elbows and knees, and the man penetrates from behind. His hands then are free to embrace her waist or caress her breasts or clitoris. Rear entry can also be done in a side-to-side position ("spoons"). In all variations, there is deep penile penetration (which could produce discomfort in some women) and the possibility of manual stimulation of the clitoris by the man. These positions are good for conception. A disadvantage, however, is the lesser opportunity of communication, and these positions are regarded as impersonal by some.

Anal Coitus

In anal coitus, the penis penetrates the anus and is moved within the rectum. This method of coitus is common in male homosexuals and in some heterosexual couples. A heterosexual couple should use a condom and never switch from anal to vaginal coitus before washing the penis since the rectum contains microrganisms that could

Figure 17-12. Two variations of the rear entry coital position.

infect the female tract (Chapter 16). The walls of the rectum are not as well lubricated as are those of the vagina, and the anal sphincter is constricted. Therefore, many lubricate the anus and penis with saliva or a sterile surgical lubricant such as K-Y Jelly. (Vaseline should never be used as it does not dissolve.)

Oral Coitus

Oral coitus is contact of the mouth with the genital organs. When the mouth of the partner caresses the genitals of a female, it is called *cunnilingus* (Latin *cunnus* = "vulva"; *lingere* = "to lick"). The clitoris, labia minora, and vaginal introitus can be kissed or licked with the mouth and tongue. Often, the clitoris is sucked, and the tongue can be inserted into the vagina. Cunnilingus is very pleasurable to many men and women (both the giver and the receiver), and some women experience orgasm as a result.[27] Cunnilingus is practiced in several cultures. The people on the islands of Ponake in the South Pacific show an interesting variation in which the man places a small fish into the woman's vulva and then licks it out prior to coitus. One danger of cunnilingus is the possibility of air being blown into the vagina, since air bubbles could enter the blood stream and could be very dangerous. Therefore, one should not blow air into the vagina.

Fellatio (Latin *fellare* = "to suck") is the oral manipulation of the penis or scrotum by a sexual partner. The glans and frenulum can be licked or nibbled, and sucking can alternate with blowing (hence the name "blow job"). A man can be induced to orgasm during fellatio. Some worry about the adverse effects of swallowing the semen. This fluid, however, is harmless. And obviously, a woman can't get pregnant from swallowing semen.

Soixante-neuf (French for "69") is when a man and woman perform simultaneous cunnilingus and fellatio. This can be done lying side-by-side or one on top of the other, facing in opposite directions.

HORMONES AND SEXUAL BEHAVIOR

In many lower animals, certain hormones need to be present for sexual behavior to be expressed. There basically are two ways that a hormone can stimulate (or inhibit) sexual behavior. First, the hormone can act directly on the brain to affect sensitivity or activity of neurons that influence *sex drive,* or *libido*. This is called a *central effect of a hormone*. Second, a hormone can affect the sensitivity or growth of peripheral tissues, such as skin or muscle, that are involved in sexual behavior. This is called a *peripheral effect of a hormone*. In both instances, however, proper erotic stimuli are necessary for the behavior to be expressed, the hormones simply increasing or decreasing sensitivity and/or response to these stimuli. The region of the human brain that influences sexual behavior is in the *limbic system,* which includes the thalamus, amygdala, hippocampus, and part of the hypothalamus and cerebral cortex. In fact, when electrical stimulation is applied to the human limbic system, a full sexual response can be elicited.[28] We now discuss the role of hormones in the human sex drive and the sexual response.

Hormones and Male Sexual Behavior

In lower animals, the androgen testosterone stimulates the male sex drive. If adults of these animals are orchidectomized (the testes are removed), the sex drive disappears. That is, the males show no behavioral response to stimuli from a female. Administration of testosterone can restore sex drive in these orchidectomized males. This androgen has a central effect on the limbic system and also increases the sensitivity of the penis to tactile stimuli. Recently, it has been discovered that testosterone does not directly stimulate male sex drive in the brain of laboratory mammals. Instead, testosterone is enzymatically converted to estradiol-17β by cells in part of the limbic system, and it is this estrogen that actually increases sex drive.[29] Whether this is true in humans is not clearly known.

Testosterone also influences the sex drive in human males, but this influence is affected markedly by learning. Orchidectomy of human males before puberty generally prevents the development of male sex drive at puberty. In ancient China and Egypt, some young males were orchidectomized. They not only failed to develop male secondary sexual characteristics but also were reported to have little or no sex drive. These men, called *eunuchs* (Greek for "guardian of the bed"), then were trusted to guard females in royal harems.

If a man is orchidectomized after he has reached puberty, the usual results are a slowly declining sex drive and a gradual loss of the ability to have an erection. The extent of the loss of libido varies from one man to the next, emphasizing the role of learning in male sexual motivation and that androgens may not be as necessary after the original behavior pattern develops. Usually, penis size and the extent of facial and body hair are not influenced by orchidectomy. Administration of an androgen to an orchidectomized man who suffers a decrease in sex drive can restore his libido and the ability to have and maintain an erection. In normal men, the higher the testosterone levels in the blood, the less time it takes to achieve maximal penile erection.[30]

If a man is given excess androgen, it does not produce an excessive sex drive. *Anabolic steroids* are androgens sometimes taken by male athletes. These can cause muscle growth, although some of their effects may be due to the expectation that they will occur.[31] This extra androgen, however, can have side effects such as hypertension, acne, headache, nausea, lowering of LH secretion, and a reduction of sperm count.[32]

Many factors can influence levels of testosterone in a man's blood and, therefore, potentially his sex drive. First, there is a daily rhythm of testosterone secretion, peaking in the morning (Chapter 4). Also, anticipation or performance of sex may increase testosterone secretion.[33] Erotic arousal stimulates LH secretion in men, which may raise testosterone levels.[34] In Rhesus monkeys, dominant males have relatively high testosterone levels, which lower when they become subordinate.[35] This also may be true in humans.[36] Finally, there is a tendency for blood testosterone levels to decrease with age.[37] However, sexually active 70-year-old men have higher testosterone levels than sexually nonactive men of the same age.[38] We discussed the subject of the "male climacteric" in Chapter 4.

Hormones and Female Sexual Behavior

In females of many lower animals, it is an estrogen (sometimes acting with progesterone) that increases the female sex drive.[39] In human females, however, the story appears to be more complicated. If a woman has her ovaries removed (ovariectomy), it generally has little or no influence on her libido, although her breasts, vagina, and uterus will regress because of the absence of estrogen. If, however, her adrenal glands are removed along with her ovaries, her sex drive decreases dramatically. Then, if this woman is given an androgen, her sex drive is restored.[40] Remember from Chapter 3 that the adrenal glands secrete "weak androgens" (predominantly dehydroepiandrosterone) and that the ovaries also secrete very low levels of testosterone. It is now generally thought that, in women, the weak androgens secreted by the adrenal glands, perhaps in concert with estradiol-17β and the low levels of testosterone, control the sex drive.[41] Secretion of adrenal androgens in women is greatest from puberty through the late twenties, declines between ages 30 and 50, and then remains at steady low levels in the later years.[42]

Other hormones also seem to affect the female sex drive. Progesterone, for example, tends to lower female libido, and combination birth control pills with high amounts of a progesterone also may have this effect.[43] Some evidence in laboratory mammals indicates that prostaglandins can influence female sex drive.[44] In addition, gonadotropin-releasing hormone (GnRH) directly stimulates female sexual behavior in some mammals.[45] We have much to learn about the role of GnRH in human sexual behavior, but there is hope expressed by some that GnRH analogs can eventually be used to stimulate human sex drive.[46]

If hormones influence female libido, there could be variation in sex drive during the menstrual cycle. The evidence for this possibility, however, is controversial. For example, some studies have used frequency of coitus during the menstrual cycle as an indicator of sex drive. It is also clear that coitus is influenced by many psychological factors, such as opportunity, fear of pregnancy, illness, and fatigue. Kinsey as well as Masters and Johnson found an increase in frequency of coitus during the three or four days before menstruation begins, which may reflect a desire to avoid conception.[47] But there also are studies showing an increase in coital frequency just after menstruation and near ovulation.[48] The general conclusion seems to be that, like other mammals that exhibit a peak in sex drive (estrous behavior) around the time of ovulation (Chapter 7), human females have retained some peak near ovulation (Fig. 17-13). However, the actual expression of this increase in sex drive can be influenced greatly by psychological factors. Therefore, the human female can be sexually receptive at any time during the menstrual cycle (Chapter 1).

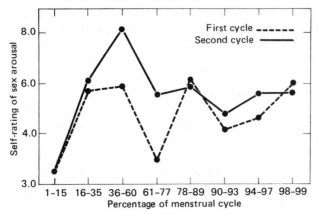

Figure 17-13. Average self-rated sex arousal of 15 women in two consecutive menstrual cycles. Note the slight peak in arousal near the time of ovulation. Reproduced from D. A. Hamburg et al., "Studies of Distress in the Menstrual Cycle and the Postpartum Period," in *Endocrinology and Human Behavior,* ed. R. P. Michael, pp. 94–118 (London: Oxford University Press, 1968). © 1968; used with permission.

PHEROMONES AND HUMAN SEXUAL BEHAVIOR

The predominant senses of humans are vision, touch, and hearing, smell being less important. To other mammals, however, the sense of smell plays a critical role in sexual interactions. When a member of a species emits an odor that changes the physiology or behavior of another member of the same species, this odor is called a *pheromone.* For example, there is a substance in the urine of sexually mature male mice that accelerates the onset of puberty in female mice and also synchronizes the estrous cycles of adult females. A male rat can tell when a female is in estrus by a pheromone in her urine. Similarly, a female Rhesus monkey that is ovulating exudes pheromones (called "copulins") that increase the sex drive in male monkeys. [49]

We are just beginning to investigate the idea that pheromones influence human behavior, a possibility that the manufacturers of perfumes and colognes have assumed for years. [50] There are some artificial musklike odors that can be smelled only by adult women, and not by children or males. What's more, adult women can smell them only near the time of ovulation. [51] Is it possible that men produce a musky pheromone that increases female sex drive, and that women differ in their ability to sense this smell in accordance with the stages of the menstrual cycle? There are glands on the male skin, axilla, penis, and scrotum that could produce pheromones.

Women also may produce pheromones. The vaginal secretion of women, especially when they are near the time of ovulation, contains several chemicals that are known to be copulins in monkeys. [52] We do not know of any behavioral effect of these potential copulins on men, but some evidence indicates that: (1) human vaginal secretions smell most pleasant to males when they are from women in the middle of the menstrual cycle, and (2) vaginal secretions applied to the bodies of women increase their sexual attractiveness. [53] Is it possible that feminine hygiene sprays are covering up something biologically meaningful?

We already discussed (Chapter 7) how women housed together in dormitories tend to have menstrual cycles that are synchronized, and a recent study suggests that an odor produced by women may cause this synchrony. In this study, a woman named Genevieve volunteered to place cotton pads under her armpits to collect secretions. Then, eight other women came in three times a week for four months to have these secretions (in alcohol) rubbed on their upper lips. Eight other women had their lips dabbed with alcohol. Well, the "essence of Genevieve" on the cotton pads caused the eight women to have synchronized menstrual cycles, whereas the cycles of those not receiving the "essence" had random menstrual cycles.[54] Much research is now being done on human skin secretions to determine their characteristics and behavioral effects.[55]

SEXUAL DYSFUNCTION

Sexual dysfunction is when an individual consistently fails to achieve sexual gratification. Such dysfunction is quite common. Masters and Johnson estimated that more than half of married couples have or will have a problem with sexual dysfunction.[56] In one study of happily married, well-educated, predominantly white couples, 40 percent of the husbands reported that they had erection or ejaculation difficulties, and 50 percent reported difficulties relating to a lack of interest in sex. Of the wives interviewed, 69 percent said that they had problems getting sexually aroused, and 20 percent were inable to achieve orgasm. About 77 percent of the wives reported difficulties relating to a lack of interest in sex.[57] About 10 to 20 percent of sexual dysfunctions have a physical or physiological cause, the remainder being caused by psychological conditions.[58] Apparently, about 40 percent of people who seek help from a sex therapist suffer from a lack of sexual desire. Obviously, physical dysfunction can influence a person's psychological well-being, and, conversely, psychological dysfunctions can influence physiology. We must also remember that if a person is not attentive to the needs of a sexual partner and does not supply effective erotic stimuli, a man or woman may be sexually dissatisfied but not have a sexual dysfunction per se.

What are some of the psychological bases for sexual dysfunction? The recent increase in sex manuals has led many people to expect too much from their sexual interactions. They read what "should" be experienced and are always judging their own and their partner's behavior, and they place unrealistic standards upon themselves. This often leads to fear of failure and rejection, and performance anxiety. These people are spectators in their sexual interactions, and they fail to really experience themselves and their partners. Also, they fail to communicate their needs, which often leads to a failure to supply effective stimuli to their partners. Sex becomes mechanistic, and love and affection take a back seat. Another psychological cause of sexual dysfunction is anxiety relating to an earlier sexual experience. A person's initial sex act may have been painful, unpleasurable, or traumatic, and he or she may have sex guilt related to parental religious or moral training (Chapter 19). Other factors inhibiting sexual enjoyment are picking the wrong time for sex, too little foreplay, the inability to relax, too little variety in the stimuli or performance of coitus, and hostility towards the sex partner.[59]

We now review some of the more common kinds of sexual dysfunction. Many of these can be treated with psychologic or behavioral therapy.[60]

Vaginismus

Vaginismus refers to painful spasmodic contractions of the outer third of the vaginal barrel or its surrounding muscles. It is not a common kind of female sexual dysfunction, occurring in only 9 percent of the women studied by Masters and Johnson.[61] A woman has a difficult time stopping these contractions voluntarily, but she can learn to control them. Vaginismus may be painful during coitus or may even prevent intromission. It can be caused by psychological states such as fear of coitus or pregnancy, or by frustration due to a partner with erectile dysfunction (see below). Also, it can be caused by the presence of scar tissue in the vagina or vulva due to childbirth, episiotomy, or infection of the vagina.

Dyspareunia

Dyspareunia is a term referring to difficult or painful coitus in either sex. In females, this can be due to sexual inhibitions or to physical conditions such as vaginismus, irritation or damage to the clitoris, failure of the vagina to lubricate, displaced or prolapsed uterus, infection in the reproductive tract, or inflammation of the Bartholin's glands. In males, the glans may be hypersensitive because of an allergic response to spermicides or urethritis. Or the foreskin may be fused to the glans, a condition called *phimosis*. Male dyspareunia also can be caused by an accumulation of smegma in the uncircumcised. Or in some cases, the penis may be bent *(chordee)*. This may be due to fibrous or calcified tissue in the top or sides of the penis, making the erect penis bend in some direction. This condition, called *Peyronie's disease,* can be the result of previous infection; it is most prevalent in older men. Also, scar tissue or infection in the male sex accessory ducts or glands can cause painful ejaculation.

Premature Ejaculation

Premature ejaculation, a common male sexual dysfunction, is when a man ejaculates too early. But what is too early? Masters and Johnson state that it is ejaculation that occurs at least 50 percent of the time before the woman reaches orgasm.[62] Some men ejaculate immediately after intromission, or even before intromission. Most studies suggest that premature ejaculation has a psychological basis. That is, it is a learned rapid response that can be unlearned.[63] In rare cases, premature ejaculation is caused by prostatitis or a disease of the nervous system. Besides being frustrating to the man, premature ejaculation can rob a woman of fulfillment. For example, women have a greater chance to achieve orgasm the longer intromission lasts without ejaculation.[64]

Ejaculatory Incompetence

Ejaculatory Incompetence is when a man is unable to ejaculate. [65] He may never have been able to ejaculate, or his problem may occur only in certain situations. A common occurrence is a man not being able to ejaculate into a vagina, while having no such problem after oral or manual stimulation. Retrograde ejaculation and coitus reservatus can be cases of learned voluntary failure to ejaculate, and are not considered to be cases of ejaculatory incompetence unless they have a physical basis. However, some men experience retrograde ejaculation after prostate surgery. [66] It should be mentioned that a man can, from time to time, ejaculate without orgasm. Most often this occurs when fear or boredom inhibits the orgasmic phase. [67]

Erectile Dysfunction

Erectile dysfunction is the failure to gain or maintain an erection. It is often called *impotence,* but this is not a desirable term because it implies lack of power, and men with this common dysfunction have enough problems without being labeled weak as well! Masters and Johnson divided erectile dysfunction into two types. [68] The first, *primary erectile dysfunction,* is when a man has never had an erection. In contrast, *secondary erectile dysfunction,* the much more common type, is when a man has had erections before but now fails to have one 25 percent of the time. Actually, many men suffer temporary erectile dysfunction at one time or another. [69]

About 10 percent of cases of erectile dysfunction have a physical basis. [70] One such cause is unusually low blood pressure in the penis, which can be corrected surgically. [71] Diabetes also is associated with erectile dysfunction in about 60 percent of cases. [72] On the average, men with erectile dysfunction do not have lower testosterone levels in their blood, but administration of androgen to men with this problem can increase their sex drive and their erection capacity. [73] Recently it has been shown that levels of the hormone prolactin are abnormally low in men with erectile dysfunction and premature ejaculation. [74] Investigations are now being conducted into the use of an implanted silicone tube into the penis of men with a physical basis of erectile dysfunction. The tube (and penis) can be made erect by the man manipulating a pump in the scrotum.

Most cases of erectile dysfunction have a psychological basis. One way of telling if the problem is psychological is if a morning erection occurs; if so, it indicates that the erection mechanisms are physiologically normal. Remember that erection requires parasympathetic dominance. Therefore, psychological conditions that activate the sympathetic system, such as fear, anxiety, or sudden intense stimuli, will tend to inhibit erection.

Orgasmic Dysfunction

Orgasmic dysfunction is the failure to achieve orgasm. It is the most common female sexual dysfunction. In Hite's survey, only 30 percent of the women achieved orgasm regularly during coitus, even though 87 percent wanted and liked vaginal coitus.[75] Masters and Johnson divided this condition into *primary orgasmic dysfunction,* when a woman never has had an orgasm, and *secondary orgasmic dysfunction,* when she fails to have orgasm only in selective situations.[76] For example, some women can have an orgasm during manual or oral stimulation but not during coitus. According to these surveys of Hite and Masters and Johnson, about 10 percent of all women have primary orgasmic dysfunction, and another 20 percent have experienced secondary orgasmic dysfunction. The term *frigidity* often is used in reference to orgasmic dysfunction, but it is also used if a woman has no interest in sex. Periodic secondary orgasmic dysfunction is a normal part of the sex life of many women[77] and is really only a problem if it interferes with the individual's happiness and sexual gratification.[78] When orgasm does not occur, however, there may be residual sexual tension that could spill over into other aspects of a woman's life. A man should not feel guilty or "less manly" if his sex partner does not have an orgasm. Often, more communication between the partners can help both to achieve more sexual gratification.

Psychological factors, such as those discussed at the beginning of this section, can inhibit orgasm. More rarely, orgasmic dysfunction has a physical basis. Illness or fatigue can be a cause, as can the absence of adequate levels of estrogen. Treatment of women who have orgasmic dysfunction with an estrogen, however, is not effective.[79] Also, about 50 percent of women with diabetes exhibit a diminished capacity to have orgasm.[80]

Many women pretend to have orgasm from time to time.[81] They may do this to maintain their feminine identity, to "get it over with," or to please the man and protect his ego. Much of the time they are good actresses in that the man is not aware of the pretense. This may be harmful to the relationship in the long run, however, and an honest talk with their partner about sexual needs and fears would be helpful.

DRUGS AND HUMAN SEXUAL BEHAVIOR

Therapeutic drugs (medicines) and nontherapeutic drugs (those taken for pleasure) can influence libido and the sexual response cycle. Before discussing some of these drugs, we should first discuss the legends about *aphrodisiacs,* substances that increase sex drive. Many of these seem to work only because one expects them to. A true aphrodisiac would be a substance that activated the part of the brain that controls sex drive, and to date there is no such substance, except, potentially, androgens and GnRH, as mentioned previously.

Many substances, however, have a reputation for being aphrodisiacs. Examples are artichokes, truffles, ginseng, and garlic. Oysters are also supposed to arouse one sexually, but there is no known chemical in oysters that has this effect.[82] This idea probably came from the supposed resemblance of oysters to human genital organs. A similar argument goes for bananas! And the Chinese have used powdered rhinocerus horn as an aphrodisiac for many years, much to the demise of the rhinocerus population. Perhaps this is the source of the term horny. Recently, Indian scientists discovered ancient Hindu formulas for aphrodisiacs made from such exotic materials as crocodile eggs, elephant dung, burnt pearls, gold dust, and lizards' eyes. People in Shakespeare's time thought prunes were an aphrodisiac, and these fruits were handed out by brothels to the clients. [83]

Nevertheless, it is true that some medicines and recreational drugs do influence sex drive and the sexual response, whereas others decrease sexual desire; the latter are called *anaphrodisiacs.* We also know more about the effects of drugs on sexual behavior of men than women, primarily because the male sexual response is easier to observe.

Therapeutic Drugs

Some drugs used to treat human illness (medicines) also influence sex drive. For example, a person with hypertension (high blood pressure) is often given an *antihypertensive drug.* These drugs lower blood pressure, and since blood pressure is important for erection, they can cause erectile dysfunction. *Spironolactone,* an antihypertensive drug, can lead to erectile dysfunction as well as to a decrease in sperm count and motility, gynecomastia, and a decrease in sex drive. This drug not only lowers blood pressure but also inhibits the action of androgens on their target tissues. Secondary amenorrhea is seen in women using spironolactone. [84] *Reserpine* is another antihypertensive drug that interferes with the action of brain neurons and can cause mental depression. Reserpine also can depress sex drive and reduce FSH and LH secretion in women. [85] In men, it can cause erectile dysfunction, ejaculatory incompetence, and gynecomastia.

Drugs that influence the sympathetic or parasympathetic nervous systems can influence the sexual response. For example, *guanethidine* (also an antihypertensive) blocks the action of some sympathetic nerves and can inhibit ejaculation and possibly produce permanent damage to the neurons in the ejaculation center. [86] Other antisympathetic drugs that can cause ejaculatory problems include ergot alkaloids, methyldopa, and rauwolfia alkaloids. Drugs that block the parasympathetic system can cause erectile dysfunction. [87] One such substance is atropine. Some *antihistamines* also can have this effect. [88]

Some minor tranquilizers that are popular prescription drugs in the United States, such as *diazepam* (Valium) and *chlorodiazepoxide* (Librium), can increase sex drive by removing inhibitions. However, chronic use of these drugs can cause ejaculatory incompetence in males and menstrual irregularities in females. [89]

Nontherapeutic Drugs

Nontherapeutic drugs are taken for their presumed pleasurable effects. Many users report that some of these drugs increase their sex drive, sexual performance, and enjoyment. Most of these effects are due to the drug acting (1) on the brain to remove sexual inhibitions, (2) on the peripheral nervous system to increase sensory acuity, or

(3) on the sympathetic or parasympathetic nervous systems to influence the sexual response cycle. Also, as mentioned, some of the effects of the drugs on sexual arousal occur simply because the person expects them to work.

Some of the *cantharides* such as "Spanish fly" have reputations as aphrodisiacs. Spanish fly is derived from a beetle *(Lytta vesicatoria)* found in an area extending from southern Europe to western Siberia and in parts of Africa. It is used to heal blisters. If this drug increases libido, one explanation is that it can irritate the urinary bladder, urethra, and digestive tract. Prolonged use of Spanish fly, however, can cause *priapism* (prolonged abnormal erection) because it dilates blood vessels in the penis. This can lead to permanent penile damage and erectile dysfunction.[90] Also, cantharides can cause vomiting and diarrhea and are poisonous at high dosages.

The active substance in marijuana and hashish from the hemp plant *(Cannabis sativa)* is tetrahydrocannabinol, or THC. This hallucinogenic substance mimics the action of the sympathetic nerves. It first stimulates euphoria, causing increased sensory awareness and distorted perception. This is followed by sleepiness and a dreamlike state. About one-quarter of marijuana and hashish users report an increase in sexual responsiveness and enjoyment;[91] THC has this effect by acting on the brain to relax sexual inhibitions and to increase sensory awareness. Some studies, however, indicate no effect of marijuana on sex drive.[92] Other studies suggest that chronic use of marijuana lowers secretion of FSH and LH in young adult males and causes a decrease in testosterone levels, sperm count, and, in a few cases, erectile dysfunction.[93] It should be mentioned, however, that one study showed no effect of chronic marijuana use on the male reproductive system.[94] Chronic THC use in females inhibits ovulation and shortens the luteal phase of the menstrual cycle if ovulation occurs.[95]

Lysergic acid diethylamide (LSD) is a hallucinogen reported to increase sex drive due to general mood change. People rate the "aphrodisiac" effects of marijuana, hashish, and LSD equally, with LSD being the best for sexual fantasies and marijuana and hashish the best for maintaining an erection.[96] LSD affects the brain by acting like serotonin (a neurotransmitter) and also mimics action of the sympathetic nervous system.[97] Mescaline has effects on sex drive similar to LSD.

Amphetamines are stimulants that decrease fatigue and activate the sympathetic nervous system. These include MDA (3, 4-methylenedioxyamphetamine), STP (2, 5-dimethoxy-4-methylamphetamine), and MMDA (3-methoxy-4, 5-methylenedioxyamphetamine). These drugs produce an aphrodisiaclike effect by increasing one's sensitivity to tactile and other stimuli.[98]

Cocaine, a widely used drug in the United States, is said to increase libido, enhance enjoyment of sex, and lower inhibitions.[99] This drug often is inhaled through the nostrils as a powder, but some women apply it to the vaginal mucosa; this is supposed to delay and enhance orgasm.[100] Chronic users of cocaine (and amphetamines), however, tend to be more interested in drugs than sex.[101]

Yohimbine is extracted from the bark of the Yohimbé tree of Africa. It is a reputed aphrodisiac, and apparently acts by stimulating the parasympathetic nervous system, causing vasodilation and erection.[102] This drug also acts on the brain and increases blood pressure, heart rate, and irritability, and can cause sweating, nausea, and vomiting. [103]

Amyl nitrite is a drug that dilates blood vessels. It is a chemical cousin of nitroglycerin, and is used to relieve heart pain. This drug is called "poppers" because of the popping sound emitted when one breaks the capsule in which it is contained. Some inhale its fumes shortly before orgam and claim it creates a more intense experience. [104] But this drug also can cause dizziness, headache, fainting, and even death. *Butyl nitrite* has similar effects. [105]

Saltpeter (*potassium nitrate*), which is the main ingredient in gunpowder and fireworks, is thought to decrease sex drive, but there is no scientific evidence for this. [106]

Alcohol (ethyl alcohol) has an anesthetic, depressive effect on the brain and thus decreases sexual inhibitions. Alcohol in repeated dosages, however, suppresses the parasympathetic nervous system and can cause erectile dysfunction. At higher dosages it suppresses the sympathetic system and can cause ejaculatory incompetence. [107] Chronic use of alcohol can destroy the neurons involved in the erection reflex. [108] In addition, male alcoholics can be feminized by the drug, developing sterility and gynecomastia. Even a single dose of alcohol lowers testosterone levels in the blood by decreasing secretion of luteinizing hormone and by increasing the activity of liver enzymes that destroy this hormone. [109]

Narcotics such as heroin, morphine, and methadone generally depress the central nervous system and decrease sex drive and responsiveness. Low doses, however, remove sexual inhibitions. [110] These narcotics lower testosterone levels in the blood of males. [111] In women, morphine, heroin, and methadone can interrupt menstrual cycles by decreasing FSH and LH secretion. [112]

Barbiturates such as *methaqualone* (Quaalude, Supor) are used as sleeping pills and hypnotic drugs. These drugs can decrease sexual inhibition, but there often is a reduced ability to perform sexually. The effects of the barbiturates on the endocrine system are similar to those described above for narcotics. [113]

CHAPTER SUMMARY

Erotic stimuli are those factors in our environment that are sexually arousing. These stimuli can be detected through vision, hearing, touching, tasting, or smelling. What is sexually arousing to one person may not stimulate another. There is cultural variation in the stimuli that are sexually arousing. Foreplay provides erotic stimuli, often as a prelude to coitus.

The human sexual response cycle is a sequence of physiological changes divided into four phases: excitement, plateau, orgasmic, and resolution. In general, this cycle in both sexes is characterized by vasocongestion of the external genitalia and skin as well as myotonia. In females, responses such as vaginal lubrication and changes in the orgasmic platform precede orgasm. In males, erection is caused by vasocongestion of the penis, and is controlled by an erection center in the spinal cord. Ejaculation occurs during the male orgasmic phase, and is controlled by an ejaculatory center in the spinal cord. In both sexes, orgasm involves release of neuromuscular tension and is experienced as an extremely pleasurable event. During the resolution phase in both sexes, the sexual systems return to an unexcited state. Women, unlike men, can experience multiple orgasms without going through resolution.

Vaginal coitus between a man and woman can utilize several positions. The more common of these are the face-to-face (man above), face-to-face (woman above), face-to-face (side-by-side), and rear entry positions. Anal coitus is penetration of the penis into the rectum, and oral coitus (oral-genital contact) can take the form of cunnilingus or fellatio.

Hormones can influence sexual behavior by acting centrally (on the central nervous system) or on peripheral tissues. In men, testosterone, both centrally and peripherally, is necessary for male sexual behavior. In women, weak androgens secreted by the adrenal glands, along with estrogens, maintain libido. A slight tendency for female sex drive to increase near the time of ovulation in the menstrual cycle is masked by psychological variables, so that the frequency of coitus during the cycle does not exhibit any consistent trend. The human male and female may produce pheromones that influence sexual behavior, but more needs to be learned about this topic.

Sexual dysfunction is present when a person consistently does not receive sexual gratification. In about 10 to 20 percent of cases, such dysfunction has a physical basis, but the majority are caused by a psychological condition. Some of the more common sexual dysfunctions in women are vaginismus and orgasmic dysfunction, whereas men can develop ejaculatory incompetence, premature ejaculation, and erectile dysfunction. Both sexes can develop dyspareunia.

A substance that increases sexual desire is an aphrodisiac, and one that decreases sexual desire is an anaphrodisiac. In reality, no true aphrodisiac exists, but some therapeutic and recreational drugs can influence human reproduction and sexual behavior. For example, some antihypertensive drugs decrease sex drive or sexual performance, as can some minor tranquilizers. Cantharides (Spanish fly) stimulate sex drive by irritating the urinary tract and dilating genital blood vessels. Tetrahydrocannabinol in marijuana and hashish can increase sexual enjoyment by stimulating euphoria, increasing sensory awareness, and causing a distortion of time perception. Chronic use of this drug, however, can decrease sexual performance by inhibiting secretion of pituitary gonadotropic hormones. Lysergic acid diethylamide (LSD) can be sexually stimulating by acting on the brain to release inhibitions. Stimulants such as cocaine, amphetamines, amyl nitrite, and yohimbine can increase sexual enjoyment, as can the narcotics heroin, morphine, and methadone. These drugs, however, can cause sexual dysfunction if used chronically. Alcohol can release sexual inhibition, but chronic use can cause sexual dysfunction, as can barbiturates. Although some of these drugs initially increase sexual drive and desire, most have long-term effects that lower sexual desire and performance as well as reproductive capacity.

NOTES

1. J. R. Heiman, "The Physiology of Erotica: Women's Sexual Arousal," *Psychology Today* (April 1975), pp. 91–94.

2. A. C. Kinsey et al., *Sexual Behavior in the Human Male* (Philadelphia: W. B. Saunders Company, 1948).

3. A. C. Kinsey et al., *Sexual Behavior in the Human Female* (Philadelphia: W. B. Saunders Company, 1953).

4. P. H. Gebhard, "Factors in Marital Orgasm," *Journal of Social Issues,* 22, no. 2 (1966), 88–95.

5. Kinsey et al. (1948; 1953), op. cit.

6. W. H. Masters and V. E. Johnson, *Human Sexual Response* (Boston: Little, Brown & Company, 1966).

7. A. Rosenfeld, "Inside the Sex Lab," *Science Digest* (November/December 1980), p. 109.

8. Masters and Johnson (1966), op. cit.

9. R. Jobaris and J. Money, "Duration of Orgasm," *Medical Aspects of Human Sexuality* (July 1976), pp. 7, 65.

10. C. A. Butler, "New Data about Female Sexual Response," *Journal of Sexual and Marital Therapy,* 2 (1976), 40–46.

11. S. Hite, *The Hite Report: A Nationwide Study of Female Sexuality* (New York: Macmillan Publishing Co., Inc., 1976).

12. E. B. Proctor et al., "The Differentiation of Male and Female Orgasm: An Experimental Study," in *Perspectives on Human Sexuality,* ed. N. Wagner (New York: Behavior Publications, 1974); and E. B. Vance and N. N. Wagner, "Written Descriptions of Orgasm: A Study of Sex Differences," *Archives of Sexual Behavior,* 5 (1976), 87–98.

13. D. E. Mould, "Neuromuscular Aspects of Women's Orgasms," *Journal of Sex Research,* 16 (1980), 193–201.

14. Masters and Johnson (1966), op. cit.

15. B. Whipple and J. D. Perry, "Pelvic Muscle Strength of Female Ejaculators: Evidence in Support of a New Theory of Orgasm, *Journal of Sex Research,* 17 (1981), 22–39.

16. J. L. Sevely and J. W. Bennett, "Concerning Female Ejaculation and the Female Prostate," *Journal of Sex Research,* 14 (1978), 1–20.

17. H. W. Masters and V. E. Johnson, *Homosexuality in Perspective* (Boston: Little, Brown & Company, 1979).

18. Kinsey et al. (1948), op. cit.

19. Masters and Johnson (1966), op. cit.

20. J. F. O'Conner and D. O. O'Conner, "Pleasureless Orgasms," *Medical Aspects of Human Sexuality* (March 1980), pp. 122–30.

21. G. D. Jensen, "Men's Unresponsive Period after Orgasm," *Medical Aspects of Human Sexuality* (March 1979), pp. 50–64.

22. Kinsey et al. (1948), op. cit.

23. M. Hunt, *Sexual Behavior in the 1970s* (Chicago: Playboy Press, 1974).

24. B. J. Sadock and V. A. Sadock, "Coital Positions," *Medical Aspects of Human Sexuality* (May 1979), pp. 115–19.

25. L. G. Barbach, *For Yourself: The Fulfillment of Female Sexuality* (New York: Anchor Books, 1976).

26. W. H. Masters and V. E. Johnson, *Human Sexual Inadequacy* (Boston: Little, Brown & Company, 1970).

27. Hite (1976), op. cit.

28. R. C. Heath, "Pleasure and Brain Activity in Man," *Journal of Nervous and Mental Disease,* 154 (1972), 3–18.

29. B. S. McEwen, "Interactions between Hormones and Nerve Tissue," *Scientific American,* 235, no. 1 (1976), 48–58.

30. J. D. Lange et al., "Serum Testosterone Concentration and Penile Tumescence Changes in Man," *Hormones and Behavior,* 14 (1980), 267–70.

31. G. Ariel and W. Saville, "Anabolic Steroids: The Physiological Effects of Placebos," *Medicine and Science in Sports,* 4 (1972), 124–26.

32. F. Murad and A. G. Gilman, "Androgens and Anabolic Steroids," in *The Pharmacological Basis of Therapeutics,* 5th ed., eds. L. S. Goodman and A. Gilman, pp. 1451–71 (New York: Macmillan Publishing Co., Inc., 1975).

33. C. A. Fox, "Studies on the Relationship of Plasma Testosterone Levels and Human Sexual Activity," *Journal of Endocrinology,* 52 (1972), 51–58.

34. J. J. LaFerla et al., "Psychoendocrine Response to Visual Erotic Stimulation in Human Males," *Psychosomatic Medicine,* 38 (1976), 62.

35. R. M. Rose et al., "Plasma Testosterone Levels in the Male Rhesus: Influences of Sexual and Social Stimuli," *Science,* 178 (1972), 643–45.

36. A. Mazur and T. A. Lamb, "Testosterone, Status, and Mood in Human Males," *Hormones and Behavior,* 14 (1980), 236–46.

37. J. G. Lewis et al., "Age-Related Changes in Serum 5α-Dihydrotestosterone and Testosterone in Normal Men," *Journal of Endocrinology,* 67 (1975), 15 p.

38. Anonymous, "Testosterone and Male Sex Drive," *Science News,* 116, no. 6 (1979), 104.

39. McEwen (1976), op. cit.

40. C. S. Carter, ed., *Hormones and Sexual Behavior: Benchmark Papers in Animal Behavior,* vol. 1 (Stroudsburg, Pa.: Dowden, Hutchenson, and Ross, 1974).

41. M. M. Grumback and J. J. Van Wyk, "Disorders of Sex Differentiation," in *Textbook of Endocrinology,* 5th ed., ed. R. H. Williams, pp. 423–501 (Philadelphia: W. B. Saunders Company, 1974).

42. W. E. Szpunar et al., "Plasma Androgen Concentrations in Diabetic Women," *Diabetes,* 26 (1977), 1125–29.

43. B. G. Everitt et al., "Humoral and Adrenergic Mechanisms Regulating Sexual Receptivity in Female Monkeys," in *Sexual Behavior: Pharmacology and Biochemistry,* eds. M. Sandler and G. L. Gessa, pp. 181–91 (New York: Raven Press, 1975).

44. B. L. Marrone et al., "Differential Effects of Prostaglandins on Lordosis Behavior in Female Guinea Pigs and Rats," *Biology of Reproduction,* 20 (1979), 853.

45. R. L. Moss and S. M. McCann, "Induction of Mating Behavior in Rats by Luteinizing Hormone-Releasing Factor," *Science,* 181 (1973), 177.

46. Anonymous, "A Chemical Key to Sexuality," *Science Digest* (November/December 1980), p. 109.

47. Kinsey et al. (1953), op. cit.; Masters and Johnson (1966), op. cit.

48. D. A. Hamburg et al., "Studies of Distress in the Menstrual Cycle and the Postpartum Period," in *Endocrinology and Human Behavior,* ed. R. P. Michael, pp. 94–116 (London: Oxford University Press, 1968); and J. R. Udry and N. M. Morris, "Distribution of Coitus in the Menstrual Cycle," *Nature,* 220 (1977), 593–96.

49. R. P. Michael et al., "Evidence for Chemical Communication in Primates," *Vitamins and Hormones,* 34 (1976), 137–86.

50. A. Comfort, "Likelihood of Human Pheromones," *Nature,* 230 (1971), 432–33; and J. L. Hopson, "Scent and Human Behavior: Olfaction or Fiction," *Science News,* 115, no. 17 (1979), 282–83.

51. J. Le Magnen, "Les Pheromones Olfactosexuals chez le Rat Blanc," *Archives des Sciences Physiologiques,* 6 (1952), 295–332.

52. R. P. Michael et al., "Human Vaginal Secretions: Volatile Fatty Acid Content," *Science,* 186 (1974), 1217–19.

53. Michael et al. (1976), op. cit; and R. L. Doty et al., "Changes in the Intensity and Pleasantness of Human Vaginal Odors During the Menstrual Cycle," *Science,* 190 (1975), 1316–18.

54. M. J. Russell, "Study on Menstrual Synchrony Using Axillary Secretion," *Science News,* 112, no. 1 (1977), 5.

55. J. A. Miller, "Whiff of Locker Room—in a Test Tube," *Science News,* 117, no. 13 (1979), 216.

56. Masters and Johnson (1970), op. cit.

57. E. Frank et al., "Frequency of Sexual Dysfunction in 'Normal' Couples," *New England Journal of Medicine,* 299 (1978), 111–15.

58. Masters and Johnson (1970), op. cit.

59. N. Roth, "Sexual Revenge," *Medical Aspects of Human Sexuality* (February 1979), pp. 8–21; and C. Nadelson, " 'Healthy' Sexual Dysfunctions," *Medical Aspects of Human Sexuality* (March 1979), pp. 106–19.

60. Masters and Johnson (1970), op. cit.

61. Ibid.

62. Ibid.

63. H. S. Kaplan, *The New Sex Therapy* (New York: Brunner/Mazel, Inc., 1974); and S. B. Levine, "Barriers to the Attainment of Ejaculatory Control," *Medical Aspects of Human Sexuality* (January 1979), pp. 32–56.

64. Gebhard (1966), op. cit.

65. Masters and Johnson (1970), op. cit.

66. R. S. Hotchkiss, "How Will an Operation on the Prostate Affect a Man's Sex Life?" *Sexual Behavior* (August 1971), p. 14.

67. P. Dormont, "Ejaculatory Anhedonia," *Medical Aspects of Human Sexuality* (February 1975), pp. 32–43.

68. Masters and Johnson (1970), op. cit.

69. Kaplan (1974), op. cit.

70. J. Marmor, "Impotence and Ejaculatory Disturbances," in *The Sexual Experience,* ed. B. J. Sadock et al. (Baltimore: The Williams & Wilkens Company, 1976).

71. Anonymous, "Do Drugs Cause Impotence by Lowering Penile Blood Flow?" *Medical World News* (February 19, 1979), pp. 85–86.

72. R. C. Kolodny et al., "Sexual Dysfunction in Diabetic Men," *Diabetes,* 23 (1974), 306–309; and M. Ellenberg, "Evaluating Sexual Impairment in Your Diabetic Patients Today," *Consultant,* 19, no. 4 (1979), 125–32.

73. F. Sicuteri et al., "Aphrodisiac Effect of Testosterone in Parachlorophenylalanine-Treated Sexually Deficient Men," in *Sexual Behavior: Pharmacology and Biochemistry,* eds. M. Sandler, and G. L. Gessa, pp. 335–39 (New York: Raven Press, 1975); and P. A. Racey et al., "Testosterone in Impotent Men," *Journal of Endocrinology,* 59 (1973), xxiii.

74. Anonymous, "The Milk Hormone and Male Sex Drive," *Science News,* 116, no. 6 (1979), 104.

75. Hite (1976), op. cit.

76. Masters and Johnson (1970), op. cit.

77. Kaplan (1974), op. cit.

78. A. M. Zeiss et al., "Orgasm during Intercourse: A Treatment Strategy for Women," *Journal of Consulting and Clinical Psychology,* 45 (1977), 891–95.

79. J. Marmor, "Frigidity, Dyspareunia, and Vaginismus," in *The Sexual Experience,* eds. B. J. Sadock et al. (Baltimore: The Williams & Wilkens Company, 1976).

80. Kolodny et al. (1974), op. cit.

81. Hite (1976), op. cit.

82. S. Neiger, "Sex Potions," *Sexology* (1968), pp. 730–33.

83. F. Donegan, "Lovable Feasts," *Viva* (February 1978), p. 70.

84. R. Caminos-Torres et al., "Gynecomastia and Semen Abnormalities Induced by Spironolactone in Normal Men," *Journal of Clinical Endocrinology,* 45 (1977), 255–60.

85. W. H. Daughaday, "The Adenohypophysis," in *Textbook of Endocrinology,* ed. R. H. Williams, pp. 31–79 (Philadelphia: W. B. Saunders Company, 1974).

86. J. Money and R. Yankowitz, "The Sympathetic-Inhibiting Effects of the Drug Ismelin on Human Male Eroticism, with a Note on Mellaril," *Journal of Sex Research,* 3 (1976), 69–82; and B. Evans et al., "Long-Lasting Damage to the Internal Genital Organs and Their Adrenic Innervation Following Chronic Treatment with the Antihypertensive Drug Guanethidine," *Fertility and Sterility,* 23 (1972), 657–67.

87. Kaplan (1974), op. cit.

88. W. W. Douglas, "Histamine and Antihistamines, 5-Hydroxytryptamine and Antagonists," in *The Pharmacological Basis of Therapeutics,* 5th ed., eds. L. S. Goodman and A. Gilman, pp. 590–629 (New York: Macmillan Publishing Co., Inc., 1975).

89. R. Byck, "Drugs and the Treatment of Psychiatric Disorders," in *The Pharmacological Basis of Therapeutics,* 5th ed., eds. L. S. Goodman and A. Gilman, pp. 152–200 (New York: Macmillan Publishing Co., Inc., 1975).

90. Kaplan (1974), op. cit.

91. E. Goode, "Drug Use and Sexual Activity on a College Campus," *American Journal of Psychiatry,* 128 (1972), 1272–76.

92. J. H. Mendelson, "Marijuana and Sex," *Medical Aspects of Human Sexuality* (November 1976), pp. 23–24.

93. R. C. Kolodny et al., "Depression of Plasma Testosterone Levels after Chronic Marijuana Use," *New England Journal of Medicine,* 290 (1974), 872–74; T. H. Mauch, "Marijuana: New Support for Immune and Reproductive Hazards," *Science,* 190 (1975), 865–67; and W. C. Hembree et al., *Marijuana: Chemistry, Biochemistry, and Cellular Effects,* eds. G. G. Nahas et al. (New York: Springer-Verlag New York, Inc., 1976).

94. J. H. Mendelson et al., "Plasma Testosterone Levels before, during, and after Chronic Marijuana Smoking," *New England Journal of Medicine,* 291 (1974), 1051–55.

95. J. Harclerode, "The Effect of Marijuana on Reproduction and Development," in *Marijuana Research Findings: 1980,* ed. R. C. Peterson, *National Institute of Drug Abuse, Monograph 31* (1980), pp. 137–66; and A. Chen, "Marijuana and the Reproductive Cycle," *Science News,* 123, no. 13 (1983), 197.

96. G. R. Gay et al., "Drug-Sex Practice in the Haight-Ashbury, or 'The Sensuous Hippie,'" in *Sexual Behavior: Pharmacology and Biochemistry,* eds. M. Sandler and G. L. Gessa, pp. 63–79 (New York: Raven Press, 1975).

97. Byck (1975), op. cit.

98. Kaplan (1974), op. cit.

99. Gay et al. (1975), op. cit.

100. D. R. Wessen and D. E. Smith, "Cocaine: Its Use for Central Nervous System Stimulation, Including Recreational and Medical Uses," in *Cocaine: 1977, National Institute on Drug Abuse Research Monograph Series,* no. 13 (1977), pp. 137–52.

101. Kaplan (1974), op. cit.

102. L. E. Hollister, "Drugs and Sexual Behavior in Man," *Life Sciences,* 17 (1975), 661–68.

103. M. Nickerson and B. Collier, "Drugs Inhibiting Adrenergic Nerves and Structures Innervated by Them," in *The Pharmacological Basis of Therapeutics,* 5th ed., eds. L. S. Goodman and A. Gilman, pp. 533–64 (New York: Macmillan Publishing Co., Inc., 1975).

104. G. M. Everett, "Effects of Amyl Nitrite ('Poppers') on Sexual Experience," *Medical Aspects of Human Sexuality,* 6, no. 12 (1972), 146–51.

105. Hollister (1975), op. cit.

106. T. G. Benedek, "Saltpeter as an Anaphrodisiac," *Medical Aspects of Human Sexuality* (March 1974), pp. 131–32.

107. V. Marks and J. Chakraborty, "The Clinical Endocrinology of Alcoholism," *Journal of Alcoholism,* 8 (1973), 94–103.

108. F. Lemere and J. W. Smith, "Alcohol-Induced Sexual Impotence," American Journal of Psychiatry, 130 (1973), 212–13.

109. R. Ylikahri and M. Huttunen, "Low Plasma Testosterone Values in Men during Hangover," *Journal of Steroid Biochemistry,* 5 (1974), 655–58; G. G. Gordon et al., "Effect of Alcohol (Ethanol) Administration on Sex-Hormone Metabolism in Normal Men," *New England Journal of Medicine,* 295 (1976), 793–97; and E. Rubin et al., "Prolonged Ethanol Consumption Increases Testosterone Metabolism in the Liver," *Science,* 191 (1976), 563–64.

110. Kaplan (1974), op. cit.

111. T. C. Cicero et al., "Function of the Male Sex Organs in Heroine and Methadone Users," *New England Journal of Medicine,* 292 (1975), 882–87.

112. R. J. Santen et al., "Mechanism of Action of Narcotics in the Production of Menstrual Dysfunction in Women," *Fertility and Sterility,* 26 (1975), 538–48.

113. L. E. Hollister, "Human Pharmacology of Drugs of Abuse with Emphasis on Neuroendocrine Effects," in E. Zimmerman et al., eds., *Drug Effects on Neuroendocrine Regulation,* vol. 39, pp. 373–81 (New York: Elsevier North-Holland, Inc., 1973).

FURTHER READING

AUSTIN, C. R., and SHORT, R. V., eds., *Reproduction in Mammals, Book 8, Human Sexuality.* Cambridge, England: Cambridge University Press, 1980.

BANCROFT, J., "Human Sexual Behavior," in *Reproduction in Mammals, Book 8, Human Sexuality,* eds. C. R. Austin and R. V. Short. Cambridge, England: Cambridge University Press, 1980.

BARBACH, L. G., *For Yourself: The Fulfillment of Female Sexuality.* New York: Anchor Books, 1976.

COMFORT, A., "Likelihood of Human Pheromones," *Nature,* 230 (1971), 432–33.

COMFORT, A., *The Joy of Sex.* New York: Crown Publishers, Inc., 1972.

HART, G., *Human Sexual Behavior, Carolina Biology Reader.* North Carolina: Carolina Biological Supply Co., 1976.

HITE, S., *The Hite Report: A Nationwide Study of Female Sexuality.* New York: Macmillan Publishing Co., Inc., 1976.

HOLLISTER, L. E., "Drugs and Sexual Behavior in Men," *Life Sciences,* 17 (1975), 661–68.

MASTERS, W. H., and V. E. JOHNSON, *Human Sexual Response.* Boston: Little, Brown & Company, 1966.

MASTERS, W. H., and V. E. JOHNSON, *Human Sexual Inadequacy.* Boston: Little, Brown & Company, 1970.

McEWEN, B. S., "Interactions between Hormones and Nerve Tissue," *Scientific American,* 235, no. 1 (1976), 48–58.

NEIGER, S., "Sex Potions," *Sexology* (1968), Pp. 730–33.

ROSENFELD, A., "Inside the Sex Lab," *Science Digest* (November/December 1980), Pp. 102–104, 124.

18 *Patterns of Human Sexual Behavior*

LEARNING GOALS

Having read this chapter, you should be able to:

1. List some cautions in interpreting "sex surveys."
2. Discuss ways in which sex is involved in the life of an infant.
3. Discuss how sex is expressed during early childhood.
4. Discuss how preadolescents express their sex drive.
5. Describe outlets for sex drive in adolescents.
6. Discuss the patterns of sexual gratification in heterosexual married adults.
7. Discuss some changes that occur in the sexual expression of older people, and some reasons for these changes.
8. Describe Kinsey's scale of homosexuality.
9. List some common myths about homosexuals.
10. Discuss psychoanalytic, behavioral, hormonal, and genetic theories about the origins of homosexuality.
11. Discuss whether or not homosexuality should be considered an illness.
12. Describe fetishism, and in particular, kleptomania, pyromania, and transvestism.
13. Describe a theory about the cause of transsexualism.
14. List the symptoms and possible causes of sadism, masochism, voyeurism, exhibitionism, pedophilia, incest, necrophilia, and bestiality.

15. Describe different kinds of male rapists.
16. Describe symptoms of the rape trauma syndrome.
17. List possible reasons why a woman would refuse to report a rape.
18. List preventive measures that a woman can take to avoid rape.
19. Discuss what a woman should do if attacked by a rapist.
20. List some reasons why a female or male would become a prostitute.
21. List some reasons why a person would seek the services of a prostitute.
22. Discuss possible causes of nymphomania, satyriasis, promiscuity, and celibacy.

INTRODUCTION

Now we embark on a discussion about patterns of human sexual behavior. In this chapter, we look at some usual and unusual patterns of sexual behavior, whereas in the next chapter we discuss some psychological and sociological influences on these patterns. In both chapters, statistics will be given indicating the frequencies of occurrence of certain sexual behaviors. These statistics are based on surveys such as those of Kinsey and his colleagues, Hunt, Levin and Levin, Athanasiou and colleagues, Westoff, Kantner and Zelnik, Sorensen, and Hite.[1-9]

When seeing these statistics, you should keep in mind some cautions. First, when your sexual behavior or that of another person does not fit with the "average" in our population, it does not mean that you or another are necessarily sick, weird, or perverted. Remember, sexual gratification is an individual thing, and there is a wide range of normal sexual behavior. Second, the above surveys are based on information supplied from a limited population of people and do not necessarily reflect our society in general. For example, Kinsey's surveys reported data only on white people, and many were from one state (Indiana). Ninety percent of Hunt's respondents were white, and all read *Playboy Magazine*. Ninety percent of the female respondents to Levin and Levin's survey were married, and most were well-educated, white, and of a high socioeconomic status. Similarly, the respondents in the Athanasiou survey were mostly young, well-educated, and well-off financially.

Also, people who would supply information about their sexual behavior in interviews or questionnaires may exhibit different patterns of behavior than those who would refuse to answer such questions.[10] Such volunteers also may exaggerate or conceal certain facts about their sex life, or their memory may be faulty. Nevertheless, this is the best information we have to date on patterns of sexual behavior in our society.

SEX AND THE HUMAN LIFE CYCLE

The sex drive and sexual behavior of an individual will vary throughout life. In this section, we discuss these age-related patterns in sexual behavior.

Sexual Behavior in Infancy

Infants (age 0–2) are not sexually indifferent people. All infants derive pleasure from parental cuddling and affection. This contact, although not directly sexual, has lasting effects on later emotional, including sexual, well-being. Infants that are rejected or lack affection can suffer adverse reactions. For instance, it is well known that infants who are separated from close human contact (in some orphanages, for example) suffer emotionally and can even develop severe physical and emotional problems.[11] If an infant experiences an emotional attachment to its parents, it later can generalize this attachment to others.

Infants also can respond sexually. Many male infants, for example, frequently have erections, and some even are born with an erection.[12] Although any sexual response in a female is more difficult to observe, such responsiveness probably is there. In the first year, infants mainly derive sexual pleasure from oral stimulation such as sucking. This is why Sigmund Freud called this the "oral stage."[13] In the second year, the infant becomes more interested in the anal region and defecation; Freud's "anal stage."[14] Many infants also *masturbate* (the act of deriving sexual pleasure from self-stimulation of the genitals). This is quite normal and is not a sign of perversion and reason for punishment, as some parents unfortunately believe. The genitals are erogenous zones in infants as in adults, and young babies often will discover and begin to explore this pleasure. Kinsey found that 32 percent of male infants were capable of having an orgasm (without ejaculation, of course) under one year of age. His studies also showed that female infants can experience orgasm through masturbation. By age 2 or 2½, infants realize their sex, mostly by comparing their genitals with those of their parents.

Sexual Behavior during Childhood

In early childhood (3 to 7 years), females and males become increasingly interested in sexually related activities, including masturbation. By the end of the third year, children begin to derive pleasure from systematically stimulating their genital region; thus, Freud's "phallic stage" begins.[15] Many children (about 60 percent) masturbate during early childhood.[16] The main difference between masturbation now and in infancy is that children learn not to do it in public.

By age 4 or 5, females and males exhibit more socially related behaviors. For example, they will hug or hold hands with the opposite sex. Much of this is an attempt to mimic their parents. Sexual games such as "playing doctor" (during which a boy and girl will secretly inspect each other's genitals) are common. Also, sexual play, usually involving handling of another's genitals, is common between children of the same sex at these ages.[17] Children in this stage often will expose their genitals to others, and also desire to view the genitals of the same and other sex. Some of this behavior is "homosexual" in nature but does not indicate any tendencies to be homosexual later in life. (Later in this chapter, we will have more to say about homosexual behavior.) By age 6 or 7, children develop more modesty and a sense of privacy about their bodies.

Sexual Behavior during Preadolescence

According to Sigmund Freud, sexual interest and behavior lie dormant during preadolescence (ages 8 to 12). This is why this stage often is called the "sexually latent stage."[18] Other studies have shown, however, that children of this age do exhibit an interest in sex.[19] Remember from Chapter 6 that physical changes associated with the process of puberty can begin during this stage. When this happens, many preadolescents tend to be uncomfortable and do not want their "new bodies" seen. Also, androgen levels in females and males of this age are increasing (Chapter 6), and sex drive is beginning to blossom. Because there usually is no outlet for this drive with another person, masturbation tends to increase during these years. According to one of the more recent surveys, by age 13, 63 percent of males had masturbated to orgasm and 33 percent of females had done so.[20] Kinsey found that many preadolescent males learn masturbation from their peers, but preadolescent females tend to discover this sexual outlet by self-exploration.[21]

The two sexes usually play separately beginning by about age 8. Therefore, between ages 8 and 12, children are becoming more interested in the opposite sex (many hear about coitus during this time) but have less contact with that sex, and this is correlated with a higher incidence of homosexual play. Many males (50 to 60 percent) engage in homosexual activity during preadolescence, as do 33 percent of females.[22] In males, this can take the form of mutual masturbation, often in a group. Preadolescent females, in contrast, tend to have these homosexual encounters in pairs, not in groups. Heterosexual activities during preadolescence consist of activities like "kissing games" at parties, casual dating, or "going steady." Children of this age also tend to form "crushes" on older individuals of the opposite sex, and they develop idols of the same sex whom they admire and try to impersonate.

Sexual Behavior during Adolescence

During adolescence (ages 13 to 19), there is a surge in sex drive; this is the "genital stage" of Freud.[23] Most individuals of this age are unmarried, and much of their time is occupied with relations (sexual and otherwise) with the opposite sex. Teenagers in America today are much more sexually active than in the past, and in the next chapter we discuss some possible reasons for this. Today, most males and about two-thirds of females experience premarital coitus before age 25. In addition, teenagers are more willing to engage in a wider variety of sexual behaviors than did previous generations. Hunt, for example, found that 72 percent of teenage males have experienced heterosexual fellatio, and 69 percent of teenage females have experienced heterosexual cunnilingus.[24] A wider variety of coital positions also are utilized by today's sexually active teenagers. In the next chapter, we discuss some reasons teenagers give for having or not having sex.

Coincident with the increase in adolescent sex drive is an increase in the frequency and incidence of masturbation. This activity increases markedly in boys at about 13 to 15 years of age (Fig. 18-1). It also increases in females, but more gradually throughout adolescence (Fig. 18-1). According to Hunt, only 15 percent of adolescents feel that masturbation is wrong,[25] and this contrasts markedly with older beliefs that masturbation is perverted and causes such maladies as warts and growth of hair on the hands, or insanity! However, adolescents who masturbate have mixed feelings about it, a holdover from these older beliefs.

Occasional homosexual behavior also occurs during adolescence. According to Sorensen, about 11 percent of adolescent males and 6 percent of adolescent females report having had some homosexual encounters during adolescence.[26] In some cases, this is a holdover from preadolescent homosexual play, but in other cases it indicates some degree of tendency toward homosexual behavior as a preference.

Sexual Behavior in Adults

In this section, we discuss patterns of adult sexual behavior in today's America. We will limit this discussion to heterosexual behavior and save a review of adult homosexual behavior for later. Since about 80 to 90 percent of Americans get married, statistics about sexual patterns usually are relevant to married couples. For detailed statistics on adult sexual behavior, you should read the original surveys mentioned at the beginning of this chapter.

Frequency and Variety of Coitus. One point that may be of interest is the frequency of coitus in married couples. According to the surveys of Hunt and

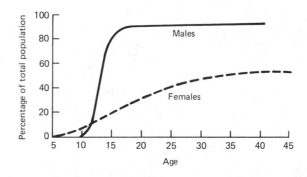

Figure 18-1. Cumulative incidence of males and females who have masturbated to orgasm at different ages. Adapted from Alfred C. Kinsey, Wardell B. Pomeroy, and Clyde E. Martin, *Sexual Behavior in the Human Male,* p. 502 (Philadelphia: W. B. Saunders Company, 1948); and Alfred C. Kinsey et al., *Sexual Behavior in the Human Female,* p. 141 (Philadelphia: W. B. Saunders Company, 1953). Reproduced by permission of Kinsey Institute for Sex Research, Inc.; © 1948 and 1953, respectively, by W. B. Saunders Company.

Westoff, the average American couple has coitus two or three times a week.[27] This frequency gradually declines with age or length of marriage (Table 18-1). The frequency of coitus can be much more or less than this in a given couple, and the final criterion for what is "normal" is what is satisfying to the couple. Married couples, on the average, use a wider variety of coital positions than in the past, and there also is a greater utilization of oral-genital contact by today's couples (Table 18-2). Over 80 percent of married couples report coitus as being pleasurable, regardless of age.

Masturbation. Masturbation often continues after a person is married. According to Hunt, 72 percent of young husbands masturbate an average of twice a month, and 18 percent of wives masturbate about once a month.[28] You may ask why this is so if so many married couples report coitus to be so pleasurable. One reason may be because masturbation can be a source of sexual outlet when the spouse is gone or is ill, or it may be used when there are wide differences in sexual desire in a couple. Also, a woman or man can derive a different kind of satisfaction from masturbation. For example, many women report a greater ease of attaining orgasm, and a greater incidence of multiple orgasms, from masturbation than from coitus.[29] Sexual fantasies usually accompany masturbation in both sexes. Both men and women have erotic dreams and can reach orgasm during these dreams. In men, these are called "nocturnal emissions."

Frequency of Coitus Related to Length of Marriage. The frequency of sexual activity for a married couple often changes as the years go by. On the average, there is a decline in coital frequency of married couples with age (Table 18-1); this decline usually can be related to time of marriage as well as to age. In some couples, the decline may be caused by sexual boredom and familiarity with the spouse. In such

TABLE 18-1. MARITAL COITUS: FREQUENCY PER WEEK (FEMALE AND MALE ESTIMATES COMBINED, 1938–1949 AND 1970S)

1938–1949 (Kinsey)		1972 (Hunt)		1970 (Westoff)	
Age	Median Frequency per Week	Age	Median Frequency per Week	Age	Median Frequency per Week
16–25	2.45	18–24	3.25	20–24	2.5
26–35	1.95	25–34	2.55	25–34	2.1
36–45	1.40	35–44	2.00	35–44	1.6
46–55	0.85	44–54	1.00		
56–60	0.50	55 and over	1.00		

Source: From J. S. Hyde, *Understanding Human Sexuality* (New York: McGraw-Hill Book Company, 1979). © 1979; used with permission. *Sources of data:* A. C. Kinsey et al., *Sexual Behavior in the Human Female* (Philadelphia: W. B. Saunders Company, 1953); M. Hunt, *Sexual Behavior in the 1970's* (Chicago: Playboy Press, 1974); and C. Westoff, "Coital Frequency and Contraception," *Family Planning Perspectives,* 6 (1974), 136–41.

TABLE 18-2. MOUTH-GENITAL TECHNIQUES IN MARITAL SEX, ACCORDING TO THE KINSEY STUDIES AND THE HUNT STUDY

Respondents	Percent of Marriages in Which Fellatio Is Used		Percent of Marriages in Which Cunnilingus Is Used	
	Kinsey (1939–1949)	Hunt (1972)	Kinsey	Hunt
Males with a high school education	15	54	15	56
Males with a college education	43	61	45	66
Females with a high school education	46	52	50	58
Females with a college education	52	72	72	72

Source: Reprinted with permission of PEI Books, Inc. From *Sexual Behavior in the 1970s* by Morton Hunt (Chicago: Playboy Press, 1974). © 1974 by Morton Hunt.

cases, experimentation with a wider variety of situations for sex and coital positions can make sex more interesting. In other couples, however, the frequency of sex increases during their marriage as they learn about each other's needs, or because they have more privacy as children grow up and leave home.

It is common for any individual to have brief periods in his or her life when because of fatigue, illness, tension, or depression he or she simply is not very interested in sex. Contrary to popular belief, this can happen to a man as easily as to a woman. When this does happen, the spouse should try to understand, realize it is temporary, and not interpret it as a decline in his or her own sexual attractiveness.

Single Adults. What happens to people's sexual activity if they are unmarried, divorced, or widowed? Hunt found that almost all divorced men and 50 percent of divorced women soon resumed sexual activity after a divorce.[30] In contrast, fewer (about half) widowed people resumed sexual activity after their spouse died. A lot of what happens after the death of a spouse depends on the age of the widow or widower, and how much grief remains. As unmarried people grow older, many find that the establishment of sexual relations becomes more difficult. It is also more difficult to meet people, and other activities consume their time. Joining "singles" social groups often helps such persons to establish relationships.

Sexual Behavior in Old Age

In Chapter 7, we discussed how the female reproductive system becomes less functional after menopause. A "male climacteric" also occurs, to varying degrees, in older men (Chapter 4). Sexual interest and activity, however, do not necessarily decline in old age, and some couples are sexually active even into their 80s. About one-third of men are still potent after age 70.[31] But many older people are widowed. This is especially true for women since the mean life expectancy in the United States for men (72 years) is less than for women (76 years). This is one reason why, in one study, about half of the women over 55 had not had coitus for some six months.[32]

There can be several reasons why sexual activity may decline in older couples.[33] The monotony and repetitiveness of sex between the couple can be a factor, and more variety and sexual experimentation may help in this regard. Some older people are preoccupied with their careers or money and "have no time for sex." Also, physical or mental illness or fatigue can interfere with sexual desires, as can overindulgence in food or drink. Finally, some older people may fear that they will not perform adequately, or they feel sex is inappropriate at this time of life or in their living situation.

VARIATIONS IN HUMAN SEXUAL BEHAVIOR

Some people exhibit patterns of sex drive or sexual behavior that are not the typical ones seen in our culture. Many of these unusual variations are considered abnormal, deviant, or perverse by the average heterosexual married American.[34] This is because such practices violate the moral standards of our culture. Often, they are also against the law (Chapter 19). For an individual, however, a given level of sex drive or kind of sexual behavior should be considered "abnormal" or "deviant" *only if* it is uncomfortable or harmful to that person, does not lead to healthy human sexual relationships, is bizarre in relation to what is usually done, harms another person, or is done within sight or sound of others.[35]

Homosexual Behavior

Homosexual behavior is the act of having sexual contact with a member of the same sex. Persons who frequently or always choose a member of the same sex in their sexual relations, or usually are sexually attracted to the same sex, are called *homosexuals* (Greek *homo* = "same"). Male homosexuals prefer to be called *gays* or *homophiles*. Female homosexuals often prefer the term *lesbian,* a name derived from a Greek homosexual poetess named Sappho, who lived on an island of Lesbos (now Mytilene) in the Aegean Sea, in 600 B.C. In contrast, a *heterosexual* person chooses to have sexual relations with the opposite sex *(heterosexual behavior).*

Kinsey contributed greatly to our understanding and communication about homosexual behavior in pointing out that there is a continuum from heterosexuality to homosexuality.[36] In his scale (Fig. 18–2), a person with a score of zero is exclusively heterosexual in his or her experience. At the other extreme are exclusive homosexuals, receiving a score of 6. These people always choose a sex partner of the same sex. Using this scale, about 2 percent of male Americans and 1 percent of female Americans have a score of 6 and are exclusively homosexual. About 63 percent of males and 87 percent of females have a score of 0 and are exclusively heterosexual.

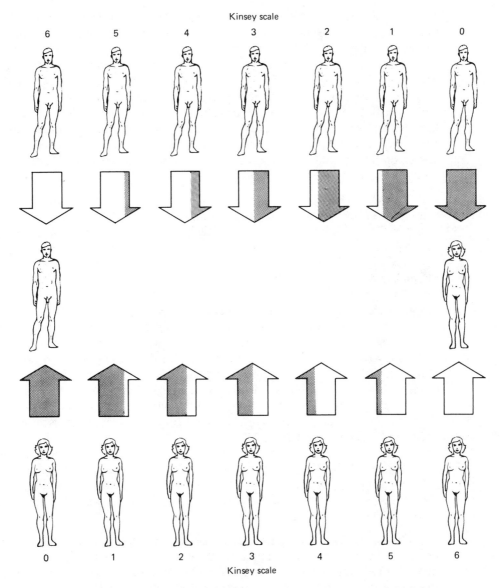

Figure 18–2. The Kinsey heterosexual-homosexual continuum: 0, exclusively heterosexual behavior; 1, mostly heterosexual, with incidental homosexual experience; 2, heterosexual, with substantial homosexual experience; 3, an individual with similar amounts of heterosexual and homosexual experience; 4, homosexual, with substantial heterosexual experience; 5, mostly homosexual, with incidental heterosexual experience; and 6, exclusively homosexual behavior. The proportion of each arrow shaded represents the percentage of heterosexual behavior. Individuals with scores of 1 through 5 could be termed *bisexual*. Source of data: A. C. Kinsey et al., *Sexual Behavior in the Human Male* (Philadelphia: W. B. Saunders Company, 1948).

The remaining people, 35 percent of males and 22 percent of females, are *bisexual* in that they have had sexual experience with their same and opposite sex at one time or another in their adult life.

Surveys more recent than Kinsey's show that these frequencies of homosexual, heterosexual, and bisexual behaviors remain similar today. It should be noted that a person may fear to express homosexual tendencies even though his or her feelings are almost exclusively homosexual. It also should be pointed out that there are many heterosexuals who have homosexual fantasies at one time or another in their lives but have no tendencies to express these in their sexual behavior. You can see from this discussion that it is difficult to classify a person as a "homosexual," and we really should talk about homosexual *behavior*.

Myths. There are some myths about homosexuals that should be put to rest. One is that all male homosexuals are effeminate, and all lesbians are masculine, in appearance. The truth is that many homosexuals look and behave (except sexually) like heterosexuals. Only about 15 percent of male homosexuals and 5 percent of female homosexuals can be identified by appearance. [37] Another misconception is that one person in a homosexual relationship always plays the "male dominant" role and the other the "female submissive" role. This certainly occurs in a few homosexual couples, especially younger ones, but in most couples the roles switch around. [38]

Another misconception is that homosexuals prey on young children. Actually, about 80 percent of cases of child molesting occur when a heterosexual man sexually molests a young girl. [39] Related to this is the idea that homosexuals in professions dealing with children (e.g., teachers, camp counselors) can introduce homosexuality to the children. The likelihood of this occurring, however, is rare and can be compared to the possibility that a heterosexual teacher or counselor will influence the direction of sexual identity taken by children. In most cases, adolescents initiate other adolescents into homosexuality. [40]

Another myth is that homosexuals are always looking around to attack heterosexuals. In reality, gays are very good at limiting their sexual approaches to other homosexuals, and most do not force their sexual attention on anyone. Finally, homosexual behavior during preadolescence or earlier does not predict that such a person will be a homosexual later in life.

Sexual Activity and Response. The sexual behaviors of homosexuals are similar to those of heterosexuals except for, of course, vaginal coitus. Kissing, hugging, and caressing can be included in foreplay of homosexuals, as can fellatio and anal coitus in homosexual males, cunnilingus in lesbians, and mutual masturbation in both sexes. Some lesbians also practice *tribadism,* which is when one woman assumes a malelike coital position. *Dildos* (artificial penises) or vibrators also are used by some lesbian couples. A recent study by Masters and Johnson indicates that the sexual responses cycles of homosexuals and heterosexuals are physiologically identical regardless of the type of stimuli present. [41] These authors note that homosexuals tended to take more time during foreplay and derive more pleasure from their sexual encounters than the average heterosexual couple. Surprisingly, they also found that homosexual and heterosexual fantasies occurred to a similar degree in heterosexuals and homosexuals. This finding, however, is not consistent with several others that show that homosexuals fantasize mostly about their same sex, and heterosexuals do the opposite. [42]

Patterns of Sexual Behavior. Male homosexuals tend to have many more sexual partners in their life than do heterosexual males. [43] They make sexual contacts by "cruising" the streets, or going to "gay" bars or bathhouses. Lesbians, on the other hand, tend to have fewer sexual partners than do male homosexuals. Some homosexual couples, however, do enter into long-lasting relationships.

Psychoanalytic Theory. What factors in a person's life predispose him or her to homosexual behavior as an adult? The answer is that we are not sure, but there are several theories. Psychoanalytic (Freudian) theory proposes that homosexuals experience an abnormal development of parental attachment. According to this theory, one- to three-year-old boys form an *Oedipus complex;* that is, they fall in love with their mother and are jealous of their father. This is based on the Greek myth where Oedipus killed his father and later married his mother—not knowing her true identity until it was too late. This conflict normally is resolved by the boy eventually identifying with his father and repressing his desire for his mother. According to this theory, male homosexuals retain a "negative Oedipus complex"; that is, they desire their father and are hostile to their mother. What would cause this reversal in a male's development? One possible answer could be the case where there is an overdominant mother who is seductive to the child, and the father is weak and detached. Thus, the child later fears relating to females because of the mother's jealous possessiveness and seductive behavior. [44] Also, he may fail to gain the love of his father and seek it in other men. [45] It is apparently true that more male homosexuals than male heterosexuals have had overcontrolling mothers [46] and emotionally distant, detached fathers. [47]

Similarly, young girls form an *Electra complex,* in which they fall in love with their father and are jealous of their mother. Later they repress their desire for their father and identify with their mother. Many lesbians are raised by rejecting mothers and distant or absent fathers. Thus, they have inadequate identification with their mother and also have not learned to relate to men. [48]

One problem with these psychoanalytic theories is that they predict that a homosexual would have a *gender identity* of the opposite sex (as we discuss in the next chapter, our gender identity is our awareness that we are male or female). As we have seen, however, many homosexuals have a normal gender identity but a reversed sex object choice.

Learning Theory. Learning theory also attempts to explain homosexual behavior. This theory assumes that all of us are born with both male and female potential. Our gender identity and sex object choice, then, are determined by what behaviors are rewarded or punished. If a boy, for example, is consistently rewarded for behavior that is typically exhibited by girls, he will develop a femalelike gender identity and prefer sexual relationships with males as an adult. There is evidence that some male homosexuals have had a childhood pattern of cross-dressing, playing with dolls, and a tendency to play with girls more than boys. [49] Similarly, lesbians would be rewarded as children when they exhibited behavior typically present in boys. Or, homosexual behavior in children may be rewarded by peers. A sociological aspect of this theory is that peer pressure can create homosexual behavior. For example, if a "sissy" boy and a "tomboy" girl are consistently labeled as "queer," they may be convinced that they are homosexual and will develop appropriate behavior for that role. One problem with these theories, as with many psychoanalytic theories, is that they predict disruptions both in gender identity and sexual object choice.

Hormonal Level in Adult Homosexuals. Does hormonal imbalance cause homosexual behavior? Some scientists have tested this idea by measuring hormone levels in the blood of adult homosexuals and comparing these levels to those in heterosexuals. One study found significantly lower testosterone levels and impaired testicular function in male homosexuals rating 5 or 6 on the Kinsey scale. [50] Another study, however, failed to find a difference in hormone levels in male homosexuals. [51] One study did find that lesbians have higher testosterone and lower estrogen levels in their blood then heterosexual females. [52] So, the jury is still out on this question. One must realize, however, that even if hormone levels in the blood of homosexuals differ from those in heterosexuals, there is no evidence that these differences can influence the direction of sex object choice. Rather, these differences probably could affect level of sex drive, either by hormones acting centrally or peripherally (Chapter 17). For example, if one administers testosterone to a male homosexual, his sex drive may increase, but it still would be directed toward other males.

Because early exposure to testosterone masculinizes the brain of male mammals (Chapter 5), some scientists theorize that homosexuals may have been exposed to abnormal levels of hormones as fetuses. Female fetuses suffering from exposure to abnormally high androgen levels, because of the adrenogenital syndrome (Chapter 5), tend to exhibit tomboyish behavior as children. However, although prenatal exposure to sex hormones can influence gender identity, the evidence that such exposure can influence sexual object choice (i.e., cause homosexuality) is not very convincing. [53] Again, this theory predicts that homosexuals would have both a disturbed gender identity and sex object choice, which often is not so.

Genetic Basis of Homosexuality. Sigmund Freud speculated that there may be a genetic basis for homosexuality. [54] One way to test for any genetic predisposition for homosexuality is to look at the sexual preferences of identical twins. If, for example, such a tendency exists, both identical twins should develop homosexuality if one of them does, even when they are raised by different families. This, however, is not the case. [55] Also, there are instances of identical twins raised in the same family, one being homosexual and the other heterosexual. [56] Therefore, there is no evidence of a genetic predisposition to homosexuality.

Situational Homosexual Behavior. Finally, there can be situational homosexual behavior. In prisons, for example, homosexual behavior is more common than in the general population. This is not because more homosexuals commit crimes and therefore end up behind bars. Rather, much of this behavior is done by heterosexuals who switch to homosexual behavior because there is no heterosexual outlet for their sex drive. Thus, the fact that homosexual rape is common in prisons can be explained by the need to establish dominance and subordinate relationships in the prison.[57]

Conclusions. In summary, we do not know what causes homosexual behavior. All human behavior is a complex interaction between inheritance, development, physiology, and learning experiences, and it is difficult to determine the relative influences of these factors. Furthermore, scientists who attempt to determine the causes of adult homosexual behavior often come up with differing conclusions. For example, remember that we discussed the general feeling that it is not always true that an adult homosexual exhibits a disturbed gender identity. Yet, a recent study by the Kinsey Institute of Sex Research suggests that a disturbance of gender identity in childhood is very common in both male and female homosexuals.[58] That is, male homosexuals exhibit feminine behaviors as young boys, and the reverse is true for female homosexuals. This study also suggests that homosexual behavior has its origins, not as a result of a disturbed family environment, but as a deep-seated predisposition, perhaps biological in origin. Thus, perhaps there are several reasons why an adult may choose to exhibit homosexual behavior, and we should speak of different kinds of homosexuality.

Many heterosexuals believe that homosexual behavior is an illness or sickness, and even 11 percent of gays feel this way.[59] Homosexuality is not a crime in most states, but certain sex acts, such as oral-genital contact and anal coitus, are considered crimes in several states (Chapter 19). In recent years, homosexuals have become more free to express their sexual preference in public, a choice often called "coming out of the closet." Many homosexuals feel better about themselves now because of support from "gay liberation" groups. Even so, there still is much prejudice against homosexuals. Homosexuality, for example, still is the basis for a dishonorable discharge from the armed forces, and gays can be fired from some government jobs because they supposedly are a security threat. However, clinicians have now agreed that homosexuality is not an illness; it is just a life style. In general, homosexuals do not have more adjustment problems than do heterosexuals.[60] Furthermore, in 1973, the American Psychiatric Association removed homosexuality from its list of "personality disorders, and certain other nonpsychotic mental disorders."

Treatment. If homosexuality is not an illness, should it be treated? A promising viewpoint is that a homosexual should seek professional help only if being gay makes him or her unhappy. Traditionally, psychologists or psychiatrists have found that it is difficult to treat a homosexual. Masters and Johnson, however, report great success not only in treating sexual dysfunction in homosexuals but also in eliminating homosexual tendencies in about two-thirds of those who wished to change their sex object choice.[61] Some critics of this finding feel that the patients studied were really mostly bisexuals. In actual fact, a majority of the ''homosexuals'' treated by Masters and Johnson were rated scores 3 or 4 on the Kinsey scale. Nevertheless, their procedures should be evaluated and tested on more people.

Fetishism

Fetishism is when a person is sexually fixated on some object (called a *fetish*) or on some part of another person's body. In most cases, this individual needs a specific object to be sexually aroused. Examples of sexual fetishes include parts of the body such as feet, hair, breasts, or buttocks. Other examples include a specific kind of physical material (e.g., leather, silk) or objects such as shoes, purses, handkerchiefs, or lingerie. Before you go too far in thinking that you have a fetishism because such an object (like part of a man's or woman's body) sexually excites you, realize that this attachment is not a sexual fetishism unless nothing else will arouse you (Fig. 18–3). *Coprophilia* and *urophilia* are rare fetishisms in which one is sexually aroused by feces or urine, respectively.

Strength of preference for fetish object

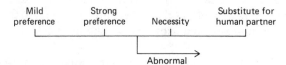

Figure 18–3. A chart showing when a fetishism becomes abnormal. Reproduced from J. S. Hyde, *Understanding Human Sexuality* (New York: McGraw-Hill Book Company, 1979). © 1979; used with permission.

Kleptomania, or compulsive stealing, is a fetishism if one derives sexual pleasure from the theft. Often the stolen object has little monetary value. Many kleptomaniacs are young females who feel unloved, but kleptomaniacs also can be young men who steal fetishes such as women's lingerie.[62]

Pyromania is when someone compulsively sets fires, and often derives sexual pleasure from this act, masturbating as he or she watches the fire.[63]

Transvestism (*trans* = ''cross''; *vest* = ''dress'') usually is considered a special case of fetishism. A transvestite is a man or woman who wears clothes typically worn by the opposite sex. Some male homosexuals (''drag queens'') will dress up as women, and some lesbians (''butches'') will wear men's clothes. Many transvestites, however, are heterosexuals who periodically put on women's clothes, often in secret. Many of these are married men with families, although some adolescents also cross-dress.[64] A true transvestite derives sexual pleasure from cross-dressing. These people often began cross-dressing as an adolescent, and in many cases the parents rewarded such behavior. Some show-business people cross-dress for entertainment value, but they usually do not derive sexual pleasure from this and are not transvestites.

Transsexualism

A *transsexual* is a person who feels that he or she is trapped in the body of the wrong sex, a condition also called *sex-role inversion*. In these people, their gender identity does not fit with their biological sex. They not only dress in the clothes of the opposite sex, but emotionally, psychologically, and sexually they want to be the opposite sex. Most transsexuals are male, and these men desire male sexual partners who are "straight," not homosexual, men. A common opinion about the cause of transsexualism is that these people were encouraged by their parents to act like, and even dress like, the opposite sex, and thus they developed the gender identity of the opposite sex.[65] Some transsexuals, however, were not treated in this way, and we are unable to rule out prenatal exposure to abnormal hormone levels as a cause of sex-role reversal, even though adult transsexuals have the normal hormone levels of their biologic sex.[66] In the next chapter, we discuss the development of sex roles in more detail.

With proper hormonal treatment and/or surgery, the individual whose sexual identity is not in tune with his or her sexual anatomy need not remain so. A transsexual man can receive estrogen treatment, which can feminize his appearance. For example, his breasts may enlarge. Or he may choose to have a *sex-change operation*. This can involve castration and surgical removal of the penis. An artificial vagina can be created by making an opening in front of the anus and lining it with skin from the penis. Part of the scrotum can be used to form labia. Such "women" can have heterosexual intercourse, and some adapt well to their new female role. Similarly, a transsexual woman can receive androgen treatment, which can produce a malelike body, facial hair, and a deep voice. There is, however, little effect of this treatment on the external genitalia other than slight enlargement of the clitoris. Sex change operations for female transsexuals are much less common and more difficult than such operations for males.

Sadism and Masochism

A *sadist* derives sexual satisfaction from inflicting physical or psychological pain on others. The term is derived from the Frenchman Marquis de Sade (1740–1814), who delighted in harming and even killing his female sex partners. Hunt found that 5 percent of males and 2 percent of females in his sample at one time or another derived sexual pleasure from inflicting pain.[67] An extreme case of sadism is "lust murder," when a psychotic person derives sexual pleasure from killing and often mutilating another person.

A *masochist,* on the other hand, derives sexual pleasure from receiving physical or psychological pain. The term is derived from the Austrian novelist Leopold von Sacher-Masoch (1836–1895), who wrote about his masochistic sexual desires. Hunt found that 2.5 percent of males and 4.6 percent of females had experienced pleasure from receiving pain.[68] This type of pleasure differs from that received from the pain experienced during gentle nibbling, pinching, or biting during foreplay or coitus (Chapter 17). In the first place, the latter causes only mild pain. Secondly, a masochist, unlike most people, needs the pain to be sexually aroused.

When a couple has a "sado-masochistic" sexual relationship, they require elaborate rituals that are necessary for the giving or receiving of pain to be arousing and pleasurable. In these couples, one person is the sadist and the other the masochist, and activities such as whipping, hard biting, pinching, and/or slapping, as well as degrading comments, are used by the sadistic partner. In extreme cases, the masochistic partner can be tied up like a slave ("bondage"). Even though most of us do not indulge in such activities, Kinsey found that 10 percent of males and 3 percent of females were sexually aroused by sado-masochistic stories.[69] How does sado-masochism develop in a person? One idea is that it is a learned association of sex with painful events. For example, a child who was spanked after being caught masturbating may begin to associate sexual arousal with physical pain. In other people, this kind of activity is a means of punishing themselves or others for having sex because they were taught that sex was dirty or a "sin."

Voyeurism

A *voyeur* (French, *voir* = "to see") is a person who derives sexual pleasure from observing a nude person or the sexual activity of others.[70] Obviously, most of us can be sexually aroused in this way, as evidenced by the sales of "skin" magazines and the popularity of X-rated movies. However, a true voyeur relies exclusively on his voyeurism to become sexually aroused. Laws often are broken in the process, such as invasion of privacy or trespassing, so that the voyeur can view the source of stimulation. The element of risk is important to the sexual experience of these people.[71] About nine males to every one female are arrested for voyeurism. Most voyeurs are young, late-blooming males, are lonely, and have had a difficult time forming meaningful sexual relationships.[72] Voyeurs often are called "peeping Toms," after Tom of Coventry, who was the only person in the town that took a peek at the naked Lady Godiva.

Exhibitionism

Exhibitionism, or indecent exposure, is when a person derives sexual pleasure from exposing his or her genital region in public. (*Triolism* is a form of exhibitionism, when two people have a third person watch their sexual activity, this third person then being a voyeur.) Usually, a male will expose himself to a female; and arrests of such males account for about 35 percent of arrests for sex offenses. Some of these men derive sexual pleasure from shocking the woman, but only in one of ten cases of exhibitionism does the man contemplate or attempt rape. Other exhibitionists are mentally deficient, or they are high on drugs or drunk during the act. Many are married, but their sex life is poor, and often they fear their masculinity.[73]

Pedophilia

Pedophilia (*pedo* = "child") is when a person desires or engages in a sexual act with a child. It also is called child molesting. In Kinsey's study, 24 percent of the females had been sexually approached between the ages of 4 and 13 by a male at least 5 years older.[74] In almost two-thirds of the cases of child-molesting, the molester is a friend, acquaintance, or relative. Violent child molesting is rare, and coitus rarely is attempted. Usually, the sexual interaction involves petting and genital fondling.[75] Most male molesters are 30- to 40-years old and heterosexual; only about 20 percent of child molesting cases involve homosexuals.[76] The actual molestation usually does not harm a child's emotional development, although these children often are frightened by adult reaction to the offense.[77] This should be kept in mind when dealing with a molested child.

Incest

Incest is sex between close relatives, including stepparents and stepchildren.[78] About 4 percent of adults have had incestual experiences as children.[79] This usually involves brothers and sisters, fathers and daughters, or, more rarely, mothers and sons.[80] The incidence of reported incest in the United States has been increasing. In 1955, only about 5000 cases were reported, but at present about 500,000 cases of incest are reported each year. In most of these cases, fathers or stepfathers are incestuous with daughters, the latter averaging 8 to 13 years of age. The actual incidence of incest probably is much higher because a majority of such incidences are not reported. Incest is against the law or is a moral taboo in many cultures; we discuss the reasons this is so in the next chapter.

Necrophilia and Bestiality

Necrophilia is having sexual relations with a corpse. It reflects an extremely rare, psychotic need to dominate. *Bestiality* is having coitus with an animal. It also is called *zoophilia* and is considered to be sodomy in some states. Kinsey found that about 8 percent of adolescent males have had some contact with animals to orgasm; this percentage was 17 percent if the adolescent males had been raised on a farm. About 4 percent of all females have had some sexual contact with animals.[81]

Rape

Rape is forced sexual relations with a person without his/her consent. Rape can occur under the threat or use of force, by inducing fear, or when a person is drugged or mentally deficient. *Statutory rape* is when a person has sexual relations with someone who is too young to make a choice in the matter. What is "too young" varies in different states, but usually this "age of consent" is from 12 to 16 years.

Rape of a female by a male is frighteningly common. There are about 50 reported cases per 100,000 women in the United States per year. About 4 of every 5 rapes, however, are not reported; if these unreported rapes were also included in the estimates, the incidence of rape would be much higher. In Los Angeles alone, for example, the chance of a woman being raped in her lifetime is about 1 in 10.[82]

Kinds of Rapists.   Is there a typical rapist? The answer is no, but these offenders can be classified to some extent. [83] The *aggressive-aim rapist* is motivated by a desire to hurt the woman. Sex has little to do with the act except as a weapon. Such men often try to harm the woman's body, especially her genital organs or breasts. The victims typically are strangers. In contrast, the *sexual-aim rapist* uses a minimal amount of violence. He usually practices fetishism or exhibitionism, and is a shy and lonely person with low self-esteem. He derives sexual pleasure from the rape. The *sexual-aggressive fusion rapist* is motivated by sadism. He derives sexual pleasure from inflicting pain on the woman. In fact, the violence is necessary for him to be sexually aroused. Often these men have a history of antisocial behavior and are very dangerous. *Impulse rapists* are those who attack a woman when the opportunity arises, for example, during a burglary of a home. So-called *date rape* usually is not formally classified as rape. This is when coitus occurs with some resistance by the woman. Such cases are quite common, often happening when the man misinterprets that a woman wants to have coitus when she really does not. [84] He then, in frustration, continues despite her resistance. The male also may be punishing her for some reason. Finally, as mentioned, homosexual rape is common in prisons, and is an expression of control or dominance.

Most rapists are young (under 25), and many repeat the crime. [85] Most do not murder their victims; only about 1 in 500 rapes ends in murder. About half of rapists are drunk. [86] And in about half of rapes, the woman knows the man, at least casually.

Rape Trauma Syndrome.   Rape can have a profound adverse effect on a woman. Physical harm, of course, can occur. There also is a somewhat typical emotional response in a woman after an attempted or actual rape called the *rape trauma syndrome.* [87] Immediately after the event, the acute phase of this syndrome begins and can last several weeks. During this phase, the woman openly expresses her fear, anxiety, anger, tension. She also can blame herself with questions like, ''Why was I so stupid?'' ''Why didn't I resist?'' This acute phase often is followed by a longer period of emotional adjustment, called the long-term reorganization phase, which may be accompanied by calmness, coldness, and jumpiness. The often subconscious tension during this phase causes symptoms like sleepless nights, fear of going outside, and fear of men in general, which can harm her sexual relationships. Sexual dysfunction can appear at this time. There also can be work disturbance and the desire to change jobs, especially if the rape was associated with the place of work. A woman may change her place of residence because of the fear of reprisal or of repetition of the rape.

Treatment of Rape.   Many women are ambivalent as to whether to report a rape and press charges if the criminal is apprehended. [88] A woman may be embarrassed or may fear reprisal if the man is let go. Besides not reporting the crime, she may even fail to discuss it with close friends. These women go through the same rape trauma syndrome as if they did report it, but usually the syndrome is more severe because they have not openly expressed their feelings and talked about the event with others. Even when a woman has the courage to report a rape to the police, the authorities have had a history of responding with little understanding and often ridicule. Sometimes it even seems as if the victim is the criminal! This is because many men still feel that a woman who is raped often "asks for it." The occurrence of occasional false rape reports, in which the woman is acting out of jealousy, blackmail, or protection of her reputation, [89] has led to suspicion on the part of authorities.

Hospital personnel where a woman is examined and treated after a rape report sometimes act in a similar manner. The woman must be examined for the presence of semen in the vulva and receive a shot of antibiotic and a test for presence of sexually transmitted disease (Chapter 16). She also may be given a "postcoital" estrogen to prevent pregnancy (Chapter 13). In fact, some women do not report a rape because they fear how they will be treated in a hospital. Recently, the generally unsympathetic behavior of law-enforcement and hospital personnel has changed to be more understanding toward rape victims.

Rape Trials. When a rape is tried in court, it often creates further trauma for the woman. She can be asked to answer questions about her past sexual experience in an attempt by the defense to show she played some voluntary role in the incident. Some states fortunately will no longer permit evidence relating to a woman's previous sexual experience. Also, she is usually asked to recall, in great detail, the events of the rape, details that she would prefer to forget!

Rape Prevention. No wonder many women fear being raped. What can a woman do to prevent rape? Some suggestions are: Stay away from dimly lit or deserted areas; walk in the middle of the sidewalk to avoid hidden recesses; drive with the car windows up and doors locked, and always lock a parked car; never let a strange male into your house; take one of several self-defense courses available for women, and carry some physical deterrent like a hat pin or mace.

What to Do If Attacked. If a woman is attacked, should she resist or give in? Unfortunately, this is a difficult question to answer. First of all, there is little time to think over the matter during an attack. Secondly, the woman will not know what kind of rapist is attacking her, so she doesn't know whether she should resist or not. As mentioned, some rapists actually are sexually aroused and spurred on by violence, so to resist could make matters worse. Obviously, if the man has a weapon or is much stronger than her, she does not have much choice but to give in. If she does resist, she should do so with as much vigor as she can muster; running away if possible, loud and continuous screaming, biting, clawing, and kicking to the genitals, and poking the eyes are all appropriate behaviors. [90]

Prostitution

Prostitution is the act of selling sex for immediate monetary reward. Most female prostitutes are heterosexual and not well-educated. [91] Some enter their trade for financial reasons, or for the glamour and excitement. Others need to punish or degrade themselves, or to rebel against their parents. Only a few are in business to support their drug habits. Despite rumors about the "white slave trade," only about 4 percent of female prostitutes are forced into their trade. About two-thirds have or have had a sexually transmitted disease, but only 5 percent of cases of venereal disease are contracted from prostitutes. [92]

Prostitutes are either in the business on their own or under the tutelage of a "pimp" who manages their affairs and protects them. Some prostitutes operate out of brothels ("whore houses") or massage parlors, or do business in bars as "call girls." Some simply solicit buyers ("Johns") on street corners.

The use of female prostitutes for a sexual outlet was once more common than it is now. Kinsey found that 69 percent of white males had had some experience with prostitutes, but Hunt found that only 19 percent of men over 35 with a college education had premarital coitus with a prostitute. [93] Why do men seek the services of a prostitute? Perhaps they have no other sexual outlet, or the prostitute supplies them with sexual acts and pleasures that are not available in their other relationships. The above-mentioned decline in use of prostitutes may be due to sexual outlets being more available in today's culture.

There can also be male prostitutes. Some young heterosexual men, called gigolos, sell their sexual services to women. Males also can be homosexual prostitutes. These, mostly young, males can operate out of homosexual brothels (called "Peg houses"), cruise the streets, or do business in "gay bars" or "tea houses" (public restrooms). About 100,000 young males, from 13- to 16-years old, are currently involved in male prostitution in the United States. [94]

Nymphomania and Satyriasis

Women with *nymphomania,* a rare variation of sex drive, possess a very high, unsatiable desire for sex. These women are compulsive and are never satisfied. Many have an obsessive need to be loved or accepted as a means to release emotional tension. In some cases, these females seek revenge against their fathers. The male equivalent of nymphomania is called *satyriasis.*

Promiscuity

Promiscuity is when a person has coitus with many people on a casual basis. This behavior, surprisingly enough, is usually not associated with a strong sex drive. Instead, sex is used to cope with nonsexual emotional problems, such as feelings of inadequacy. Promiscuous men often are called "Don Juans."

Celibacy

Celibacy is the total abstention from sexual activity. Celibacy is practiced by some religious groups, such as Roman Catholic priests. In other cases, celibacy can be situational. An example would be the absence or illness of a spouse. Also, some people choose to be celibate, preferring to express their sexual drive through other channels. Celibacy is not harmful to one's health and should be considered an alternative life style.

CHAPTER SUMMARY

Statistics from "sex surveys" about patterns of sexual behavior may not be indicative of such patterns in the population as a whole, but they are the only information we have.

Infants are sexually responsive and derive sensual pleasure from oral, anal, and genital stimulation. During childhood, sexual expression becomes more social, and sex play with the same or opposite sex is common, as is masturbation. The preadolescent represses sexual tendencies to some extent, but children of this age frequently masturbate and exhibit homosexual and heterosexual interactions. Puberty occurs at the start of adolescence, and a person's sex drive increases. Heterosexual activity begins around the time of puberty, and masturbation remains another form of sexual outlet. Most heterosexual married adults derive pleasure from coitus and masturbation. In some older people, sexual activity may decline, but many older people are physically capable of, and desire, sexual interactions.

Adult homosexual behavior occurs to various degrees in about 37 percent of males and 23 percent of females. Most of the people exhibiting homosexual behavior, however, are bisexual, and only about 2 percent of adult males and 1 percent of adult females are strictly homosexual. Some myths about homosexuals include: (1) male homosexuals are always effeminate and lesbians always masculine, (2) in homosexual couples, one person always plays the "male" role, (3) homosexuals prey on children, (4) homosexuals are looking to attack heterosexuals, (5) homosexuals should not be in jobs dealing with children, and (6) homosexual behavior during preadolescence or childhood means a person will be a homosexual as an adult.

Sexual behavior and response in homosexuals are similar to those of heterosexuals, except for the absence of vaginal coitus. Although most (especially male) homosexuals have several sexual partners, some homosexual couples develop long-lasting relationships. Theories about what causes adult homosexuality are numerous, and include retention of a negative Oedipus complex (psychoanalytic theory), reward for homosexual behavior as a child (learning theory), exposure to abnormal hormone levels (hormonal theory), presence of a genetic influence on homosexuality (genetic theory), and situations where no members of the opposite sex are available (situational theory). None of these theories fully accounts for the behavior of homosexuals. Homosexuality is not an illness, but simply an unusual sexual life style.

Other unusual patterns of sexual object choice include fetishism (including kleptomania, pyromania, and transvestism), transsexualism, sado-masochism, voyeurism, exhibitionism, pedophilia, incest, necrophilia, and bestiality.

Rape of a female by a male is alarmingly common. Men can be classified as aggressive-aim, sexual-aim, sexual-aggressive fusion, impulse, or homosexual rapists. A woman who is raped may exhibit a typical sequence of reactions called the rape trauma syndrome. Many women do not report a rape because of embarrassment or fear or because of the trauma they go through from unsympathetic treatment by police, hospitals, and the courts. A woman can take several steps to avoid rape and should prepare to defend herself if necessary.

Female prostitutes are in the business for money, excitement, or emotional reasons. Men seek the services of these women because they need sexual outlet or new sexual experiences. Homosexual or heterosexual male prostitutes also are common.

Sex drive can be at an extreme level in some people. A high, insatiable sex drive is present in nymphomania and satyriasis. Promiscuous people, on the other hand, have a low sex drive but have many indiscriminate sexual encounters for nonsexual emotional reasons. Finally, some people may choose celibacy as a life style.

NOTES

1. A. C. Kinsey et al., *Sexual Behavior in the Human Male* (Philadelphia: W. B. Saunders Company, 1948).

2. A. C. Kinsey et al., *Sexual Behavior in the Human Female* (Philadelphia: W. B. Saunders Company, 1953).

3. M. Hunt, *Sexual Behavior in the 1970s* (Chicago: Playboy Press, 1974).

4. R. J. Levin and A. Levin, "The Redbook Report on Premarital and Extramarital Sex," *Redbook* (October 1975), p. 38.

5. R. Athanasiou et al., "Sex," *Psychology Today,* 4, no. 2 (1970), 39–52.

6. C. Westoff, "Coital Frequency and Contraception," *Family Planning Perspectives,* 6 (1974), 136–41.

7. J. K. Kantner and M. Zelnik, "Sexual Experience of Young Unmarried Women in the United States," *Family Planning Perspectives,* 4 (1972), 9–18.

8. R. C. Sorenson, *Adolescent Sexuality in Contemporary America* (New York World Publications, 1973).

9. S. Hite, *The Hite Report* (New York: Macmillan Publishing Co., Inc., 1976).

10. G. Katts and K. Davis, "Effects of Volunteer Biases in Studies of Sexual Behavior and Attitudes," *Journal of Sex Research,* 17 (1971), 26–34.

11. R. A. Spitz, *The First Year of Life* (New York: International Universities Press, 1965); and S. Provence and R. C. Lipton, *Infants in Institutions* (New York: International Universities Press, 1963).

12. H. M. Halverson, "Genital Sphincter Behavior of the Male Infant," *Journal of Genetic Psychology,* 43 (1940), 95–136.

13. S. Freud, *Three Essays on the Theory of Sexuality,* J. Strachey, ed. and trans. New York: Basic Books, 1963.

14. Ibid.

15. Ibid.

16. G. V. Ramsey, "The Sexual Development of Younger Boys," *American Journal of Psychiatry,* 56 (1943), 217–33.

17. F. M. Martinson, *Infant and Child Sexuality: A Sociological Perspective* (St. Peter, Minn.: Book Mark, 1973).

18. Freud (1963), op. cit.

19. Martinson (1973), op cit.

20. Hunt (1974), op. cit.

21. Kinsey et al. (1948; 1953), op. cit.

22. W. R. Reevy, "Child Sexuality," in *The Encyclopedia of Sexual Behavior,* eds. A. Ellis and A. Abarbanil (New York: Hawthorn Books, Inc., 1967).

23. Freud (1963), op. cit.

24. Hunt (1974), op. cit.

25. Ibid.

26. Sorensen (1973), op. cit.

27. Hunt (1974), op cit; Westoff (1974), op. cit.

28. Hunt (1974), op. cit.

29. Hite (1976), op. cit.

30. Hunt (1974), op. cit.

31. J. Rubin, *Sexual Life after Sixty* (New York: Basic Books, Inc., 1965).

32. W. C. Wilson, "The Distribution of Selected Sexual Attitudes and Behaviors among the Adult Population of the United States," *Journal of Sex Research,* 11 (1975), 46–64.

33. W. H. Masters and V. Johnson, *Human Sexual Inadequacy* (Boston: Little, Brown & Company, 1970); and W. S. Appleton, "Why Marriages Become Dull," *Medical Aspects of Human Sexuality* (March 1980), pp. 73–85.

34. J. Marmor, " 'Normal' and 'Deviant' Sexual Behavior," *Journal of the American Medical Association,* 217 (1971), 165–170.

35. A. Ellis, *Sex without Guilt* (Secaucus, N. J.: Lyle Stuart, Inc., 1958); and A. Buss, *Psychopathology* (New York: John Wiley & Sons, Inc., 1966).

36. Kinsey et al. (1948), op. cit.

37. W. B. Pomeroy, "Parents and Homosexuality," *Sexology* (March 1966), pp. 508–11.

38. D. Martin and P. Lyon, *Lesbian/Woman* (San Francisco: Glide Publications, 1972).

39. C. H. McCaghy, "Child Molesting," *Sexual Behavior,* 1 (1971), 16–24.

40. Sorenson (1973), op. cit.

41. W. Masters and V. Johnson, *Homosexuality in Perspective* (Boston: Little, Brown & Company, 1979).

42. J. Marano, "New Light on Homosexuality," *Medical World News,* 20, no. 9 (1979), 8–19.

43. M. S. Weinberg and C. Williams, *Male Homosexuals: Their Problems and Adaptations* (New York: Oxford University Press, 1974).

44. I. Bieber et al., *Homosexuality: A Psychoanalytic Study of Male Homosexuals* (New York: Basic Books, Inc., 1962).

45. I. Bieber, "A Discussion of Homosexuality: The Ethical Challenge," *Journal of Consulting and Clinical Psychology,* 44 (1976), 163–66.

46. M. T. Sager and E. Robins, *Male and Female Homosexuality. A Comprehensive Investigation* (Baltimore: The Williams & Wilkins Company, 1973).

47. Ibid.

48. C. Wolff, *Love between Women* (New York: Harper & Row, Publishers, Inc., 1971).

49. F. L. Whitman, "Childhood Indicators of Male Homosexuality," *Archives of Sexual Behavior,* 6 (1977), 89–96.

50. R. C. Kolodny et al., "Plasma Testosterone and Semen Analysis in Male Homosexuals," *New England Journal of Medicine,* 285 (1971), 1170–74.

51. W. L. Jaffee et al., "Plasma Hormones and the Sexual Preferences of Men," *Psychoneuroendocrinology,* 5 (1980), 33–38.

52. J. A. Loraine et al., "Patterns of Hormone Excretion in Male and Female Homosexuals," *Nature,* 234 (1971), 552–54.

53. A. A. Ehrhardt and H. F. L. Meyer-Bahlburg, "Effects of Prenatal Sex Hormones on Gender-Related Behavior," *Science,* 211 (1981), 1312–18; J. D. Yalom et al., "Prenatal Exposure to Female Hormones," *Archives of General Psychiatry,* 28 (1973), 554–61; and G. Dorner, "Sex-Hormone Dependent Brain Differentiation and Reproduction," in *Handbook of Sexology,* eds. J. Money and H. Musaph, pp. 227–43 (New York: Elsevier North-Holland Biomedical Press, 1978).

54. Freud (1963), op. cit.

55. L. Heston and J. Shields, "Homosexuality in Twins: A Family Study and a Registry Study," *Archives of General Psychiatry,* 18 (1968), 149–60.

56. J. Marmor and R. Green, "Homosexual Behavior," in *Handbook of Sexology,* eds. J. Money and H. Musaph, pp. 1051–68 (New York: Elsevier North-Holland Biomedical Press, 1978).

57. S. Brownmiller, *Against Our Will* (New York: Simon & Schuster, 1975).

58. A. Bell et al., *Sexual Preference: Its Development in Men and Women* (Bloomington, Indiana: Indiana University Press, 1981).

59. Weinberg and Williams (1974), op. cit.

60. R. Green, "Homosexuality as a Mental Illness," *International Journal of Psychiatry,* 10 (1972), 77–98; and D. H. Rosen, *Lesbianism: A Study of Female Homosexuality* (Springfield, Ill.: Charles C. Thomas, Publisher, 1974).

61. Masters and Johnson (1979), op. cit.

62. M. Alexander, "Sex and Stealing," *Sexology* (April 1965), pp. 410–12.

63. E. S. Robbins et al., "Sex and Arson: Is There a Relationship?" *Medical Aspects of Human Sexuality* (October 1969), pp. 57–64.

64. V. Prince and P. M. Butler, "Survey of 504 Cases of Transvestism," *Psychological Reports,* 31 (1972), 903–17.

65. J. P. Driscoll, "Transsexuals," *Transaction* (March-April 1971), pp. 28–31.

66. J. R. Jones, "Plasma Testosterone Concentrations in Female Transsexuals," *Archives of Sexual Behavior,* vol. 2 (1972).

67. Hunt (1974), op. cit.

68. Ibid.

69. Kinsey et al., (1948; 1953), op. cit.

70. R. S. Smith, "Voyeurism: A Review of the Literature," *Archives of Sexual Behavior,* 5 (1976), 585–608.

71. E. Sagarin, "Power to the Peep-Hole," *Sexual Behavior,* 3 (1973), 2–7.

72. Smith (1976), op. cit.

73. J. C. Coleman, *Abnormal Psychology and Modern Life,* 4th ed. (Glenview, Ill.: Scott, Foresman & Company, 1972).

74. Kinsey et al., (1953), op. cit.

75. P. H. Gebhard et al., *Sex Offenders: An Analysis of Types* (New York: Harper & Row, Publishers, Inc., 1965).

76. McCaghy (1971), op. cit.

77. Ibid.

78. E. Sagarin, "Incest: Problems of Definition and Frequency," *Journal of Sex Research,* 13 (1977), 126–35; and N. H. Greenberg, "The Epidemiology of Childhood Sexual Abuse," *Pediatric Annals,* 8 (1979), 16–28.

79. Gebhard et al. (1965), op. cit.

80. D. Lester, "Incest," *Journal of Sex Research,* 8 (1972), 268–85.

81. Kinsey et al. (1948; 1953), op. cit.

82. J. Selkin, "Rape," *Psychology Today,* 8, no. 8 (1975), 70–76.

83. M. L. Cohen et al., "The Psychology of Rapists," *Seminars in Psychiatry,* 3 (1971), 307–27.

84. E. J. Kanin, "Selected Dyadic Aspects of Male Sex Aggression," *Journal of Sex Research,* 5 (1969), 12–18.

85. Cohen et al. (1971), op. cit.

86. R. T. Rada, "Alcoholism and Forcible Rape," *American Journal of Psychiatry,* 132 (1975), 444–46.

87. A. W. Burgess and L. L. Holmstrom, "Rape Trauma Syndrome," *American Journal of Psychiatry,* 131 (1974), 981–86.

88. Brownmiller (1975), op. cit.

89. J. M. MacDonald, "False Accusations of Rape," *Medical Aspects of Human Sexuality* (May 1973), pp. 170–93.

90. K. Kollias and J. Tucker, "Interview: Woman and Rape," *Medical Aspects of Human Sexuality* (May 1974), pp. 183–97.

91. B. E. Bess and S. S. Janus, "Prostitution," in *The Sexual Experience,* eds. B. J. Sadock et al. (Baltimore: The Williams & Wilkins Company, 1976).

92. J. James, "Prostitution: Arguments for Change," in *Sexuality Today and Tomorrow: Contemporary Issues in Human Sexuality,* eds. S. Gordon and R. W. Libby (N. Scituate, Mass.: Duxbury Press, 1976).

93. Kinsey et al. (1948), op. cit.; Hunt (1974), op. cit.

94. R. Lloyd, *For Money or Love: Boy Prostitution in America* (New York: Vanguard Press, Inc., 1976).

FURTHER READING

BROWNMILLER, S., *Against our Will.* New York: Simon & Schuster, 1975.

GREEN, R., "Variant Forms of Human Sexual Behavior," in *Reproduction in Mammals, Book 8, Human Sexuality,* ed. C. R. Austin and R. V. Short. Cambridge, England: Cambridge University Press, 1980.

HITE, S., *The Hite Report: A Nationwide Study on Female Sexuality*. New York: Macmillan Publishing Co., Inc., (1976).

HUNT, M., *Sexual Behavior in the 1970s*. Chicago: Playboy Press, 1974.

KINSEY, A. C., W. B. POMEROY, and C. E. MARTIN, *Sexual Behavior in the Human Male*. Philadelphia: W. B. Saunders Company, 1948.

KINSEY, A. C., et al., *Sexual Behavior in the Human Female*. Philadelphia: W. B. Saunders Company, 1953.

MARANO, H., "New Light on Homosexuality," *Medical World News,* 20, no. 9 (1979), 8–19.

MARMOR, J., " 'Normal' and 'Deviant' Sexual Behavior," *Journal of the American Medical Association,* 217 (1971), 165–170.

MARTIN, D., and P. LYON, *Lesbian/Woman*. San Francisco: Glide Publications, 1972.

MASTERS, W., and V. JOHNSON, *Homosexuality in Perspective*. Boston: Little, Brown & Company, 1979.

SORENSON, R. C., *Adolescent Sexuality in Contemporary America*. New York: World Publications, 1973.

WESTOFF, C., "Coital Frequency and Contraception," *Family Planning Perspectives,* 16 (1974), 136–41.

19 The Psychology and Sociology of Sexual Behavior

LEARNING GOALS

Having read this chapter, you should be able to:

1. List the differences between femaleness and maleness, sex role and gender identity, and femininity and masculinity.
2. List evidence for and against the theory that the development of sex role is an interaction between our biological nature and our environment.
3. Describe the traditional female and male stereotypes in the United States; discuss what is currently happening to these stereotypes; and list some reasons for these changes.
4. Present some ideas about what has caused the "sexual revolution" in the United States, stating evidence that indicates the extent of this revolution.
5. List the nine basic points of the new sexual code as proposed by leading scholars of human sexuality.
6. Be able to answer the following questions about forces that can influence our sexual behavior:
   (a) What are the major sexuality differences between restrictive and permissive societies?
   (b) What are ways that religious training can influence sexuality?
   (c) Why are statements that racial differences in sexuality are inherited invalid?
   (d) How do people with different educational background and social class differ in their sexuality?
   (e) How do the media present a distorted picture of human sexuality?
   (f) Why are most laws relating to sex between consenting adults outmoded?

(g) Why are parents often more conservative than their children about sexuality, and why do they rarely communicate with their children about sex?

(h) How do traditional female and male stereotypes influence sexual behavior?

(i) What are ways that sexual guilt influences sexual behavior?

(j) How does peer pressure influence sexual behavior?

7. Define the different kinds of love, and discuss why romantic love must be combined with other forms of love for a relationship to endure.

8. List ways in which sex is used other than as an expression of love.

9. Discuss what goes on with you and your peers when initiating sexual relationships, and how these practices differ from past dating practices.

10. Compare and contrast the goals of past and present marriages in the United States.

11. List legal restrictions on marriage.

12. Discuss how homogamy and heterogamy relate to marital stability.

13. List legal grounds for annulment of a marriage.

14. Discuss factors that may be responsible for the rising divorce rate in the United States.

15. Explain what is meant by "legal grounds for divorce," and discuss how divorce laws are changing.

16. Differentiate among common-law marriage, contract marriage, group marriage, renewable marriage, and cohabitation.

INTRODUCTION

In the previous chapter, we saw that the patterns and frequency of sexual behavior vary remarkably among different individuals. Now, we look in more detail at some forces that influence our *sexuality,* in terms of our sexual attitudes, feelings, and behavior. All of human behavior is an interaction among biological, psychological, sociological, and cultural factors. Sexual behavior is no exception, and it is difficult to distinguish the relative influences of these different factors on the manifestation of our sexuality. Nevertheless, we will try to discuss here some of these influences and their impact on our sexual behavior and relationships. You must realize that our sexuality influences every aspect of our lives, and many books and hundreds of articles have been written on this subject. It would be impossible to cover every aspect of the psychology and sociology of human sexuality, so this chapter serves only as an introduction to this topic. You can explore these topics further by consulting the "Further Reading" section at the end of this chapter.

SEX ROLES

Factors Influencing Sex Roles

Sex roles are a product of our biological nature, how we perceive this nature, and how we present our sexuality to others. First, our biological sex determines our anatomical and physiological *femaleness* or *maleness.* If our cells have two X chromosomes, we develop ovaries, female sex accessory structures, and female secondary sexual characteristics (Chapter 5). If our cells have one X and one Y chromosome, we develop testes, male sex accessory structures, and male secondary sexual characteristics (Chapter 5). Besides these anatomical and physiological sex differences, our biological maleness and femaleness also include sex differences in brain function that could influence our sexual behavior (Chapter 5). *Gynandry* refers to genetic females with malelike biological characteristics, whereas *androgyny* refers to males with femalelike biological characteristics.

Gender Identity. Gender identity is the psychological belief or awareness that one is biologically female or male. It is the private experience of one's biological sex. *Sex role* (or *gender role*), on the other hand, is the outward expression of gender identity. It is the way we present our gender identity publicly through our behavior (sexual or otherwise), speech, dress, etc. We are deemed *feminine* if our sex role fits what society defines as the female role, and *masculine* if the form of our sex role fits society's definition of what is male. Our sex role is greatly influenced by our culture, economic status, social life, nature of home and family, and religious beliefs.

Nature versus Nurture. Great controversy has existed over whether our sex role is mainly a product of our biology (our "nature") or of our learning and experience (our "nurture"). Are we feminine or masculine because of our biological make-up or because of our rearing and present environment? In the past, most favored the theory that nature plays a more important role, but recent information suggests that how we are nurtured has a major influence on the development of our sex role.

Anthropological studies have shown that sex role is highly influenced by social training or culture. The famous anthropologist Margaret Mead, in an influential book called *Sex and Temperament in Three Primitive Societies,* beautifully illustrated the role of culture in sex role development.[1] She studied three primitive tribes in New Guinea. In one tribe, the Arapesh, both sexes were "feminine" according to Western standards. Males and females were warm, gentle, tender, and sensitive. Both sexes initiated sexual encounters in a nonaggressive manner. In fact, an aggressive individual was considered deviant. In the Mundugumor, however, both sexes were aggressive headhunters and cannibals. Males and females were equally "masculine," being competitive, independent, and aggressive sexually. Mild and meek behavior was considered deviant behavior. Finally, the sex roles in the Tchambuli were similar to those in Western cultures, only reversed! In other words, females were "masculine" according to Western standards in that they were the breadwinners and rulers of the family. The men, in contrast, were emotional, dependent, and vain. Women in this tribe were the aggressive sex. Thus, sex role is not fixed in our biology, and social factors and training can even reverse the situation from that found in Western culture. This point is complicated, however, by studies of sex role development in hermaphrodites.

Hermaphrodites. Hermaphrodites are born with sexual organs of both sexes (Chapter 5). These individuals are usually assigned a sex role shortly after birth. As adults, their sex role is consistent with their rearing as boys or girls. Thus, their biological make-up appears to have little influence on the development of their sex role.[2] However, the biological forces operating in hermaphrodites may be abnormal. Recently, a group of investigators studied some selected people in three small villages in the Dominican Republic.[3] In these villages, inbreeding has led to a high incidence of pseudohermaphrodites (Chapter 5). Because of an inherited deficiency in androgen metabolism, genetic males are born with femalelike external genitalia but normal male internal organs. Of 38 such people studied, 17 were reared as girls. However, when these "girls" reached puberty, all but one of them developed normal adult male genitalia, and regardless of how they were raised, they all (but one) then behaved as normal adult males in their sex roles. In the villages, these teenage boys are called *Guevedoces,* which means "penis at 12" (Chapter 5). Although some authorities have questioned whether the 17 raised in the female role were actually treated as normal girls,[4] this study suggests that testosterone primed these individuals' behavior in a male direction while they were fetuses (Chapter 5), and that we cannot eliminate biological factors as having an influence on the development of gender role.

Modern Theory of Sex Role Development. A modern theory of gender identity and sex role describes a continuous interaction of biological and social influences from conception. This theory holds that chromosomal and hormonal make-up determine whether the genitals develop in a male or female direction. This "genital dimorphism stage," a biologically determined event (Chapter 5), is followed by the increasing influence of psychological and social input. How individuals perceive their bodies (that is, do they have male or female sex structures) leads to a body image of maleness or femaleness (gender identity). Generally, this awareness of one's maleness or femaleness begins around 18 months of age. The expression of this gender identity through sex role develops later and is greatly influenced by familial and social factors. That is, one's sexual behavior, appearance, and even the way one thinks are a product of training.

Children tend to imitate the parent of the same sex, and parents and peers behave differently toward boys and girls, reinforcing or punishing certain behaviors.[5] Although there are sex differences in behavior at birth,[6] any role of these differences in development of sex role is conjectural at this time (Chapters 5 and 18). Thus, sex role is the result of some undefined interaction between biological sex, environmental input, and a person's awareness of self and his or her relationship to others. This theory is still undergoing revision,[7] and much remains to be learned about this fascinating and important subject.

Changing Sex Roles in the United States

Not long ago, the United States was an *androcentric society.* That is, the society was characterized by domination by the masculine sex role, with its assignments of strength, competition, power, joviality, courage, independence, aggressiveness, success, and lack of sensitivity. This view was based on religious and philosophical beliefs that the male was superior. The female, on the other hand, was believed to be physically and mentally inferior. It was thought that women were more gentle, confused, fearful, vulnerable, vain, compassionate, and meek. Also, a woman's place was in the home where her supposed motherly attributes of tenderness, compassion, and giving of self to others could best be expressed. Various religious doctrines assumed that man was created in God's image, and God was masculine. Most religious and political leaders were (and still are) men. This androcentric social system was perpetuated by teaching children sex roles that supported this system. Thus, little girls were taught to avoid masculine behaviors and dress, and little boys were allowed and often forced to be aggressive, dominant, and "manly." But, we have seen that the development of sex role is highly subject to social change, and this is happening to some extent, in the United States today.

Female Liberation. The recent resurgence of the *feminist movement* (or the *women's liberation movement*) was sparked by Betty Friedan's book, *The Feminine Mystique,* published in 1963.[8] In this book, Frieden attacked the belief systems leading to male dominance. The feminist movement has greatly influenced sex roles in our society and has begun to free women (and men) from previous stereotyped sex roles. This movement seeks to free women from economic, social, behavioral, and sexual restraints that have limited their freedom, and women are participating more in traditionally masculine areas, such as sports, business, and politics.[9]

Because of the development of this freedom, sexual behavior of women is also changing. For example, it was formerly believed that only men should initiate sex, experiment with sex, and enjoy the "pleasures of the flesh"; this double standard was based on the stereotyped sex roles of the androcentric society. Now that these stereotypes are being attacked, many women have developed more self-esteem and are not as inhibited and self-conscious about sex. Women with self-esteem are now more likely to initiate sex, experiment more with sex, and are more open to sexual experience than in the past. More specifically, several studies suggest that confident women are more likely to masturbate, participate in oral-genital coitus, find the male genitals attractive, and experiment with various coital positions.[10]

Male Liberation. The women's liberation movement may also be a "men's liberation movement." Our traditional concept of masculine behavior is changing. There is more participation by men in traditionally feminine activities such as housekeeping and child care. Some men also are allowing themselves to be more sensitive and are more honest about their feelings. More specifically, they are showing more tenderness in public, cry more to release feelings, and are willing to be more passive at times in sexual encounters. It is not clear, however, how well men and women are adjusting to these new roles,[11] and the high incidence of unhappy marriages and divorces is partly related to the fact that our society is in a transitional stage of defined sex roles.

In the past, sexual behavior in the United States was restricted because of our cultural background. Religious and philosophical training taught people that the main purpose of coitus was procreation. The joys of sex, although of course present, were clouded with guilt and secrecy. A man's right (and privilege) was to have, enjoy, and even demand sex, whereas a woman was to be passive and serve her mate. However, sexual attitudes and behavior are changing at such a rapid rate that some say the United States is in the midst of a "sexual revolution."[12]

Nonmarital Coitus

Attitudes and beliefs concerning *nonmarital coitus* (coitus outside a marital union) are changing at a rapid pace. In the past, *premarital coitus* (coitus before marriage) was denounced for women. Because of sex role stereotyping, it was more acceptable if males had premarital coitus. Recently, however, there has been a dramatic change in this regard, and premarital coitus has increased markedly in both sexes.[13] In 1959, a Gallup Poll showed that 68 percent of Americans thought that premarital coitus was wrong; but by 1970, this percentage was down to less than 10 percent, and about 75 percent had experienced premarital sex.[14] Among college students surveyed in 1973, about 73 percent of women and 84 percent of men approved of premarital coitus for men, and about 68 percent of women and 81 percent of men approved of this activity for women.[15] In the early 1940s, about 45 percent of college men had experienced premarital coitus, but this percentage was up to 75 percent in the early 1970s.[16] Although about 5 to 15 percent of college women approve of premarital coitus without any emotional commitment, most feel that a loving commitment is a prerequisite for this sexual experience.[17] As discussed in Chapter 18, one must be aware of the limitations of such statistics, but they do indicate general trends.

Premarital coitus also is on the rise in young teenagers. In the late 1940s and early 1950s, Kinsey and his co-workers reported that only 3 percent of females and 10 percent of males had experienced coitus by age 16.[18] In contrast, Hass reports that among the 15- to 16-year-old males surveyed in the late 1970s, 43 percent had experienced coitus by then, and 18 percent by age 13.[19] Among 15- to 16-year-old females, 31 percent had experienced coitus, some of these (7 percent) before age 13. About two-thirds of teenage females and four-fifths of teenage males are sexually active (have had coitus at least once) in the United States today. Figure 19–1 indicates the sexual activity and fate of pregnancies in teenage females.

This liberalization of views about premarital coitus is only part of a broader change in attitudes about and practice of sexual behavior. For example, a survey by Hunt in 1972 showed that almost 60 percent of both sexes approved of anal coitus, up to four-fifths of college students approved of fellatio, and up to 95 percent of cunnilingus.[20] In America today, over 80 percent of adults think that either the male or female can initiate sex. In other words, many people in the United States now have

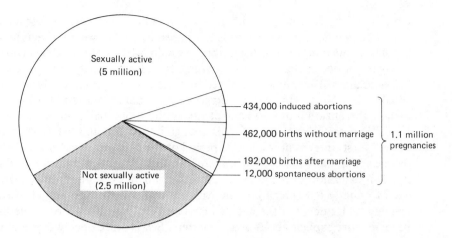

Figure 19-1. Current sexual activity in American teenage females, based on a survey by the Alan Guttmacher Institute (1981; see Further Reading section). This figure also shows the fate of teenage pregnancies.

more open feelings about sexual expression and view sex as a source of pleasure for both women and men, as well as a means of reproducing. It should be kept in mind, however, that whereas people may approve of certain behaviors, it does not necessarily mean that they participate in them.

Views about sexual expression also are changing in married couples. *Extramarital coitus* (adultery) has been long a taboo in most cultures because of its potential adverse affects on family life. However, extramarital coitus is on the rise in the United States. For example, Kinsey and his co-workers found in the midtwentieth century that 27 percent of American women thought that extramarital sexual activity of their spouse was sufficient grounds for divorce, and 51 percent of men felt this way.[21] In 1970, however, about 80 percent of men and women approved of extramarital sex under certain circumstances, although only about 40 percent said that they had experienced extramarital intercourse.[22] Another form of extramarital sex practiced by a few couples in the United States today is *mate-swapping* or *swinging,* in which married couples exchange sex partners to gain new sexual experience.[23] Although some couples experience enrichment of their own marriage as a result of mate-swapping, relationships often suffer because of jealousy. In swinging circles, strict rules on the degree of emotional involvement and depth of communication are followed to help prevent jealousy.

The Sexual Revolution

Thus, attitudes and beliefs about sexual behavior are becoming less conservative in the United States. But, I must emphasize two points in relation to this statement. First, this sexual revolution is occurring more rapidly about some aspects of sexuality (e.g., premarital coitus) than others (e.g., extramarital coitus, homosexuality).[24] Second, what we *say* we approve of is not necessarily what we *do*. Even though we are becoming much more permissive about certain aspects of sexual behavior, America today is not dominated by a "sex is fun" morality. In fact, some aspects of sexual behavior are still very much restricted by social values, personal inhibitions, and personal ethics. Thus, what we can expect is a gradual revolution of behavior that lags behind liberalization of views and attitudes about that behavior.[25]

What factors are causing this change in attitude and behavior in the United States? First, there is now a longer period of social adolescence, during which young people are sexually mature but still are learning about the expression of their sexuality. This extension of the time of sexual learning has resulted from a decrease in the age of puberty (Chapter 6), along with a lengthening of the period of educational training and a delay in the age of marriage (Chapter 20). Second, the media have become more explicit about sexual matters. Sex is more openly depicted in movies, books, and television, and this may influence the development of sexual expression. Third, traditional religious and social training is becoming more liberal, and youth are questioning, evaluating, and often rejecting traditional teachings or adapting them according to their own and peer value systems. Availability of effective contraceptives also has influenced our sexual behavior by freeing people from the fear of pregnancy. In addition, an increase in our knowledge about human sexuality through psychological, sociological, and biological research has changed our views and understanding of sexuality. Finally, the changing ideas about sex roles have resulted in at least a partial convergence of male and female sexuality, with women more free to enjoy sex and with men more free to express the "feminine" sides of their nature.

New Sexual Code

In 1976, a new humanistic sexual code was described, and it has been adopted by several leading sociologists. It describes several rights and responsibilities that one should consider in today's sexual world. Briefly, the following points are made:

1. The boundaries of human sexuality must be expanded so that both women and men can be enriched and fulfilled. Sex should be an experience of intimacy as well as a relief from sexual tension and a means of reproducing.
2. Both sexes must have equal opportunities to develop to their fullest potential, sexually and otherwise.
3. Repressive taboos about sexuality must be replaced by new views based on an awareness of human needs.
4. Each person is obliged to be fully informed on issues of human sexuality in the community.
5. Potential parents have the right and responsibility to plan the number and time of birth of their children in accordance with personal desire and social need (Chapter 20).

6. Sexual morality should come from a sense of caring and respect for the needs of others.

7. Physical pleasure has a place as a moral value and not as an evil or sinful experience.

8. It must be recognized that each person has the right to respond positively to his or her own sexuality throughout all stages of life.

9. In all sexual encounters, there must be a commitment to humane and humanistic values.

For a more thorough discussion of these points, you should read the original essay by Kirkendahl.[26]

FACTORS THAT INFLUENCE OUR SEXUALITY

Culture

Basically, there are three kinds of societies in relation to attitudes about sexual behavior: restrictive, semirestrictive, and permissive. In *restrictive societies,* coitus mainly is limited to marital life, and its primary purpose is procreation. Usually, men initiate sex in these societies, and women are passive and unresponsive. In restrictive societies, fewer women reach orgasm,[27] which may show that cultural restrictions can influence sexual experience. In *permissive societies,* women are as sexually active as men, and the joys of sex are as important as procreation. *Semirestrictive societies* (such as the United States today) are somewhere in between the restrictive and permissive situation.

Regardless of the society, there are some general cultural restrictions on certain kinds of sexual activities. For example, about 20 percent of societies do not approve of premarital coitus for either sex, 45 percent approve, and the remainder approve of this activity only for men.[28] Adultery is taboo in about 80 percent of societies, based mainly on the potential disruption of family life.[29] Incest also is restricted in many cultures (see below).

Religion

Judeo-Christian teachings emphasize that the main purpose of sex is reproduction, and that pleasures of the flesh are either a byproduct of procreation or not sanctioned. These traditional teachings primarily repress sexual behavior through the underlying mechanism of guilt.[30] A woman with strong religious beliefs is more likely to be a virgin when married, masturbate less, and have lower sexuality in general.[31] Nevertheless, recent liberalization of religious views about human sexuality and such issues as contraception and induced abortion has freed many people to experience their sexuality more fully.

Race

It is a general belief of many that there are racial differences in patterns of sexual behavior. Implicit in these beliefs is that biological differences among races produce different sexual behaviors. When one remembers, however, the great influence of social factors on sexuality, it is clear that racial differences in sexual behavior disappear when these social factors are eliminated. In general, for example, more black Americans practice premarital coitus than whites, and they begin sexual activity at an earlier age.[32] However, when one compares people of the two races with similar educational, economic, and religious backgrounds, the "racial" differences in sexual behavior are not there.[33]

Socioeconomic Class and Education

Socioeconomic class and education appear to influence patterns of sexual attitudes and behavior.[34] For example, in the early 1970s, four-fifths of people with a college education approved of fellatio, but three-quarters of those without a college education disapproved of this form of sexual expression.[35] Our feeling about sexual behavior of others generally is related to our social class.[36] For example, among people less than 35 years of age, 65 percent of white-collar workers thought homosexuality should be legalized, but only 41 percent of blue-collar workers approved of this form of sexuality.[37] Thus, people with less educational background or a lower social class tend to be more conservative about sexual matters, although there are, of course, exceptions.

The Media

We constantly are bombarded with sexual messages from the media. Sexuality expression in the media often is based on past stereotypes of sex roles and sexual behavior. Unfortunately, these stereotyped sexual messages are a major way that many young people learn about sex roles in our society. Thus, being a "macho man" or a female fashion model are standards to be achieved. Movies and books support the sex role of men as hard-driving, high-performance sex machines, and women as seduced victims. According to movies and books, we get married and live happily ever after in a life of high romance and leisure. This places unattainable expectations on many relationships. The influence of figures in the media is so great on some people's lives that they experience deeply involved, romantic, artificial relationships with them.[38] We must realize that the media in general give us a distorted picture of sex roles, sexual behavior, and personal relationships, and should base our sexual behavior and feelings on reality and not fantasy.

Sex Laws

There are many variations in sexual behavior, and laws have been established to define what is "deviant." Unfortunately, many sex laws are based on religious dogma, taboo, superstition, or ignorance.[39] In most states, for example, the only legal sexual activities are heterosexual hugging, kissing, contact of the mouth with other parts of the body except the genitals, and touching another's body with the hands. Coitus is legal only between married people. In addition, homosexual activities, anal coitus, oral-genital coitus, and sexual contacts with a prostitute or animals are illegal in many states. If many of these sex laws were enforced, there would be a lot of adults in jail!

In some states, there have been legal changes that align sex laws with modern views about sexuality, but many states still lag far behind. The purpose of sex laws should be to protect people from being forced into sexual acts, to protect children from sexual abuse, and to prevent flagrant display of sexual acts (Chapter 18). Sex laws should not enforce morality or attempt to regulate sex between consenting adults.[40]

Family Life

Most of our training about sex roles and sexual behavior comes from our parents. Some of the early training in relation to sexuality is on the subconscious level of both parent and child. For example, children are taught to take baths by washing the face first but the genitals last because they are "dirty." Other messages come across loud and clear. When a child asks a question about sex, the reply from some parents is "later" or "not now," and it does not take long for the child to realize that communication about sex with his or her parents is not available. To be sure, some parents feel free to discuss such topics as male and female reproductive anatomy, but when it comes to a need for communication about critical concerns of sexual behavior and sex roles, silence is the rule.[41]

The general family environment (happy or destructive) also can have an effect on the sexual behavior of children in the family. For example, females from happy families tend to have better sexual relationships and more sexual partners than do females from unhappy families.[42] Perhaps this is because of better communication about sex in happy families. Communication about children's and parents' sexuality should be part of parental responsibilities.

The attitudes and beliefs about sexual matters often can differ between parents and their children. This is especially true in a society such as the United States where a change in sexual attitudes is in progress. Parents growing up before this change may have different, more conservative views on the subject than do their children. In 1969, Wake reported that 30 percent of mothers practiced premarital coitus, but that only 3 percent of them would approve of their daughters, and 9 percent their sons, engaging in premarital coitus. About 50 percent of fathers surveyed had experienced premarital coitus, but only 10 percent approved of this for their daughters, and 9 percent for their sons.[43] Thus, parents at that time tended to disapprove of their children doing something they may have done themselves.

There is some indication that today's parents, however, are becoming more liberal in their views on sexuality of their children. For example, Levin found that of the parents surveyed in 1975, only 12 percent objected to their sons having premarital coitus, and only 24 percent of mothers objected to their daughters experiencing sex in this way.[44] Any age difference in attitudes about sex between parents and their children may not be related to age at all, but rather to the degree of responsibility for another's behavior. Parents with older children are more conservative about their children experiencing premarital sex than are similar-aged parents with younger children or those with no children.[45]

Changing Sex Roles

As discussed, there has been a change in the United States in male and female sex roles. Convergence of sex roles in the United States appears to be at least partially eliminating the double standard of sexual attitudes and behavior. For example, Kinsey and his colleagues found, in the late 1940s, that women approved more of their husbands having extramarital relationships than the reverse.[46] In addition, it was felt by many that only men should initiate sexual encounters. Such sex differences in attitudes about sex have, to some extent, disappeared. In 1970, about 80 percent of men and women approved of extramarital coitus under certain circumstances, and 40 percent of men and 36 percent of women had experienced extramarital affairs.[47] In addition, about 80 percent of both sexes felt that either the man or woman may initiate sex, although many men still feel uncomfortable when a woman initiates sex.[48]

It is interesting to note that even if both men and women may approve of and participate in certain forms of sexual behavior more often than in the past, there still can be a sex difference in the reasons for choosing to participate or not. For example, in a study of college students by Frede in 1970, reasons listed for not participating in premarital coitus were similar between men and women but differed in order of importance.[49] For men, the reasons in priority order were lack of opportunity, fear of pregnancy, religious or moral considerations, fear of jeopardizing a future marriage, loss of partner's respect, and loss of self-respect. For women, the reasons were similar but in a different order of priority. The first reason was religious or moral considerations, followed by loss of self-respect, fear of jeopardizing a future marriage, fear of pregnancy, and loss of partner's respect. Lack of opportunity was near the bottom of the female list!

Stereotyped sex roles can lead to difficulties in sexual relationships. For example, women who feel that men should be strong and protective sometimes are impressed by, and attracted to, "macho" behavior. Women who have guilt about their sexual behavior may even purposefully choose sex partners who are aggressive and even abusive. And male students who feel that being masculine is to drop out of school, take drugs, or drive fast cars are often displaying these behaviors because they suffer from a lack of self-confidence. Also, males with the macho self-image often think of women in a stereotyped way as being weak and submissive. After "conquering" a female, these men feel that they have seduced the woman and feel guilty about it. Thus, they tend to be attracted to "bad" women (i.e., those who do not fit their stereotype of a "good" woman).

Sexual Guilt

Sexual guilt is a strong factor influencing human sexual behavior.[50] In fact, guilt plays a role in the sequence of development of one's sexual experience.[51] For example, as young persons begin dating, their first sexual experience (kissing, for example) often is followed by guilt. When this guilt subsides, they then go to a further stage (e.g., petting), acquire the associated guilt, rid themselves of that guilt, and then proceed on to a further stage of sexual experience. New sexual relationships tend to begin where previous ones left off, and the older and more experienced a person is, the less sexual guilt is present.[52]

Guilt comes from doing or thinking something that is against one's conscience (social, moral, ethical, or religious training). For example, young women may feel guilty about their first experience with coitus because they were taught that women should be chaste before marriage. And young men may feel guilty because of seducing such women. Guilt can also interfere with sexual desire; women who feel guilty about sex tend to have fewer orgasms and are less responsive sexually.[53] Furthermore, guilt can lead to difficulties in marital relationships. For example, a married man, believing that he should love only his wife, may be attracted to another woman. Then, because of the attraction, he feels that he must not love his wife, and feels guilty about this. He thus develops hostility toward the source of his guilt, his wife, and his hostility harms their relationship.

Peer Pressure

During adolescence and adulthood, there is a great need to be accepted by one's peer group, and this means, among other things, that a person feels he or she should behave sexually as do others in the group. Today in the United States, young people often get the message that something is wrong with them if they do not experiment sexually.[54] Some level of rejection of parental, religious, and ethical training always has been a common facet of adolescent development, and this includes evaluating and often rejecting parental views about sexual matters. This, plus the increased availability of contraceptives to teenagers and the influence of the media, has resulted in an increase in the incidence of premarital coitus in teenagers. Many young women still are taught that they should be chaste before marriage, but they are convinced that they must be sexy to be attractive. They also learn that sex is pleasurable. This can lead to a conflict about their behavior that W. A. Layman has called the "saint or sinner" syndrome.[55] Thus, young women (and men) can experience their first sex as dissatisfying and filled with guilt. In addition, young men are still told by society that being masculine is being sexually aggressive and experienced, and communication among adolescent males about sexual conquests often is a means of keeping score on one's progress to manhood.

Regardless of these very real pressures, a young person should answer some important questions before engaging in premarital coitus. Why do you want to experience sexual intercourse? Do you love your proposed partner, or will coitus satisfy another desire such as curiosity or satisfaction of an intense need? Do you understand your and your partner's needs and feelings? Would having a sexual relationship agree with your value systems and those of your partner? Have you dealt with your fears about coitus? Are you influenced by peer pressures? What do you expect sexual intercourse to do for you? Are you and your partner aware of methods of contraception, and can you get adequate protection? Furthermore, will you use this protection? Finally, are you and your partner willing and able to deal with pregnancy if it occurs?

SEXUAL RELATIONSHIPS

Love and Sex

It is unfortunate that the English language has only one word for *love*, because we use this word to describe very different feelings and situations. For example, we say, "I would love to go to a movie," "I'm in love with that man," "I love chocolate ice cream," "I love my child," or "I have a loving relationship with my wife." Psychologists and philosophers have identified different kinds of love.[56] *Sexual love* (or "lust") is the desire to experience sex with another person and is part of *erotic love,* or the desire for complete fusion with another person. *Philia* (or "brotherly love") is the kind of love one has for a friend or sibling, and is not exclusive to any one person. *Agape* (or "caritas") is selfless love for people, as in the love of a parent for a child. More than one kind of love can exist in a single relationship, and the importance of each varies with the situation.

In our initial romantic relationship with another person, we are "madly in love," and this form of love usually is dominated by the passion of erotic and sexual love. In essence, a major force behind this experience is that we feel affirmed and worthwhile in the presence of the lover, and sexual encounters during this time are new and exciting. This romantic love, however, often is based on distorted views of the true nature of the person, and it is usually the case that the passion of this love must develop into another form of love in order for the relationship to endure. That is, philia and agape must combine with sexual and erotic love; and friendship and commitment to the personal growth and well-being of the partner become important components of a lasting relationship.

Maslow distinguishes "deprivation love" from "being love."[57] Deprivation love is a thirst to fill a need. In this form of love, the partner is seen as a source of fulfillment of one's needs. In a way, the identity and self-esteem of the receiver are dependent on the feelings and behavior of the lover. If these messages are withdrawn, the love dies and the relationship ends. Being love, on the other hand, rises not out of one's needs but out of the experience of giving. In this form of love, satisfaction of one's needs are self-generated and are not dependent on others. Thus, the lover is free to give and to support the personal growth of the partner as one of many persons in the universe. We all need to be cared for, and in a "being love" relationship, this need is satisfied.

Many people wonder if sex leads to love, or love leads to sex. The answer to both questions is yes, and it depends on the particular relationship or situation. Of 20,000 Americans surveyed in 1970, most felt that love greatly improves sex but is not necessary for the enjoyment of sex.[58] Another theory holds that sex, being pleasurable, reinforces a relationship, which then leads to love and marriage.[59] We have seen in Chapter 17 that affection in the form of touching and cuddling is a vehicle for expression of love, not only during sex but in other situations. But, sex can be performed without affection or love. It can even be used as a solution for aloneness, as a desire to dominate or be dominated, or out of vanity.[60] Sex (or the refusal of it) can also be used to communicate moods and feelings. For example, a woman may refuse sex with her male partner to "pay him back" or to say "pay more attention to me." On the other hand, a man may demand sex from his female partner to prove his masculinity. In a workable relationship, sex is integrated into life and each person's needs are satisfied, their commitments fulfilled, and sex is consistent with goals in other areas of life.

Dating

In the restrictive American culture before 1920, a young man would not date until he was in his upper teens. Any dating activities required approval of parents, and kissing or petting usually occurred in the woman's home. The woman then would become the man's exclusive partner, and this would lead to engagement and marriage, with the goal being to have a home and family. In America today, people are having sexual relationships at an earlier age, and sexual activity often is not restricted to one other person. Formal dating has been replaced to some extent by casual get-togethers, and some women are becoming as aggressive as men in initiating sexual relationships. Usually, these relationships are short-lived and are more for the purpose of immediate pleasure than serious commitment for marriage. As discussed, premarital coitus is very common today, and some feel that this experience will lead to a more stable relationship in a marriage. Even so, it is still acceptable in our society that a female, and to a lesser extent a male, be chaste before marriage.

Marriage

Purposes of Marriage. In the nineteenth century, marriages in America were strengthened by legal, economic, and religious bonds. Formal weddings were the rule, which served not only to legally unite the couple but as a social function for family and friends. Today, many couples still opt for a formal wedding, although some prefer a more simple ceremony, and others marry in secret (*elopement*).

In the past, the part of the marriage ceremony saying "until death do us part" often came true. This marriage usually was dominated by the husband, and economic survival, emotional security, and children were its goals. Today, however, the sexual revolution has influenced the goals of marriage so that personal growth and freedom, companionship, satisfaction, and free sexual expression are some major aspirations of marriage. Still, successful marriages today require a sense of loving commitment, responsibility, and sacrifice, and having a spouse who also is a good friend.

Marriage can have several advantages in today's world, including an opportunity for abundant sexual expression, trust, self-confidence, companionship, love, and fidelity. However, many marriages today are unhappy, and the divorce rate is climbing (Fig. 19–2). About half of marriages in the United States today end in divorce. Also, the incidence of extramarital sex is increasing in the United States. The reasons given for experiencing extramarital sex may reflect some problems with marriage today. According to G. Neubeck, reasons given for having an extramarital affair include a desire for sexual variety, retaliation against the spouse, more satisfaction of emotional needs, and just plain boredom.[61] Some other factors that may be influencing the rising divorce rate in the United States will be discussed later in this chapter. Suffice it to say that what is needed today is a reevaluation of the marital

Figure 19-2. First marriage, remarriage, and divorce rates in the United States from 1920 to 1980. Adapted from "Marital Instability: Past, Present and Future," by A. J. Norton and P. C. Glick in *Divorce and Separation: Context, Causes and Consequences,* eds. G. Levinger and O. C. Moles (New York: Basic Books, Inc., 1979). © 1979 by the Society for the Psychological Study of Social Issues; used with permission.

relationship along the lines of a partnership, with commitments to and support of the personal growth and experience of each partner.

Marriage Laws.   Marriage is a legal contract, and there are laws governing this union in all states. Marriages can be performed by licensed or ordained clergy or by authorized judicial or public officials. Every state has a minimal age for marriage without parental consent, ranging from 18 to 21 for males and 14 to 18 for females. With parental consent, this age is 16 to 18 for males and 14 to 18 for females. Most states make exceptions to these minimal ages if the woman is pregnant.

A marriage license is required in all states, and there is a waiting period (about 3 to 5 days) between applying for a license and issuance of the license, supposedly to help prevent "spur of the moment" marriages. Most states require a physician's report saying that the proposed couple are free of communicable (especially sexually transmitted) diseases. Blood tests are needed for blood typing and testing for syphilis. In a few states, the partners must be free of mental disorders.

Marriage between members of the immediate family (incest) is prohibited by law and by religious or social taboos,[62] even though incest has been permitted in some past cultures. For example, brother-sister matings were considered highly desirable by the ancient Egyptian Pharaohs, and Cleopatra was the last in a long line of such matings. Some states will not permit coitus or marriage between first cousins, and some preclude even distant cousins. Taboos against incest are common in most cultures and are based on several facts and beliefs. For example, incest laws prevent inbreeding between close relatives, which could lead to an increase in genetic disease in children (Chapter 12). Also, incest taboos may have evolved to protect the family unit and the personal growth of the children. Even though against the law, incest in the United States is more common than most people realize (Chapter 18).

Marriage between a single man and woman (*monogamy*) is the accepted norm in the United States. *Polygamy* is the situation where either sex has more than one marriage at one time, and *bigamy* is when either sex has two spouses at one time. Polygamy or bigamy can exist in the form of *polygyny* (a man having multiple wives at one time) or *polyandry* (a woman having multiple husbands at one time). Only monogamy is legal in the United States.

Factors Influencing Marital Success.   Several factors seem to be correlated with marital success. For example, we tend to be sexually attracted to and marry people with similar economic, social, ethnic, and attitudinal backgrounds.[63] Also, when the female is less than 18 years old, or the man is less than 21 years old, the marriage tends to be less stable than when the participants marry at a later age. Great age differences between the partners may offer special problems, although many of these can be overcome. Generally, similar social class of partners can lead to greater marital stability. If there is a class difference between partners, marriages seem to be longer lasting if the man has a higher social position than the woman. Marriages between lower-class people tend to be less successful than those between middle- or upper-class people. Marriages between people of similar educational background tend to be more stable than those between people with differing educational backgrounds. When there is a difference in education, a marriage tends to work better if the man has more education.

Intermarriages are marital unions between people with different cultural, religious, or racial backgrounds. More specifically, *interfaith marriages* are those in which the partners differ in religious belief. Marriages between people of similar faith are more successful than those where faith differs. Similarly, marriages between people from different cultures (*interethnic marriages*) are less successful than are those between people of similar cultures. Finally, marriages between people of different racial backgrounds (*interracial marriages*) tend to present more problems than do those between people of a similar race.

From the above discussion, you will gather that marriages between individuals of a similar age, social class, culture, education, religion, and race are more likely to be successful than other marriages. Even people with synchronized daily rhythms of body temperature have more stable marriages.[64] To put it another way, marriages between partners with like characteristics (*homogamy*) are more likely to be successful than those between partners with unlike characteristics (*heterogamy*). However, for women in a second marriage after a divorce, heterogamy tends to be prevalent for age, education, and religious belief, and it appears that many of these women are prone to choosing men unlike themselves.[65] The tendency that homogamous marriages work better has other exceptions. For example, people with differing personality types (e.g., subordinate vs. dominant) may have happier marriages than do those with similar personality types. Finally, please be aware that the above correlations represent trends only, and there are many exceptions. A good dose of love, commitment, and understanding can produce heterogamous marriages that are both joyful and lasting.

In the past, there has been evidence that marriages were less happy in families with low income, with husbands who had low-prestige occupations, when both spouses attended church less regularly or not at all, and in those where the wife was employed outside the home. A recent study, however, found that the stability of modern marriages is not related to any of these factors, but only to whether or not young children are present or if the wife is middle-aged (less stable in each case).[66] It appears that unhappy marriages are more readily ended nowadays, so that those remaining are mostly happy and are not enduring simply because of the stigma of divorce.

Marital Dissolution. When a marriage is not working, and means of helping this situation (such as counseling) have not worked, the relationship can be dissolved by annulment, separation, or divorce.

Marriages can be *annuled,* which means that a court order says that the marriage never legally occurred. Annulment is the only form of marriage dissolution presently recognized by the Roman Catholic Church. An annulment can occur for the following legal reasons:

1. Concealment by one of the partners of a previous marriage or divorce.
2. Misrepresentation of chastity.
3. One or both of the partners were under legal age when married.
4. There was fraudulent intent not to perform the marriage vows.
5. There can be no sexual consummation of the marriage.
6. One or both of the partners are sterile.
7. The marriage was performed under force or duress.

Often a couple will decide to live apart for awhile but will keep their marriage legally intact to see if being apart improves their relationship. This *separation* is by mutual consent, and can be followed by divorce, annulment, or reconciliation. When one partner leaves the other without consent, it is called *desertion*.

A decree of *divorce,* when final, legally terminates a marriage. The rate of divorce in the United States has risen about 250 percent in the last 20 years (Fig. 19–2), and about one-half of marriages end in this way. The practice of divorce and remarriage (80 percent of divorcees marry again) is becoming so common nowadays that some say a new life style of *serial monogamy* (having several marriages in a sequence) is developing.[67]

The present high divorce rate has been attributed to many causes, some of which are the following:

1. Greater independence and freedom of women.
2. Liberalization in religious doctrines about marriage and divorce.
3. Urbanization of our society, which results in less dependence on a patriarchal male as the breadwinner.
4. An increase in mobility and change of residence.
5. Liberalization of divorce laws and simplification of divorce proceedings.
6. Less dependence on the family as a vehicle for ethical and educational training of children.
7. An increase in financial problems.

In most states, divorce requires proof of a marital offense, such as adultery, insanity, criminal offense, use of alcohol or drugs, desertion, lack of economic support, impotence or frigidity, or mental or physical cruelty. The last reason accounts for about 50 percent of all divorce grounds, followed by desertion (23 percent). Several states have "no-fault" divorces in which divorce is granted based on "irreconcilable differences that have led to irreparable harm" to the marriage. This is the most humane divorce procedure since other "grounds" usually are based on these differences. Traditional agreements reached during divorce are also being questioned. For example, custody of children usually was awarded to one parent (usually the mother) in the past. Recently, however, joint or "alternating" custody, with each parent sharing equally in child-rearing and its responsibilities, is becoming more common.[68]

NEW LIFE STYLES

The sexual revolution and other aspects of our changing society have produced alternative forms of marriage. Besides the traditional heterosexual monogamy, in which a male and female are married for a theoretically indefinite period, we see some alternate life styles becoming more common, as discussed below.

A *consensual arrangement* (also called sexual cohabitation or living together) is common among college students and the divorced.[69] In fact, 20 to 30 percent of college students experience this form of relationship. Most of these arrangements are temporary, but some consider this life style as a form of "trial marriage."

Common-law marriages are recognized as a legal form of union in many states. In this form of marriage, a male and female must live together usually for seven years. Then, they can be considered married under common law, and they have all rights available that are available to traditionally married couples. No license or ceremony is required. In some states, parental consent is needed if the male is less than 21 and the woman less than 18.

A *contract marriage* is a form of traditional marriage in which formal agreements are reached concerning such things as allocation of housework, relations with friends, career activities, and sexual behavior within and outside the marriage. This kind of marriage has also been called an *open marriage*.[70] Some contracts are for a defined period (e.g., five years), after which there is an option to renew.

In a *group marriage* (or *commune*), small groups of men, women, and children live together under one roof or in a small community.[71] In some communes, individual couples are married traditionally. In others, polygyny or polyandry are common. Many of these communes practice open or group sex and collective parenting of children. Most group marriages do not last; the most enduring and successful are those that have a single leader, selective membership, and some regimentation of duties within the commune.

A single parent of either sex may raise a family. This arrangement often occurs when a child is born out of wedlock, or an unmarried person adopts a child, or (most frequently) a parent is divorced or widowed.

Another type of marriage might be the *renewable marriage*. A marriage license that can be renewable after a certain amount of time has been suggested. In fact, it has been suggested that two kinds of marriage licenses could be issued. The first would be a trial marriage license where the couple would be legally married but would not be allowed to have children. After a certain amount of time, the couple would then be eligible to apply for the second license, which would allow raising of a family. Considering the high incidence of divorce and the population problem in the United States (Chapter 20), this may not be such a bad idea.

Marriage between members of the same sex (homosexual marriages) are legal in some states but not others. Marriages of this kind theoretically are subject to the same legal requirements as are heterosexual marriages.

CHAPTER SUMMARY

Human sexuality refers to our sexual attitudes, feelings, and behavior. We are aware of our biological maleness and femaleness, and this awareness is our gender identity. We express this identity to others as our sex role or gender role. The development of sex role reflects an interaction between our biological nature and our environment. In the United States, traditional definitions of what is masculine and feminine are changing, partly because of the women's liberation movement.

In the United States today, people's sexual attitudes and behavior are becoming more liberal, and this sexual revolution is caused by biological, social, and medical changes. The type of society we live in (restrictive or permissive) influences our sexuality, as does our religious training, social class, educational background, sex in the media, sex laws, family life, sexual guilt, peer pressure, and sex role. Racial differences in sexuality are a reflection of the above influences and not of an inherited biological difference.

Erotic love is the desire for fusion with another person, and sexual love is a main force in erotic love. The passionate, erotic love of a new relationship must acquire new forms of love to endure, such as *agape* (selfless love) and *philia* (friendship or brotherly love). Sex can be a vehicle for expression of love but can also be used for other purposes.

Dating practices in the United States are changing, and each sex now often shares equally in initiating sexual relationships. Marriage is a legal union, and laws restricting marriage relate to age, health, genetic relatedness (incest), and whether one is already married. Marriages between like individuals (homogamy) tend to be more successful than are those between individuals differing in age, race, culture, social class, or religion (heterogamy). Marriages can be ended temporarily by separation or desertion, invalidated by annulment, or ended by divorce. The divorce rate in the United States is rising, and several sociological factors may be decreasing the stability of marriages. New forms of life style include common-law marriage, consensual arrangement, contract marriage, group marriage, renewable marriage, single-parent families, and homosexual marriage.

NOTES

1. M. Mead, *Sex and Temperament in Three Primitive Societies* (New York: William Morrow, 1935).
2. J. Money and A. Ehrhardt, *Man and Woman: Boy and Girl. The Differentiation and Dimorphism of Gender Identity from Conception to Maturity* (Baltimore: Johns Hopkins Press, 1972).

3. J. Imperato-McGinley et al., "Steroid 5α-Reductase Deficiency in Man: An Inherited Form of Male Pseudohermaphroditism," *Science,* 86 (1974), 1213–15.

4. E. Sagarin, "Sex Rearing and Sexual Orientation: The Reconciliation of Apparently Contradictory Data," *Journal of Sex Research,* 11 (1975), 329–34; and R. T. Rubin et al., "Postnatal Gonadal Steroid Effects on Human Behavior," *Science,* 211 (1981), 1318–24.

5. A. Constantinople, "Sex-Role Acquisition: In Search of the Elephant," *Sex Roles,* 5 (1979), 121–33.

6. M. Lewis and M. Weintraub, "Origins of Sex-Role Development," *Sex Roles,* 5 (1979), 135–49.

7. M. B. Parlee, "The Sexes under Scrutiny: From Old Biases to New Theories," *Psychology Today,* 12, no. 6 (1978), 62–69; and A. A. Ehrhardt and H. F. L. Meyer-Bahlburg, "Effects of Prenatal Sex Hormones on Gender-Related Behavior," *Science,* 211 (1981), 1312–18.

8. B. Friedan, *The Feminine Mystique* (New York: Norton, 1963).

9. J. B. Rohrbaugh, "Femininity on the Line," *Psychology Today,* 13, no. 3 (1979), 30–42.

10. A. H. Maslow, "Self-Esteem (Dominance Feeling) and Sexuality in Women," in *Sexual Behavior and Personality Characteristics,* ed. M. F. DeMartino (New York: Grove Press, Inc., 1966).

11. S. Bear et al., "Even Cowboys Sing the Blues: Difficulties Experienced by Men Trying to Adopt Nontraditional Sex Roles and How Clinicians Can Be Helpful to Them," *Sex Roles,* 5 (1979), 191–98.

12. J. L. McCary, "Human Sexuality: Past, Present, and Future," *Journal of Marriage and Family Counseling,* 4 (1978), 3–12.

13. K. E. Bauman and R. R. Wilson, "Premarital Sexual Attitudes of Unmarried University Students: 1968 vs. 1972," *Archives of Sexual Behavior,* 5 (1976), 29–37.

14. R. Athanasiou et al., "Sex," *Psychology Today,* 4 (1979), 39–59.

15. M. Hunt, *Sexual Behavior in the Seventies* (Chicago: Playboy Press, 1974).

16. F. W. Finger, "Changes in Sex Practices and Beliefs of Male College Students over 30 Years," *Journal of Sex Research,* 11 (1975), 304–17.

17. K. E. Davis, "Sex on Campus: Is There a Revolution?" *Medical Aspects of Human Sexuality* (January 1971), pp. 128–42.

18. A. C. Kinsey et al., *Sexual Behavior in the Human Male* (Philadelphia: W. B. Saunders Company, 1948); and A. C. Kinsey et al., *Sexual Behavior in the Human Female* (Philadelphia: W. B. Saunders Company, 1953).

19. A. Hass, *Teenage Sexuality* (New York: Macmillan Publishing Co., Inc., 1979).

20. Hunt (1974), op. cit.

21. Kinsey et al. (1948; 1953), op. cit.

22. Athanasiou et al. (1970), op. cit.

23. R. R. Bell, "Swinging, The Sexual Exchange of Marriage Partners," *Sexual Behavior* (May 1970), pp. 70–79.

24. N. D. Glenn and C. N. Weaver, "Attitudes toward Premarital, Extramarital, and Homosexual Relations in the U.S. in the 1970s," *Journal of Sex Research,* 15 (1979), 108–18.

25. C. W. Wilson, "The Distribution of Selected Attitudes and Behaviors among the Adult Population of the United States," *Journal of Sex Research,* 11 (1975), 46–64.

26. I. Kirkendahl, "A New Bill of Sexual Rights and Responsibilities," *The Humanist* (January-February 1976), pp. 4–6.

27. P. Kronhausen and E. Kronhausen, *The Sexually Responsive Woman* (New York: Ballantine Books, Inc., 1963).

28. G. P. Murdock, *Social Structure* (New York: The Free Press, 1956); W. F. Stephens, *The Family in Cross Cultural Perspective* (New York: Holt, Rinehart and Winston, 1969).

29. Ibid.

30. D. J. Ogren, "Sexual Guilt, Behavior, Attitudes, and Information," Ph.D. dissertation, University of Houston, 1974.

31. Maslow (1966), op. cit.

32. I. L. Reiss, "The Influence of Contraceptive Knowledge on Premarital Sexuality," *Medical Aspects of Human Sexuality* (February 1970), pp. 71–86.

33. R. L. Delcampo et al., "Premarital Sexual Permissiveness and Contraceptive Knowledge: A Biracial Comparison of College Students," *Journal of Sex Research,* 12 (1976), 180–92.

34. A. C. Kerchoff, "Social Class Differences in Sexual Attitudes and Behavior," *Medical Aspects of Human Sexuality* (November 1974), pp. 10–31.

35. Hunt (1974), op. cit.

36. R. R. Bell, "Female Sexual Satisfaction as Related to Levels of Education," *Sexual Behavior* (November 1971), pp. 8–14.

37. Hunt (1974), op. cit.

38. J. L. Caughey, "Media Mentors," *Psychology Today,* 12, no. 4 (1978), 44–49.

39. H. G. Beigel, "Outmoded Sex Laws Should Be Changed," *Sexology* (December 1965), pp. 341–43.

40. D. W. Cory and J. R. LeRoy, "A Radically New Sex Law," *Sexology* (January 1964), pp. 374–76.

41. C. T. Cory, "Parents' Sexual Silence," *Psychology Today,* 12, no. 8 (1979), 14, 84.

42. I. L. Reiss, "Adolescent Sexuality," in *Sex and the Life Cycle,* eds. W. W. Oaks et al. (New York: Grune, 1976).

43. F. R. Wake, "Attitudes of Parents toward the Premarital Sex Behavior of Their Children and Themselves," *Journal of Sex Research,* 5 (1969), 170–77.

44. R. J. Levin, "The Redbook Report on Premarital and Extramarital Sex," *Redbook* (October 1975), pp. 38–44.

45. N. Uddenberg, "Mother-Father and Daughter-Male Relationships: A Comparison," *Archives of Sexual Behavior,* 5 (1976), 69–79.

46. Kinsey et al. (1948; 1953), op. cit.

47. Athanasiou et al. (1970), op. cit.

48. Hunt (1974), op. cit.

49. M. C. Frede, "Sexual Attitudes and Behavior of College Students at a Public University in the Southwest," Ph.D. dissertation, University of Houston, 1970.

50. D. L. Mosher and H. J. Cross, "Sex Guilt and Premarital Sexual Experiences of College Students," *Journal of Consulting and Clinical Psychology,* 36 (1971), 27–32.

51. W. Ehrmann, *Premarital Dating Behavior* (New York: Holt, Rinehart and Winston, 1959).

52. Reiss (1970), op. cit.

53. S. J. Kutner, "Sex Guilt and the Sexual Behavior Sequence," *Journal of Sex Research,* 7 (1971), 107–15.

54. R. W. Menninger, "Decisions on Sexuality: An Act of Impulse, Conscience or Society," *Medical Aspects of Human Sexuality* (June 1974), pp. 56–85.

55. W. A. Layman, "The 'Saint or Sinner' Syndrome: Separation of Love and Sex by Women," *Medical Aspects of Human Sexuality* (August 1976), pp. 46–53.

56. E. Fromm, *The Art of Loving* (New York: Bantam Books, Inc., 1970).

57. A. Maslow, *Beyond the Farthest Reaches of Human Nature* (New York: The Viking Press, 1971).

58. Athanasiou et al. (1970), op. cit.

59. J. Kelley, "Sexual Permissiveness: Evidence for a Theory," *Journal of Marriage and Family,* 40 (1979), 455–68.

60. Fromm (1956), op. cit.

61. G. Neubeck, *Extra-Marital Relationships* (Englewood Cliffs, N.J.: Prentice-Hall, Inc., 1969).

62. F. Cutter, "The Crime of Incest," *Sexology* (June 1963), pp. 744–46.

63. D. Byrne, *The Attraction Paradigm* (New York: Academic Press, Inc., 1971); and S. G. Vandenberg, "Assortive Mating, or Who Marries Whom?" *Behavior Genetics,* 2 (1972), 127–58.

64. M. B. Parlee, "The Temperatures of a Marriage," *Psychology Today,* 13, no. 1 (1979), 29.

65. G. Dean and D. T. Gurak, "Marital Homogamy, The Second Time Around," *Journal of Marriage and Family,* 40 (1978), 559–70.

66. N. D. Gleen and C. N. Weaver, "A Multivariate, Multisurvey Study of Marital Happiness," *Journal of Marriage and Family,* 40 (1978), 269–82.

67. L. Tiger, "Omnigamy: The New Kinship System," *Psychology Today,* 12, no. 2 (1978), 14–17.

68. M. Roman and W. Haddad, "The Case of Joint Custody," *Psychology Today,* 12, no. 4 (1978), 96–105.

69. E. D. Macklin, "Cohabitation in College: Going Very Steady," *Psychology Today,* 8, no. 6 (1974), 53–59.

70. N. O'Neill and G. O'Neill, *Open Marriage* (New York: Avon Books, 1972).

71. R. M. Kanter, "Communes," *Psychology Today,* 4 (1970), 53–57.

FURTHER READING

ALAN GUTTMACHER INSTITUTE, *Teenage Pregnancy: The Problem That Hasn't Gone Away.* New York: The Alan Guttmacher Institute, 1981.

ATHANASIOU, R., P. SHAVER, and C. TRAVIS, "Sex," *Psychology Today,* 4 (1970), 39–59.

BARBACH, L. G., *For Yourself: The Fulfillment of Female Sexuality.* New York: Anchor Books, Inc., 1976.

FROMM, E., *The Art of Loving*. New York: Bantam Books, Inc., 1970.

HATCHER, R. A., and J. B. ADAMS, "Solving the Teenage Pregnancy Problem," *Medical Aspects of Human Sexuality* (March 1980), Pp. 10–23.

HITE, S., *The Hite Report: A Nationwide Survey on Female Sexuality*. New York: Macmillan Publishing Co., Inc., 1976.

HUNT, M., *Sexual Behavior in the Seventies*. Chicago: Playboy Press, 1974.

KIRKENDAHL, L., "A New Bill of Sexual Rights and Responsibilities," *The Humanist* (January-February 1976), Pp. 4–6.

KRANTZLER, M., *Creative Divorce*. New York: Signet Books, 1973.

LEVIN, R. J., "The Redbook Report on Premarital and Extramarital Sex," *Redbook* (October 1975), Pp. 38–44, 190.

MACCOBY, E. E., and C. N. JACKLIN, *The Psychology of Sex Difference*. Stanford, Calif.: Stanford University Press, 1974.

MACKLIN, E. D., "Cohabitation in College: Going Very Steady," *Psychology Today,* 8, no. 6 (1974), 53–59.

MASTERS, W. H., and V. E. JOHNSON, *The Pleasure Bond: A New Look at Sexuality and Commitment*. Boston: Little, Brown & Company, 1975.

MONEY, J., "The Development of Sexuality and Eroticism in Humankind," *The Quarterly Review of Biology,* 56 (1981), 379–404.

MONEY, J., and A. EHRHARDT, *Man and Woman: Boy and Girl*. Baltimore: Johns Hopkins Press, 1972.

SCHULZ, D. A., *Human Sexuality*. Englewood Cliffs, N.J.: Prentice-Hall, Inc., 1979.

SORENSON, R. C., *Adolescent Sexuality in Contemporary America*. New York: World Publishing Company, 1973.

WHEELER, M., *No-Fault Divorce*. Boston: Beacon Press, 1974.

20 *Human Population Growth and Family Planning*

LEARNING GOALS

Having read this chapter, you should be able to:

1. Describe how birth rate, death rate, emigration, and immigration interact to determine population size.
2. Describe how the age distribution of a population reflects its growth rate.
3. Discuss how environmental resistance limits the biotic potential of a population to the carrying capacity of the environment.
4. Calculate the doubling time of a population with a given rate of increase.
5. Differentiate between exponential and arithmetic growth.
6. Discuss the contribution of Thomas Malthus to present theories about human population biology.
7. State the present earth's population and its rate of increase and doubling time.
8. Discuss the reasons for, and implications of, a 1.0 percent population growth rate in developed countries as opposed to a 2.0 percent growth rate in underdeveloped countries.
9. List the steps that would have to occur to reach zero population growth in the United States.
10. Explain why it still takes several years to reach zero population growth after each couple has only two children.
11. Explain why the post-World War II "baby boom" did not result in a great increase in the U.S. birth rate in the 1970s.

12. Compare and discuss differences in age structure of populations in developed and underdeveloped countries.

13. Discuss the cause and effect relationships of overpopulation and illiteracy.

14. List reasons why foreign aid to underdeveloped countries in the form of food has not solved the problem of hunger and starvation in these countries.

15. List ways in which increasing population density may influence our health, psychological well-being, and political situation.

16. Discuss the past and possible future effects of the Agricultural and Scientific-Industrial Revolutions on our population.

17. Explain why family planning does not necessarily result in population control.

18. Describe the activities and interests of family planning programs, and tell how these relate to population control.

19. List various educational, psychological, cultural, economical, religious, and racial barriers to family planning programs.

20. List possible ways in which governments could attempt to limit family size.

INTRODUCTION

In France, schoolchildren are given a riddle. They are told that a lily pond contains a single leaf. Each day thereafter the number of leaves doubles. They then are asked, "If the pond is full of leaves on the thirtieth day, at what point is the pond half full?" The answer is the twenty-ninth day![1] This riddle teaches about a reality of population growth; it dispels the myth that it will take as long as people have been on earth to double our present population. The previous chapters in this book have presented what we know about human reproductive anatomy, physiology, and behavior, and something about how science and medicine are providing us with tools to influence these phenomena. This chapter deals with problems of choice in how we will go about our reproduction, which after all means "producing offspring."

THE BIOLOGY OF POPULATION GROWTH

Basic Principles of Population Biology

The size of any given population of organisms is influenced by four factors: (1) *birth rate,* or natality, (2) *death rate,* or mortality, (3) *immigration,* or movement of new individuals into a population from another population, and (4) *emigration,* or movement of individuals out of a defined population. That is:

Population size = (birth rate + immigration) − (death rate + emigration)

Birth Rate

Birth rate is the number of individuals born in a population in a given amount of time. Human birth rate is stated as the number of individuals born per year per 1000 in the population. For example, if 35 births occur per year per 1000 individuals, the birth rate is 35. Often this rate is expressed as a percentage, in this case 3.5 per 100, or 3.5 percent. Populations can be subdivided into *juveniles* (before puberty), *reproductive adults,* and *postreproductive adults* (those too old to have offspring). The more younger individuals in a population, the faster that population is growing because the birth rate is higher and the death rate lower (Fig. 20-1). When birth rate is expressed per age group, it is called the *standardized birth rate,* as opposed to the *crude birth rate* of the total population.

Death Rate

Death rate is the number of individuals dying in a population in a given amount of time. As with birth rate, death rate is the number of humans in the population dying per year per 1000, and death rate also can be expressed as a percentage. The *crude death rate* can be broken down further into a *standardized death rate* for certain age groups. The higher the percentage of older individuals in a population, the higher is the death rate and the lower the birth rate; and the population will decline (Fig. 20-1).

Emigration and Immigration

The balance between emigration and immigration of individuals would determine the net addition or deletion of individuals in the population if birth rate and death rate were equal. We consider emigration and immigration to some extent when we talk about specific countries, but we ignore these when talking about the earth's popula-

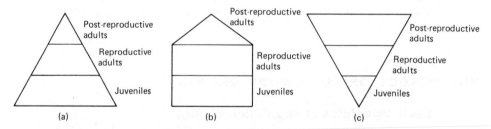

Figure 20-1. Three types of age pyramids. Population (a) is growing because of a large proportion of juveniles. In (b), the population is stable; note the similar proportion of reproductive adults and juveniles and the relatively large proportion of postreproductive adults as compared to (a). Population (c) is declining because the preponderance of postreproductive adults means the death rate is high and the birth rate is low. Adapted from *Life Science,* 2nd ed., by Gerard J. Tortora and Joseph F. Becker (New York: Macmillan Publishing Co., Inc., 1978). © 1978 by Macmillan Publishing Co., Inc; used with permission of Macmillan Publishing Co., Inc.

tion since there is no good evidence of anyone coming or leaving our planet permanently!

Population Growth Rate

In general, and ignoring immigration and emigration, the *population growth rate* is the number of individuals added to or subtracted from a population in a given amount of time. Thus, growth rate is related to birth rate, and death rate is described in the following equation:

$$\text{Population growth rate} = \text{birth rate} - \text{death rate}$$

The growth rate has a positive sign in a growing population and a negative sign in a declining population, depending on whether the birth rate exceeds or is less than the death rate. When the two rates are equal, the population is *stable*.

Reproductive Potential

Theoretically, every species has a maximal birth rate defined by its genes. This maximal birth rate is the *reproductive potential* of the species. For example, humans usually have one child at a time with a minimal spacing between babies of about a year, assuming about three months of breast-feeding (which tends to inhibit ovulation, Chapter 11) and no use of contraceptives. Thus, the reproductive potential of humans is about one child per year per reproductive female. Assuming a reproductive life span of women to be 13 to 50 years of age, a woman theoretically is capable of having 37 children!

Biotic Potential

The *biotic potential* of a population is the ability of that population to grow under optimal environmental conditions. When populations are expressing their biotic potential, birth rate is maximal and deaths are few and are due to the aging process. Population growth under these conditions is at first slow but gets faster and faster. This kind of increase, called *exponential growth,* occurs because additions to the population are increasing like money does in a compound interest savings account. At a given amount of interest (percent population growth rate), the total amount of money in the account (population size) increases at a greater and greater rate because the interest is calculated on an ever-increasing amount in the account. The population growth curve of a population expressing its biotic potential looks like Fig. 20-2.

Environmental Resistance

Usually, natural populations do not express their biotic potential because environmental factors curb the birth rate and increase the death rate. This *environmental resistance* to population growth can take the form of limited food and habitat for

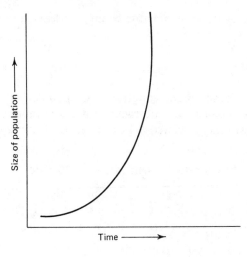

Figure 20–2. The exponential growth curve of a population under optimal environmental conditions. Adapted from *Life Science,* 2nd ed., by Gerard J. Tortora and Joseph F. Becker (New York: Macmillan Publishing Co., Inc., 1978). © 1978; reprinted with permission of Macmillan Publishing Co., Inc.

breeding and deaths due to disease, parasitism, and accidents. Many of these limiting environmental factors have a greater impact on a population, the greater the population size. In other words, more individuals will suffer, the higher the population density. In summary, populations are prevented from expressing their biotic potential as defined in this equation:

$$\text{Population growth} = \text{biotic potential} - \text{environmental resistance}$$

The *carrying capacity* of the environment for a given population is defined as the number of individuals that can be supported by that environment. Thus, environmental resistance usually causes a population previously expressing its biotic potential to level off at the carrying capacity; death rate increases, birth rate lowers, and population size stabilizes (Fig. 20-3). If the population is below carrying capacity, the opposite occurs. Thus, populations are regulated by a feedback system, and the carrying capacity is the "set point" (Chapter 2). In some lower animals, behaviors such as aggressive defense of territories keep the population below the carrying capacity, but the extent to which this occurs in humans is not clear.

Population Crashes

When a population grows far beyond its carrying capacity, the environment reprimands this error with disaster! That is, the population reaches a level where an important environmental resource such as food is destroyed by the individuals themselves. The result is a population crash to a level far below the carrying capacity (Fig. 20-4). After a crash, a population then would express its biotic potential for a time but soon would be regulated at carrying capacity if the environment has recovered. If the environment has not recovered, the carrying capacity and stable population size

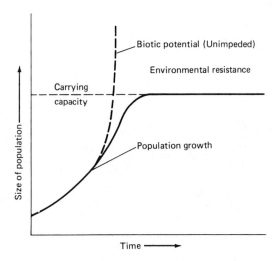

Figure 20-3. Environmental resistance causes a population previously expressing its biotic potential to level off at the carrying capacity of the environment. Adapted from *Life Science,* 2nd ed., by Gerard J. Tortora and Joseph F. Becker (New York: Macmillan Publishing Co., Inc., 1978). © 1978; used with permission of Macmillan Publishing Co., Inc.

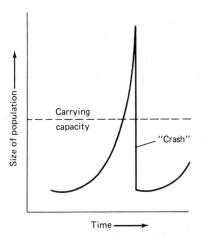

Figure 20-4. When population growth exceeds the carrying capacity of the environment, components of the environment are destroyed and the population crashes. Adapted from *Life Science,* 2nd ed., by Gerard J. Tortora and Joseph F. Becker (New York: Macmillan Publishing Co., Inc., 1978). © 1978; used with permission of Macmillan Publishing Co., Inc.

would be lower. Some animal populations repeat this overpopulation error and exhibit cycles of population build-ups followed by crashes.

Doubling Time

If one knows the birth rate and death rate of a population, the future of that population can be predicted with some accuracy. For example, if the birth rate is 35 per 1000 per year and the death rate is 25 per 1000 per year, the growth rate of the population is + 10 per 1000 per year. At this 1.0 percent rate of population increase, it will take 70 years for the population to double. Thus the number of years it will take for a popula-

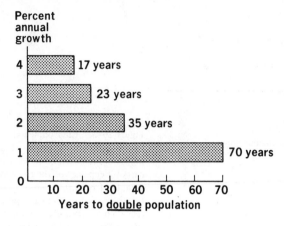

Percent
annual
growth

4 | 17 years
3 | 23 years
2 | 35 years
1 | 70 years
0

10 20 30 40 50 60 70
Years to <u>double</u> population

Figure 20-5. Even extremely small percent annual growth rates produce rapid population increases. Reproduced from P. Corfman and E. Huyck, *Population Research at the U.S. National Institutes of Health: A Response to a National and International Problem,* DHEW Publ. No. (NIH) 75-781, 1978. © 1978; used with permission.

tion *(doubling time)* can be calculated by dividing the rate of population inrease into 70:

$$\text{Doubling time in years} = \frac{70}{\text{Percent annual rate of increase}}$$

Thus, a 2.0 percent annual rate of increase would produce a doubling time of 35 years.[2] Surprisingly, low growth rates still have relatively brief doubling times (Fig. 20-5).

HUMAN POPULATION GROWTH

The Prediction of Thomas Malthus

In 1798, Thomas R. Malthus, an English economist, historian, and clergyman, wrote an article entitled "An Essay on the Principle of Population."[3] In his discussion, he proposed that human populations tend to grow in an exponential manner, whereas food supply increases only arithmetically. For example, suppose we begin with a human population of 1 and a food abundance of 1. The growth of each would then proceed in the following manner:

Exponential human population growth: $1 \rightarrow 2 \rightarrow 4 \rightarrow 8 \rightarrow 16 \rightarrow 32$
Arithmetic increase in food: $1 \rightarrow 2 \rightarrow 3 \rightarrow 4 \rightarrow 5 \rightarrow 6$

Malthus concluded that "the power of population is infinitely greater than the power in the earth to produce subsistence of man."[4] That is, the "passion of the sexes" will drive the human population to a level that exceeds the capacity of the environment to support it. The end result, in his words, will be "misery and vice."

Malthus failed to predict two things: first, that people may have a choice whether or not to express their reproductive potential, and second, that they would be able to increase artificially the carrying capacity, including food supply, of the earth's ecosystem.[5] Nevertheless, there have been limitations to what we have done and can do about this Malthusian dilemma, and his prediction looms over the horizon.

Human Population Growth on Earth

Let us look at what has happened to the human population in the past, and what may happen in the future. One way we can view this is by looking at the numbers of people on earth from ancient times to the present.[6] About 25,000 years ago, people relied on hunting and food gathering for subsistence, and the total population on earth was about 3 million. At the beginning of the Agricultural Revolution (8000 B.C.), when people discovered how to grow food, the population was 5 million. Between that time and the birth of Christ (A.D. 1), the population doubled about six to seven times, resulting in about 300 million people. In 1650, there were 500 million people, and it took only 200 years (to 1850) for this number to double again to 1 billion. By this time, the Scientific-Industrial Revolution, which is still going on in developed countries and has not even begun or is just beginning in underdeveloped countries, assisted the population to double in only 80 years, to a population of 2 billion in 1930. By 1980, it had more than doubled again to 4.4 billion, and this took only about 50 years. Thus, the human population has not only increased but has increased faster and faster in an exponential manner (Fig. 20-6). This history of human population growth and doubling time is summarized in Table 20-1.

The human population in late 1982 was about 4.6 billion, and was growing at a rate of about 2.0 percent per year, which translates to more than 211,000 people a day! At that rate of increase, the doubling time would be 35 years, so the population in the year 2000 will be about 6.8 billion. Of this number, about 1.45 billion will be in developed countries, whereas 5.35 billion will reside in underdeveloped countries. No country that desires social and economic progress can survive this rate of population growth.

TABLE 20-1. CHANGE IN WORLD POPULATION AND DOUBLING TIME

Date	Estimated World Population	Time It Took For Population to Double
8000 B.C.	5 million	
A.D. 1	300 million	
A.D. 1650	500 million	
A.D. 1850	1,000 million (1 billion)	200 years
A.D. 1930	2,000 million (2 billion)	80 years
A.D. 1980	4,000 million (4 billion)	50 years
A.D. 2015	8,000 million (8 billion)	35 years

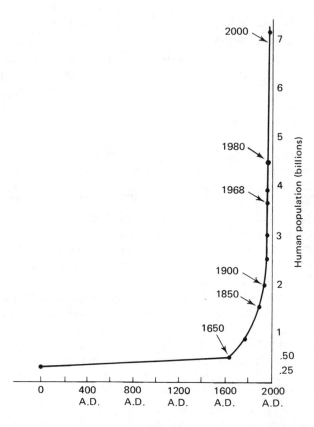

Figure 20-6. Growth of the world population. At the beginning of the Christian era, the world population is estimated to have been 300 million. The 1650 population of 500 million had doubled by the year 1850 to 1 billion people. In the past 100 years, the growth rate has accelerated greatly. It is estimated that by the year 2000, the earth will have approximately 6.8 billion inhabitants. Note the similarity of this curve to that in Fig. 20-2. Adapted from *Life Science,* 2nd ed., by Gerard J. Tortora and Joseph F. Becker (New York: Macmillan Publishing Co., Inc., 1978). © 1978; used with permission of Macmillan Publishing Co., Inc.

Population Growth in Developed Countries

Several developed regions, such as the Western European countries, the USSR, Canada, the United States, and Japan, have reduced their population growth rate to approximately 1.0 percent or less, and East Germany, Luxembourg, Austria, and Belgium have stable or declining populations.[7] At present, the population of West Germany is declining because the birth rate has gone below the death rate while the latter has remained stable (Fig. 20-7). In Japan, legal abortion supplied by the government has resulted in about one-third of pregnancies being terminated in the first trimester, and the result has been a marked decline in Japan's population growth rate. Thus, use of birth control methods such as IUDs, condoms, oral contraceptives, and sterilization in some cases has succeeded in lowering the rate of population growth in certain developed countries. A few countries in Europe have a low birth rate and a low population growth rate not because of the use of "modern" birth control methods but because of extensive use of coitus interruptus.[8] However, recent disturbing news is that the birth rate in England and Wales has been increasing in the past few years.[9]

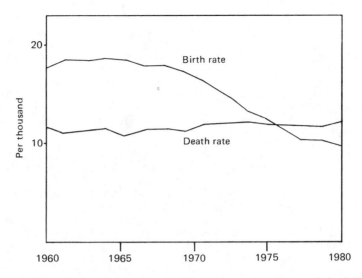

Figure 20-7. Birth and death rates in West Germany. As in several developed countries, the birth rate has decreased while the death rate has remained constant. At present, the population of West Germany is declining slightly. Adapted from L. J. Brown, "World Population Trends: Signs of Hope, Signs of Stress," *Population Reports,* ser. J, no. 13 (1977), p. 240. Used with permission of the Population Information Program, Johns Hopkins University, Baltimore, Md.

Age Distribution. When one looks at the distribution of ages in developed countries with a relatively low rate of population growth, different ages are fairly evenly distributed (Fig. 20-8), so the population has relatively less potential for producing children than would a population with a predominance of younger people. However, you must remember that even a very small population growth rate (1.0 percent per year, for example) can result in doubling of the population in a relatively short time (Fig. 20-5).

U.S. Population Growth. The birth rate in the United States exhibited a steady decrease until 1940. However, the so-called "baby boom" after World War II reflected an increase in the birth rate, from around 2.0 percent in 1940 to 2.7 percent in 1947. Since then, the birth rate again declined to a low in the late 1970s, but recently it has shown a slight increase (Fig. 20-9). The baby boom resulted in a bulge in the pyramid of U.S. age structure in the 1960 population in the 0–14 age class, and these young individuals were predicted to increase greatly the birth rate in the 1970s. But this prediction has only slightly been borne out. [10] This is because several factors inhibited women from having as many children in the 70s as in the past. [11] The age of marriage increased, desired family size decreased, use of birth control methods increased, abortion laws were liberalized, and there were changes in attitude about the role of women in our society.

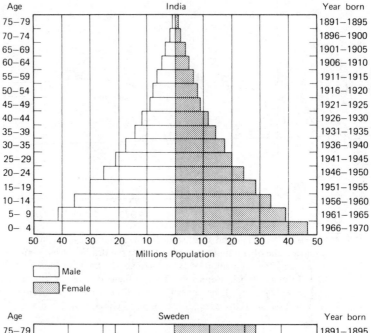

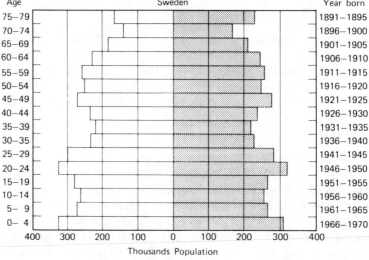

Figure 20-8. Age structure in India and Sweden. Width of bars shows proportion of total population in age group. Note the predominance of youth in India, a rapidly growing population, as opposed to the more equal age distribution in more stable, developed Sweden. Compare these figures to those in Fig. 20-1. Reproduced from C. J. Avers, *The Biology of Sex* (New York: John Wiley & Sons, Inc., 1974). © 1974; used with permission.

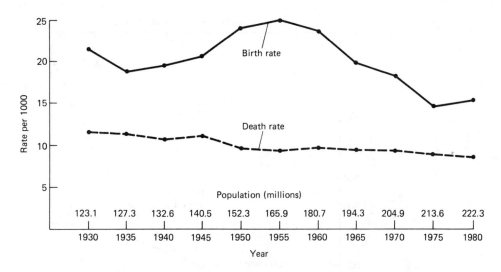

Figure 20-9. Birth rates and death rates in the United States from 1930 to 1980. Whenever birth rate exceeds death rate, a population is growing. Note that the death rate has slowly been declining, but the birth rate has fluctuated widely. The peak in births in the 1950s represents the post-World War II "baby boom." Note the recent slight increase in birth rate. From *Statistical Abstract of the United States, 1980,* 101st ed. (Washington, D.C.: U.S. Bureau of the Census, 1980).

The U.S. population in 1980 was about 200 million. At the rate of population growth at that time (slightly less than 1.0 percent), the doubling time was 70 years, and the U.S. Census Bureau predicted about 300 million people in the United States in the year 2000. To reach zero population growth in the United States at the present death rate of 9.6/1000/year would require limiting family size to 2.0 to 2.2 children. But even if we averaged 2.0 children per female, the population would continue to increase because of the 400,000 immigrants entering our country each year. Stabilization of the U.S. population at this rate of immigration and with a family size of 2.0 children still would require 50 to 60 years (Fig. 20-10) because it would take that long for the age structure to shift to favor middle-aged instead of younger people. The average age in the United States in 1979 was about 28 years, and this would increase to about 37 in these 50 to 60 years.

Population Growth in Underdeveloped Countries

The world population is growing most rapidly in underdeveloped countries, many of which are still growing at a rate of over 2.0 percent a year. For example, in the late 1970s, the population growth rate in India was 2.5 percent, and the population would double in only 28 years. The growth rate was 3.0 percent in tropical South America, 3.4 percent in middle American countries, and 3.5 percent in the Philippines! The age

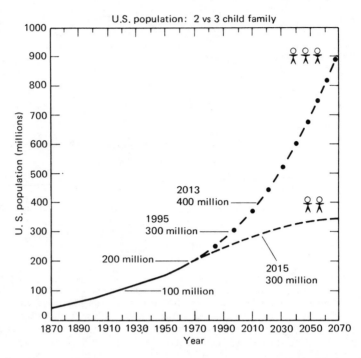

Figure 20-10. Differing predictions for population size in the United States, depending on a family size of two children (- - -) or three children (—·—·—). These projections assume small future reductions in mortality and continuation of immigration at the present level. Population would stabilize at 350 million in the year 2070 if each family had only two children now. Adapted from C. J. Avers, *The Biology of Sex* (New York: John Wiley & Sons, Inc., 1974). © 1974; used with permission.

structure of these countries favors the young (Fig. 20–8). In fact, about half of the people in underdeveloped countries are under 15 years of age.

These marked growth rates are present even though many of these countries have introduced formal family planning programs and receive aid from advanced countries. In 1979, two decades after oral contraceptives were first introduced, birth control pills delivered for one year to underdeveloped countries amounted to a supply for 25 million women.[12] However, cultural, educational, and religious factors have been a barrier to the effectiveness of family planning programs in these countries. For example, 85 percent of the world's educated women use the combination pill, but only less than 10 percent of uneducated women use this birth control method. Most women in underdeveloped countries are uneducated and either use no birth control measures or their mates practice coitus interruptus. Thus, because the birth rate did not decrease as rapidly as hoped for and the death rate has declined, the end result has been rapid population growth (Fig. 20-11). A report in 1979 showed that birth rate is declining in 14 of 15 underdeveloped countries (but not Latin American countries).[13]

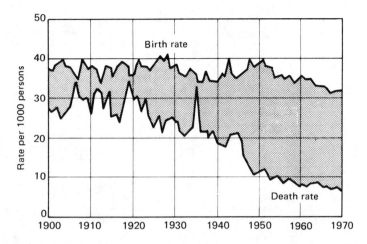

Figure 20–11. Birth and death rates in Ceylon. Note that although the birth rate is slightly decreasing, death rate is dropping. Thus, as in many underdeveloped countries, the population is growing rapidly not because of an increase in birth rate but because the death rate is decreasing as a result of better nutrition and medicine. Adapted from C. J. Avers, *The Biology of Sex* (New York: John Wiley & Sons, Inc., 1974). © 1974; used with permission.

The reasons for this decline are: (1) women are marrying at a later age, (2) families want fewer children, and (3) there is greater utilization of birth control methods. The population growth rate, however, is still high in these countries because birth rate remains higher than death rate. About 800 million people on earth still live in absolute poverty, and more than 30 million children under the age of five died of starvation in 1978!

Life Expectancy

Another way of looking at the population situation is life expectancy, which relates to the death rate. Upper-class Egyptians 2000 years ago had a mean life expectancy of 25 to 30 years, as determined by examination of mummies. Europeans in the Middle Ages lived an average of 35 years, which not long ago was the life expectancy of people in India. Tombstones suggest that only 18 percent of people in ancient Greece reached 40 years of age. Better nutrition and improved health care have now increased mean life expectancy to over 70 years in many developed countries such as the United States and United Kingdom. In these countries, most deaths occur in the very old or very young (Fig. 20-12). In some underdeveloped countries, infant mortality is very high. Although this is deplorable, it decreases birth rate more effectively than does mortality of postreproductive adults because these children do not survive to reproduce. Surprisingly, the United States ranks only fifteenth in the world in prevention of infant mortality because health care to the poor is limited in this country.

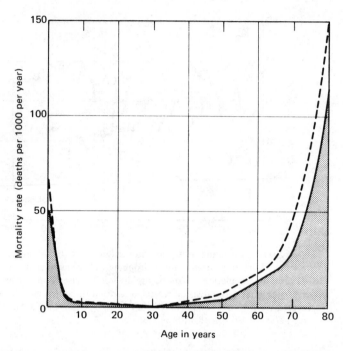

Figure 20–12. The mortality rate in the United States varies with age and is higher among males at older ages. The female mortality rate is shaded, and the dashed line is that of males. Adapted from C. J. Avers, *The Biology of Sex* (New York: John Wiley & Sons, Inc., 1974). © 1974; used with permission.

Overall View of Human Population Growth

The world's population is increasing at an alarming rate due to an increase in "death control" without enough birth control. This phenomenon is most evident in underdeveloped countries, but it is also present to a lesser degree in most developed countries. Even if every family on earth would limit its reproduction to two children now, the earth's population would still increase about 10.5 billion in the year 2110 because it would take many years for the age structure to equalize.

EFFECTS OF OVERPOPULATION

When a human population exceeds the carrying capacity of the earth to support it, as is happening now in some regions, many miseries are inflicted on each individual. These miseries take many forms, and it is beyond the scope of this book to discuss them in detail; read the book *Population, Resources, Environment,* by Paul R. Ehrlich and Anne H. Ehrlich (see Further Reading list at end of chapter), for a more thorough discussion of the impact of population growth on our environment. We now discuss briefly some of these population-related problems.

Illiteracy

It may come as a surprise to you as you are reading this book that about one-third of the people on earth are illiterate (can't read or write); and two-thirds of these are women, which reflects social attitudes about women and their role in different societies.[14] And, it appears that the higher the birth rate in a country, the greater the illiteracy. This is because of the large number of young in these populations and the inability to educate most of them. Conversely, illiteracy and poor education in general can lead to an increase in birth rate. In the United States, couples with lower intelligence scores have more children, not because they want more but because they fail to use contraception.[15] Accompanying this widespread illiteracy is unemployment, low income, poor housing, and poor health care.

Food Production and Hunger

Hunger and starvation are serious problems in many underdeveloped countries, while a major concern in developed countries is obesity! The Scientific-Industrial Revolution has not reached many areas in which malnourishment is predominant. Much of the land in these countries is overcultivated, and this, along with overgrazing by domestic livestock and deforestation, has resulted in land erosion and destruction of land potentially available for crops.[16] Aid from advanced countries, involving shipping of food and technological assistance, has and is helping, but this aid has not worked as well as planned. This is because: (1) this aid does not keep up with population growth, (2) such aid is only a stop-gap measure and usually does not motivate the people to produce more food for themselves or to decrease their birth rate, (3) the food and aid, once delivered to the country, are not fairly distributed, and (4) the availability (and cost) of food in developed countries is changing.

Unless population growth is curbed, the ultimate result in underdeveloped countries can be devastating famine.[17] In fact, massive starvation and death have occurred quite recently. Four million died of starvation in China in 1920 and 1921, and 2 to 4 million in West Bengal, India, in 1943. Even without widespread famine, about half of the people on earth are undernourished. And even in the United States, about 20 million people do not have proper nutrition to live a healthy life.

Some people believe that more food production and better technology in underdeveloped countries would lower the birth rate in these countries. This is because some of the reasons for having large families are economical, such as a family wanting several children to help farm their land. Theoretically, if tractors or other technologies to help people farm were available, the need for large families would disappear. In countries such as Brazil, Mexico, and Kuwait, however, rapid economic and technological development in the 1960s was not accompanied by a decline in birth rate.

Natural Resources and Energy

There is a myth that the earth's resources are infinite. For example, it is a widespread belief that food from the sea will save us. This hope is probably unfounded, however, because most of the sea is a biological desert. Proper management may be able to double or even triple the yield of protein from the sea, but this will not help in the long run. Even now, ocean fisheries, which supply much protein to the world's population, are in trouble. The catch of fish, although tripling between 1950 and 1970, has declined recently. As a result, the price of fish protein is rising. It is true that fish are a renewable resource, but harvesting must balance with production if this resource is to remain stable. Fish have their own population problem due to excessive human predation.

Energy is at the forefront of world concern. We are already seeing that most of our sources of energy are not renewable, and some are running out.[18] Minerals and fossil fuels (oil, coal, natural gas) are limited in some countries (such as the United States), and these countries must rely on imports from other countries or must develop alternative energy sources. In this situation, the "haves" can and are putting economic and political pressures on the "have-nots." In the late 1970s, virtually one-half of the oil supplies of the United States was imported; and Americans consumed 35 percent of the fossil fuels of the world, yet made up only 6 percent of the world's population. Some countries, like Japan, import *all* of their oil. Increased pressures from oil-producing nations can only result in higher gas prices and gas rationing in "have-not" countries and a forced change in life style of their citizens. Nuclear reactors are an important source of electricity and may become more so in the future. But this source of energy has drawbacks, including high costs and the possibility of nuclear accidents. Alternative energy sources, such as solar energy (which potentially is unlimited), must be developed.

POLLUTION AND ENVIRONMENTAL ILLNESS

Pollution of our ecosystems is a major problem that is related directly to both population number and life style.[19] Our waters often contain pesticides, insecticides, and industrial and human waste. Our air is becoming saturated in some places with automobile and industrial emissions; many of these pollutants are not biodegradable. Some of these pollutants even affect our rainfall, making it more acidic, and this could have profound effects on the earth's plant and animal life. Noise pollution is common in cities. It is clear that we must curb population growth and change our life styles in order to avoid poisoning ourselves. Much public health literature suggests that severe human ailments, such as cancer and respiratory disease, can be traced to environmental contamination.

Crowding and Stress

John Calhoun has studied the behavior of crowded rats and found that high densities cause, among other things, decreased reproduction, reduced maternal care, inceased fighting, and aberrant sexual behaviors such as rape.[20] In addition, crowding in animals produces physiological changes that cause stress-related conditions such as ulcers and cardiovascular disease. Although we have little reliable information on the effects of crowding on our own physiology and behavior, it is evident that the incidence of poor health and crime is higher in cities than in less densely populated areas. Humans are a gregarious species and tend to congregate in clumps. For example, most of the people in the State of California are on less than 10 percent of the land in the state. And about 50 percent of the world population are on 5 percent of the land surface. Many Americans are moving away from cities, so the rural areas are also becoming crowded. As our total population increases, the effects on the environment will be accentuated by the tendency for people to clump together.

Quality of Life

It is a basic human right to be well-nourished, have ample housing and clothing, live in a clean environment, and be able to commune with nature. These rights in some underdeveloped countries are almost totally unfulfilled. Even in the United States, the effects of overpopulation and poor management are widespread. In this affluent society, there are millions of undernourished and poor people. Air, soil, and water pollution are widespread, highways are clogged, schools and jails are overcrowded, and "urban blight" is common. Some natural areas are visited by reservation only, and some have been damaged or destroyed. The list of endangered animal and plant species is getting longer and longer. Even if you are not bothered by the present situation, you must realize that the population of the United States will double in 70 years, and that about 1.5 million people are added each year. Even if we improve our quality of life and decrease our rate of population growth, many other regions of the world will not, and these regions will be of constant humanistic, political, and economic concern.

WILL SCIENCE AND TECHNOLOGY SAVE US?

As mentioned, Malthus did not predict that human beings, through their creative genius, could increase the carrying capacity of the earth to support an increasing population. The Agricultural Revolution by increasing food production and the Scientific-Industrial Revolution by providing new technologies have greatly increased the earth's capacity to support more people in some regions, but these advances have reached only some people, and many have not reaped any benefits. Furthermore, along with the increase in some components of the earth's carrying capacity, came a decrease in the death rate around the world due to mass vaccination and inoculation, improved sanitation, antibiotics, hormones, and other wonder drugs. Thus, science, medicine, and technology have increased components of the carrying capacity but at the same time have decreased the death rate, so the population continues to grow. We are now at the limits of some components of the earth's carrying capacity and are beyond others.

There are a variety of suggestions as to how science and technology can save us in the future. For example, synthetic foods could be produced, we could farm the algae in the sea, we could desalinate the seawater, we could develop better fertilizers, and we could cultivate unused land. It is true that we either now have or will have the capability to do many of these things. [21] But the results will only be a temporary solution. [22] For example, if the usable land that now is not cultivated is changed to crop land, the food produced may still not keep pace with the 80 million people added to the earth each year. Similarly, increased food production from the sea may not keep up with the population explosion. Some have even suggested that we will be able to ship people to other planets in our solar system. However, we would need more than 2000 ships per day, each carrying 100 people and their baggage, to keep the earth's population stable! [23] Even if this were possible, the planets in our solar system probably are uninhabitable.

FAMILY PLANNING AND POPULATION CONTROL

The messages of the previous discussion are that: (1) the human population is subject to basic biological laws, (2) the human population is increasing at an alarming rate, especially in underdeveloped countries, (3) the carrying capacity of the earth's ecosystem for people is finite, and (4) unless our population growth is curbed, the earth will not be able to support our population, and it will crash with, in the words of Malthus, much "misery and vice." Thus, our goal must be to achieve zero population growth as soon as possible before nature does this for us, and to achieve harmony of our activities with the earth's ecosystems. Two factors can decrease the rate of population growth: decreasing the birth rate and/or increasing the death rate. Since we do not want to increase the death rate, we must decrease birth rate to about two children (or less) per female. To do this, we must overcome educational, economic, cultural, psychological, religious, and racial barriers to curbing the birth rate.

Family Planning Programs

For individual freedom to continue, it is necessary for each person to have a choice as to how many children he or she will have. Twelve heads of state, in a United Nations Declaration in December 1966, declared that the opportunity to decide the number and spacing of children is a basic human right. This declaration inferred the right to control excess offspring. Thus, the goal of many *family planning programs* is to eliminate unwanted children. Family planning programs can be supported by private, federal, state, or local governments, and often are staffed by volunteers. Personal discussions with clients center around age of marriage, choice of being married or not, and the availability of and use of birth control methods. Economic, educational, and health advantages of well-spaced and limited numbers of children are emphasized. Advice on pregnancy and child care often is provided. Complete physical examinations, blood tests, a Pap cervical smear test, and a chest X-ray usually are available.

Margaret Sanger, a nurse, is considered to be the instigator of family planning programs in the United States.[24] Her interest stemmed from an experience, in 1912, of a young mother who nearly killed herself by attempting to abort her fourth child. Sanger's activities, in first disseminating birth control literature (which at that time was against federal and state law, being considered pornographic) and later setting up family planning clinics, got her arrested and jailed several times. In the end, however, her efforts had great impact on the use of birth control, resulting in a reduction of birth rate in the United States. Several agencies in the United States are now involved in family planning programs, including Planned Parenthood-World Population, The Family Planning Association, The Population Council, and Zero Population Growth.

In 1952, Ms. Sanger, along with Ms. Attesen-Hensen of Sweden, Lady Ramu Rau of India, and others, succeeded in forming the International Planned Parenthood Federation (IPPF), a combination of family planning groups in several nations. Many countries, most of which are underdeveloped, are now members of IPPF. Other agencies of the United Nations, such as the World Health Organization (WHO) and the United Nations Population Division, assist in governmental family planning programs, distribution of birth control devices, and advertising in the media or from village to village about the value of family planning. Family planning programs are active in every country that has over 100 million people. Recent statistics show that in countries with formal family planning programs for five years or more, the birth rate has declined markedly. In contrast, the birth rate has not declined in countries without effective family planning programs.

Barriers to Family Planning

Limiting the number of children in a family is unacceptable for many people because of economic, social, and/or religious reasons. The family is the fundamental organization unit of human societies, and the size of this unit varies from region to region and culture to culture. One must remember that family planning doesn't necessarily mean population control since family planning simply implies elimination of *unwanted* children. And the family may want and plan for more than two children. In fact, the number of children desired in families of underdeveloped countries is 3.5 to 5.5, as opposed to 2.0 to 3.3 in Europe and the United States. Thus, even if the underdeveloped countries eliminated unwanted births, it would not reduce the marked rate of population growth in these countries. Family planning programs must change the attitudes of some societies and individuals about desired family size if the overall goal of curbing population growth is to succeed. This, however, is a "tough nut to crack."

Cultural Barriers.   Some people want large families for economic security. Consider a couple in a village in India.[25] They may want five or more children to ensure enough people to work their small plot of land. In addition, when they get old and can't work, they want children around to take care of them, taking into account the fact that some of their children may die from malnutrition or disease. How would you go about convincing this couple to have only two children? The population of India in the late 1970s was about 516 million, and it was growing at 2.5 percent per year in spite of a governmental family planning program that began in the 1950s. Besides these attitudinal barriers, consider that there are 550,000 Indian villages with varying languages and customs, and you have got a real problem in reducing population growth in this country.

Cultural traditions have a strong influence on birth rate in some regions. For example, in Bangladesh, the high population growth rate is due mainly to the young age of marriage and the virtual absence of contraceptive use.[26] Most women marry when they reach menarche, which is at age 16 or 17 because of poor nutrition. An average woman is pregnant nine times. Thus, Bangladesh women spend much of their married life either pregnant or nursing. Only 14 percent of married couples in Bangladesh use any form of birth control, including withdrawal and abstinence. In many underdeveloped countries, up to one-third of women do not want more children but also do not use contraceptives.[27] Nevertheless, there is some evidence that family planning programs are causing the birth rate to decline in some underdeveloped countries.[28]

Religious Barriers.   Religious attitudes and beliefs may also limit the effectiveness of family planning programs. Consider this quote from the Bible (Genesis 1:26-28, King James Version):

> And God said, Let us make man in our image, after our likeness: and let them have dominion over the fish of the sea, and over the fowl of the air, and over the cattle, and over the earth, and over every creeping thing that creepeth upon the earth.
> So God created man in his own image, in the image of God created he him; male and female created he them.
> And God blessed them, and God said unto them, Be fruitful, and multiply, and replenish the earth, and subdue it: and have dominion over the fish of the sea, and over the fowl of the air, and over every living thing that moveth upon the earth.

The Roman Catholic Church has proclaimed that any "artificial" means of birth control (that is, anything except rhythm methods or abstinence) is a sin. Even with this position, there is recent evidence that the usage of all forms of birth control in the United States is similar between Catholics and Protestants, although family size is slightly higher among U.S. Catholics as opposed to Protestants.[29] The Anglican (Episcopal) Communion in England and the United States approved all methods of contraception in 1958, as did the Central Conference of American Rabbis (Reformed) in 1960 and the National Council of Churches in 1961. Religious beliefs, however, still influence usage of contraception and family size in many other countries.

Other Barriers. Racial and economic factors sometimes influence population control.[30] For example, many family planning programs in the United States are attempting to reach the poor, and there is a misconception that most poor are racial minorities and that most minorities are poor, which is not true. However, some minority groups believe that attempts to limit family size in the poor, which in their minds means a specific race, is a form of genocide in that it permits whites to breed at the expense of other races.[31] What is true is that the poor in the United States comprise several races, including many whites. And, about 42 percent of families with greater than five children are poor in the United States, whereas only about 10 percent of families with one or two children are poor. The incidence of unwanted children also is higher in the poor (40 percent) as compared to nonpoor (14 percent).

Other factors influencing family size include established patterns of sexual behavior, tolerance of women working, social status of women, and cost of raising and educating children. All of these factors should be of interest to family planners. Tables 20-2 and 20-3 summarize the reasons people have for limiting or not limiting family size. You may want to discuss these with your instructor, classmates, and friends. Another major barrier to effectively reaching zero population growth is that humans tend not to believe in something that is not at present having a serious effect on their lives. Thus, education about the present and future effect of the population explosion should be at the forefront of family planning programs. Not only must these programs deal with the practices and attitudes of a target population about contraception, but they must educate people, especially young men and women, about the potentially disastrous consequences of population growth.

Family Planning in the United States

The Congress of the United States, in 1969, established the Commission of Population Growth and the American Future. This commission recommended several changes for controlling population size in the United States.[32] For example, the government should develop a national plan for population stabilization, and schools should educate children about the population problem. Additional recommendations included that child care services be facilitated so that more women could work, that adoption procedures should be simplified, and that women have equal rights compared to men. Finally, access to contraceptives to minors should be greater, there should be no restrictions on sterilization services, and development of better contraceptives should be emphasized in government-sponsored research programs. Another recommendation, that abortion laws be liberalized, was accomplished by the U.S. Supreme Court (Chapter 14).

Many of the above recommendations have been put into action, resulting in

TABLE 20-2. SOME REASONS WHY A FAMILY WOULD WANT TO HAVE SEVERAL CHILDREN

High Fertility Motives

HEALTH
   Children often die. It is necessary to have large families in order to get living children who grow to adulthood.

ECONOMIC
   Children are an economic advantage. They are needed or are useful in helping the family earn a living. They pay for themselves by working as they grow. Social security in old age. If you have many children, one will be able to take care of you in old age.

FAMILY WELFARE
   Can help with work around the house. Older children help the younger.
   Big families are happy families. Family life is more enjoyable; they have a good time together. Children from big families have better personalities. They are better adjusted, better able to get along with other people, not so spoiled or egotistical.
   Continue the family name. It is necessary to have many children to be sure to have a son to carry on the family name.
   Strength of the clan. The family is stronger; sons can help you fight your battles; family can be upheld.

MARRIAGE ADJUSTMENT
   Large families promote good marriage adjustment. Couples get along with each other better, marriage is better.

PERSONALITY NEEDS
   Ego support. A demonstration of virility, manliness.

COMMUNITY AND NATIONAL WELFARE
   Large families are good for the community or nation. They promote population growth.

MORAL AND CULTURAL
   Large families are God's will. It is against religious belief to limit fertility.
   Large families promote morality. Help prevent divorce or infidelity.
   Tradition. The community, village, family, and clan expect large families.
   You have high status in the community. If you have a big family, you are more important, are looked up to.

DISLIKE FOR CONTRACEPTION
   Dislike use of contraception for aesthetic or health reasons or because it interferes with sex.

Source: From D. J. Bogue, *Mass Communication and Motivation for Birth Control* (Chicago: University of Chicago Community and Family Study Center, 1967). Used with permission.

smaller families and a reduction in the number of unwanted births in the United States. However, a significant portion of unmarried young American men and women are still sexually active, and many of these do not practice effective contraception. A survey by the Alan Guttmacher Institute indicates that two-thirds of the 5 million sexually active teenage females and 7 million males in the United States do not use effective contraceptive protection. In this same survey, 51 percent of the females said that they didn't think they could get pregnant. [33] In sexually active teenagers, 58 percent of those not using contraception get pregnant as opposed to only 6 percent

TABLE 20-3. SOME REASONS WHY A FAMILY WOULD WANT TO LIMIT THEIR OFFSPRING NUMBER

HEALTH
  Preserve health of mother.
  Assure healthy children.
  Reduce workload of father and mother.

ECONOMIC
  Everyday, general expenses are less.
  Avoid worsening present (poor) economic condition.
  Gain a higher standard of living, more comfort, afford better house.
  Permit saving for future, for retirement.
  Desire to avoid subdividing property or savings among many children.
  Family able to have money for recreation, vacation.

FAMILY WELFARE
  Improve children's lot in life, give them good education, help get started in a career.
  Opportunity to do a better job of rearing children; able to devote more time to each, better able to socialize with child.
  Avoid overcrowding of house; more opportunity for individual expression.
  Easier to find a more desirable house or apartment.

MARRIAGE ADJUSTMENT
  Provides husbands and wives more leisure opportunity to enjoy each other's companionship.
  Improves the sexual adjustment by eliminating or reducing fear of unwanted pregnancy.

PERSONALITY NEEDS
  Facilitates realization of ambitions. Permits either husband or wife or both to pursue occupational or vocational objectives.
  Facilitates self-development. Permits an intelligent or talented wife to express herself outside the home, yet have a normal family and married life.
  Facilitates realization of social needs. Permits the person (especially wife) to have contacts and friendships outside the home and participate in neighborhood activities.
  Reduces worry of the future. Avoids danger of childbearing when one is really too old—danger of dying and leaving behind orphan children.

COMMUNITY AND NATIONAL WELFARE
  Helps avoid overpopulation, overcrowding.
  Helps community meet demands for education, other community services.
  Helps nation with economic development.
  Helps keep down delinquency, social problems of youth.
  Helps reduce welfare burden of the community.

Source: From D. J. Bogue, *Mass Communication and Motivation for Birth Control* (Chicago: University of Chicago Community and Family Study Center, 1967). Used with permission.

that do use contraception.[34] And the U.S. population still is growing at a rate of about 1.0 percent a year!

Americans must be aware that it is their reproduction that has the most profound impact on our population in pure numbers, and it is their life style that is polluting, damaging, and consuming our environment. An average middle-class American uses about 26,000 gallons of water, 21,000 gallons of gasoline, and 10,000 pounds of beef in his or her lifetime. As noted earlier, Americans use 35 percent of the world's supply of minerals and fossil fuels but comprise only 6 percent of the world's population. It may be time for a voluntary change in life style before such a change is required for survival. In addition, the ideas that "I'll get mine while I'm here," and "Why should I save it for somebody to use and enjoy after I'm dead?" must be discarded.

Individual Freedom and Family Planning

Individual freedom dictates that family planning should be on a voluntary basis. However, if voluntary reduction in family size is not achieved, governments could initiate measures to control population size.[35] The Chinese government, for example, recently passed a law raising the legal age of marriage and making birth control compulsory for newlyweds. Table 20-4 lists some possible ways that governments could attempt to limit our reproduction. What you should do is look at each suggestion, think about it, and discuss its advantages and disadvantages. Which of these is favorable to voluntary limitation of family size?

CHAPTER SUMMARY

The subject of this chapter has been human population growth, the effects of this growth upon us and our environment, and the importance of family planning and population control. Population size and growth rate are determined by interactions among birth rate, death rate, emigration, and immigration. Rapidly growing populations are characterized by a predominance of the young, whereas stable populations have similar numbers of all ages except the very old. Declining populations have a predominance of older individuals.

A population expresses its biotic potential when the individuals are exhibiting their reproductive potential, which is their maximal rate of reproduction. However, populations do not reproduce at their biotic potential because of environmental resistance. Population growth must stabilize eventually at the carrying capacity of the environment. Populations expanding far above carrying capacity will crash. Populations with a high rate of population growth have a short doubling time.

Thomas Malthus predicted that the human population would outgrow its environmental requirements, and this may well occur. The earth's population, currently in excess of 4 billion people, has exploded in the last 75 years. At the current growth rate (2.0 percent/year), the population will double again in 35 years. In underdeveloped countries, death rates have lowered and birth rates have remained stable or decreased only slightly, so the population growth rate in these countries is still alarmingly high (>2.0 percent). In many developed countries, the growth rate has decreased to 1.0 percent or less because of a marked decrease in birth rate. In fact, some developed countries have stable or even declining populations. However, in the United States, it will still take 50 to 60 years to reach a stable population even if each

TABLE 20-4. POSSIBLE PROPOSALS WHEREBY A GOVERNMENT COULD CONTROL POPULATION GROWTH

Birth Control	Taxing	Marriage	Social and Political
1. Sterilize poor people	1. Reduce taxes for single people	1. Charge a high fee for a marriage license	1. More support of contraceptive research
2. Liberalize abortion and sterilization laws	2. No tax exemptions for parents	2. Incentive payments for late marriages	2. Pressure on religious and other groups
3. Sterilize men or women with more than two living children	3. Tax families with greater than two children	3. Issuance of two kinds of marriage licenses:	3. Reduce new housing, transportation, and water supplies to certain regions
4. Temporary sterilization of women; reversal only by government consent	4. Free schooling for up to two children	(a) family marriage, licensed for children; more difficult to dissolve; expensive	4. Make foreign aid inversely proportional to population growth rate in a given country
5. More contraceptive devices available without prescription	5. Tax benefits for working wives	(b) couple marriage, licensed for no children; easy to dissolve; cheap	
6. Payment for cost of abortion			
7. Mandatory abortion of illegitimate pregnancies			
8. Selective placing of contraceptive chemicals in water supplies			

couple only has two children, largely because the population has a disproportionate amount of young people.

Many human populations are plagued by illiteracy, hunger, and starvation, depletion of natural resources and energy supplies, pollution and environmental illness, stress due to crowding, political conflict, and an overall reduction in the quality of life. Science and technology, which in the past have increased the earth's carrying capacity for people, will make further advances. These improvements, however, are not likely to keep pace with the population growth if it continues at the present rate.

Family planning programs have as their goal to decrease the birth of unwanted children and to limit family size. These programs deal with subjects such as age of marriage, availability and use of contraceptive methods, and the economic, educational, and health advantages of well-spaced children and small families. The International Planned Parenthood Federation and the United Nations offer assistance to family planning programs in various countries. Family planning programs in the United States and some other developed countries as well as in a few underdeveloped countries have helped reduce family size. Economic, social, cultural, psychological, racial, and religious barriers still confront the proponents of family planning, and these barriers vary in different regions. Many people, especially in underdeveloped countries, want more than two children. If voluntary limitation of birth rate does not occur in the future, governments may seek to control famly size through legislation.

NOTES

1. D. H. Meadows et al., *The Limits of Growth: A Report for the Club of Rome's Project on the Predicament of Mankind* (New York: Universe Books, 1972).
2. R. O. Greep, "Prevalence of People," *Perspectives in Biology and Medicine,* 12 (1969), 332–43.
3. G. Hardin, ed., *Population, Evolution, and Birth Control* (San Francisco: W. H. Freeman & Company, Publishers, 1969).
4. Ibid, p. 4.
5. E. S. Deevey, Jr., "The Human Population," *Scientific American,* 203, no. 3 (1960), 195–204.
6. A. Desmond, "How Many People Have Ever Lived on Earth?" *Population Bulletin,* vol. 18, no. 1 (1962); Deevey (1960), op. cit; and G. P. Coale, "The History of the Human Population," *Scientific American,* 231 (1975), 41–51.
7. C. F. Westoff, "The Population of the Developed Countries," *Scientific American,* 231 (1974), 109–20; and L. C. Brown, "The Twenty-Ninth Day: Accommodating Human Needs and Numbers to the Earth's Resources," *Population Reports,* ser. E, no. 5 (1978), pp. 53–72.
8. P. R. Ehrlich and A. H. Ehrlich, *Population, Resources, Environment* (San Francisco: W. H. Freeman & Company, Publishers, 1970).

9. Anonymous, "Births in 1980," *Research in Reproduction,* 12, no. 3 (1980), 1.

10. L. R. Brown, "World Population Trends: Signs of Hope, Signs of Stress," *Population Reports,* ser. J, no. 13 (1977), pp. 237-51.

11. R. H. Weller and F. B. Hobbs, "Unwanted and Mistimed Birth in the United States: 1968-1973," *Family Planning Perspectives,* 10 (1978), 168-72.

12. Rinehart and P. T. Piotrow, "OCs—Update on Usage, Safety, and Side Effects," *Population Reports,* ser. A, no. 5 (1979), pp. 136-85.

13. M. Kendall, "The World Fertility Survey: Current Status and Findings," *Population Reports,* ser. M, no. 3 (1979), pp. 75-103.

14. L. R. Brown et al., "Twenty-Two Dimensions of the Population Problem," *Population Reports,* ser. J, no. 11 (1976), pp. 177-203.

15. J. R. Udry, "Differential Fertility by Intelligence: The Role of Birth Planning," *Social Biology,* 25 (1978), 10-14.

16. Brown et al. (1976), op. cit.

17. P. R. Ehrlich, "Paying the Piper," *New Scientist,* 36 (1967), 652-55.

18. Brown et al. (1976), op. cit.

19. Ibid.

20. J. B. Calhoun, "Population Density and Social Pathology," *Scientific American,* 206, no. 2 (1962), 139-48.

21. A. H. Boerma, "A World Agricultural Plan," *Scientific American,* 223, no. 2 (1970), 54-69.

22. Ehrlich and Ehrlich (1970), op. cit.

23. Ibid.

24. Ibid.

25. L. E. Taylor, "Population Trends in an Indian Village," *Scientific American,* 223, no. 1 (1970), 106-14.

26. L. T. Ruzicka and A. K. M. Alauddin Chowdhuru, "Marriage and Fertility in Rural Bangladesh," *International Planned Parenthood Medical Bulletin,* vol. 12, no. 4 (1978).

27. C. F. Westoff, "The Immediate Need for Birth Control in Five Asian Countries," *Family Planning Perspectives,* 10 (1978), 175-81.

28. B. Berelson and R. Freedman, "A Study in Fertility Control," *Scientific American,* 210, no. 5 (1964), 29-37.

29. P. Corfman and E. Huyck, *Population Research in the U.S. National Institute of Health: A Response to a National and International Problem* (Washington, D.C.: U.S. Government Printing Office, 1978) No. (NIH) 75-781, pp. 1-26.

30. Ibid.

31. Ehrlich and Ehrlich (1970), op. cit.

32. Corfman and Huych (1978), op. cit.

33. Alan Guttmacher Institute, *Teenage Pregnancy: The Problem That Hasn't Gone Away.* New York: Alan Guttmacher Institute, 1981.

34. M. Zelnik and J. F. Kantner, "Contraceptive Patterns and Premarital Pregnancy among Women Aged 15-19 in 1976," *Family Planning Perspectives,* 10 (1978), 135-42.

35. B. Berelson, "Beyond Family Planning," *Science,* 163 (1969), 533-43.

FURTHER READING

BOERMA, A. H., "A World Agricultural Plan," *Scientific American,* 223 (1970), 54–69.

BUMPASS, L., AND C. F. WESTOFF, "The 'Perfect Contraceptive' Population," *Science,* 169 (1970), 1177–82.

CALHOUN, J. B., "Population Density and Social Pathology," *Scientific American,* 206 (1962), 139–48.

COALE, A. J., "The History of the Human Population," *Scientific American,* 231 (1974), 41–51.

DAVIS, K., "The Migration of Human Populations," *Scientific American,* 231 (1974), 93–105.

DAVIS, W. H., ed., *Readings in Human Population Ecology.* Englewood Cliffs, N.J.: Prentice-Hall, Inc., 1971.

DEMENY, P., "The Population of the Underdeveloped Countries," *Scientific American,* 231 (1974), 149–59.

EHRLICH, P. R., and A. H. EHRLICH, *Population, Resources, Environment.* San Francisco: W. H. Freeman and Co., 1970.

FREEDMAN, R., and B. BERELSON, "The Human Population," *Scientific American,* 231 (1974), 31–39.

FREJKA, T., "The Prospects for a Stationary World Population," *Scientific American,* 228 (1973), 15–23.

HARDIN, G., ed., *Population, Evolution and Birth Control.* San Francisco: W. H. Freeman & Company, Publishers, 1969.

HAWTHORN, G. P., "The Changing World Population," *Research in Reproduction,* 7, no. 6 (1975), map.

LANGER, W. L., "Checks on Population Growth: 1750–1850," *Scientific American,* 226 (1972), 92–99.

MYRDAL, G., "The Transfer of Technology to Underdeveloped Countries," *Scientific American,* 231 (1974), 173–82.

NORTMAN, D., "Population and Family Planning: A Fact Book," *Reports on Population/Family Planning,* no. 2 (December 1974).

REVELLE, R., "Food and Population," *Scientific American,* 231 (1974), 161–70.

RYDER, N. B., "The Family in Developed Countries," *Scientific American,* 231 (1974), 123–32.

TAYLOR, C. E., "Population Trends in an Indian Village," *Scientific American,* 223 (1970), 106–14.

WESTOFF, C. F., "The Population of Developed Countries," *Scientific American,* 231 (1974), 109–20.

WESTOFF, C. F., and L. BUMPASS, "The Revolution of Birth Control Practices of U.S. Roman Catholics," *Science,* 1979 (1973), 41–44.

Glossary

Abdominal pregnancy An ectopic pregnancy wherein implantation has occurred in the abdominal cavity.

ABO blood type An inherited presence of specific antigens on the surface of red blood cells; exposure of blood to a foreign ABO antigen causes red blood cells to agglutinate.

Abruptio placenta Premature detachment of the placenta from the uterus.

Acetaminophen A nonaspirin pain reliever.

Acidophils Cells present in the adenohypophysis that stain with acidic dyes. There is an acidophil that secretes growth hormone and one that secretes prolactin.

Acne A condition in which sebaceous glands become clogged and infected, producing pimples and blackheads; common in pubertal teenagers.

Acquired immune deficiency syndrome (AIDS) A viral, sexually transmitted disease found in some male homosexuals, Haitians, drug users, and those who have received blood transfusions.

Acrosin An enzyme (protease) present in the sperm acrosome that, when released during the acrosome reaction, breaks down protein of the zona pellucida.

Acrosin inhibitor Constituent of ovum cortical granules that inhibits acrosin action and thus prevents more than one sperm from penetrating the zona pellucida.

Acrosomal space Region in the sperm head, between the outer and inner acrosomal membranes, that contains enzymes important to fertilization.

Acrosome Thick cap over the head of a sperm that contains enzymes important in sperm penetration and fertilization.

Acrosome reaction When the outer acrosomal membrane of the sperm head fuses with the sperm plasma membrane, forming a composite membrane with openings through which pass acrosomal enzymes necessary for sperm penetration; see also **Outer acrosomal membrane; Composite membrane.**

Acupuncture   The blocking of sensation by inserting needles into specified body regions.

Adaptive trait   An inherited feature that makes individuals better suited to live and reproduce in their environment.

Adenine   A nitrogenous base in DNA and RNA.

Adenohypophysis   Epithelial lobe of the hypophysis that consists of the pars distalis, pars intermedia, and pars tuberalis.

Adolescence   The socially defined period of youth between puberty and social adulthood.

Adrenalectomy   Surgical removal or destruction of the adrenal glands.

Adrenogenital syndrome   An inherited condition characterized by excessive secretion of androgens from the adrenal glands, producing precocious development in both sexes and masculinization of the female; **Congenital adrenal hyperplasia.**

Adultery   See **Extramarital coitus.**

A-frame orgasm   See **Uterine orgasm.**

Afterbirth   Maternal and fetal tissue (amniotic membrane, placenta) expelled from the mother in stage 3 of labor.

Afterpains   Painful uterine contractions occurring for a few days after delivery of a child.

Agamaglobulinemia   A sex-linked, recessive, inherited disorder characterized by the inability to form antibodies.

Agape   Nonsexual, selfless love, as in the love of a parent for a child.

Aggressive-aim rapist   A rapist who is motivated by a desire to harm a woman.

Albinism   Inherited, homozygous recessive condition producing an absence of pigment (melanin).

Albumin   A protein found in the human body; when in oviductal fluid, it may serve to activate sperm and cause the acrosome reaction.

Aldosterone   A steroid hormone secreted by the adrenal glands that causes salt and water retention by the kidneys.

Allantois   Extraembryonic membrane that contributes to the formation of the placenta.

Alleles   Two or more genes that occupy the same position (locus) on homologous chromosomes.

Alpha-fetoprotein   An estrogen-binding protein, present in the blood of newborn rats, that keeps estrogen from masculinizing the female brain.

Amenorrhea   Absence of menstruation; see also **Primary amenorrhea, Secondary amenorrhea,** and **Oligomenorrhea.**

Amines   A group of nitrogen-containing compounds formed from ammonia.

Amniocentesis   Method of sampling amniotic fluid, and the fetal cells within this fluid, to test for chromosomal and genetic fetal disorders.

Amnion   Extraembryonic membrane that forms a closed, fluid-filled sac about the embryo and fetus.

Amniotic fluid   Fluid secreted by the amnion that bathes the embryo and fetus.

Amphetamines   A group of stimulant drugs.

Amyl nitrite   A fruity-smelling volatile liquid, fumes of which can be inhaled to intensify an orgasm.

Anabolic steroids   Synthetic androgens utilized to promote muscle growth.

Anal coitus   Insertion of the penis through the anus into the rectum.

Analgesic   Substance that diminishes sense of pain.

Anaphrodisiac   An external substance that lowers sexual desire.

Androcentric society Male-centered society dominated by the masculine sex role.

Androgen-binding protein (ABP) A protein, secreted by the Sertoli cells, that binds to androgens present in the testicular fluid.

Androgenesis When a sperm penetrates an ovum and does not fuse with the oocyte pronucleus but instead develops into an embryo with haploid cells derived from the father.

Androgens Substances that promote development and maturation of male sex structures; secreted by the testes and ovaries as well as the adrenal glands.

Androgyny When a male has femalelike biological characteristics.

Androstenedione A weak androgen that serves as a precursor for estrogen synthesis in the ovaries; present in male and female blood.

Anencephaly A congenital defect resulting in a stillborn fetus with no brain.

Anesthetic Substance that diminishes all sensations.

Anestrus A condition in which estrogen levels are low and a female is not sexually receptive.

Aneuploidy Embryos with either one too few (45; monosomy), or one too many (47; trisomy) chromosomes.

Annulment Legal declaration that a marriage was never valid.

Anorexia nervosa Emotional disturbance characterized by decrease of food intake and resultant emaciation as well as amenorrhea.

Anovulatory cycle An infertile menstrual cycle with no ovulation; often occurs in the first months after menarche.

Anteflexion Referring to the uterus, it is the normal forward bend to this organ.

Anterior hypothalamic-preoptic-suprachiasmatic region (APS) Area in hypothalamus, activity of which generates a surge of GnRH release immediately before ovulation.

Anterior pituitary gland See **Adenohypophysis.**

Anticodon A triplet of nitrogenous bases in tRNA that recognizes and binds to a specific codon in mRNA.

Antidiuretic hormone (ADH) See **Vasopressin.**

Antihistamines Drugs that block the production or action of histamine, a potent vasodilator.

Antihypertensive drugs Drugs that lower blood pressure.

Antiprogesterone pill A new chemical developed in France that is a contraceptive by rendering cervical mucus hostile to sperm.

Antral cavity Cavity in tertiary ovarian follicles that is filled with antral fluid.

Antral fluid Fluid in antral cavity; secreted by granulosa cells, but mainly is a transudate of blood plasma; **Follicular fluid.**

Apgar score Rating given to the respiration, muscle tone, reflex irritability, heart rate, and skin color of a newborn.

Aphrodisiac Any substance that stimulates sexual desire.

Apnea Temporary cessation of respiration.

Arcuate uterus Uterus with a slight dent on the fundus.

Areola The pigmented region of the skin surrounding the nipple.

Arginine vasotocin Octapeptide hormone, secreted by the fetal neurohypophysis, that may regulate secretion of amniotic fluid; also found in adult pineal gland.

Artificial insemination The artificial introduction of semen into the vagina or uterus; sperm can be provided by the husband or another male (donor).

Asexual reproduction Any reproductive process, such as budding, that does not involve fusion of gametes.

Atherosclerosis Deposition of fatty or crystalline deposits in the lining of blood vessels.

Atresia Process by which ovarian follicles die and degenerate.

Autoallergic orchiditis   Inflammation of the testes due to a man's allergic reaction to his own sperm that leak into his tissues after a vasectomy.

Autosomes   Any chromosomes that are not sex chromosomes.

Axon   Extension of a neuron that conducts nerve impulses and secretory products away from the cell body.

Azoospermia   Absence of sperm in semen.

B$_1$-Glycoprotein   A placental protein that can be used to detect the future possibility of spontaneous abortion.

Back labor   Back pain during stage 1 of labor, caused by the back of the fetal head pressing against the mother's sacrum.

Bacterial prostatitis   See **Prostatitis.**

Barbiturates   Drugs, such as methaqualone, that are used as hypnotics and sleeping pills.

Barr body   A condensed mass of chromatin in female cells, resulting from inactivation of all X chromosomes but one. Thus, a Barr body is not present in male cells, but there is one in normal female cells and more than one in female cells with greater than two X chromosomes; **Sex chromatin.**

Bartholin's glands   See **Greater vestibular glands.**

Basal body temperature method   Determining the safe and fertile periods in the menstrual cycle utilizing the rise in body temperature during the luteal phase; See also **Rhythm methods.**

Basement membrane   The membrane encasing each seminiferous tubule.

Basophils   Cells present in the adenohypophysis that stain with basic dyes.

Bendictin   A drug commonly used to treat morning-sickness in pregnant women. There is circumstantial evidence that it has teratogenic effects.

Benign tumor   A tumor in which the cells are not cancerous.

Bestiality   Sexual contact with an animal.

Bicornuate uterus   A forked uterus that is the result of a developmental error.

Bifid penis   A forked penis produced by a developmental error.

Bigamy   The act of marrying or being married to another person while still legally married to someone else.

Bilirubin   Yellowish breakdown product of destroyed red blood cells.

Bioassay   Technique that measures levels of a specific hormone in blood or tissue using a biological response to that hormone.

Biotic potential   Ability of a population to increase under optimal environmental conditions.

Birth canal   The cavities of the cervix and vagina through which the fetus passes during parturition.

Birth defects   See **Congenital defects.**

Birth rate   The rate at which young are born into a population.

Birthing room   Hospital room that provides a homelike setting for labor and delivery.

Bisexual   A person who has had sexual contact with both sexes as an adult.

Blastocoel   Fluid-filled cavity within the blastocyst; **Blastocyst cavity.**

Blastocyst   Stage in embryonic development after the morula stage; consists of an inner cell mass, an outer trophoblast layer, and a blastocoel.

Blastocyst cavity   See **Blastocoel.**

Blastomeres   The cells produced by mitotic division of a zygote.

Bloody show   Small amount of pinkish fluid discharged from the vagina during early labor; made up of mucus and blood.

Body image   The awareness or perception of one's body.

Body stalk An embryonic structure that connects the embryo to the chorion. It contains the allantois and blood vessels that become part of the umbilical cord.

Bougie A cylindrical instrument, used by the Japanese, that induces abortion after it is placed into the uterus.

Braxton-Hicks contractions Mild and irregular uterine contractions during pregnancy.

Breaking of the bag of waters Natural rupture of the amniotic sac, usually in the initial stages of labor; can be done artificially by the physician.

Breakthrough bleeding See **Spotting.**

Breast-feeding When an infant suckles the breast to obtain milk.

Breasts See **Mammary glands.**

Breech presentation When the feet or buttocks of the fetus are presented at the cervix.

Broad ligaments Paired ligaments that suspend the body of the uterus from the pelvic wall.

Bromocriptine A drug that inhibits prolactin secretion; used to treat male and female infertility and to inhibit lactation.

Brown fat Fat present in the newborn that, when broken down, gives off heat to help the baby maintain body temperature.

Buccal smear test When the cells lining the newborn's mouth are examined for the presence of X and Y chromosomes.

Bulbocavernosus muscle Muscle at the base of the penis that contracts rhythmically during ejaculation.

Bulbourethral glands Paired male sex accessory glands that produce a small amount of mucus; **Cowper's glands.**

Butyl nitrite A drug that has similar sexual and side effects to amyl nitrite; See **Amyl nitrite.**

Calendar method Determining the safe and fertile periods in the menstrual cycle utilizing a woman's variation in cycle length and the predicted time of ovulation; see also **Rhythm methods.**

Calmodulin A protein in seminal plasma that may render sperm in the epididymis capable of being capacitated.

Calymmatobacterium granulomatis A gram-negative bacterium that causes granuloma inguinale; *Donovania granulomatis.*

Cancer A condition in which certain cells lose the ability to control their growth and multiplication. The cells of a cancerous (malignant) tumor can remain in place or can spread (metastasize) to other regions of the body.

Candida Albicans A yeastlike fungus, normally present in the mouth, intestines, and vagina, which causes monilia vaginitis.

Cantharides Drugs such as "Spanish fly" that irritate the urinary tract and thus may have an aphrodisiaclike effect.

Caput Thick, soft swelling on the part of a newborn's head that rested against the cervix.

Cardiovascular system disease Disease that results in death due to myocardial infarction, pulmonary embolism, or cerebral hemorrhage; more common in combination pill users, especially those who are over 35 or who smoke.

Carnitine A constituent of seminal plasma from the epididymis that breaks down fatty acids, the products being used as nutrients for sperm.

Carrier A person who is heterozygous for dominant-recessive alleles.

Carrying capacity The number of individuals who can be supported by a given environment.

Caudal A conduction anesthetic injected into the outside membrane of the spinal cord low in the back.

Cavernous urethra See **Spongy urethra.**

Celibacy   Abstention from sexual activity.

Central effect of a hormone   The action of a hormone on the central nervous system.

Central nervous system   The brain and spinal cord.

Cephalohematoma   Blood clot between the skull bone and bone membrane in a newborn.

Cerebral palsy   Spastic muscle paralysis due to brain damage.

Cerebrospinal fluid   Viscous fluid in the ventricles of the brain and the central canal of the spinal cord.

Cervical biopsy   Surgical sampling of a small piece of cervical tissue for further examination.

Cervical canal   A narrow passage in the uterine cervix that connects the uterine cavity with the vagina.

Cervical cap   A mechanical barrier to sperm transport through the cervix; similar to but smaller than a diaphragm.

Cervical crypts   Deep recesses in the wall of the cervix that serve as reservoirs for sperm.

Cervical cysts   Small, noncancerous, pimplelike growths in the cervical lining.

Cervical dilatation   Widening of the external cervical os from 0.3 cm to 10 cm in diameter during stage 1 of labor. Also, artificial dilatation of the cervical canal before an induced abortion or other medical procedures involving the uterus.

Cervical dysplasia   Noncancerous cellular changes in the cervix.

Cervical effacement   Thinning of the wall of the cervix during stage 1 of labor.

Cervical effacement and dilatation   Stage 1 of labor.

Cervical mucous method   Determining the safe and fertile periods in the menstrual cycle using changes in cervical mucous volume and consistency; see also **Rhythm methods.**

Cervical polyps   Tear-shaped, noncancerous growths from the cervical lining.

Cervicitis   Inflammation or infection of the cervix.

Cesarean delivery   Surgical incision into the uterus through the abdominal wall to remove the fetus.

Chancre   An ulcerlike sore seen, for example, in the primary stage of syphilis and chancroid.

Chancroid   A sexually transmitted disease caused by the bacterium *Hemophilus ducreyi.*

Chlamydia trachomatis   A microorganism that causes some cases of nonspecific urethritis as well as lymphogranuloma venereum.

Chloasma   Brown facial spotting seen in some users of the combination pill and in pregnant woman.

Chlordiazepoxide   A minor tranquilizer (Librium).

Cholesterol   Precursor steroid for the synthesis of all steroid hormones.

Chordee   When the penis is bowed or bent.

Chorion   Extraembryonic membrane that, along with the allantois, forms the fetal portion of the placenta.

Chorionic villi   Fingerlike projections of the chorion that come in contact with maternal blood in uterine sinusoids of the placenta.

Chromosomal aberrations   Abnormal chromosome number or structure.

Chromosomal deletion   Absence of a piece of chromosome; see also **Chromosomal translocation.**

Chromosomal translocation   The breaking off of a piece of a chromosome and its attachment to a nonhomologous chromosome.

Circumcision   Surgical removal of the prepuce of the penis in the male or of the clitoris, sometimes along with the labia, in the female.

Classical hemophilia   A sex-linked, recessive, hereditary disease characterized by failure of blood to clot.

Clitoral glans Tip of the clitoris; homologous to the glans of the penis in the male.

Clitoral orgasm An orgasm resulting from stimulation of the clitoris.

Clitoral prepuce Flap of skin partially covering the clitoris; homologous to the prepuce (foreskin) of the penis.

Clitoral shaft Main body of the clitoris; partially homologous to the shaft of the penis in the male.

Clitoris A small, erectile structure embedded in tissue at the upper junction of the labia minora; homologous to the penis.

Clomiphene An antiestrogenic drug that commonly is used to treat female infertility.

Cloning Production of a new individual by removing the haploid nucleus from an unfertilized ovum, or the male and female pronuclei from a fertilized ovum, and replacing these structures with a nucleus from a diploid embryonic or adult cell from another individual.

Cocaine A recreational drug used as a stimulant.

Codominance When two or more alleles are dominant over recessive alleles for a single trait.

Codon Three adjacent nucleotides in mRNA that form a code for a specific amino acid.

Coitus Strictly speaking, sexual intercourse between a male and female, during which the penis is inserted into the vagina; **Vaginal coitus** or **Sexual intercourse.**

Coitus interruptus Withdrawal of the penis so that ejaculation occurs outside the vagina and vulva.

Coitus obstructus Avoidance of ejaculation by pressing on the penile urethra, thus causing retrograde ejaculation into the male's bladder.

Coitus reservatus When a male stops short of ejaculation and allows penile detumescence to occur within the vagina.

Collagenase Enzyme that destroys collagen fibers in connective tissue.

Colostrum The secretion of the mammary glands during late pregnancy and the first two days after delivery. Contains important nutrients and maternal antibodies for the newborn.

Colpotomy Incision through the vaginal wall; used for tubal sterilization.

Combination pill An oral contraceptive containing a synthetic estrogen and progestogen.

Common-law marriage Marriage not requiring a license or formal ceremony; requires living together for several years.

Commune See **Group marriage.**

Composite membrane See **Acrosome reaction.**

Conception See **Fertilization.**

Condom Barrier method of contraception using a rubber or skin sheath fitted over the penis.

Conduction anesthetic Anesthetic injected into tissue.

Condylomata acuminata See **Genital warts.**

Congenital adrenal hyperplasia See **Adrenogenital syndrome.**

Congenital defects Physiological or anatomical defects present at birth; **Birth defects.**

Congenital syphilis A condition in which a child is born with syphilis.

Conjoined twins Identical twins in which, during their development, parts of their bodies remain fused; **Siamese twins.**

Consensual arrangement An unmarried couple living together in a sexual relationship; sexual cohabitation.

Contraception Prevention of pregnancy by abstinence or the use of substances, devices, or surgical procedures that prevent ovulation, fertilization, or implantation.

Contract marriage A marriage with formal agreements about the rights, responsibilities, and behavior of the partners; may be renewable after a certain defined period; **Open marriage.**

Copper IUD An intrauterine device made of flexible plastic with a wrapping of copper wire.

Coprophilia   When sexual arousal is associated with defecation or feces.

Corona glandis   Rounded ridge on the back edge of the glans penis.

Corona-penetrating enzyme   Enzyme from acrosome that dissolves material between cells of the corona radiata.

Corona radiata   A thin layer of granulosa cells that surrounds the oocyte in tertiary follicles and the ovum after ovulation.

Corpora atretica   Structures formed from atretic follicles.

Corpora cavernosa   Paired columns of spongy, erectile tissue present in the shaft of both the clitoris and penis.

Corpus albicans   Connective tissue-filled structure that is the remnant of the regressed corpus luteum.

Corpus luteum   A yellowish endocrine gland formed from the wall of an ovulated follicle.

Corpus hemorrhagicum   Empty ovarian follicle existing after ovulation and before its transformation into a corpus luteum.

Corpus spongiosum   Spongy erectile tissue in the shaft of the male penis; it is homologous to the labia minora in the female.

Cortical cords   Sex cords in the cortex of indifferent ovaries.

Cortical granules   Granules in cortical vesicles lying at the periphery of the ovum cytoplasm; sperm penetration causes them to release acrosin inhibitor and a protease, both of which prevent more sperm from penetrating the egg.

Cortical vesicles   Membrane-bound bodies in the ovum that release cortical granules after sperm penetration.

Corticosteroids   Steroid hormones, such as cortisone and cortisol, that are secreted by the adrenal glands.

Corticotropin (ACTH)   Polypeptide hormone secreted by basophils of the adenohypophysis that stimulates secretion of steroid hormones (corticosteroids) from the adrenal glands.

Corticotropin-releasing hormone (CRH)   Polypeptide from the hypothalamus that increases synthesis and secretion of corticotropin.

Cortisol   Steroid hormone secreted by the cortex of the adrenal glands that raises blood sugar levels.

Countercurrent heat exchange   Cooling system present in the spermatic cords, involving heat transfer between the testicular artery and veins.

Courtship behaviors   Behaviors that serve to bring males and females together for mating.

Cowper's glands   See **Bulbourethral glands.**

Cremaster   A muscle layer in the scrotum that contracts upon sexual stimulation.

Cremasteric reflex   Contraction of the cremaster muscle of the scrotum when the inner thighs are stroked.

Crib death   See **Sudden-infant-death syndrome.**

Cri-du-chat syndrome   Chromosomal deletion on chromosome 5, resulting in a syndrome of physical abnormalities; affected newborns cry like a hungry kitten.

Critical body fat hypothesis   Theory that a minimal ratio of fat weight to whole body weight is necessary for menarche to occur.

Crossing-over   The exchange of genetic material between homologous chromosomes during meiosis.

Crowning   When the top of the fetal head appears in the birth canal and does not recede between contractions.

Crown-rump length   The distance between the crown of the head and the rump of the embryo.

Crude birth rate   The number of individuals born in a population in a given amount of time.

Crude death rate The number of individuals dying in a population in a given amount of time.

Cryptorchid testes Testes that remain in the body cavity during childhood; undescended testes.

Culdoscopy Incision through the vaginal wall that is used in tubal sterilization.

Cumulus oophorus Column of granulosa cells that attaches the oocyte to the wall of tertiary follicles; leaves with the ovulated ovum.

Cunnilingus When the tongue or mouth is used to stimulate or caress the female external genitalia.

Curettage Scraping the lining of the uterus using a metal instrument (curette).

Curette Metal instrument used to scrape the lining of the uterus.

Cyanosis When the blood of a newborn does not contain enough oxygen, leading to a bluish tint to the skin; can be caused by patent ductus arteriosus or open foramen ovale ("blue babies"), or may be present briefly in normal newborns.

Cyproterone acetate An antiandrogen that blocks binding of androgens to receptors in target cells.

Cystic fibrosis An inherited, homozygous recessive disorder characterized by overproduction of mucus.

Cystic follicle A large fluid-filled or solid sac derived from an unovulated Graafian follicle.

Cystic mastitis Development of one or more fluid-filled sacs (cysts) in the breasts.

Cystitis Inflammation or infection of the urinary bladder.

Cytomegalovirus The virus that cause cytomegalic inclusion disease.

Cytosine A nitrogenous base present in DNA and RNA.

Cytotrophoblast Cellular component of the trophoblast that gives rise to the chorion.

Date rape A phrase to describe coitus when a woman resists to some degree (not formally classified as rape).

Death rate The rate at which individuals die in a population.

Deciduoma response Tumorlike growth of uterine tissue over the implanted embryo.

Dehydroepiandrosterone (DHEA) Weak androgen serving as a precursor for estrogen and testosterone synthesis; synthesized by ovaries, testes, and "fetal zone" of fetal adrenal glands.

Delayed puberty When puberty has not begun by age 14 (testicular growth) or 18 (skeletal growth) in males, or by age 14 (breast development) or age 15 (skeletal growth) in females.

Dendrite Extension of a neuron that conducts a nerve impulse toward the cell body.

Deoxyribonucleic acid (DNA) The carrier of genetic information, composed of two complementary chains of nucleotides wound in a double helix.

Dermoid cyst Lump of tissue within an ovarian follicle, containing embryonic organs; possibly derived from fertilization of an oocyte while it still is in the ovary.

Desertion The abandonment of spouse or children without the spouse's consent.

Diabetes mellitus Condition in which lack of insulin produces high blood sugar and copious urination.

Diaphragm A mechanical barrier to passage of sperm into the cervix; fits into vagina to cover the cervix.

Diazepam A minor tranquilizer (Valium).

Diencephalon Portion of the brain containing the hypothalamus, thalamus, epithalamus, and third ventricle.

Diesterase An enzyme in the uterus that breaks down glycerylphosphocholine, thus supplying nutrients for sperm.

Diethstilbestrol (DES) A potent synthetic estrogen.

Differential gene action A phrase describing the fact that, even though every cell of one individual has the same genes, different cells exhibit differences in gene activity.

Dilatation and Curettage (D & C) Surgical procedure of first dilating the cervix with a metal instrument and then scraping out the contents of the uterus; used for induced abortion during the late first trimester.

Dilatation contractions Contractions in stage 1 of labor that dilate (as well as efface) the cervix.

Dilator Metal instrument used to dilate the cervical canal before uterine curettage.

Dildo An artificial penis.

Diploid Containing a double complement of chromosomes; in human cells, there are 23 pairs (46 chromosomes); 2*N*.

Diuretics Drugs used to increase urine flow.

Divorce decree Legal dissolution of a marriage.

Dizygotic twins Twins resulting from fertilization of two separate ova by two separate sperm; **Fraternal twins** or **nonidentical twins.**

DNA-replication When the two strands of a **DNA** molecule separate and each strand copies itself.

Döderlein's bacillus A gram-positive bacterium normally found in the vagina; *Lactobacillus acidophilus.*

Dominant Refers to one allele for an inherited trait preventing expression of other alleles for that trait.

Donovania granulomatis See *Calymmatobacterium granulomatis.*

Dose-response relationship When a response to a given hormone is predictably related to the amount of the hormone present.

Double helix Spiral-staircase, three-dimensional shape of the DNA molecule.

Double penis Two penises produced by a developmental error.

Double uterus Two uteri, with a single vagina or two vaginas being present, produced by a developmental error.

Doubling time The time in years it will take to double population size at a given population growth rate.

Douching Rinsing the vagina with a solution; is not an effective contraceptive measure.

Down's syndrome An affliction caused by trisomy of chromosome 21. These individuals have physical disorders and mental retardation; **Mongolism.**

Duchenne's muscular dystrophy Sex-linked, recessive, inherited disorder leading to degeneration of skeletal muscle and early death.

Ductus arteriosus A fetal blood vessel that shunts blood from the pulmonary trunk to the descending aorta, bypassing the fetal lung; closes after birth.

Ductus deferens See **Vas deferens.**

Ductus epididymis Main duct of the body of the epididymis.

Ductus venosus Fetal blood vessel that directs oxygenated blood away from the fetal liver.

Due date Expected date of birth as calculated from first day of last menstrual period or day of conception; **Term.**

Dysmenorrhea Painful menstruation caused by uterine cramps.

Dyspareunia Difficult or painful coitus.

Eclampsia See **Toxemia.**

Ectoderm Outermost of the three primary germ layers of the embryo; gives rise to the nervous system, the external sense organs (ear, eye, etc.), the skin, and the mucous membranes of the mouth and anus.

Ectopic pregnancy   Implantation of an embryo outside the uterus.

Ectromelia   Fetal condition characterized by total absence of limbs.

Edema   Excessive accumulation of extracellular water in tissues.

Effacement contractions   Contractions in stage 1 of labor that efface (thin out) the wall of the cervix.

Ejaculation   Expulsion of semen from the male sex accessory ducts; See also **Female ejaculation.**

Ejaculation reflex   Reaction when stimuli associated with sexual arousal activate the ejaculatory center in the spinal cord, which in turn causes ejaculation.

Ejaculatory center   A group of neurons in the spinal cord that controls ejaculation.

Ejaculatory ducts   Small paired ducts in the male that receive the ampulla of each vas deferens and the duct from each seminal vesicle. The ejaculatory ducts empty into the prostatic urethra.

Ejaculatory incompetence   When a man is unable to ejaculate, even though he may have an orgasm.

Electra complex   The stage in a female child's life when she is in love with her father and jealous of her mother.

Embryo   Stage of prenatal development between the blastocyst and fetus, or from the second through the eighth week after fertilization.

Embryonic disc   Early embryo, derived from the inner cell mass of the blastocyst, that contains the primary germ layers.

Embryonic period   Stage of pregnancy from the second through the eighth week of pregnancy.

Emigration   Leaving a place of residence in a defined population for elsewhere.

Emission stage of ejaculation   The first stage of the ejaculatory response involving contractions of the male sex accessory ducts and glands.

Endemic syphilis   See **Nonvenereal syphilis.**

Endocrine glands   Ductless glands that secrete hormones into the blood.

Endocrine system   System consisting of the endocrine glands; includes the hypophysis, pineal gland, gonads, placenta, thyroid and parathyroid glands, adrenal glands, digestive tract, kidneys, pancreas, and thymus.

Endocrinology   The scientific discipline involving the study of endocrine glands and their secretions.

Endoderm   The innermost of the three primary germ layers; gives rise to the digestive and respiratory systems.

Endometrial biopsy   Surgical sampling of a small piece of endometrium for further examination.

Endometrial hyperplasia   Unusually thick uterine endometrium due to prolonged estrogenic stimulation and consequent cell multiplication.

Endometrial polyps   Mushroomlike growths of the endometrium.

Endometriosis   Abnormal growth of uterine cells outside the uterus; can cause infertility if present in the oviducts.

Endometritis   Inflammation or infection of the endometrium.

Endometrium   The uterine lining, which consists of the stratum functionalis and stratum basalis.

Endoplasmic reticulum   An extensive system of membranes in the cell, dividing the cytoplasm into compartments; often coated with ribosomes.

Endorphins   Natural, opiatelike substances that decrease pain.

Engagement of the presenting part   Dropping of the fetus between the pelvic bones so that its head rests against the neck of the cervix; occurs near term.

Enkaphalins   Natural, opiatelike substances that decrease pain.

Ensoulment   Supposed event wherein a developing human acquires a soul and thus becomes a person.

Environmental resistance   Limitation of population growth by factors in the environment.

Epidermal-growth factor   Substance present in human milk and other tissues that promotes tissue growth.

Epididymis (pl., Epididymides)   Comma-shaped organ on the posterior surface of each testis; its ducts transport sperm from the testis to the vas deferens.

Epidural   A conduction anesthetic injected into the outside membranes of the spinal cord.

Episiotomy   Surgical incision of the perineum during parturition to reduce the possibility of tearing the perineal tissues.

Epstein-Barr virus   The virus that causes infectious mononucleosis.

Erectile dysfunction   The failure to gain or maintain an erection; **Impotence.**

Erection   The stiffening and enlargement of the penis (or clitoris), usually as a result of sexual arousal.

Erection center   A group of neurons in the lower end of the spinal cord that controls erection of the penis.

Erection reflex   When an erotic stimulus causes nervous activation of the erection center, which in turn activates the erection mechanisms.

Ergonamine   A drug derived from a fungus growing on rye that causes uterine contractions; used to stop uterine hemorrhage after expulsion of the placenta.

Erogenous zones   Areas of the body that are especially sexually sensitive to tactile stimuli.

Erotic love   The desire for complete fusion with another; sexual love is a major component of erotic love.

Erotic stimuli   Stimuli that cause sexual arousal.

Erythroblastosis fetalis   Fetal condition in which immature red blood cells are predominant in the blood; caused by Rhesus disease.

Escherichia coli   A bacterium normally present in the human intestinal tract.

Estradiol-17β   The major natural estrogen, secreted by the ovaries, testes, and placenta.

Estriol   An estrogen secreted in large amounts by the placenta.

Estrogens   Substances that promote maturation of female reproductive organs and secondary sexual characteristics; secreted by the ovaries, testes, adrenal glands, and placenta.

Estrone   An estrogen secreted in small amounts by the ovaries and placenta.

Estrous behavior   A cyclic period of sexual receptivity ("heat") in females occurring near the time of ovulation and related to high levels of estrogen in the blood.

Estrous cycle   A reproductive cycle in which the female exhibits sexual receptivity (estrus) only near the time of ovulation.

Eugenics   The artificial manipulation of gene frequencies in a population by selective breeding.

Eunuch   An orchidectomized (castrated) male.

Excitement phase   The initial stage in the human sexual response cycle that follows effective sexual stimulation.

Exhibitionism   When a person derives sexual pleasure from exposing his or her genital region to others.

Exocrine glands   Glands that secrete substances into ducts that empty into body cavities and onto body surfaces.

Exponential growth   Growth at an increasing rate based on a "compound-interest" principle.

Expulsion of the fetus   Stage 2 of labor.

Expulsion of the placenta   Stage 3 of labor.

Expulsion stage of ejaculation   The second stage of the ejaculation process, during which semen is expelled.

External cervical os   Opening of the cervical canal into the vagina.

External fertilization   When fertilization occurs *in vitro* (in a dish or test tube).

External genitalia   In the female, the mons pubis, labia majora and minora, vaginal introitus, hymen, and clitoris (also called the **vulva**); in the male, the penis and scrotum.

Extirpation   Removal of tissue or organs.

Extraembyronic membranes   Four membranes (yolk sac, amnion, allantois, chorion) derived from but not part of the developing embryo.

Extramarital coitus   Coitus between a married person and someone other than his or her spouse; **Adultery.**

Fallopian tubes   See **Oviducts.**

False labor   Contractions that occur near term; can be rhythmic, but do not last or cause much cervical dilatation or effacement.

False pregnancy   See **Pseudocyesis.**

Family planning programs   Programs with the goals of eliminating birth of unwanted children and limiting family size.

Feedback loop   When production of a product either increases or decreases further production of that product.

Fellatio   Taking the penis into the mouth and stimulating it for erotic purposes; **Oral coitus.**

Female climacteric   See **Menopause.**

Female ejaculation   Expulsion of fluid from the lesser vestibular glands into the vestibule during orgasm.

Femaleness   Biological characteristics of the female sex.

Feminine   Social definition of what are female characteristics; often used to mean "womanly."

Feminist movement   See **Women's Liberation Movement.**

Femoral coitus   Insertion of the penis between the thighs of another person.

Fern test   Laboratory method of determining phase of the menstrual cycle using degree of crystallization in dried cervical mucus.

Fertilization   When the pronuclei of a haploid ootid and sperm fuse to form a diploid zygote; **Conception.**

Fetal alcohol syndrome   Condition of newborn caused by chronic ingestion of alcohol during pregnancy, characterized by small size and retardation.

Fetal distress   Abnormal cardiovascular signs of the fetus during labor and delivery.

Fetal ejection reflex   Stimulation of oxytocin release by mechanical stimulation of the uterus, cervix, or vagina, resulting in further uterine contractions.

Fetal monitoring   Continuous evaluation of vital signs of the fetus during labor and delivery.

Fetal zone   Region in fetal adrenal glands that secretes large amounts of a weak androgen (dehydroepiandrosterone).

Fetish An inanimate object or part of a person's body that is necessary for sexual arousal.

Fetishism When an inanimate object or part of a person's body is necessary for sexual arousal.

Feto-placental unit Integrated system of fetus and mother that combines to produce secretion of estrogens from the placenta.

Fetoscopy Method of fetal scanning using direct observation with an optical instrument inserted through an abdominal incision.

Fetus Offspring at stage in prenatal development from the ninth week until term.

Fibrin See **Fibrinogenase.**

Fibrinogen See **Fibrinogenase.**

Fibrinogenase Enzyme in seminal plasma that converts fibrinogen to fibrin, and thus causes semen coagulation.

Fibrinolytic enzyme Enzyme in seminal plasma that breaks down fibrin and thus causes semen liquification.

Fibroids Benign tumors of the uterine smooth muscle.

Fimbriae Fingerlike projections on the edge of the infundibulum of each oviduct.

First meiotic arrest When the first meiotic division in primary oocytes is stopped before completion; condition present in all primary oocytes.

First meiotic division First phase of meiosis, during which reduction division occurs; produces a secondary oocyte in females and a secondary spermatocyte in males.

First polar body Small haploid cell produced by unequal first meiotic division of oocyte; degenerates before or after dividing again.

5-Alpha dihydrotestosterone (DHT) A potent androgen secreted by the testes and adrenal glands and synthesized from testosterone in some target tissues.

5-Alpha reductase An enzyme, present in some androgen target tissues, that converts testosterone to 5-α dihydrotestosterone.

Follicle-stimulating hormone (FSH) Glycoprotein hormone secreted by basophils of the adenohypophysis that, together with LH, maintains gonadal function in both sexes.

Follicle-stimulating hormone-releasing hormone (FSHRH) See **Gonadotropin-releasing hormone.**

Follicular fluid See **Antral fluid.**

Follicular phase Period in menstrual cycle beginning at the end of menstruation and ending with ovulation.

Fontanels Soft spots in the skull of newborns.

Foramen ovale Opening between the right and left atria of the fetal heart that allows blood flow to equalize between the right and left sides of the heart; closes soon after birth.

Forceps delivery Assisting expulsion of fetus by using a tonglike instrument called a *forceps.*

Foreplay The preliminary stages of coitus, involving kissing, touching, and caressing between partners.

Foreskin See **Penile prepuce.**

Fornication Coitus between two unmarried people.

Fornix A circular recess at the upper end of the vagina that encircles the uterine cervix.

Fraternal twins See **Dizygotic twins.**

Frigidity When a woman is disinterested in sex or is unable to achieve sexual gratification.

Galactorrhea Condition where prolactin levels are high in a nonpregnant female, producing milk secretion and amenorrhea.

Gametes Spermatozoa in the male, and an ovulated ovum in the female; are haploid.

Gay A male homosexual.

Gender identity The awareness of or belief about one's maleness or femaleness.

Gender role See **Sex role.**

Gene A unit of heredity in the chromosome; a sequence of nucleotides in a DNA molecule that codes for a polypeptide or protein.

Gene expressivity The degree of phenotypic expression of a gene.

General adaptation syndrome A set of physiological responses to any long-term stressor, mainly involving increased adrenal gland activity; proposed by Hans Selye.

Genetic counseling Advice given by trained individuals to potential parents about the possible genetic characteristics of potential offspring.

Genetic mosaic When populations of cells of an individual differ genotypically.

Genital ridges Embryonic ridges on the dorsal wall of the abdominal cavity that will form the gonads.

Genital tubercle Structure of the indifferent external genitalia of the early embryo that will form the penis or clitoris; **Phallus.**

Genital warts Virus-induced warts in the genital region that can be sexually transmitted; **Condylomata acuminata.**

Genotype The genetic make-up of an individual's cells.

Germinal vesicle breakdown Disintegration of the nuclear membrane of a secondary oocyte at the start of final oocyte maturation; LH-induced.

Gestation See **Pregnancy.**

Glans penis The soft bulbous end of the penis.

Glycerylphosphocholine Constituent of seminal plasma, secreted by the epididymis, that is a nutrient source for sperm.

Glycogen A long-chain polysaccharide made up of a series of glucose molecules.

GnRH analogs Synthetic molecules similar in structure to gonadotropin-releasing hormone (GnRH). Some inhibit, and others stimulate, gonadotropin secretion.

Gonadostat Area in hypothalamus that regulates feedback control of GnRH secretion.

Gonadotropic hormones Pituitary hormones that affect gonadal function, including FSH and LH.

Gonadotropin-releasing hormone (GnRH) Hormone from the hypothalamus that increases synthesis and release of both LH and FSH; same as **Luteinizing hormone-releasing hormone** and **Follicle-stimulating hormone-releasing hormone.**

Gonads Structures that produce the male gametes (testes) and female gametes (ovaries).

Gonococcal ophthalmia neonatorum A condition in which a newborn's eyes are infected with *Neisseria gonorrhoeae.*

Gonococcus A common name for the bacterium *Neisseria gonorrhoeae,* the cause of gonorrhea.

Gonorrhea A sexually transmitted disease caused by the bacterium *Neisseria gonorrhoeae.*

Gossypol A new male contraceptive chemical derived from the cotton plant.

Grafenberg spot A small region in the anterior vaginal wall that, when stimulated, causes sexual arousal.

Granuloma inguinale A rare, sexually transmitted disease caused by *Calymmatobacterium granulomatis.*

Granulosa cells Cells of the membrana granulosa of ovarian follicles.

Grasp reflex Flexion of a newborn's fingers when its palm is touched.

Greater vestibular glands Mucus-secreting glands in the female vestibule, homologous to the Cowper's glands of the male; **Bartholin's glands.**

Group marriage Marriage involving several males and females, often including indiscriminate sex and collective childrearing; **Commune.**

Growth hormone (GH) Protein hormone secreted by acidophils of the adenohypophysis that stimulates incorporation of amino acids into protein (tissue growth) and raises blood sugar.

Guanethidine An antihypertensive drug that blocks the sympathetic nervous system and can cause erectile dysfunction.

Guanine A nitrogenous base present in **DNA** and **RNA.**

Gubernaculum Ligament connecting the testes with the scrotum that plays a role in testicular descent in the late fetus.

Guevedoces Inherited form of male pseudohermaphroditism resulting from an absence of the enzyme 5-α reductase in external genitalia tissues, so that they are unable to change testosterone to 5-α DHT and grow.

Gumma A soft, gummy tumor, as seen in cases of tertiary stage syphilis.

Gynandromorph A true hermaphrodite in which one side of the reproductive tract is male and the other side is female.

Gynandry When a female has malelike biological characteristics.

Gynecomastia Excessive development of the mammary glands in the male.

Gynogenesis When a sperm penetrates an egg, the male pronucleus does not form, and a haploid (N) embryo develops from the unfertilized ovum.

Haploid Containing only a single complement of chromosomes; in man, 23 chromosomes, or *N;* present in gametes.

Hegar's sign Softening of the lower part of the uterus occurring about four weeks after a missed menses; a probable sign of pregnancy.

Hemochorial placenta Type of placenta, present in the human, in which maternal blood in sinusoids directly bathes the chorionic villi.

Hemophilus ducreyi A gram-negative bacterium that causes chancroid.

Hemophilus vaginalis A gram-negative bacterium that is one cause of vaginitis and non-specific urethritis.

Hemophilus vaginitis Inflammation of the vagina caused by *Hemophilus vaginalis;* can be sexually transmitted.

Hepatitis Inflammation or infection of the liver.

Herpes genitalis A sexually transmitted disease caused by herpes simplex virus, type 2.

Herpes simplex virus, type 1 The virus that causes cold sores.

Herpes simplex virus, type 2 The virus that causes herpes genitalis.

Heterogametic sex The sex with two dissimilar sex chromosomes, which is the male in humans and other mammals.

Heterogamy Marriage of unlike individuals; compare **Homogamy.**

Heterosexual A person who is sexually attracted to and engages in sexual activity primarily with members of the opposite sex.

Heterosexual behavior Sexual activity with a member of one's opposite sex.

Heterozygous Condition when a cell contains two different alleles on homologous chromosomes.

Heterozygous dominant genotype When cells have the dominant and recessive allele for a given trait.

Hilus cells Sparse, testosterone-secreting cells present in the hilus of the ovary; homologous to the Leydig cells of the testes.

Hirsutism Abnormal hairiness, especially in women.

Histones NonDNA proteins in chromosomes.

Home birth Labor and birth at home, with or without medical supervision.

Homeostasis Regulation of body functions at a steady state.

Homogametic sex The sex with two similar sex chromosomes, which is the female in humans and other mammals.

Homogamy Marriage between people of the same race, religion, social class, or personality type; compare **Heterogamy.**

Homophile Term used for a male homosexual.

Homosexual A person who is sexually attracted to and engages in sexual activity primarily with members of the same sex.

Homosexual behavior Sexual activity with a member of one's same sex.

Homozygous A condition when a cell contains two identical alleles on homologous chromosomes.

Homozygous dominant genotype When the two alleles for a given trait are both dominant.

Homozygous recessive genotype When the two alleles for a given trait are both recessive.

Hormone receptors Proteins on the cell membrane or in the cytoplasm of cells in target tissues for hormones. Each kind of receptor binds to a specific hormone.

Hormones Substances secreted by endocrine glands into the blood, which travel to specific target tissues and affect the growth and function of those tissues.

Hot flashes Intense feeling of warmth and profuse sweating that occur periodically in some menopausal and postmenopausal women.

Human chorionic gonadotropin (hCG) Protein hormone secreted by the blastocyst and placenta; similar in biological activity to luteinizing hormone (LH).

Human chorionic somatomammotropin (hCS) Protein hormone secreted by the placenta; similar in biological activity to both prolactin (PRL) and growth hormone (GH).

Human menopausal gonadotropin (hMG) The combination of FSH and LH found at high concentrations in the blood and urine of postmenopausal women.

Huntington's chorea See **Huntington's disease.**

Huntington's disease An inherited, autosomal dominant disorder that leads to neurological malfunction beginning in midlife; **Huntington's chorea.**

Hyaluronidase Enzyme present in the sperm acrosome that breaks down acid that cements cells of the corona radiata together.

H-Y Antigen Protein coded by a gene or genes on the Y chromosome that is present on the cell membrane of male cells and induces formation of testes from the indifferent gonads.

Hydatidiform mole Tumorlike uterine growth resulting from implantation of a trophoblast without an embryo. The cells of these moles are diploid, but the chromosomes are derived entirely from the father.

Hydrocele Accumulation of body fluid in a saclike cavity, especially in the scrotum when the inguinal canal does not close after testicular descent.

Hydrocephaly A congenital defect characterized by excessive fluid accumulation in the brain.

Hymen A membrane that partially closes the vaginal introitus.

Hypophysectomy Extirpation or destruction of the hypophysis.

Hypophysial portal veins Small veins in the hypophysis that carry hypothalamic neurohormones to the adenohypophysis.

Hypophysis Gland lying in the sphenoid bone and connected to the hypothalamus; consists of the adenohypophysis and neurohypophysis; **Pituitary gland.**

Hypophysiotropic area (HTA) Region in the hypothalamus; contains cell bodies of the neurosecretory neurons that control the function of the adenohypophysis.

Hypospadias Failure of the embryonic urethra to close, so that the urethral meatus opens on the lower surface of the penis or into the vagina.

Hypothalamo-hypophysial portal system Vascular system connecting the median eminence of the hypothalamus with the adenohypophysis; consists of the primary and secondary capillary plexi connected by the hypophysial portal veins, see also **Primary capillary plexus** and **secondary capillary plexus.**

Hypothalamus Floor of the diencephalon; contains cells that regulate function of the hypophysis and some aspects of sexual behavior.

Hysterectomy Surgical removal of the uterus.

Hysterotomy Surgical method of abortion used after the twentieth week of pregnancy, during which the fetus is removed through an incision in the abdomen and uterus; also called a **Mini-cesarean.**

Identical twins See **Monozygotic twins.**

Immature newborn When the newborn birth weight is between 0.50 kg (1 lb, 1.5 oz) and 0.99 kg (2 lb, 3 oz).

Immigration Moving to a new place of residence from another defined population.

Immunoassay pregnancy test Pregnancy test combining anti-hCG with a woman's urine.

Imperforate hymen Condition in which tissue completely blocks the vaginal introitus; must be removed surgically.

Implantation Apposition and attachment to, and finally invasion of, the endometrium by a blastocyst; **Nidation.**

Impotence See **Erectile dysfunction.**

Impulse rapist One who rapes a woman when the opportunity arises.

Inborn error of metabolism An inherited disorder characterized by the absence of an enzyme.

Incest Sexual activity or coitus between close relatives.

Incompetent cervix Condition in which the cervix is unable to support a growing fetus to term; more common after induced abortions that used a metal dilator.

Incomplete dominance When different alleles of a gene are equally expressed in the phenotype.

Incubation period Time between when an infectious microorganism enters the body and the appearance of disease symptoms.

Independent assortment The mendelian principle that unlinked genes assort into gametes independently of one another.

Indifferent gonads Embryonic gonads, consisting of a cortex and medulla, that will give rise to testes or ovaries.

Indomethacin A potent antiprostaglandin drug.

Induced abortion Termination of pregnancy by artificial means.

Induced labor Labor artificially induced by breaking the bag of water (amnion) and/or by administering pitocin or prostaglandins.

Induced ovulation Ovulation that only occurs as a result of copulation or stimuli associated with the presence of a male.

Infanticide Killing a newborn or infant; used in the past to control population and to eliminate undesired children.

Infertility When a couple fails to produce pregnancy after about a year of unprotected coitus.

Injectable progestogen Female contraceptive method using an injected progestogen.

Inguinal canals Openings in the lower pelvic wall through which pass the testes during their descent; vasa deferentia pass through these canals in the adult male, as do the round ligaments in females.

Inguinal hernia   Extension of abdominal contents through an open inguinal canal into the scrotal sac.

Inhibin   Substance, found in fluid of testes and associated ducts as well as in antral fluid in the ovaries, that inhibits FSH secretion.

Inner acrosomal membrane   Inner membrane of acrosome.

Inner cell mass   Portion of the blastocyst that gives rise to the embryo as well as the amnion, yolk sac, and allantois.

Interethnic marriage   Marriage between people of different cultures.

Interfaith marriage   Marriage between people of different religions.

Intermarriage   Marriage between people of different race, culture, or religion.

Internal cervical os   Opening of the uterine cervix into the uterine cavity.

Internal fertilization   When fertilization takes place within the female reproductive tract.

Interracial marriage   Marriage between people of different races.

Intersexuality   When a person has ambiguous reproductive structures.

Interstitial cells   In the male, see **Leydig cells.** In the female, see **Ovarian interstitial cells.**

Interstitial cell-stimulating hormone (ICSH)   See **Luteinizing hormone.**

Intraamniotic saline   Abortion method used to terminate second trimester pregnancies, also called salting out. Besides (or in addition to) saline, the substances urea, glucose, or prostaglandin can be injected into the amniotic fluid to induce abortion.

Intracervical ring   A new contraceptive device, containing progesterone, that is placed within the cervix.

Intradermal progestogen implant   Female contraceptive method using implantation of a progestogen-containing capsule within the skin.

Intramural oviduct   Region of oviduct within the uterine wall.

Intrauterine device (IUD)   Foreign object, usually flexible plastic, placed into the uterus for contraception; some can include copper or progestogen.

Intromission   Penetration of the penis into the vagina during vaginal coitus.

Jaundice   Yellowish skin color due to the deposition of bilirubin.

Juveniles   Individuals that have not reached puberty.

Kegel exercise   Voluntary contraction of the pubococcygeus muscle, used to strengthen its tone.

Kleptomania   Compulsive stealing, often associated with sexual arousal.

Klinefelter's syndrome   One of the more common developmental disorders of the reproductive tract in which a male has a chromosomal abnormality (47; XXY), undeveloped testes, and gynecomastia.

Labia majora   Outer, major lips of female vulva; homologous to the male scrotum.

Libia minora   Inner, minor lips of the female vulva; homologous to the corpus spongiosum of the male penis.

Labioscrotal swelling   Part of the indifferent external genitalia of the early embryo that gives rise to the scrotum in males or the labia majora and mons pubis in females.

Labor   Process leading to delivery that involves uterine contractions and fetal movement; divided into three stages: (1) cervical effacement and dilatation; (2) expulsion of fetus; and (3) explusion of placenta.

Lactation   Secretion of milk by the mammary glands.

Lactiferous duct   Duct of mammary tissue lying between the mammary ampulla and nipple.

Lactobacillus acidophilus   See **Döderlein's bacillus.**

Lamaze method   Prepared childbirth method involving knowledge of biological processes, controlled breathing and relaxation, and assistance by a partner.

Laminaria tent Cylinders of dried, sterilized seaweed used to dilate the cervix slowly by absorbing moisture.

Lanugo Fine downy hair covering the fetal skin.

Laparoscopy Examination of abdominal cavity with a tiny telescope and illuminator inserted through its wall.

Laparotomy Incision through the abdominal wall; used for tubal sterilization.

Larynx Voice box.

Latent stage of syphilis The third stage of syphilis, characterized by a relatively long-lasting absence of symptoms.

Lateral cervical ligaments Paired ligaments that attach the cervix to the pelvic wall.

Latex agglutination inhibition test Pregnancy test utilizing inhibition of antibody reaction to hCG.

Leboyer method Method of childbirth and newborn care in which the delivery environment is made as nonstressful as possible.

Lectins Proteins, derived from certain plants, that bind to sugars in the zona pellucida and prevent sperm penetration.

Lesbian A female homosexual.

Lesch-Nyhan syndrome A sex-linked, recessive, inherited disorder characterized by physical abnormalities and self-mutilation.

Lesser vestibular glands Fluid-secreting glands, opening into the female vestibule, that are homologous to the prostate gland of the male; **Skene's glands.**

Leucocytes White blood cells.

Leukorrhea A whitish, sticky vaginal or uterine discharge containing white blood cells.

Leydig cells Cells in the interstitial spaces between seminiferous tubules that secrete androgens; **Interstitial cells.**

Libido See **Sex drive.**

Lightening Movement of the fetus down into the pelvic cavity two or three weeks before labor begins.

Limbic system The part of the brain that influences sex drive.

Linea nigra A dark line that appears on the lower abdominal wall from the pubic area upward toward the navel of some pubertal females.

Linkage The tendency of certain genes to be inherited together because they are located on the same chromosome.

Lipotropin (LPH) Polypeptide hormone, secreted by basophils of the adenohypophysis, that causes fat to be metabolized to fatty acids and glycerol.

Lithopedion Calcified remnants of an abdominal ectopic pregnancy; "stone baby."

Lochia The vaginal discharge characteristic of women after they have delivered a child.

Loci (sing., **Locus**) In genetics, the position of a gene in a chromosome.

Long-loop feedback Feedback affecting function of the hypophysis that is mediated through products produced elsewhere than the brain or hypophysis.

Low tubal ovum transfer A method for treating tubal blockage by removing the ovum from a point above the block and inserting it back into the oviduct below the block so that it can be fertilized.

Luteal cells Endocrine cells of the corpus luteum.

Luteal phase Period in menstrual cycle between ovulation and the onset of menstruation.

Luteinization Process by which granulosa cells of the Graafian follicle switch from being primarily estrogen-secreting to primarily progesterone-secreting.

Luteinized cysts Abnormal lumps of luteal tissue in the ovary.

Luteinizing hormone (LH) Glycoprotein hormone secreted by basophils of the adeno-hypophysis that, together with FSH, maintains gonadal function in both sexes; **Interstitial cell-stimulating hormone.**

Luteinizing hormone-releasing hormone (LHRH) See **Gonadotropin-releasing hormone.**

Luteolytic factor Substance that causes the corpus luteum to regress.

Lymphogranuloma venereum A sexually transmitted disease caused by a strain of *Chlamydia trachomatis;* see also **Reiter's syndrome.**

Lysergic acid diethylamide (LSD) A hallucinogen that arouses brain centers and mimics sympathetic nervous system arousal.

Male climacteric Physiological and psychological changes in older men associated with a decline in testicular function.

Maleness Biological characteristics of the male sex.

Malignant Adjective applying to cancerous growth; dangerous, likely to cause death if not treated.

Mammae See **Mammary glands.**

Mammary alveoli Glandular structures of the mammary glands that secrete milk.

Mammary ampulla Wide storage tube for milk existing between a mammary duct and a lactiferous duct in the mammary gland.

Mammary coitus When a man inserts his penis between a women's breasts.

Mammary duct Duct in mammary gland between the secondary mammary tubule and the mammary ampulla.

Mammary glands Paired structures on the thoracic wall of both sexes; in the female, they secrete milk; **Breasts** or **Mammae.**

Mammography X-ray examination of the mammary glands.

Marfan's syndrome A dominant, autosomal disorder characterized by longer than normal limbs and digits.

Masculine Having the qualities and characteristics of a male; often used to mean "manly."

Masochist A person who derives sexual pleasure from experiencing physical or psychological pain.

Mastectomy Surgical removal of a breast alone (**simple mastectomy**) or with underlying muscle and axillary lymph nodes (**radical mastectomy**).

Masturbation Self-stimulation of the genitals to produce sexual arousal.

Mate-swapping Exchange of sex partners between two or more married couples; **Swinging.**

Mating behaviors Behaviors associated with coitus.

Meconium Greenish fecal material present in the fetal large intestine.

Median eminence Region in floor of the hypothalamus, near the infundibulum, where neurohormones are released from neurosecretory neurons of the hypophysiotropic area of the hypothalamus.

Medullary cords Cords in the medulla of the indifferent gonad of the embryo that give rise to the seminiferous tubules in males and the rete ovarii in females.

Meiosis A type of cell division found in the gonads, in which the chromosome number is reduced from 46 (2*N*) to 23 (*N*).

Melanin Brown pigment that is synthesized from the amino acid tyrosine.

Melanophores Pigment cells that contain melanin.

Melanophore-stimulating hormone (MSH) Protein hormone, secreted by basophils of the pars intermedia of the adenohypophysis, that stimulates melanin synthesis.

Melatonin Hormone, synthesized and secreted by the pineal gland during darkness, that inhibits reproductive function.

Membrana granulosa Single or multiple layer of granulosa cells between the theca and the zona pellucida of the ovarian follicle.

Membranous urethra The portion of the urethra passing through the urogenital diaphragm.

Menarche The onset of menstruation at puberty.

Menopause The cessation of menstruation in the human female occurring between the ages of 35 and 55; **Female climacteric.**

Menses See **Menstruation.**

Menstrual cramps Painful, spasmodic contractions of the uterus, possibly caused by unusually high uterine prostaglandin levels during the menstrual phase.

Menstrual cycle A monthly reproductive cycle in which periodic uterine bleeding occurs, as in humans and some other primates.

Menstrual extraction See **Menstrual regulation.**

Menstrual phase Period of menstrual cycle when menstruation occurs; **Menses.**

Menstrual regulation Removal of embryo by sucking it out with a syringe attached to a plastic tube; used to induce abortion in the early first trimester; **Menstrual extraction.**

Menstruation Discharge of blood along with sloughed cells of the uterine lining from the uterus through the vagina that normally occurs at monthly intervals.

Mesoderm Middle of the three primary germ layers of the embryo; gives rise to the muscular, skeletal, circulatory, excretory, and reproductive systems.

Mesonephric ducts See **Wolffian ducts.**

Mesonephric tubules Structures of the embryonic kidney (mesonephros) that form the vasa defferentia of the testes.

Mesonephros Embryonic kidney, which regresses before birth, except that its tubules and ducts (wolffian ducts) develop into portions of the adult male reproductive tract.

Messenger RNA (mRNA) The single-stranded RNA that carries genetic information from the gene to the ribosome, where it determines the order of amino acids in the formation of a polypeptide or protein.

Metabolic rate Rate at which tissues use oxygen.

Metastasis Spread of malignant cells from a tumor to other regions of the body.

Methaqualone A barbiturate (Quaalude; Supor).

Methoxyindoles Small molecules present in the pineal gland; of unknown biological importance.

Metreurynter A rubber bag used by the Japanese to induce abortion when placed within the uterus.

Microcephaly A congenital defect characterized by a relatively small brain.

Micropenis An extremely small penis, often due to an underdeveloped fetal anterior pituitary gland.

Milk-ejection reflex Stimulation of oxytocin secretion by suckling, oxytocin then causing ejection of milk from the nipple.

Milk line Paired lines on the ventrum of the embryo containing potential pairs of mammary glands.

Mini-cesarean See **Hysterotomy.**

Minilaparotomy "Band-aid" method of tubal sterilization using a small abdominal incision just above the pelvic bone.

Minimata's disease Cerebral palsy caused by mercury poisoning.

Minipill Female oral contraceptive containing only a low amount of progestogen.

Miscarriage See **Spontaneous abortion.**

Mitosis Process of cell division in which a diploid cell divides to produce two diploid progeny.

Mittelschmerz Mild abdominal pain associated with ovulation.

Molluscum contagiosum A virus-induced disease characterized by pinkish skin warts.

Monestrus Having only one estrous cycle a year.

Mongolism See **Down's syndrome.**

Monilia vaginitis Inflammation of the vagina caused by *Candida albicans;* can be sexually transmitted.

Monogamy Marriage of two people for a theoretically indefinite period; in biology, when a male has a single mate for an extended period.

Monosomy A condition of aneuploidy in which an individual has one too few (45) chromosomes.

Monozygotic twins Twins usually resulting from separation of an inner cell mass of a blastocyst into two, after which two identical individuals develop; **Identical twins.**

Mons pubis An elevated cusion of fat in the female, covered with skin and pubic hair; **Mon veneris.**

Mons veneris See **Mons pubis.**

Morning-after pill See **Postcoital pill.**

Morning sickness Nausea experienced by some women in the first few weeks of pregnancy.

Moro reflex Reflexive movement of a newborn's arms and legs in response to being startled; **Startle reflex.**

Morula Stage in prenatal development when there is a solid ball of eight or more cells.

Mucous plug Barrier of mucus in the cervix that is lost during early labor.

Mucus Viscous substance on surface of membranes that moistens, lubricates, and protects; made up of mucin (a mucopolysaccharide) and water; serves as a barrier to foreign agents and a partial barrier to sperm in the cervix.

Müllerian duct-inhibiting factor Polypeptide secreted by the Sertoli cells of the embryonic testes that causes regression of the müllerian ducts in males.

Müllerian ducts Embryonic ducts that form the sex accessory ducts of the adult female; **Paramesonephric ducts.**

Multiparous Adjective applied to a woman who has previously borne children.

Mumps A viral infection of the salivary glands that also can infect the testes of an adult male and cause infertility.

Mutagen A chemical or other factor that changes the genetic components of cells.

Myoepithelial cells Cells surrounding each alveolus of the mammary gland; their contraction leads to milk ejection.

Myometrium The smooth muscle layer of the uterine wall.

Myotonia Increased muscular tension.

Narcotics Addictive drugs that induce stupor or arrest activity, such as heroin, morphine, and methadone.

Natural selection The differential reproduction of certain individuals in a population because they are better adapted to the environment.

Necrophilia Sexual attraction to, or activity with, a human corpse.

Negative feedback When production of a product decreases further production of that product.

Neisseria gonorrhoeae A gram-negative bacterium, also called gonococcus, that causes gonorrhea.

Neonatal death Infant death within the first day of life.

Neonate   A newborn child.

Neoplasm   See **Tumor.**

Neural tube   The embryonic structure that forms the brain and spinal cord.

Neural tube defects   Congenital abnormalities in neural tube development, such as spina bifida, microencephaly, and hydroencephaly.

Neuramidase   An enzyme produced by sperm that, when released during the acrosome reaction, breaks down neuraminic acid in the zona pellucida.

Neurohormones   Hormones synthesized and secreted by neurosecretory neurons.

Neurohypophysis   Neural portion of the hypophysis; **Posterior pituitary gland** or **Pars nervosa.**

Neurons   Nerve cells.

Neurosecretory neuron   Nerve cell that is specialized for synthesis and release of neuro-hormones.

Neurosecretory nuclei   Paired clusters of neurosecretory neuron cell bodies in the hypo-thalamus.

Neurotransmitter   Substance secreted by a regular neuron that crosses the synapse between two neurons and causes nerve impulses to be generated in the next neuron.

Nidation   See **Implantation.**

Nipple   A conical structure on the breast that bears the opening of the lactiferous ducts through which milk is ejected.

Nitrogenous base   A nitrogen-containing molecule having basic properties; part of a nucleo-tide in DNA and RNA.

Nocturnal emissions   Involuntary ejaculation of semen, usually at night during sleep.

Nonbacterial prostatitis   See **Prostatitis.**

Nondisjunction   The failure of homologous chromosomes to separate during meiosis.

Nongonococcal urethritis   See **Nonspecific urethritis.**

Nonidentical twins   See **Dizygotic twins.**

Nonmarital coitus   Coitus between two people, neither of whom is married; see also **Pre-marital coitus.**

Nonspecific urethritis   A sexually transmitted inflammation or infection of the urethra that is not due to the presence of gonorrhea.

Nonspecific vaginitis   Inflammation or infection of the vagina for which the cause is not known.

Nonvenereal syphilis   A disease of children in arid regions of the world caused by a bacterium similar or identical to *Treponema pallidum*. It is transmitted by direct (nonsexual) body con-tact or when drinking or eating; **Endemic syphilis.**

Norepinephrine   A neurotransmitter secreted by neurons of the sympathetic nervous system and central nervous system.

Nucleotide   Molecule containing a nitrogenous base connected to a phosphate and sugar; present in DNA and RNA.

Nulliparous   Adjective applied to a woman who has never borne a child.

Nurse midwife   Trained nurse that assists in home deliveries.

Nymphomania   When a female has an extremely high, insatiable sex drive.

Oedipus complex   In Freudian theory, the sexual attraction of a little boy to his mother.

Oligomenorrhea   When a woman periodically misses a menstrual period.

Oligospermia   When sperm count is below the fertile level in a male.

Oneida community   Religious community of the last century in upstate New York that practiced group sex, including among prepubertal children.

Oocyte Female germ cell.

Oocyte maturation inhibitor Proposed substance in antral fluid that inhibits oocyte maturation until the LH surge.

Oogenesis The process of meiosis that results in a female gamete.

Oogonia (sing., **Oogonium**) Immature, mitotically dividing female germ cells; present only in fetal ovaries.

Ootid Haploid female germ cell after completion of the second meiotic division.

Ootid pronucleus Nucleus of female germ cell after completion of meiosis and before fusion with sperm pronucleus.

Open marriage See **Contract marriage.**

Oral coitus Contact of the mouth or oral cavity with the genitals; see also **Cunnilingus** and **Fellatio.**

Orchidectomy Surgical removal of the testes.

Orchiditis Inflammation or infection of the testes.

Orgasm The peak of arousal in sexual activity.

Orgasmic dysfunction Failure to achieve orgasm.

Orgasmic phase The third stage in the human sexual response cycle, during which orgasm occurs.

Orgasmic platform The area, including the outer third of the vaginal barrel and the labia minora, which displays vasocongestion during the female plateau phase.

Osteoporosis Loss of calcium and phosphorus from bones; common in older people, especially postmenopausal women.

Ostium Internal opening of the oviductal infundibulum.

Outer acrosomal membrane Outer membrane of acrosomal cap on sperm head; see also **Acrosome reaction.**

Ovarian cortex Outer, more dense layer of the ovarian stroma.

Ovarian cyst An abnormal growth of ovarian follicular tissue.

Ovarian follicle A bag or sac of tissue in the ovary that contains the oocyte.

Ovarian interstitial cells Clumps of endocrine cells in the ovarian stroma that are derived from the theca of atretic follicles.

Ovarian medulla Central, less dense region of the ovarian stroma.

Ovarian stroma Connective tissue framework of the ovary.

Ovariectomy Surgical removal of ovaries.

Ovaries The gonads of the female.

Oviductal ampulla Widened portion of the oviduct between the infundibulum and the isthmus.

Oviductal infundibulum Widened end of the oviduct that captures the ovulated egg.

Oviductal isthmus Narrow portion of the oviduct between the ampulla and the intramural region.

Oviductal muscularis Layer of smooth muscle in the oviductal wall.

Oviductal serosa Thin external covering of the oviduct.

Oviducts Paired tubes that lead from the ovaries to the upper portion of the uterus; same as **Fallopian tubes** or **Uterine tubes.**

Ovotestis Gonad containing both testicular and ovarian tissue.

Ovulation Process by which an ovum is extruded from a mature ovarian follicle.

Ovum An ovulated, haploid secondary oocyte, in second meiotic arrest, before it has been penetrated by a sperm.

Ovum activation Re-initiation of meiosis in the ovum as a result of sperm penetration.

Oxytocin   Octapeptide hormone, secreted by the neurohypophysis, that causes contraction of smooth muscle in reproductive tissues of both sexes.

Papule   A small elevation on the skin.

Papanicolaou test   Test by which cells of the uterine cervix are removed with a swab and examined for the presence of cervical cancer; **Pap smear.**

Pap smear   See **Papanicolaou test.**

Paracervical   A local anesthetic injected into both sides of the cervix.

Paramesonephric ducts   See **Müllerian ducts.**

Pars distalis   Major portion of the adenohypophysis.

Pars intermedia   Portion of adenohypophysis between the pars distalis and the neuro-hypophysis; often absent in adult humans.

Pars nervosa   See **Neurohypophysis.**

Pars tuberalis   Portion of adenohypophysis existing as a collar of cells around the pituitary stalk.

Parthenogenesis   Development of an embryo from an unfertilized egg; virgin birth.

Partial hydatidiform mole   Implantation of a trophoblast and a dead embryo; cells of these moles are triploid.

Parturition   Birth.

Patent ductus arteriosus   When the ductus arteriosus fails to close after birth.

Paternal behavior   Care-giving behavior of the father for his child.

Pearl's formula   Method of expressing failure rate of contraceptive measures per 100 woman-years.

Pediculosis pubis   A condition in which one is parasitized by the louse *Phthirus pubis;* also called "crabs"; can be sexually transmitted.

Pedigree   A flow chart that depicts the presence or absence of an inherited trait in a family line.

Pedophilia   When an adult is sexually attracted to children.

Pelvic inflammatory disease   Widespread infection of the female pelvic organs.

Penicillinase   An enzyme produced by certain bacteria that inactivates penicillin.

Penile agenesis   Congenital absence of a penis, due to the lack of a genital tubercle in the embryo.

Penile body   See **Penile shaft.**

Penile prepuce   The loose flap of skin that partially covers the penile glans; **Foreskin.**

Penile shaft   The cylindrical, erectile portion of the penis; **Penile body.**

Penis   Portion of the male external genitalia that serves as an intromittent and urinary organ.

Peptides   Small molecules that contain amino acids.

Perimetrium   Thin membrane covering the outside of the uterus.

Perineum   Pelvic floor; the space between the anus and scrotum in the male, and the anus and vulva in the female.

Peripheral effect of a hormone   Action of a hormone on tissues outside the central nervous system.

Perivitelline space   Region between the zona pellucida and the vitelline membrane of an oocyte.

Permissive society   A society in which both sexes are free to express their sexuality.

Persistent pulmonary hypertension   A condition in the newborn in which blood flow through the lungs is resisted by the presence of thickened pulmonary arteries; can be caused by fetal exposure to aspirin or indomethacin.

Peyronie's disease A condition in which the penis develops a fibrous ridge along its top or side, causing curvature.

Phagocytosis One process by which cells engulf external substances.

Phallus See **Genital tubercle.**

Phenotype The observable properties of the body; the physical expression of the genotype.

Phenylalanine An amino acid, levels of which are high in the blood and urine of an infant with phenylketonuria.

Phenylketonuria (PKU) An inherited disorder in protein metabolism that can cause mental retardation.

Pheromone A chemical produced by one individual of a species that changes the behavior or physiology of another member of the same species.

Philia Love as for a friend or sibling.

Phimosis Tightness of the penile prepuce.

Phocomelia Condition in which a child is born with hands and feet but no arms or legs.

Phthirus pubis A blood-sucking crab louse that parasitizes humans and causes pediculosus pubis (crabs).

Physiological jaundice Yellowing of the skin and eyes of a newborn because of excessive deposition of the pigment bilirubin, a breakdown product of red blood cells.

Pineal gland Single glandular outpocketing of the epithalamus that synthesizes and secretes melatonin.

Pitocin Trade name of an artificial oxytocin used to induce or hasten labor.

Pituitary gland See **Hypophysis.**

Pituitary stalk The tissue connecting the hypophysis with the brain.

Placenta A structure, composed of fetal and maternal tissue, attached to the inner surface of the uterus and connected to the fetus by the umbilical cord; serves as a respiratory, excretory, and nutritive organ for the fetus, and also secretes hormones important in pregnancy.

Placenta previa Uncommon condition in which the placenta is formed low in the uterus and partially or completely covers the opening of the cervix; Cesarean delivery often is required.

Placentophagia Eating of the placenta (afterbirth).

Plantar reflex When a newborn flexes its toes after someone touches the sole of its foot.

Plateau phase The second stage in the human sexual response cycle that immediately precedes orgasm.

Point mutation The spontaneous or induced change of a gene from one form to another; an inheritable change in DNA of a chromosome.

Polyandry When a woman is married to more than one man at the same time.

Polychlorinated biphenyls (PCBs) Environmental pollutants that have been associated with lowering sperm production in American males.

Polycystic ovarian syndrome Condition in which several large abnormal follicles exist in each ovary.

Polyestrous Having more than one estrous cycle a year.

Polygamy When the spouse of either sex is married to more than one person at one time.

Polygenic inheritance The determination of a given trait, such as body weight or height, by the interaction of several genes.

Polygyny When a man is married to more than one woman at the same time; in biology, when a single male has more than one mate.

Polypeptide Small chain of amino acids.

Polyspermy Fertilization of an ovum by more than one sperm.

Polythelia When more than two mammary glands persist in an adult.

Population growth rate Change in number of individuals in a population in a given amount of time.

Positive feedback When production of a product increases further production of that product.

Positive signs of pregnancy Signs that definitely indicate pregnancy, including detection of a fetal heartbeat, visualization of the fetus with ultrasound or fetoscopy, and feeling the fetus move.

Postcoital pill High dose of estrogen (oral or injected) given soon after coitus to prevent pregnancy; **Morning-after pill.**

Posterior pituitary gland See **Neurohypophysis.**

Postmature newborn A baby born more than two weeks after the due date.

Postpartum After birth.

Postpartum depression Depression (fits of crying, anxiety) in a woman appearing soon after she has had a child.

Postpartum psychiatric syndrome Severe psychosis developing in a woman soon after she has had a child.

Postreproductive adults Those too old to have offspring.

Potassium nitrate See **Saltpeter.**

Precocious puberty When puberty occurs before age eight in females or age ten in males.

Preeclampsia See **Toxemia.**

Pregnancy The condition of carrying a developing embryo or fetus in the uterus; **Gestation.**

Pregnancy tests Tests that utilize detection of human chorionic gonadotropin (hGC) in blood or urine to determine if a woman is pregnant.

Pregnanediol A metabolite of progesterone found in urine.

Pregnenolone Precursor steroid for synthesis of steroid hormones; derived from cholesterol.

Premarital coitus Coitus before one's first marriage; see also **Nonmarital coitus.**

Premature ejaculation Ejaculation prior to, just at, or soon after intromission.

Premature newborn When the newborn birth weight is between 0.99 kg (2 lb, 3 oz) and 2.49 kg (5.5 lb).

Premenstrual syndrome A group of physical and emotional symptoms associated with the late luteal phase of the menstrual cycle.

Premenstrual tension Emotional components of the premenstrual syndrome.

Prepared childbirth Term used for several kinds of "natural" childbirth procedures.

Presumptive signs of pregnancy Possible signs of pregnancy, including secondary amennorhea, morning sickness, and breast changes.

Priapism Persistent, abnormal erection of the penis, usually without sexual desire.

Primary amenorrhea When a female has not reached menarche by age 16.

Primary capillary plexus Cluster of capillaries in the median eminence that receives neurohormones secreted in that region; see also **Hypothalamo–hypophysial portal system.**

Primary erectile dysfunction When a man is unable to gain an erection, now or in the past.

Primary follicles Ovarian follicles consisting of a primary oocyte surrounded by a single layer of cuboidal granulosa cells and a thin theca.

Primary germ layers The layers of cells (ectoderm, mesoderm, endoderm) in the early embryo that give rise to all adult tissues.

Primary oocyte An oocyte in first meiotic arrest.

Primary orgasmic dysfunction When a woman is unable to have an orgasm now or in the past.

Primary sex ratio Ratio of male to female embryos determined at, or shortly after, conception.

Primary sexual characteristics Those reproductive structures that distinguish male from female, including the gonads, sex accessory ducts and glands, and external genitalia.

Primary spermatocyte Male germ cell before the first reduction division of meiosis.

Primary stage of syphilis Initial stage of syphilis, during which a sore (chancre) develops.

Primiparous Giving birth for the first time.

Primordial follicles Ovarian follicles consisting of a primary oocyte surrounded by a single layer of flattened granulosa cells.

Primordial germ cells Cells that migrate from the yolk sac to the indifferent gonads of the embryo, where they give rise to the oogonia of the ovaries or the spermatogonia of the testes.

Probable signs of pregnancy Indications that a woman, in all probability, is pregnant; include Hegar's sign and a positive pregnancy test.

Proctitis Inflammation or infection of the rectum.

Progestasert An IUD that contains a progestogen.

Progesterone The major natural progestogen.

Progestogens Substances that promote secretory function of the uterus; secreted by the ovaries and placenta.

Prolactin (PRL) Protein hormone, secreted by acidophils of the adenohypophysis, that stimulates milk synthesis in the mammary glands.

Prolactin release-inhibiting hormone (PRIH) Hormone from the hypothalamus that decreases synthesis and secretion of prolactin.

Prolapsed uterus When the uterus falls down from the normal position.

Promiscuity When a person engages in frequent sexual activity with many people, often for nonsexual reasons.

Prophylactic measure A measure taken to prevent occurrence of a disease or disorder.

Prostaglandins (PG) Family of small 20-carbon atom molecules derived from fatty acids that have various effects on the reproductive system of both sexes.

Prostate gland An unpaired male sex accessory gland that produces 13 to 33 percent of seminal plasma.

Prostatic urethra The portion of the urethra enclosed by the prostate gland.

Prostatic utricle Small blind pouch, present in the adult male prostate gland, which is a remnant of the müllerian duct system.

Prostatitis Inflammation of the prostate gland; can be the result of a bacterial infection (**Bacterial prostatitis**) or due to unknown nonbacterial causes (**Nonbacterial prostatitis**).

Prostitution Indulging in sexual activity for payment.

Protease An enzyme that hydrolyzes proteins to peptides and/or amino acids.

Proteins Long chains of amino acids linked by peptide bonds.

Pseudocyesis When some presumptive signs of pregnancy appear in a nonpregnant woman; has a psychosomatic origin; **False pregnancy.**

Pseudohermaphrodite An individual who has gonads that agree with chromosomal sex but has ambiguous external genitalia.

Puberty The biological transformation that takes a person from being a sexually immature child to a sexually mature adult; **Sexual maturation.**

Pubescence   The state of a child between the onset of pubertal changes and the completion of sexual maturation.

Pubococcygeus muscle   Muscle in the pelvic floor that forms support for the pelvic organs.

Pudendal block   When an anesthetic is injected into the pudendal nerves in the vaginal wall to numb the vagina.

Puerperium   The period of confinement of a woman after she has given birth.

Purulent   Pus-filled.

Pyromania   Compulsive fire-setting, often for sexual gratification.

Quinacrine   A dye used to label the Y chromosome; used as an aid in separation of X and Y sperm.

Radical mastectomy   See **Mastectomy**.

Radioimmunoassay   Procedure for hormone assay utilizing competitive binding of nonradioactive and radioactive hormone to an antibody to that hormone.

Rape   Forced sexual relations with an individual without that person's consent.

Rape trauma syndrome   A typical series of emotional reactions in a woman following an attempted or actual rape.

Raphe   Ridge in the midline of the scrotum.

Recessive   An allele that is not expressed phenotypically when in the heterozygous condition.

Recombinant DNA   Synthetic DNA that is spliced onto the natural DNA of an organism's cells.

Recombination   The combining of maternal and paternal alleles at fertilization.

Reduction division   Process of the first meiotic division during which the diploid ($2N$) condition is changed to a haploid (N) condition.

Refractory period   A temporary state after orgasm when a male is not responsive to sexual stimuli.

Regulatory genes   Genes that control the activity of other genes by coding for synthesis of repressor molecules.

Reiter's syndrome   Serious, advanced symptoms of Lymphogranuloma venereum.

Relaxin   Small polypeptide hormone, secreted by the corpus luteum and placenta during pregnancy, that relaxes the pubic symphysis, inhibits uterine contractions, and helps dilate the cervix.

Release-inhibiting hormone (RIH)   A hormone, secreted by neurosecretory neurons of the hypothalamus, that decreases synthesis and secretion of a hormone of the adenohypophysis.

Releasing hormone (RH)   A hormone, secreted by neurosecretory neurons of the hypothalamus, that increases synthesis and secretion of a hormone of the adenohypophysis.

Renewable marriage   When a marriage license would be renewed after a defined length of time.

Replacement therapy   Administration of a gland extract or pure hormone that reverses the effects of extirpation of an endocrine gland.

Repressor   In genetics, the substance produced by a regulatory gene that inhibits mRNA formation.

Reproductive adults   Sexually mature individuals.

Reproductive potential   Maximal number of offspring an individual is capable of producing in a given amount of time.

Reserpine   A drug that lowers blood pressure; can reduce fertility and cause sexual dysfunction.

Resolution phase   The last stage in the human sexual response cycle, during which the sexual system returns to its unexcited state.

Respiratory distress syndrome   When a newborn's lungs do not have surfactant, and it is unable to breath properly.

Restrictive society   Society in which the sexuality of one sex (usually female) is repressed, and the pleasures of sex are restricted.

Rete ovarii   Vestigial remnants of the medullary cords present in the medulla of the adult ovary.

Rete testis   Tiny tubules within the testis that transport sperm from the seminiferous tubules to the vasa efferentia.

Retroflexion   Referring to a uterus that is tilted backwards.

Retrolental fibroplasia   Disorder of the newborn leading to partial or complete blindness; caused by exposure to too much oxygen.

Rhesus disease   Condition of fetus in which the mother's cells are producing antibodies to the fetal red blood cells.

Rhythm methods   Contraceptive methods based on prediction of safe and fertile periods in the menstrual cycle; see also **Calendar, Basal body temperature, Cervical mucous,** and **Sympto-thermal methods.**

Ribonucleic acid (RNA)   A class of nucleic acids characterized by the presence of ribose sugar and the nitrogenous base uracil.

Ribosomes   Cellular organelles composed of protein and RNA; sites of translation in protein synthesis.

Rivanol   A chemical (ethacridine lactate) used intraamniotically to induce second trimester abortion in some countries.

Rooting reflex   When the hungry newborn keeps its head in contact with an object that touches its mouth or cheek.

Round ligaments   Paired, cordlike ligaments that connect the top of the uterus with the pelvic wall.

Rubella   German measles virus.

Saddle block   A conduction anesthetic injected into the subdural space of the spinal cord low in the back.

Sadist   A person who derives sexual pleasure from inflicting physical or psychological pain on someone else.

Salpingectomy   Surgical removal of the oviducts.

Salpingitis   Inflammation or infection of the oviducts.

Saltpeter   A chemical falsely accused of being an anaphrodisiac; **Potassium nitrate.**

Sarcoptes scabiei   A mite that parasitizes human skin and causes scabies.

Satyriasis   An extraordinarily high, insatiable sex drive in a male.

Scabies   A skin condition caused by the mite *Sarcoptes scabiei;* can be sexually transmitted.

Scrotum   The pouch, suspended from the groin, that contains the testes and its accessory ducts.

Sebaceous glands   Skin glands at the base of hair follicles that secrete an oily substance called sebum; their activity increases during puberty.

Sebum   Oily secretion of sebaceous glands.

Secondary amenorrhea   When a woman, who had menstruated previously, fails to menstruate for at least six consecutive months; see also **Amenorrhea.**

Secondary capillary plexus   Capillaries of the adenohypophysis from which neurohormones leave and affect function of the endocrine cells in the adenohypophysis; see also **Hypo-thalamo–hypophysial portal system.**

Secondary erectile dysfunction When a man is unable to gain an erection at least 25 percent of the time that the opportunity is present.

Secondary follicles Ovarian follicles consisting of a primary oocyte surrounded by several layers of granulosa cells and a single theca.

Secondary mammary tubules Small ducts in the mammary gland that carry milk from the alveoli to the mammary ducts.

Secondary oocyte A haploid oocyte produced by completion of the first meiotic division.

Secondary orgasmic dysfunction When a woman fails to have an orgasm in selective situations.

Secondary sex ratio Ratio of males to females at birth.

Secondary sexual characteristics External, sex-typical male or female structures (excluding the external genitalia) that are not directly involved in coitus or gamete production and transport.

Secondary spermatocyte Haploid male germ cell that is the product of reduction division of primary spermatocytes.

Second meiotic arrest Occurs after the secondary oocyte has begun the second meiotic division; ends only after sperm penetration of the ovum.

Second meiotic division The phase of oocyte meiosis in which a haploid secondary oocyte divides to produce a haploid ootid and second polar body.

Second polar body Small haploid cell that is a product of the second meiotic division of an oocyte. Degenerates before or after dividing.

Secondary stage of syphilis The second stage of syphilis, characterized by a widely distributed rash.

Segregation The separation of alleles into different gametes during meiosis.

Sella turcica Cup-shaped depression in the sphenoid bone at the base of the skull in which lies the hypophysis.

Semen Mixture of sperm and secretions of male sex accessory structures (seminal plasma) that leaves the male urethra during ejaculation; **Seminal fluid.**

Semen coagulation Increase in viscosity of semen within one minute of ejaculation.

Semen liquification Increase in semen fluidity several minutes after ejaculation.

Seminal fluid See **Semen.**

Seminal plasma Fluid portion of semen secreted by male sex accessory glands.

Seminal vesicles Paired male sex accessory glands that produce 46 to 80 percent of seminal plasma rich in fructose and prostaglandins.

Seminiferous epithelium Layer of cells lining each seminiferous tubule.

Seminiferous tubules The small tubes within the testes that contain male germ cells; sites of sperm production.

Semirestrictive society When a society's sexual attitudes lie between those in a restrictive and a permissive society.

Separation A married couple living apart by mutual consent.

Septic abortion Spontaneous abortion (miscarriage) caused by uterine infection.

Septic pregnancy Condition in which infection occurs within the uterus of a pregnant woman.

Sequential pill Type of oral contraceptive in which an estrogen alone is taken, followed by a combination of an estrogen and progestogen; now withdrawn from the market.

Serial monogamy Having several monogamous marriages in succession.

Sertoli cells   Cells in seminiferous tubule that produce androgen-binding protein and support male germ cells; **Sustentacular cells.**

Sex accessory ducts   Those internal structures in the male or female that transport or support germ cells or embryos.

Sex accessory glands   Those glands that secrete into sex accessory ducts in either sex.

Sex-change operation   Operation that constructs artificial external genitalia into those characteristic of the opposite sex.

Sex chromatin   See **Barr body.**

Sex chromosomes   Pair of homologous chromosomes that are XX in female somatic cells and XY in male somatic cells.

Sex drive   The motivation or desire to behave sexually; **Libido.**

Sex flush   The vasocongestion of the skin that begins during the excitement phase of the sexual response cycle.

Sex-limited trait   An inherited trait whose expressivity is influenced by sex hormones.

Sex-linked trait   An inherited trait controlled by an allele located on a sex chromosome (usually the X chromosome).

Sex role   The public manifestation of one's gender identity; **Gender role.**

Sex-role inversion   See **Transsexual.**

Sexual-aggressive fusion rapist   One who derives sexual pleasure from harming a woman.

Sexual-aim rapist   A rapist who derives sexual pleaure from his crime.

Sexual dimorphism   When the male and female of a species differ in appearance.

Sexual dysfunction   Any psychological or physical condition that inhibits sexual gratification.

Sexual intercourse   See **Coitus.**

Sexuality   One's sexual attitudes, feelings, and behavior.

Sexual love   Desire for sexual intercourse; part of erotic love.

Sexually indifferent stage   Stage in the early embryo when the gonads, sex accessory glands, and external genitalia are similar in both sexes.

Sexually transmitted diseases   Infectious diseases that can be transmitted from one person to another during coitus or other genital contact; **Venereal diseases.**

Sexual maturation   See **Puberty.**

Sexual reproduction   Reproduction involving meiosis and fertilization.

Sexual response cycle   The sequence of four phases of sexual response after effective erotic stimuli are present.

Short-loop negative feedback   Feedback in which a hormone of the adenohypophysis reaches the median eminence and directly influences secretion of its specific releasing or release-inhibiting hormone.

Siamese twins   See **Conjoined twins.**

Sickle-cell anemia   An inherited recessive disorder of hemoglobin in blacks that produces clumping of hemoglobin molecules and respiratory difficulties.

Sickle-cell trait   The heterozygous condition of the sickle-cell genotype, in which individuals have both normal and abnormal hemoglobin.

Simple mastectomy   See **Mastectomy.**

Skene's glands   See **Lesser vestibular glands.**

Smegma   Cheesy secretion from glands present under the male prepuce.

Sodomy   A legal term, defined differently in various states, but usually referring to "acts against nature" such as anal or oral coitus, or coitus with animals.

Soixante-neuf   A French term for simultaneous oral-genital contact between a man and a woman ("69").

Somatostatin Another name for growth hormone release-inhibiting hormone.

Sperm See **Spermatozoon.**

Sperm activation Increase in sperm motility, probably caused by contact with oviductal fluid.

Spermatic cords The paired structures in males that suspend the testes and scrotum from the pelvic wall and contain the vas deferens, nerves, and blood vessels.

Spermatid Haploid male germ cell that matures into a spermatozoon, a process called spermiogenesis.

Spermatogenesis The process of sperm formation occurring in the testis, whereby spermatogonia are transformed by meiosis to spermatids.

Spermatogonium (pl., Spermatogonia) Diploid male germ cell that divides mitotically before becoming a primary spermatocyte.

Spermatozoa See **Spermatozoon.**

Spermatozoon (pl., Spermatozoa) A mature male germ cell; **Sperm.**

Sperm capacitation Process by which sperm acquire the ability to penetrate the zona pellucida of a recently ovulated ovum; probably occurs in the uterus or oviduct.

Sperm forward mobility protein Protein from epididymis that causes sperm to swim straight instead of in circles.

Sperm granuloma Lump formed after a vasectomy because of an autoallergic response to a man's sperm leaking into surrounding tissues.

Spermiation Process by which mature sperm are released into the lumen of the seminiferous tubule from the Sertoli cells in which they were embedded.

Spermicide Agent placed in vagina that acts as a barrier to sperm transport and kills sperm.

Spermiogenesis Maturation of a spermatid to a spermatozoon.

Sperm midpiece Portion of sperm that contains the mitochondria; between the sperm neck and tail.

Sperm neck Portion of the sperm between the head and midpiece.

Sperm plasma membrane The outer membrane of the sperm head.

Sperm pronucleus Sperm nucleus immediately before fusion with the haploid ovum pronucleus.

Spina bifida A congenital defect in which the spinal cord is exposed or protrudes through the back.

Spinal anesthetic Injection of an anesthetic into the subdural space around the spinal cord; used during labor and delivery.

Spinnbarheit Threadability of cervical mucus.

Spironolactone An antihypertensive drug that inhibits the action of androgens on target tissues.

Sponge barrier contraceptive A new form of barrier contraceptive that is placed in the external cervical os.

Spongy urethra The portion of the urethra in the penis; **Cavernous urethra.**

Spontaneous abortion Spontaneous loss of an embryo or fetus; **Miscarriage.**

Spontaneous erections Penile erections that occur with increasing frequency during puberty.

Spontaneous ovulation Ovulation that occurs cyclically, not requiring copulation or presence of a male.

Spotting Slight blood loss from the uterus around the time of ovulation; **Breakthrough bleeding.**

Stable population A population in which input (births plus immigration) equals output (deaths plus emigration).

Standardized birth rate The number of individuals born to a specific age group of a population in a given amount of time.

Standardized death rate The number of individuals dying within a specific age group of a population in a given amount of time.

Staphylococcus aureus Bacterium present in the vagina of some women that is associated with toxic shock syndrome.

Startle reflex See **Moro reflex.**

Status orgasmus A sustained orgasm in a woman, lasting 20 seconds or longer.

Statutory rape Sexual relations with a person who is below the age of consent.

Stein-Leventhal syndrome Female condition caused by inability of the ovaries to convert androgens to estrogens; characterized by polycystic ovaries, infertility, obesity, and excessive body hair.

Steroid A molecule with a basic structure similar to cholesterol.

Steroidogenesis The biosynthesis of steroid hormones in the gonads, adrenal glands, and placenta.

Stigma Small avascular region on the surface of a graafian follicle; future site of ovulation.

Stillbirth When fetus is dead at birth.

Stratum basalis Outer layer of the uterine endometrium that is not shed during menstruation.

Stratum functionalis Inner layer of uterine endometrium that is shed during menstruation.

Strawberry marks Small reddish blotches on the skin of a newborn that go away in due time.

Stress The physiological response to a stressor.

Stressor Any set of circumstances that disturbs the normal homeostasis of the body.

Suckling reflex Sucking movements of a newborn's mouth that are elicited by touching its lips.

Sudden-infant-death syndrome (SIDS) Sudden unexpected death of an infant, probably due to failure of its respiratory mechanisms; **Crib death.**

Superfetation Development of a second fetus after one has already started development in the uterus.

Surface epithelium Thin epithelial covering on each ovary.

Surfactant Lipoprotein present in the newborn's lungs that allows the air sacs to fill by affecting surface tension.

Surrogate pregnancy When a woman contracts to carry the child of another couple to term, after which she relinquishes the infant to the couple.

Suspensory ligaments of Cooper Ligaments in breasts that provide support.

Sustentacular cells See **Sertoli cells.**

Sweat glands Skin glands; these can be apocrine, causing a typical odor, or eccrine, secreting a salty fluid involved in temperature regulation.

Swinging See **Mate-swapping.**

Sympto-thermal method Method of determining safe and fertile periods in the menstrual cycle using cervical mucus, basal body temperature, and other signs of ovulation; see also **Rhythm methods.**

Synapse Small space between two connecting neurons through which pass neurotransmitters.

Syncytiotrophoblast Syncytial component of the trophoblast.

Syncytium Tissue with cells that have communicating cytoplasm.

Syndrome A group of symptoms relating to a single cause or event.

Syphilis A sexually transmitted disease caused by the bacterium *Treponema pallidum.*

Target tissues Tissues containing cells that have specific receptors for a certain hormone and exhibit a growth or physiological response to that hormone.

Tay-Sach's disease An autosomal recessive, inherited disorder of the nervous system, most common in Jews of Central European ancestry.

Tenting effect Expansion of the vagina during the female sexual response cycle; caused by elevation of the uterus.

Teratogen A chemical or other substance that causes abnormalities in embryonic or fetal growth and development.

Term See **Due date.**

Tertiary follicles Large ovarian follicles in which there is an antral cavity; the theca of these follicles is divided into a theca interna and externa.

Tertiary stage of syphilis The life-threatening final stage of syphilis.

Testes (sing., **Testis**) The male gonads; site of spermatogenesis and androgen secretion; **Testicles.**

Testicles See **Testes.**

Testicular artery Artery in each spermatic cord that supplies blood to the testis and epididymides.

Testicular biopsy Surgical sampling of a small piece of testicular tissue.

Testicular feminization syndrome Form of male pseudohermaphroditism that is caused by an inherited absence of androgen receptors in target cells.

Testicular lobules Compartments within the testis that contain the seminiferous tubules.

Testicular veins Two veins in the spermatic cord that drain blood from the testis and epididymides.

Testosterone The major natural androgen.

Test-tube baby Baby conceived outside the mother, the resultant embryo then being implanted into the mother's uterus; does not imply complete embryonic and fetal development outside of the body.

Tetrahydrocannabinol (THC) The active, hallucinogenic chemical in marijuana and hashish.

Thalassemia A rare inherited disorder of hemoglobin that relates to ineffective "punctuation" of mRNA messages.

Thalidomide Drug formerly prescribed for pregnant women as a sedative or to treat morning sickness until it was learned that it had severe teratogenic effects.

Theca Connective-tissue covering of the ovarian follicle and corpus luteum.

Theca externa External, dense, vascular, connective-tissue layer of the theca of tertiary ovarian follicles.

Theca interna Internal glandular layer of the theca of tertiary ovarian follicles.

Therapeutic abortion When an abortion is induced in consideration of the woman's physical health.

Thermography Method of detecting tumors using the heat they produce.

Thymine A nitrogenous base in the DNA molecule.

Thyrotropin (TSH) Glycoprotein hormone, secreted by basophils of the adenohypophysis, that stimulates the thyroid glands to synthesize and secrete thyroid hormones.

Thyrotropin-releasing hormone (TRH) Hormone of the hypothalamus that increases synthesis and secretion of thyrotropin.

Thyroxin A major thyroid hormone.

Toxemia Sometimes life-threatening complication of pregnancy. Early stage (**preeclampsia**) is characterized by hypertension, edema, and the presence of protein in the urine; late stage (**eclampsia**) is accompanied by coma, convulsion, and sometimes death.

Toxic shock syndrome A disease caused by the presence of certain strains of the bacterium *Staphylococcus aureus* in the vagina, especially in tampon users.

Transcription The process by which the genetic information in DNA specifies a complementary RNA molecule.

Transfer RNA (tRNA) A class of RNA attached to a specific amino acid and having a specific anticodon that recognizes a codon in mRNA.

Transition contractions Strong contractions during stage 1 of labor that dilate the cervix from 7 cm to 10 cm.

Transition dilatation The final phase of stage 1 labor, during which the cervix dilates from 7 cm to 10 cm.

Translation The synthesis of a polypeptide or protein from amino acids carried by tRNAs to the ribosome-bound mRNA specifying the polypeptide or protein.

Transsexual One who feels that he or she is trapped in the body of the wrong gender; **Sex-role inversion.**

Transverse presentation When the shoulders and arms of the fetus are engaged in the cervix.

Transvestite One who derives sexual pleasure from dressing in clothes of the opposite sex.

Treponema pallidum The spirochaete bacterium that causes venereal syphilis and also may cause nonvenereal syphilis.

Treponema pertenue The spirochaete bacterium that causes yaws.

Tribadism When lesbians assume coital positions.

Trichomonas vaginalis A pear-shaped flagellate protozoan, normally found in the vagina, which causes trichomonas vaginitis.

Trichomonas vaginitis Inflammation of the vagina due to *Trichomonas vaginalis;* can be sexually transmitted; **Trichomoniasis.**

Trichomoniasis Infection with *Trichomonas vaginalis.*

Trimester Pregnancy is divided into three-month intervals, and each interval is a trimester.

Triolism When three people engage together in sexual activity.

Triploidy An embryo with an extra set of chromosomes in each cell (3N).

Trisomy A condition of aneuploidy in which an individual has one too many (47) chromosomes.

Trophoblast Outer layer of blastocyst that plays a role in implantation and secretes hCG.

True hermaphrodite A person who has an ovary on one side and a testis on the other, or an ovotestis on one or both sides.

Trypsin Protease from the pancreas that blocks fertilization in some laboratory mammals.

Tubal pregnancy Ectopic pregnancy within an oviduct.

Tubal sterilization Surgical blockage of the oviducts; used as a contraceptive measure.

Tubuli recti Straight ducts within the testis that transport sperm from the seminiferous tubules to the rete testis.

Tumescent The condition of being swollen.

Tumor A lump or mass containing cells that have lost their control of growth and multiplication; **Neoplasm.**

Tunica albuginea Thin connective-tissue covering of ovary between surface epithelium and the ovarian cortex. Also, the thin shiny covering of each testis.

Tunica dartos Muscle in the scrotum that contracts in response to a lowering of scrotal temperature.

Tunica vaginalis The thin membrane lying directly over the tunica albuginea of each testis.

Turner's syndrome Females with cells having only one sex chromosome (XO condition); individuals are short and sterile, with undeveloped genitalia and sex accessory structures.

Ultrashort-loop negative feedback   Feedback in which a releasing hormone reaches the median eminence and directly influences further secretion of the same releasing hormone.

Ultrasound   High-frequency sound used for fetal scanning and other purposes.

Umbilical arteries   Two fetal blood vessels in the umbilical cord that carry deoxygenated blood to the placenta.

Umbilical cord   Cord connecting the fetal circulation to the placenta.

Umbilical vein   Fetal blood vessel in the umbilical cord that carries oxygenated blood from the placenta to the fetus.

Uracil   A nitrogenous base in the RNA molecule.

Ureaplasma urealyticum   An organism with viral and bacterial properties that may cause some cases of nonspecific urethritis.

Urethra   The duct that carries urine from the urinary bladder to the outside.

Urethral bulb   The enlarged end of the spongy urethra in the male.

Urethral meatus   The external opening of the urethra in the penis.

Urethral orifice   Opening of the urethra into the vestibule in the female, through which urine passes.

Urethritis   Inflammation or infection of the urethra.

Urinary incontinence   Uncontrollable leakage of urine from the urethra.

Urogenital folds   Structure of the embryonic external genitalia that gives rise to the labia minora in the female and the ventral aspect of the penile shaft in the male.

Urogenital sinus   Embryonic structure that gives rise to the urinary bladder and urethra in both sexes, as well as the lower two-thirds of the vagina in the female.

Urophilia   When one derives sexual pleasure from urine or urination.

Uterine cervix   Lower regions of the uterus between the uterine corpus and the vagina.

Uterine corpus   Body of the uterus, between the uterine fundus and cervix.

Uterine fundus   Dome-shaped top of the uterus.

Uterine glands   Tubular glands in the uterine endometrium that secrete fluid and nutrients during the luteal phase of the menstrual cycle.

Uterine isthmus   External constriction of the uterus that demarks the junction between the uterine corpus and cervix.

Uterine ligaments   Bands or cords of tissue that support the uterus.

Uterine orgasm   Orgasm that results from stimulation of the Grafenberg spot in the vaginal wall.

Uterine stroma   The connective tissue framework of the endometrium.

Uterine tubes   See **Oviducts.**

Uterosacral ligaments   Paired ligaments attaching the lower uterus to the sacrum.

Uterotubal junction   Muscular constriction at the entrance of the intramural oviduct into the uterine cavity.

Uterus   Pear-shaped female organ located in the pelvic cavity; receives the oviducts and connects with the vagina; source of menstrual flow and site of pregnancy; also known as the womb.

Vacurette   Tube used to suck out the uterine contents during a vacuum aspiration abortion.

Vacuum aspiration   Method of inducing abortion during the late first trimester by suction applied through a tube (vacurette) inserted into the cervix; **Vacuum curettage.**

Vacuum curettage   See **Vacuum aspiration.**

Vacuum extraction   Assistance in expulsion of fetus by use of a suction device attached to the top of the fetal head.

Vagina The tube extending from the uterine cervix to the vestibule.

Vaginal coitus See **Coitus.**

Vaginal introitus Opening of the vagina into the vestibule.

Vaginal lubrication When a clear fluid, leaking from blood vessels in the vaginal wall, appears in the vagina a few seconds after sexual arousal.

Vaginal orgasm Orgasm resulting from stimulation of the vagina; see **Uterine orgasm.**

Vaginal pessary A device placed in the vagina to support a sagging vagina and uterus.

Vaginal ring A doughnut-shaped, progestogen-containing contraceptive device that is placed within the vaginal cavity.

Vaginal rugae Transverse folds in the vaginal wall that allow vaginal expansion during coitus or childbirth.

Vaginal smear A small sampling of vaginal tissue used for detection of changes in vaginal lining during the menstrual cycle.

Vaginismus Strong, spasmodic contractions of the vagina or its surrounding muscles.

Vaginitis Inflammation or infection of the vagina.

Varicella-zoster virus The virus that causes chicken pox in children and shingles in adults.

Varicocele Enlarged testicular veins in the spermatic cord, which can press on the vasa deferentia and reduce fertility.

Varicose veins When veins enlarge and swell on the skin surface.

Vasa efferentia Series of ducts in the testis and head of the epididymis that convey sperm from the rete testis to the ductus epididymis.

Vas deferens (pl., Vasa deferentia) Duct carrying sperm from the epididymis to the ejaculatory duct; **Ductus deferens.**

Vasectomy Surgical sterilization by interruption of the vasa deferentia.

Vasocongestion Pooling or engorgement of blood in a tissue.

Vasopressin Octapeptide hormone secreted by the neurohypophysis that causes the kidneys to retain water; **Antidiuretic hormone.**

Venereal diseases See **Sexually transmitted diseases.**

Veratrum californicum A plant containing an alkaloid that harms the fetus when eaten by a pregnant sheep. This alkaloid causes delayed parturition.

Vernix caseosa Protective coating on skin of fetus, consisting of oil and sloughed skin cells.

Vestibule The space in the female between the labia minora, into which open the vaginal introitus and urethral orifice.

Viral hepatitis, type B Infection of the liver by a virus that can be sexually transmitted.

Vitamin B₆ Member of the B complex of vitamins that influences the function of the nervous system.

Voyeur A person who requires viewing nudes or watching others have coitus for sexual arousal.

Vulva The female external genitalia.

Vulvitis Inflammation or infection of the vulva.

Wharton's jelly Jellylike supporting substance within the umbilical cord.

Whitten effect Induction and synchronization of rat estrous cycles by the presence of a male rat.

Witch's milk Fluid secreted by the mammary glands of male and female infants.

Wolffian ducts Embryonic precursors of the male sex accessory ducts and some accessory glands; **Mesonephric ducts.**

Woman-year Use of a contraceptive by one woman for one year; used to express failure rates of contraceptive measures; see also **Pearl's formula.**

Women's Liberation Movement Movement dedicated to the sexual, economic, and political freedom of women, and to their personal growth; **Feminist Movement.**

X-chromosome inactivation When, in female cells, all but one X chromosome condenses into a clump (Barr body) and is genetically inactive.

Xeroradiography X-ray technique by which the image is expressed on paper, not film.

Yaws A skin disease common in children in tropical regions, caused by the bacterium *treponema pertenue;* usually not sexually transmitted.

Yohimbine A reputed aphrodisiac that stimulates the parasympathetic nervous system.

Yolk sac Extraembryonic membrane in embryo that contains a slight amount of yolk and gives rise to the primordial germ cells.

Yolk sac cavity Cavity, within the yolk sac, that is derived from the blastocoel.

Zona pellucida Translucent membrane between the oocyte and membrana granulosa of an ovarian follicle.

Zona reaction Change in chemical nature and structure of the zona pellucida caused by ovum cortical granule secretion; prevents polyspermy.

Zoophilia See **Bestiality.**

Zygote Diploid, fertilized cell produced at conception.

Index